This book addresses the issue of designing the microstructure of fiber composite materials in order to obtain optimum performance. Besides the systematic treatment of conventional continuous and discontinuous fiber composites, the book also presents the state-of-the-art of the development of textile structural composites as well as the nonlinear elastic finite deformation theory of flexible composites.

The author's experience during twenty years of research and teaching on composite materials is reflected in the broad spectrum of topics covered, including laminated composites, statistical strength theories of continuous fiber composites, short fiber composites, hybrid composites, two- and three-dimensional textile structural composites and flexible composites. This book provides the first comprehensive analysis and modeling of the thermomechanical behavior of fiber composites with these distinct microstructures. Overall, the inter-relationships among the processing, microstructures and properties of these materials are emphasized throughout the book.

The book is intended as a text for graduate or advanced undergraduate students, but will also be an excellent reference for all materials scientists and engineers who are researching or working with these materials.

Microstructural design of fiber composites

Cambridge Solid State Science Series

EDITORS:

Professor R. W. Cahn
*Department of Materials Science and Metallurgy,
University of Cambridge*

Professor E. A. Davis
Department of Physics, University of Leicester

Professor I. M. Ward
Department of Physics, University of Leeds

TSU-WEI CHOU

Jerzy L. Nowinski Professor of Mechanical Engineering
University of Delaware

Microstructural design of fiber composites

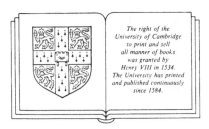

The right of the
University of Cambridge
to print and sell
all manner of books
was granted by
Henry VIII in 1534.
The University has printed
and published continuously
since 1584.

CAMBRIDGE UNIVERSITY PRESS

Cambridge
New York   Port Chester
Melbourne   Sydney

CAMBRIDGE UNIVERSITY PRESS
Cambridge, New York, Melbourne, Madrid, Cape Town, Singapore, São Paulo

Cambridge University Press
The Edinburgh Building, Cambridge CB2 2RU, UK

Published in the United States of America by Cambridge University Press, New York

www.cambridge.org
Information on this title: www.cambridge.org/9780521354820

First published 1992
This digitally printed first paperback version 2005

A catalogue record for this publication is available from the British Library

Library of Congress Cataloguing in Publication data
Chou, Tsu-Wei
Microstructural design of fiber composites/Tsu-Wei Chou.
   p. cm.—(Cambridge solid state science series)
ISBN 0-521-35482-X
1. Fibrous composites. 2. Microstructure. I. Title.
II. Series.
TA481.5.C48 1992
620.1'18—dc20   90-43347   CIP

ISBN-13 978-0-521-35482-0 hardback
ISBN-10 0-521-35482-X hardback

ISBN-13 978-0-521-01965-1 paperback
ISBN-10 0-521-01965-6 paperback

To Mei-Sheng, Helen, Vivian and Evan

Contents

Preface

The science and technology of composite materials are based on a design concept which is fundamentally different from that of conventional structural materials. Metallic alloys, for instance, generally exhibit a uniform field of material properties; hence, they can be treated as homogeneous and isotropic. Fiber composites, on the other hand, show a high degree of spacial variation in their microstructures, resulting in non-uniform and anisotropic properties. Furthermore, metallic materials can be shaped into desired geometries through secondary work (e.g. rolling, extrusion, etc.); the macroscopic configuration and the microscopic structure of a metallic component are related through the processing route it undergoes. With fiber composites, the co-relationship between microstructure and macroscopic configuration and their dependence on processing technique are even stronger. As a result, composites technology offers tremendous potential to design materials for specific end uses at various levels of scale.

First, at the microscopic level, the internal structure of a component can be controlled through processing. A classical example is the molding of short-fiber composites, where fiber orientation, fiber length and fiber distribution may be controlled to yield the desired local properties. Other examples can be found in the filament winding of continuous fibers, hybridization of fibers, and textile structural forms based upon weaving, braiding, knitting, etc. In all these cases, the desired local stiffness, strength, toughness and other prespecified properties may be achieved by controlling the fiber type, orientation, and volume fraction throughout the structural component.

Second, the external geometrical shape of a structural component can also be designed. Advances in the technology of filament winding enable the automated production of components with complex contours. It is now also feasible to fabricate three-dimensional fiber preforms using advanced textile technology. As the ability to fabricate larger and more integrated structural components of net shape is further enhanced, the need to handle and join a large number of small parts, as is currently done with metallic materials, diminishes.

The integrated and system approach, ranging from microstructure to component net shape, offers almost unlimited opportunity in composites processing and manufacturing. The figure below depicts the interdependence of processing, microstructure, properties, responses to external fields (physical, chemical and mechanical), and performance of composites.

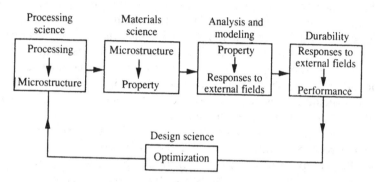

The purpose of this book is to address the issue of designing the microstructure of composites for optimum performance. This is achieved through the selection of fiber and matrix materials as well as the placement of both continuous and discontinuous fibers in matrix materials. Continuous fibers can assume straight or wavy shapes; they can also be hybridized or woven into textile preforms. The wide range of microstructures available offers tremendous versatility in the performance of composites; the ability to design microstructures enables performance to be optimized.

The book is intended as an intermediate-level textbook for students and a reference for research scientists and engineers. Readers need some background and preparation in materials science and applied mechanics. The first chapter examines the driving forces for advances in fiber composites, as well as the trends and opportunities of this rapidly developing field. Besides providing a concise summary of the linear elastic laminate theory, Chapter 2 examines some of the recent developments in the mechanics of laminated composites. Particular emphasis is given to thick laminates, hygrothermal effects and thermal transient effects. Chapter 3 treats the strength of continuous-fiber composites. Analyses of the local load redistribution due to fiber breakages are presented first. They are followed by a fairly comprehensive treatment of the statistical tensile strength theories which encompasses the behavior of individual fibers, fiber bundles, unidirectional fiber composites, cross-ply composites and laminates of multi-directional plies. Various modes of failure of laminated composites are examined. Section 3.4.6.2 is contributed by S. L. Phoenix, and Sections 3.4.7.4 and 3.4.8 are contributed by A. S. D. Wang. Chapter 4 deals with the elastic, physical and viscoelastic properties as well as the strength and fracture behavior of short-fiber composites. The effects of variations in fiber length and orientation are examined using a probabilistic approach. In Chapter 5, fiber hybridization serves as a vivid example of how the performance of composites can be controlled through the selection of material systems and their geometric distributions. The synergistic effects between the component phases with low elongation and high elongation fibers are of particular interest. Chapter 6 is devoted to two-dimensional textile structural composites based on woven, knitted and braided preforms. A comprehensive treatment of the techniques for analyzing and modeling the thermomechanical behavior of two-dimensional textile composites is presented. Chaper 7 introduces recent developments in the processing of three-dimensional textile preforms based on braiding, weaving, stitching and knitting. The processing–microstructure relationship is demonstrated by the establishment of 'processing windows' for a specific forming technique. Then the microstructure–property relationship is exemplified through the construction of 'performance maps'. Mechanical properties of polymer- and metal-based composites using three-dimensional textile preforms are reviewed. Chapters 8 and 9, in contrast to the earlier chapters, treat the topic of finite elastic deformation of flexible composites. The fundamental characteristics of flexible composites and the technique for analyzing them are presented in Chapter 8. A rigorous treatment of the constitutive relations of flexible composites is developed in Chapter 9 based upon both the Lagrangian and Eulerian descriptions of finite elastic deformation. Overall, the inter-relationship among processing, microstructure, property, responses to external fields, and performance of composites is emphasized throughout this text.

The contents of this book have evolved from my experience during two decades of teaching and research of composite materials at the University of Delaware. Stimulation from students and co-workers was indispensable to the preparation of this book. The contributions of the individuals with whom I had the privilege and pleasure to interact are too numerous to cite here. However, this book serves as a tribute to the intellectual achievements of them all. The generous support provided by the National Science Foundation, Department of Energy, Department of Transportation, Army Research Office, Office of Naval Research, Naval Research Laboratory, Air Force Office of Scientific Research, NASA, industrial companies and the Center for Composite Materials of the University of Delaware for conducting the research reported in this book is greatly appreciated. Ding-Guey Hwang, Shen-Yi Luo, Joon-Hyung Byun and Wen-Shyong Kuo read the manuscript and gave critical comments. Te-Pei Niu, Yih-Cherng Chiang, Mark Deshon and Alison Gier provided valuable assistance in the preparation of the manuscript.

Lastly, I should like to express my deep appreciation to the following persons. The late Prof. Alan S. Tetelman of Stanford University first pointed out to me the technological potential of fiber composites. As a colleague of mine at Delaware, Prof. R. Byron Pipes has greatly enriched my perspective on the subject matter. The scholarship and guidance of Prof. Anthony Kelly have always been a source of inspiration to me. Prof. Jerzy L. Nowinski encouraged me throughout the course of this endeavor.

1 Introduction

1.1 Evolution of engineering materials

Compared to the evolution of metals, polymers and ceramics, the advancement of fiber composite materials is relatively recent. Ashby (1987) presented a perspective on advanced materials and described the evolution of materials for mechanical and civil engineering. The relative importance of four classes of materials (metal, polymer, ceramic and composite) is shown in Fig. 1.1 as a function of time. Before 2000 BC, metals played almost no role as engineering materials; engineering (housing, boats, weapons, utensils) was dominated by polymers (wood, straw, skins), composites (like straw bricks) and ceramics (stone, flint, pottery and, later, glass). Around 1500 BC, the consumption of bronze might reflect the dominance in world power and, still later, iron. Steel gained its prominence around 1850, and metals have dominated engineering design ever since. However, in the past two decades, other classes of materials, including high strength polymers, ceramics, and structural composites, have been gaining increasing technological importance. The growth rate of carbon-fiber composites is at about 30% per year – the sort of growth rate enjoyed by steel at the peak of the Industrial Revolution. According to Ashby the new materials offer new and exciting possibilities for the designer and the potential for new products.

1.2 Fiber composite materials

Fiber composites are hybrid materials of which the composition and internal architecture are varied in a controlled manner in order to match their performance to the most demanding structural or non-structural roles. The fundamental characteristics of fiber composites have been summarized by Vinson and Chou (1975), Chou and Kelly (1976), Chou, Kelly and Okura (1985), Kelly (1985), and more recently by Chou, McCullough and Pipes (1986), from which the following is excerpted.*

On the face of it a composite might seem a case of needless complexity. The makings of ideal structural materials would appear

* From 'Composites', Chou, McCullough and Pipes. Copyright © (1986) by Scientific American, Inc. All rights reserved.

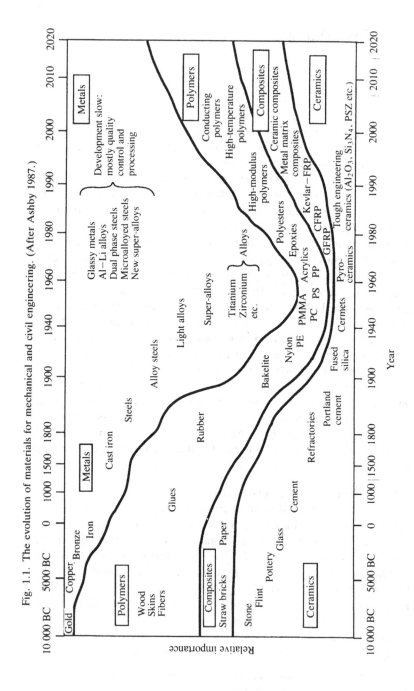

Fig. 1.1. The evolution of materials for mechanical and civil engineering. (After Ashby 1987.)

to be at hand, in the midsection of the periodic table. Those elements, among them carbon, aluminum, silicon, nitrogen and oxygen, form compounds in which the atoms are joined by strong and stable bonds. As a result, such compounds, typified by the ceramics, for instance, aluminum oxide, silicon carbide and silicon dioxide, are strong, stiff and resistant to heat and chemical attack. Their density is low and furthermore their constituent elements are abundant.

Yet because of a serious handicap these substances have rarely served as structural materials. They are brittle and susceptible to cracks. In bulk form the substance is unlikely to be free of small flaws, or to remain free of them for long in actual use. When such a material is produced in the form of fine fibers, its useful strength is greatly increased. The remarkable increase in strength at small scales is in part a statistical phenomenon. If one fiber in an assemblage does fail, moreover, the crack cannot propagate further and the other fibers remain intact. In a similar amount of the bulk material, in contrast, the initial crack might have led to complete fracture.

Tiny needlelike structures called whiskers, made of substances such as silicon carbide and aluminum oxide, also contain fewer flaws and show greater strength than the material in bulk form. Whiskers are less likely to contain defects than the bulk material, not only for statistical reasons but also because they are produced as single crystals that have a theoretically perfect geometry. The notion that many materials perform best as fibers also holds for certain organic polymers. Composites are a strategy for producing advanced materials that take advantage of the enhanced properties of fibers. A bundle of fibers has little structural value. To harness their strength in a practical material the designer of a composite embeds them in a matrix of another material. The matrix acts as an adhesive, binding the fibers and lending solidity to the material. It also protects the fibers from environmental stress and physical damage that could initiate cracks.

The strength and stiffness of the composite remain very much a function of the reinforcing material, but the matrix makes its own contribution to properties. The ability of the composite material to conduct heat and current, for example, is heavily influenced by the conductivity of the matrix. The mechanical behavior of the composite is also governed not by the fibers alone but by a synergy between the fibers and the matrix.

The ultimate tensile strength of a composite is a product of the synergy. When a bundle of fibers without a surrounding matrix is stressed, the failure of a single fiber eliminates it as a load carrier. The stress it had borne shifts to the remaining intact fibers, moving them closer to failure. If the fibers are embedded in a matrix, on the other hand, fracture does not end the mechanical function of a fiber. The reason is that as the broken ends of the fiber pull apart, elastic deformation or plastic flow of the matrix exerts shear forces, gradually building stress back into the fragments. Because of such load transfer the fiber continues to contribute some reinforcement to the composite. The stress on the surrounding intact fibers increases less than it would in the absence of the matrix, and the composite is able to bear more stress without fracturing. The synergy of the fibers and the matrix can thus strengthen the composite and also toughen it, by increasing the amount of work needed to fracture it.

Although the general requirement that the matrix be ductile provides some guidance for choosing a matrix material, the most common determinant of the choice is the range of temperatures the composite will face in its intended use. Composites exposed to temperatures of no more than between 100 and 200°C usually have a matrix of polymer. Most composites belong to this group.

Polymer matrices are often thermosets, that is polymers in which bonds between the polymer chains lock the molecular structure into a rigid three-dimensional network that cannot be melted. Thermosets resist heat better than most thermoplastics, the other class of polymeric materials, which melt when they are heated because no bonds cross-link the polymer chains. Epoxies are the most common thermosetting matrix for high-performance composites, but a class of resins called polyimides, which can survive continuous exposure to temperatures of more than 300°C, have attracted considerable interest. If the resin is a thermoset, the structure must then be cured, subjected to conditions that enable the polymer chains to cross-link. Often the composite must be held at high temperature and pressure for many hours.

In part to shorten the processing time, thermoplastic matrix materials are attracting growing interest; one promising example is a polymer called PEEK (polyetheretherketone). Consolidating a composite that has a thermoplastic matrix requires only relatively short exposure to a temperature that is sufficient to soften the plastic. The melting temperature of some thermoplastic matrices is so high that they rival thermosets in heat resistance: PEEK, for example, melts at 334°C. Thermoplastics have the additional advantage of being tougher than most of the thermosets.

Temperatures high enough to melt or degrade a polymer matrix call for another kind of matrix material, often a metal. Along with temperature resistance a metal matrix offers other benefits. Its higher strength supplements that of the reinforcing fibers, and its ductility lends toughness to the composite. A metal matrix exacts two prices: density that is high in comparison with polymers, even though the light metals such as aluminum, magnesium and titanium are the most common matrices, and complexity of processing. Indeed, whereas the production of many advanced polymer matrix composites has become routine, the development of metal matrix composites has progressed more slowly, in part because of the extreme processing conditions needed to surround high strength fibers with a matrix of metal.

Metal matrix composites might assume a place in the cooler parts of the skin of a hypersonic aircraft, but at the nose, on leading edges of the wings and in the engines temperatures could exceed the melting point of a metal matrix. For those environments, there is growing interest in a class of composites that have matrices as resistant to heat as the fibers themselves, and also as lightweight and potentially as strong and stiff, namely, ceramics. Because they are brittle, ceramics behave differently from other matrices. In metal and polymer matrix composites the fibers supply most of the strength, and the ductile matrix acts to toughen the system. A ceramic matrix, in contrast, is already abundantly stiff and strong, but to realize its full potential it needs toughening. The fibers in a ceramic matrix composite fill that need by blocking the growth of cracks. A growing crack that encounters a fiber may be deflected or may pull the fiber from the matrix. Both processes absorb energy.

The ceramic matrix gives such composites great temperature resistance. Borosilicate glass reinforced with carbon fibers retains its strength at 600°C. Such matrices as silicon carbide, silicon nitride, aluminum oxide or mullite (a complex compound of aluminum, silicon and oxygen) yield composites that remain serviceable at temperatures well above 1000°C. The heat resistance of a ceramic matrix composite, however, complicates its fabrication.

The characteristics of these three classes of composites can be exemplified by the relation of stress and strain for the unreinforced polymer, metal and ceramic as compared with curves for the corresponding composites. Whereas unreinforced epoxy stretches easily, an epoxy matrix composite containing 50% by volume of silicon carbide fibers is far stiffer (Fig. 1.2a). In an aluminum matrix the same volume of reinforcement, in this case aluminum oxide fibers, also improves stiffness dramatically (Fig. 1.2b). Because the

Fig. 1.2. Stress–strain curves for (a) SiC/epoxy, (b) Al_2O_3/aluminum, and (c) SiC/borosilicate glass composites. (From 'Composites,' Chou, McCullough and Pipes). Copyright © (1986) by *Scientific American, Inc.* All rights reserved.

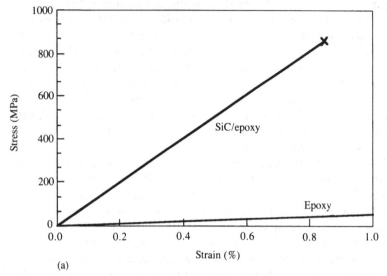

(a)

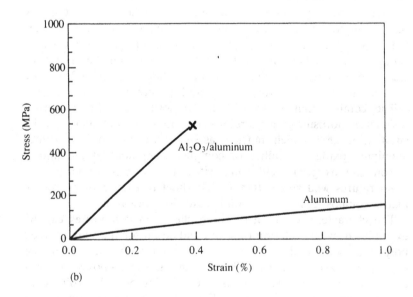

(b)

fibers are brittle, the composite fails at a much lower strain than unreinforced aluminum does. A similar fraction of silicon carbide fibers stiffens a matrix of borosilicate glass only slightly but toughens it considerably, increasing the percentage by which it can be strained without breaking (Fig. 1.2c). The fibers do so by restraining the growth of matrix cracks that might otherwise lead to fracture.

Related to ceramic matrix composites in character but distinctive in manufacture is a composite in which both the matrix and the reinforcing fibers consist of elemental carbon. Carbon–carbon composite is reinforced by the element in a semicrystalline form, graphite; in the matrix the carbon is mostly amorphous. A carbon–carbon composite retains much of its strength at 2500°C and is used in the nose cones and heat shields of re-entry vehicles. Unlike most ceramic matrix composites, it is vulnerable to oxidation at high temperatures. A thin layer of ceramic is often applied to the surface of a carbon–carbon composite to protect it.

The combination of fiber and matrix gives rise to an additional constituent in composites: an interface (or interphase) region. Chemical compatibility between the fibers and the matrix is most crucial at this region. In polymer and metal matrix composites a bond must develop between the reinforcement and the matrix if they are to act

Fig. 1.2. (*cont.*)

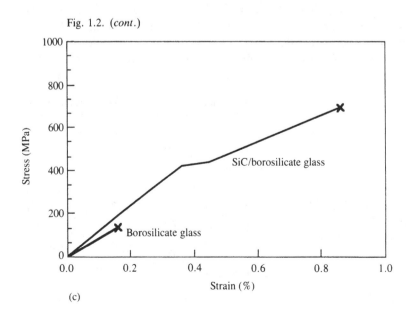

(c)

in concert. A prerequisite for adhesion is that the matrix, in its fluid form, be capable of wetting the fibers. Fibers that would otherwise not be wetted by their matrix can be given a coating that fosters contact by interacting with both the fibers and the matrix. In some cases varying the matrix composition can also promote the process. Once the matrix has wetted the fibers thoroughly, intermolecular forces or chemical reactions can establish a bond.

The properties of an advanced composite are shaped not only by the kind of matrix and reinforcing materials it contains but also by a factor that is distinct from composition: the geometry of the reinforcement. Reinforcing geometries of composites can be grouped roughly by the shape of the reinforcing elements: particles, continuous fibers or short fibers (Fig. 1.3). Sets of parallel continuous fibers are often embedded in thin composite layers, which are assembled into a laminate. Alternatively, each ply in a laminate can be reinforced with continuous fibers woven or knitted into a textile 'preform'. Recently developed geometries dispense with lamination: the fibers are woven or braided in three dimensions (Fig. 1.4), a strategy that in some cases enables the final shape of the composite to be formed directly.

Progress toward managing the many variables of composite design has encouraged investigators to contemplate new complexities. An ordinary composite reinforced with stiff, straight fibers usually displays a nearly constant value of stiffness. New composites designed to display specific non-linear relations of strain and stress are now attracting interest. One such example, a flexible composite consisting of undulating fibers in an elastomeric matrix, can

Fig. 1.3. Particle- and fiber-reinforced composites. (From 'Composites' Chou, McCullough and Pipes.) Copyright © (1986) by *Scientific American, Inc.* All rights reserved.

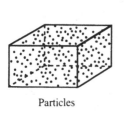

Particles

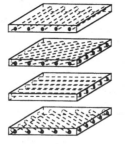

Continuous fibers

Short fibers

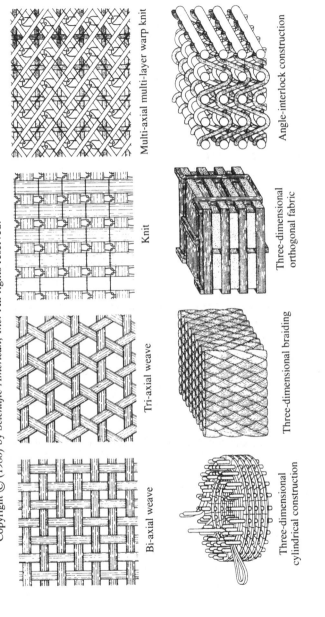

Fig. 1.4. Preforms of textile structural composites. (From 'Composites' Chou, McCullough and Pipes.) Copyright © (1986) by *Scientific American, Inc.* All rights reserved.

Bi-axial weave

Tri-axial weave

Knit

Multi-axial multi-layer warp knit

Three-dimensional cylindrical construction

Three-dimensional braiding

Three-dimensional orthogonal fabric

Angle-interlock construction

elongate readily at low stresses but stiffens when the fibers become fully extended. A hybrid composite strengthened with two kinds of fibers, some of them brittle and inextensible and the others ductile and tough, can display the opposite behavior. The stiff fibers cause stress to increase very sharply at low strains, but when the strain is sufficient to break the stiff, brittle fibers, the curve of stress over strain flattens. The ductile fibers come into play, and as a result the composite becomes more extensible. The hybrid design can yield a material that combines much of the stiffness of an ordinary composite containing only stiff fibers with increased toughness.

Overall, the opportunity in the engineering of fiber composites is the potential to control the composition as well as internal geometry of the materials for optimized performance.

1.3   Why composites?

The question of 'Why composites?' was raised in the 1975 text by Vinson and Chou (1975). The rationale provided then focussed on

(a)   the limitations in strength and ductility for metallic alloys from the viewpoints of theoretical cohesive strength of solids and the arrangement of crystalline defects,

(b)   the need of a balanced pursuit in strength and ductility and the potential of achieving both in fiber composites, and

(c)   the strength limitation of metallic alloys at elevated temperatures and the potential of carbon–carbon composites and refractory metal wire reinforced super-alloys.

The field of fiber composites has witnessed drastic changes and advancement since the mid-1970s because of the availability of several ceramic fibers, high-temperature thermoplastics, glass–ceramic matrices, and intermetallic solids for composites. Although the fundamental physical principles governing the synergism of the component phases in composites should not change, the advancement in materials technology coupled with that in processing, surface science and instrumentation has greatly changed the perspective of composite technology. In the following, the answer to the question of 'why composites?' is re-examined from both economic and technological points of view.

1.3.1   *Economic aspect*

For the discussion of the economic aspect of advanced materials in general and fiber composites in particular, it is worthwhile referring to a recent survey entitled *Problems and Opportunities in Metals and Materials: An Integrated Perspective* by the U.S. Department of the Interior (Sousa 1988). The report asserts that the future growth prospects seem best not in tonnage commodities but rather in materials that are more technology-intensive and more high-value-added. As the economy grows and matures, the rate of growth in consumption of tonnage metals first exceeds, eventually parallels, and finally trails that of the economy as a whole.

Figure 1.5 shows the estimated current relative market maturity of the major metals and other materials. The vertical dimension indicates intensity-of-use (amount/GNP). The potential of polymer, metal and ceramic based composites is obvious. This figure also demonstrates a hard fact of life that eventually catches up with virtually any product – that of market saturation and, as the inexorable evolution of technology proceeds, eventual displacement and decline.

By incorporating different materials into composites, the synthetic class of materials can thus draw on the essential characteristics of diverse materials: the high strength, ductility, thermal–electrical conductivity and formability of metals, the low cost fabrication, light weight and corrosion resistance of polymers, and the strength, corrosion resistance and high-temperature performance of ceramics.

Fig. 1.5. Relative market maturity of materials. (After Sousa (1988).)

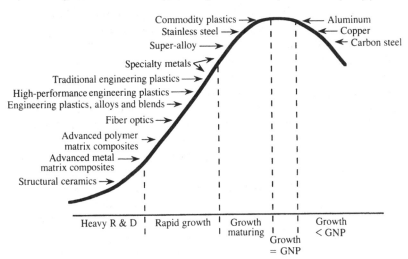

The survey of the U.S. Department of the Interior forecasts the total demand for advanced materials in the U.S. in the year 2000 to be approximately $55 billion annually, roughly the same magnitude as the current U.S. steel market. By comparison, a Japanese Ministry of International Trade and Industry report showed that the Japanese annual demand for advanced materials is expected to be about $34 billion. The breakdown of the market in terms of material categories is (1) advanced polymer composites: 22% (U.S.), 7.6% (Japan); (2) advanced metal alloys and composites: 35% (U.S.), 28.3% (Japan); (3) advanced ceramics: 30% (U.S.), 35.9% (Japan); (4) engineering plastics: 13% (U.S.), 28.3% (Japan). Although the rudimentary nature of such forecasts cannot be overemphasized, the transition from a metals economy to a materials economy, and the importance of composite materials to the economy of advanced materials, is unmistakable.

1.3.2   *Technological aspect*

From the technological viewpoint, advanced composite materials can offer a competitive edge in many products, including aircraft, automobile, industrial machinery and sporting goods, provided their overall production costs can be reduced and their performance improved. According to the study *New Structural Materials Technologies* made by the Congress of the United States, Office of Technology Assessment (1988), the broader use of advanced structural materials requires not only solutions to technical problems but also changes in attitudes among researchers and end-users. The traditional approach based upon discrete design and manufacturing steps for conventional structural materials needs to be replaced by an integrated design and manufacturing process which necessitates a closer relationship among researchers, designers, and production personnel as well as a new approach to the concept of material costs. A fully integrated design process capable of balancing all of the relevant design and manufacturing variables requires an extensive database on matrix and fiber properties, the ability to model fabrication processes, and three-dimensional analysis of the properties and behavior of the resulting structure. Knowledge of the relationships among the constituent properties, microstructure and macroscopic behavior of the composite is basic to the development of an integrated design methodology.

To further understand the impacts of advanced structural materials on manufacturing, this report examines the following two possibilities: substitution by direct replacement of metal components in existing products and the use in new products that are made possible by the new materials. Direct substitution of a ceramic or composite part for a metal part is not likely to take full advantage of the superior properties and design flexibility of advanced materials. Substitution of conventional structural metals such as steel and aluminum alloy by composites is highly unlikely. Because of their low cost and manufacturability, these metals are ideally suited for applications in which they are now used. On the other hand, the metal industry has responded to the potential of direct substitution by developing new alloys with improved properties, such as high-strength, low-alloy steel and aluminum–lithium. According to this assessment, significant displacement of metals could occur in four potential markets: aircraft, automobiles, containers and constructions.

In the choice of material substitution, a variety of factors need to be taken into account. Compton and Gjostein (1986) analyzed the weight saving and cost for material substitution for ground transportation. Weight reduction that can be achieved in designing a part by substituting a light-weight material for a conventional one depends critically on the part's function. A unit volume of cast aluminum weighs 63% less than an equal volume of cast iron. Cast iron, however, is stiffer than cast aluminum. Therefore if a hypothetical cast-aluminum part is to be as stiff as a cast-iron one, more aluminum would have to be used and the weight saving would be reduced to 11%. If equal loading carrying capacity is required in the hypothetical aluminum part, the weight saving would be 56%. (In actual design situations the weight saving offered by the substitution of aluminum for cast iron ranges from 35 to 60%.) Similarly, aluminum and fiber-reinforced plastics are much lighter than mild (ordinary) steel by volume. The weight savings, however, are much smaller if equal stiffness or equal collapse load and bending stiffness (a measure of structural strength) is needed. High-strength steel is no lighter by volume than mild steel, nor is it stiffer. Where structural strength is the main concern, however, high-strength steel does offer a weight saving: 18% in the example discussed by Compton and Gjostein.

In terms of innovative designs and new products based upon advanced composites, the automotive industry undoubtedly provides an excellent paradigm. The use of polymer matrix composites for primary body structures and chassis/suspension systems is under evaluation by the major automobile manufacturers. The potential advantages of using composites are: weight reduction and resulting

fuel economy; improved overall quality and consistency in manufacturing; lower assembly costs due to parts consolidation; lower
investment costs for plant, facilities, and tooling; improved corrosion resistance; and lower operating costs. The major barriers to the
large-scale applications of composites are the lack of high-speed,
high-quality, low-cost manufacturing processes; uncertainties regarding crash integrity and long-term durability; and lack of
adequate technologies for repair and recycling of polymer composite structures. According to Compton and Gjostein, glass fiber
reinforced composites are capable of meeting the functional requirements of the most highly loaded automotive structures. Candidate fabrication methods include resin transfer molding, compression molding, and filament winding. Among these methods, resin
transfer molding seems the most promising, although none of these
methods can satisfy all of the production requirements at this time.
There is no doubt that the large-scale adoption of polymer matrix
composites for automotive structures would have a major technological impact on the fabrication and assembly of automobiles.

Fig. 1.6. Temperature capabilities of polymer, metal and ceramic matrix
materials. (After Mody and Majidi 1987, with permission from the Society
of Manufacturing Engineers.)

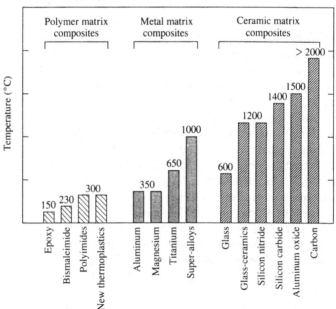

Another technological aspect that motivates the use of fiber composites pertains to the demand of an elevated temperature environment (Steinberg 1986). Temperature capabilities of polymer, metal and ceramic matrix materials are shown in Fig. 1.6 (Mody and Majidi 1987). The demand for high-temperature applications of composites is best exemplified by the need for aerospace materials. The U.S. goals for subsonic, supersonic and hypersonic flight and for space explorations require alloys and composites with superior strength, light weight and resistance to heat. According to Steinberg, the evolution of aircraft has required continual improvements in materials because increased speed raises the heating of the skin from friction with the air and increased power raises the temperature of the engine. Figure 1.7 shows the changes in skin temperatures from aircraft of the 1930s to the proposed Orient Express which is a transatmospheric craft capable of cruising at great speed in space. The skin materials have progressed from wood and fabric to advanced alloys of aluminum, nickel and titanium and graphite fiber reinforced polymer composites.

Figure 1.8 shows the changes in engine temperature from engines cooled by water to those of scramjets. The need for composites in engine components can be understood from the evolution in engine

Fig. 1.7. Evolution of aircraft skin temperatures. (From 'Materials for Aerospace', Steinberg). Copyright © (1986) by *Scientific American, Inc.* All rights reserved.

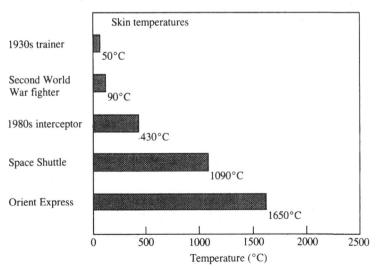

performance. According to Steinberg, the thrust delivered by a big jet engine for transport and cargo aircraft has increased about six fold over the past 30 years, approaching 294 000 newtons (66 000 pounds) now. During the same period the weight of the engine has increased by a factor of only two or three. The thrust-to-weight ratio of the military aircraft may approach 15:1 by the year 2000. The performance of jet engines has been made possible partially with improvements in turbine blades. It is predicted that with the further improvements in blades and other aspects of aircraft propulsion, a typical propulsion system in the year 2000 will be likely to contain about 20% each of composites, steel, nickel and aluminum, 15% titanium, 2% ordered alloys (aluminides, e.g. titanium–aluminum or nickel–aluminum) and 1% ceramics (Steinberg 1986).

Clark and Flemings (1986) have also examined the present and future material systems for meeting the engine operating temperature requirements. In Fig. 1.9 the lowest band on the graph indicates the temperature increase that has been achieved so far through improvements in nickel-based super-alloys, the standard turbine material. It is believed that in the coming decades alloy turbine blades made of metal strengthened by directional crystal structures, and blades protected by a coating of ceramics or special

Fig. 1.8. Evolution of aircraft engine temperatures. (From 'Materials for Aerospace' Steinberg). Copyright © (1986) by *Scientific American, Inc.* All rights reserved.

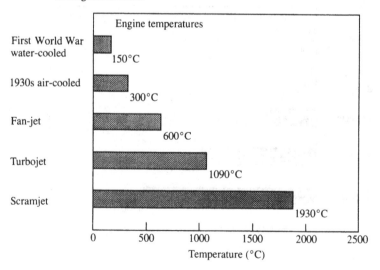

alloys, will allow an increase in turbine-inlet temperatures. However, ultimately, the demand for very-high-temperature material
can only be met by ceramic matrix composites and carbon–carbon
composites.

1.4   Trends and opportunities

Kelly (1987a&b), in a recent outline of the trends in
materials science and processing, examined the status of fiber
composites. It was concluded that the development of this field has
been mainly driven by the aerospace industry. This development
has contributed to the growth of a relatively small body of new
science which related the colligative properties of fiber composites
to the properties of the individual components. There have been
interesting combinations of properties not hitherto available in
single phase materials, for example, a negative thermal expansion
and a negative Poisson's ratio. However, there have not been large
non-linear synergistic effects. There is perhaps not much new
science of the colligative properties of composites. However, in
Kelly's view, the studies of design of fiber composites are critical for

Fig. 1.9. Rise in the operating temperature of jet engines with time.
(From 'Advanced Materials and Economy' Clark and Flemings). Copyright © (1986) by *Scientific American, Inc.* All rights reserved.

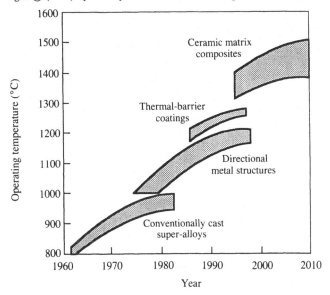

their applications. Furthermore, there may be much new science on how to produce composites.

The significant trends in structural composites point to the direction of low-temperature metal matrix, resin matrix, metal–resin matrix, rubber matrix, cement–ceramic matrix and elevated-temperature composites. Non-structural composites are increasingly being recognized for their unique opportunities in electric, magnetic, superconducting and biomedical applications. A brief summary of those trends follows (see Kelly 1987a). A major motivation behind the development of low-temperature metal matrix composites in the U.S. has been for the utilization of high-stiffness continuous fibers in a matrix material without the disadvantage of thermosetting resins of low thermal conductivity, high thermal expansion, dimensional instability, hygrothermal degradation, material loss in high vacuum, susceptibility to radiation damage, and lower temperature brittleness. The lighter metals do not possess these disadvantages; their low atomic number (Z) is important in a neutron-rich environment. It is useful to bear in mind that five out of the 13 lowest-Z solids are metals. Some of these metals, together with their atomic number and density, are listed below: lithium $(Z = 3$, density $= 0.53 \, \mathrm{Mg \, m^{-3}})$, sodium $(11, 0.97)$, potassium $(19, 0.86)$, calcium $(20, 1.55)$, magnesium $(12, 1.741)$, beryllium $(4, 1.85)$, and aluminum $(13, 2.7)$.

Reinforcement of a light metal, e.g. aluminum and magnesium, is attractive in the automobile industry in reducing creep at moderate temperatures and improving wear resistance. Coating for carbon fibers is necessary for incorporation into aluminum and magnesium matrices.

Thermoplastic resins have certain advantages over thermosets in their infinite shelf life, good resistance to water and solvents, and ductility. Thermoplastics are attractive particularly from the viewpoint of composites manufacture because they are rapidly processable, and are better adapted to automated manufacturing. Also, they can be recycled and joined by welding.

Laminates formed by bonding metal sheets to fiber–resin composites take advantage of the synergistic effects of hybrid composites. For instance, the combination of aluminum foil with Kevlar/epoxy composite results in enhanced fatigue resistance and compressive strength.

Rubber (elastomeric) matrix can be reinforced with short and continuous fibers and can provide the capability of large non-linear elastic deformation. Automobile tires and coated fabrics are examples in this category.

Contrary to the large deformation of rubber type flexible composites, ceramic based composites offer the other extreme on the scale of deformation. The brittle nature of ceramic solids requires a new way of thinking in 'reinforcement'. Fibers are added for the purpose of improving toughness against fracture and ductility in terms of energy absorption and deformation range.

Ceramic matrix composites, directionally solidified eutectics, intermetallic solids, certain types of metal based composites, and carbon–carbon composites are the candidate materials for elevated-temperature applications. Among these, carbon–carbon composites present the ultimate in high-temperature materials under reducing conditions. They have many tribological applications. Protection against oxidation and densification of the matrix are major challenges to carbon–carbon composites.

Finally, the potential of non-structural composites has not been fully explored. Kelly (1987a) cited the examples in making special devices. For example, a magnetoresistive device obtained by coupling a metal rod with a semiconductor matrix provides a contactless potentiometer or a fluxmeter, or coupling a piezoelectric and magnetostrictive material gives a magnetoelectric material. The potential for biomedical applications of flexible composites also exists (see Chou 1989).

1.5   Microstructure–performance relationships

Chapters 2–9 examine the stiffness, strength and failure behavior of several types of composites: laminated composites composed of continuous fibers; composites reinforced with short fibers in biassed or random orientations; composites with two types of fibers in intermingled, interlaminated or interwoven forms; composites reinforced with textile preforms; and flexible composites exhibiting large deformations. The mathematical tools for analyzing their thermomechanical properties have been presented. Most significantly, an effort has been made to delineate the relationship between the behavior and these composites.

In the following, a comparison is first made among the stress–strain behaviors of three composite systems. The purpose is to demonstrate the versatility in composite performance through the design of microstructure. This is followed by specific examples of tailoring the material performance through microstructural design. Lastly, the emerging field of 'intelligent composites' is introduced.

1.5.1 *Versatility in performance*

For the purpose of demonstrating the versatility of the performance of composites, the stress–strain relationships of three types of composites are examined. Figure 1.10 shows the stress–strain curves of a unidirectional carbon fiber reinforced glass matrix composite (Nardone and Prewo 1988). The behavior is typical for brittle matrix composites based upon polymer and glass/ceramic matrices. The knee phenomenon of the stress–strain curve resembles the yield behavior of metallic alloys.

Figure 1.11 gives the stress–strain curves of interlaminated carbon/glass hybrid composites. The ability of the low elongation phase (carbon) in developing multiple fractures enables the hybrid composites to sustain deformations at a level much higher than that of the all-low elongation fiber composite. The energy absorption capability as indicated by the area under the stress–strain curve is also much higher than that of the all-carbon fiber composite. The shape of this stress–strain curve resembles those of ductile metals with strain-hardening behavior.

The stress–strain data of a flexible composite (Fig. 1.12) show rapid increase in stress and stiffness at large deformation (Chou 1989). It resembles the behavior of certain biological materials such as soft animal tissues (Humphrey and Yin 1987; Gordon 1988).

Fig. 1.10. Tensile stress–strain curves of a carbon/borosilicate glass composite. (After Nardone and Prewo 1988).

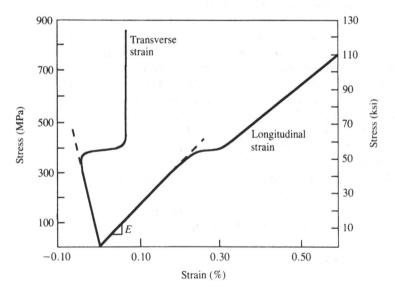

Fig. 1.11. Tensile stress–strain curves of a carbon/glass/epoxy interlaminated composite. (After Bunsell and Harris 1974).

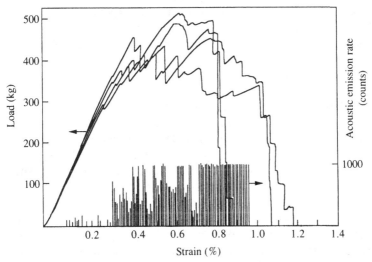

Fig. 1.12. Tensile stress–strain data of a flexible composite. (After Chou 1989.)

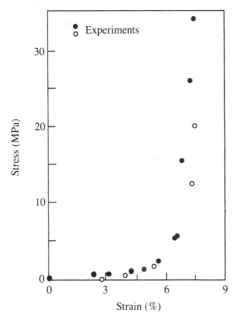

It is interesting to note that through the selection of fiber and matrix materials, as well as their geometric arrangements, a broad spectrum of material performance can be accomplished. It is feasible to design the physical and mechanical properties of composites which not only duplicate the performance of some existing materials but also fulfil the most demanding structural roles not envisioned before.

1.5.2   *Tailoring of performance*

The structure–performance relationships of the various types of fiber composites are further demonstrated in this section. First, for continuous fiber composites, the problem of edge delamination is used as an example. Next the variation of composite electric properties with the configuration of reinforcements is demonstrated.

Consider the $[\pm 45°/0_2°/90_2°]_s$ laminate. The effect of fiber orientation on the deformation of each individual lamina is highly anisotropic (Fig. 1.13). The compatibility of displacements among the laminae induces interlaminar stresses through the thickness direction of the laminate. Sun (1989) has demonstrated that the opening mode of delamination can be minimized through fiber hybridization, stitching, the use of adhesive layers, ply termination, and modification of edge geometry.

Figure 1.14 shows the free-edge interlaminar normal stresses in the all-carbon composite and the hybrid composite formed by replacing 90° plies with a glass/epoxy composite. A significant

Fig. 1.13. Effect of fiber orientation on the deformation of composite laminae. (After Sun 1989.)

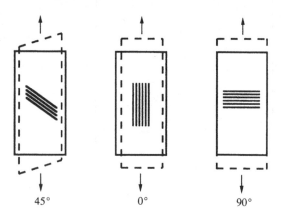

45°   0°   90°

reduction in interlaminar normal stress is achieved with hybrid laminates. The experimentally measured delamination initiation stress and failure stress are 324.3 MPa and 800.4 MPa, respectively. The corresponding stresses for the hybrid laminate are 800.4 MPa and 883.2 MPa, respectively. Thus, the addition of the glass/epoxy plies significantly improves the delamination stress. The gain in failure stress is not as significant since the 0° plies in both laminates dominate the ultimate strength.

Reinforcements in the thickness direction can suppress interlaminar failure. Figure 1.15 shows the X-ray radiographs of $[\pm 45°/0_2°/90°]_s$ laminates under uniaxial tension. The specimen with through-the-thickness stitches along the free edges experiences much less delamination than the specimen without stitches.

Besides relying on textile performing techniques such as stitching, weaving and braiding, delamination in brittle resin matrix composites can be remedied by adding a ductile matrix in the form of thin adhesive layers. The resulting composite has a hybridized matrix. It has been demonstrated in $[0°/90°/45°/-45°]_s$ carbon/epoxy laminates that by reducing the free-edge effect the laminate strength can be greatly improved. Furthermore, the laminate strength becomes an isotropic property which can be predicted by the classical failure theory. The use of adhesive layers in laminates subject to low-velocity impact also proves to be effective in suppressing the development of matrix cracking and delamination.

Fig. 1.14. Free-edge interlaminar normal stresses on the mid-surface in carbon/epoxy and carbon/glass/epoxy laminates. (After Sun 1989).

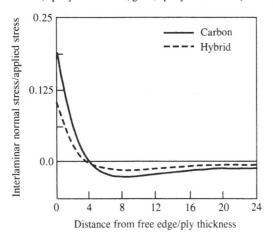

The transport properties, e.g. electrical conductivity, thermal conductivity, dielectric constants, magnetic permeability and diffusion coefficients of composites, are also sensitive to the microstructure of the reinforcements. McCullough (1985) has demonstrated the importance of structural features that promote transport along the preferred path, i.e. percolative mechanisms. Consider, for instance, the electrical behavior of metal-filled polymers. The effective resistivity changes sharply from non-conducting to conducting behavior upon crossing a 'percolation threshold'. Figure 1.16 illustrates such a transition for a composite containing conductive fillers ($\rho_f = 10^{-6}\,\Omega\,cm$) in an insulating polymer matrix ($\rho_m = 10^{16}\,\Omega\,cm$). The decrease in resistivity with the increase in filler volume fraction is attributed to the enhancement in probability of particle–particle contact. McCullough has concluded that these contacts promote the formation of continuous conduction paths that mimic the behavior of conducting fibers.

1.5.3 *Intelligent composites*

Traditionally, fiber composites have been designed and manufactured with the purpose of serving very specific functional

Fig. 1.15. X-ray radiographs showing delamination in unstitched (left) and stitched (right) $[\pm45°/0_2°/90°]_s$ laminates under uniaxial tension. (After Mignery, Tan and Sun 1985.)

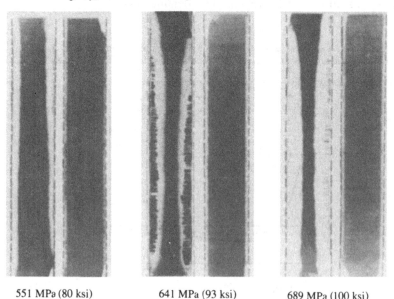

551 MPa (80 ksi)   641 MPa (93 ksi)   689 MPa (100 ksi)

goals. Such goals and considerations may include stiffness, fracture toughness, fatigue life, impact resistance, electromagnetic shielding, corrosion resistance, and biocompatibility, just naming a few. With the expansion in available material systems for composites, advancements in fabrication technologies, and improvements in analysis and design techniques, it becomes increasingly feasible for developing *multi-functional* fiber composites for which a number of functional goals are satisfied simultaneously, and the performance can be optimized.

A new breed of multi-functional composites is dubbed 'smart composites' or 'intelligent composites'. Takagi (1989) has defined intelligent materials as 'those which can manifest their own functions intelligently depending on environmental changes'. Thus, intelligent composites can react to the thermal, electrical, magnetic, chemical or mechanical environment and adjust their performance accordingly. It should be borne in mind that intelligent composites are made possible only through the design of their microstructures.

There are two basic requirements for intelligent composites to 'think' for themselves. First, the ability to detect the change in the environment, such as pressure, strain, temperature, and electro-

Fig. 1.16. Illustration of chain formation in a particulate filled composite. Open circles and closed circles indicate, respectively, isolated particles and contacting particles participating in chain formation. ρ and V_f denote resistivity and filler volume fraction, respectively. (After McCullough 1985.)

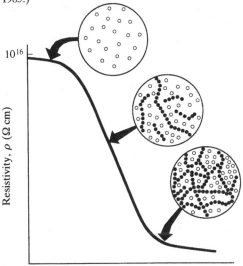

Resistivity, ρ (Ω cm)

10^{16}

Filler volume fraction, V_f

magnetic radiation is necessary. Next, the ability in feedback and control is also needed so corrective actions can be taken.

An example of intelligent composites under consideration by researchers is the skin of an aircraft wing (see Port, King and Hawkins 1988). The resin-based composite skin in this case has built-in optic-fiber sensors which through the pulses of laser light can detect internal defects and damages, the weight of ice or incoming electromagnetic radar waves. Signals from the sensors would be analyzed by patches of chips mounted on a flexible printed circuit board bonded over the skin.

It has been suggested that implanting monolithic microwave integrated circuit chips around an airplane's surface would produce a huge, omnidirectional antenna that would be far more effective than the small forward looking units now mounted on its nose. Other applications of intelligent composites have been envisioned for the purpose of in-flight damage assessment capability on airplanes and orbiting spacecrafts, prelaunch checks for leaks and structural integrity of the casing around rockets, altering the stiffness of sporting equipment such as golf club and fishing rods in response to the changing operating conditions, and monitoring the sway of high-rise buildings induced by hurricane winds or earthquakes so measures to compensate such deformations can be activated (Port, King and Hawkins, 1988). Some of the issues of intelligent structures have been discussed by Rogers (1988).

In summary, the challenges of intelligent composites are manifested by the following factors: (a) development of sensing, feedback and control systems as well as the technologies for fabricating composites imbedded with such devices, (b) implementation of the required changes in the shapes of the structural components, for example the change of the angle and shape of an airplane's wing, and (c) perhaps the most challenging task, the ability of a material to change its performance, for example the stiffness or transport properties.

The combination of the structural and non-structural roles of a composite in an integrated manner will undoubtedly change the performance of fiber composites in a way not envisioned in the past.

1.6   Concluding remarks

Having examined the evolution of engineering materials, and the role of fiber composites in materials technology, it is perhaps useful to put in perspective the research and economic opportunities of advanced composites.

First, from the viewpoint of materials research, it is important to recognize that the distinction between the three classes of materials, i.e. metals, ceramics and plastics, is disappearing. As observed by Kelly (1987a), there are now plastics as strong as metals which show some electrical conductivity. Metals are being made which are super-plastic and can be subjected to deformations in processing like conventional polymeric materials. Also the three classes of materials are beginning to show the same limits of strength and stiffness; fibers made from all three can attain stiffness and strength close to the theoretically predicted values. Furthermore, the properties of all three classes of materials can be modified and improved by the use of surface coatings.

As the distinction between the three classes of materials disappears, new possibilities and opportunities arise. One of these, according to Kelly, is the possibility of designing materials not so much for final properties but equally in terms of processability. These thoughts have profound implications for the future technology of fiber composites:

(1) The commonality in processing shared by the three classes of materials, e.g. super-plastic forming of metal and polymers, injection molding of polymers and ceramic powders, will enable more extensive and effective transfer of knowhow among the three basic disciplines and effect efficient processing technology for fiber composites.

(2) The commonality in performance shared by the three classes of materials, e.g. stiffness, strength, thermal expansion, enables the material scientist to engineer composites with a broad spectrum of component materials. Consequently, hybridizations of materials, e.g. glass and low-melting-point metal, ceramics and thermoplastics, and polymer and metal in laminates or other interdispersed composite forms can be achieved and the properties optimized (e.g. composites composed of metal and polymer components of nearly the same stiffness but different fatigue resistance, or thermal expansion coefficient).

(3) The similarity in material property and behavior implies that analytical and design methodologies originally developed for a specific class of composites may be transferable to others. A notable example is the fracture and failure behavior of ceramic and polymer based composites.

(4) The complex task inherent in conceiving components and their materials and developing the proper design methodology will grow increasingly dependent on computers and multi-disciplinary teams. Such an approach will harness the full potential of composites for the technologies of the future.

2 Thermoelastic behavior of laminated composites

2.1 Introduction

Laminated composites are made by bonding unidirectional laminae together in predetermined orientations. The basis for analysis of thin laminated composites is the classical plate theory. When the thickness direction properties significantly contribute to the response of the laminate to an externally applied elastic field, the classical plate theory breaks down.

Fundamental to the treatment of thin laminates is the knowledge of the thermoelastic properties of a unidirectional lamina. These properties are predictable from the corresponding properties of constituent fiber and matrix materials as well as the fiber volume fraction. Having established the elastic response of a unidirectional lamina, the behavior of laminated composites is then analyzed from the strain and curvature of the mid-plane of the laminate as well as the force and moment resultants acting on its boundary edges. Because of the complexity of the constitutive equations for a general anisotropic laminated plate, simplifications of the stress–strain relations are accomplished through the manipulation of the geometric arrangement of the laminae. The lamination theory is a relatively mature subject; its treatment can be found in text books of, for instance, Ashton, Halpin and Petit (1969), Jones (1975), Vinson and Chou (1975), Christensen (1979), Tsai and Hahn (1980), Carlsson and Pipes (1987), and Chawla (1987), and in the review articles of Chou (1989a and b). A modification of the classical plate theory is in the inclusion of higher order terms in the displacement field expansion to account for the transverse shear deformation. An outline of such modifications adopted by various researchers is presented.

The classical thin laminated theory has been extended to take into consideration the effects of thermal and moisture diffusions, with particular emphasis on the transient behavior. Because of the large differences in the magnitudes of the thermal conductivity and moisture diffusion coefficients, the thermal and hygroscopic problems can be solved separately and their linear elastic fields can be superposed. Stress concentrations due to transient thermal effects are of particular interest in the study of laminate thermal shock resistance.

The mechanics of the thermoelastic behavior of laminated composites is fundamental to the understanding of the strength, fracture and fatigue behavior of all continuous-fiber composites including those reinforced with textile preforms.

2.2   Elastic behavior of a composite lamina

2.2.1   *Elastic constants*

It is well known that for a homogeneous isotropic material (i.e. the material properties are independent of the location and direction), two independent material elastic constants are sufficient to specify the constitutive relations. These could be any two of the five constants commonly used: E (Young's modulus), v (Poisson's ratio), G (shear modulus), K (bulk modulus), and k (plane strain bulk modulus). The relations among these constants are

$$G = E/2(1 + v)$$
$$K = E/3(1 - 2v) \tag{2.1}$$
$$k = E/2(1 - v - 2v^2)$$

Twenty-one independent constants are necessary to describe the elastic stress–strain relation of a generally anisotropic material (i.e. the material properties are different in different directions). However, due to the material symmetries, the number of the independent constants can be greatly reduced. Consider a lamina (Fig. 2.1) composed of unidirectional straight fibers in a matrix. Assume that

Fig. 2.1. A unidirectional fiber composite lamina.

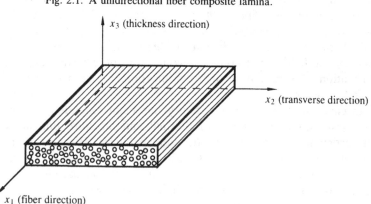

it is homogeneous on a scale much larger than that of the inter-fiber spacing. Then, the unidirectional lamina can be treated as a homogeneous orthotropic continuum (i.e. having three mutually perpendicular planes of symmetry). The coordinates $x_1-x_2-x_3$ shown in Fig. 2.1 are known as the material principal coordinates, where x_1 is parallel to the fibers and x_2 lies in the plane of the lamina. For circular cross-section fibers randomly distributed in a unidirectional lamina, the lamina can be further assumed macroscopically as transversely isotropic, namely the material properties in planes transverse to the fiber direction are isotropic. Then, there are only five independent constants. The commonly used engineering elastic constants for the transversely isotropic lamina, referring to the fiber (x_1) and in-plane transverse (x_2) directions, are denoted by E_{11} (longitudinal Young's modulus), E_{22} (transverse Young's modulus), v_{12} (Poisson's ratio due to loading in the x_1 direction and contraction in the x_2 direction), and G_{12} (in-plane shear modulus). These four independent elastic constants can be determined experimentally by three simple tensile tests of composite specimens with fiber orientations of $0°$, $90°$ and $[\pm 45°]_{2s}$; the relevant testing standards are ASTM D3039-76 and ASTM D3518-76. The fifth independent constant, representing the transverse isotropic properties, could be either v_{23} (transverse Poisson's ratio) or G_{23} (transverse shear modulus); the two are related by

$$G_{23} = \frac{E_{22}}{2(1 + v_{23})} \tag{2.2}$$

The other engineering constants are:

$$v_{21} = \frac{E_{22}}{E_{11}} v_{12}$$

$$E_{33} = E_{22}$$

$$G_{13} = G_{12} \tag{2.3}$$

$$v_{32} = v_{23}$$

$$v_{13} = v_{12}$$

$$v_{31} = v_{21}$$

Various micromechanical models are available for predicting the elastic properties of unidirectional laminae from their constituent properties. Most of the matrices and some of the fibers used in composites can be considered as isotropic. Let the elastic constants of Eq. (2.1) for the isotropic fiber and matrix materials be denoted by the subscripts f and m, respectively. Also, the fiber volume fraction of the composite is indicated by V_f. Assuming no void in the composite, the volume fraction of matrix is

$$V_m = 1 - V_f \tag{2.4}$$

The following relations due to Hashin and Rosen (see Rosen 1973) are quoted for their concise forms and, hence, ease in application.

$$E_{11} = E_f V_f + E_m V_m + \frac{4V_f V_m (v_f - v_m)^2}{\dfrac{V_m}{k_f} + \dfrac{V_f}{k_m} + \dfrac{1}{G_m}}$$

$$E_{22} = \frac{4k_t^* G_t^*}{k_t^* + G_t^* \left(1 + \dfrac{4k_t^* v_{12}^2}{E_{11}}\right)}$$

$$v_{12} = v_f V_f + v_m V_m + \frac{V_f V_m (v_f - v_m)\left(\dfrac{1}{k_m} - \dfrac{1}{k_f}\right)}{\dfrac{V_m}{k_f} + \dfrac{V_f}{k_m} + \dfrac{1}{G_m}} \tag{2.5}$$

$$G_{12} = G_m \frac{V_m G_m + (1 + V_f) G_f}{(1 + V_f) G_m + V_m G_f}$$

$$v_{23} = \frac{E_{22}}{2G_t^*} - 1$$

where

$$k_f = E_f / 2(1 - v_f - v_f^2)$$

$$k_m = E_m / 2(1 - v_m - v_m^2)$$

$$k_t^* = \frac{k_m k_f + (V_f k_f + V_m k_m) G_m}{V_m k_f + V_f k_m + G_m}$$

$$G_t^* = G_m \frac{(\alpha + \beta_m V_f)(1 + \rho V_f^3) - 3V_f V_m^2 \beta_m^2}{(\alpha - V_f)(1 + \rho V_f^3) - 3V_f V_m^2 \beta_m^2} \tag{2.6}$$

$$\alpha = (\gamma + \beta_m)/(\gamma - 1)$$

$$\beta_m = \frac{1}{3 - 4v_m}, \qquad \beta_f = \frac{1}{3 - v_f}$$

$$\rho = (\beta_m - \gamma \beta_f)/(1 + \gamma \beta_f)$$

$$\gamma = G_f / G_m$$

Fibers such as carbon and Kevlar exhibit anisotropic behavior; their thermoelastic properties along and transverse to the fiber axis are significantly different. These fibers are considered to be transversely isotropic, and thus five independent constants are needed to describe their elastic properties, namely, E_{1f}, E_{2f}, G_{12f}, v_{12f} and G_{23f}. The following expressions, due to Chamis (1983), describe the elastic properties of a unidirectional lamina composed of anisotropic fibers in an isotropic matrix:

$$E_{11} = E_{1f}V_f + E_m V_m$$

$$E_{22} = E_{33} = \frac{E_m}{1 - V_f(1 - E_m/E_{2f})}$$

$$G_{12} = G_{13} = \frac{G_m}{1 - V_f(1 - G_m/G_{12f})} \tag{2.7}$$

$$G_{23} = \frac{G_m}{1 - V_f(1 - G_m/G_{23f})}$$

$$v_{12} = v_{13} = v_{12f}V_f + v_m V_m$$

$$v_{23} = \frac{E_{22}}{2G_{23}} - 1$$

Halpin and Tsai (1967) have developed some semi-empirical relations for the laminar elastic properties. These expressions contain certain parameters which are influenced by the geometry of the reinforcing phases, their packing in the composite, and the loading conditions. Estimates of the values of these parameters can be obtained by comparing the Halpin–Tsai equation predictions with the numerical solutions employing formal elasticity theory (Halpin 1984). The effect of interfacial debonding on elastic properties has been discussed by Takahashi and Chou (1988).

2.2.2 *Constitutive relations*

Consider a unidirectional lamina exhibiting orthotropic symmetry. The constitutive relations, referring to the material principal coordinates x_1–x_2–x_3, assume the general form (Vinson and Chou 1975):

$$
\begin{pmatrix} \varepsilon_{11} \\ \varepsilon_{22} \\ \varepsilon_{33} \\ 2\varepsilon_{23} \\ 2\varepsilon_{13} \\ 2\varepsilon_{12} \end{pmatrix} = \begin{pmatrix} S_{11} & S_{12} & S_{13} & 0 & 0 & 0 \\ S_{12} & S_{22} & S_{23} & 0 & 0 & 0 \\ S_{13} & S_{23} & S_{33} & 0 & 0 & 0 \\ 0 & 0 & 0 & S_{44} & 0 & 0 \\ 0 & 0 & 0 & 0 & S_{55} & 0 \\ 0 & 0 & 0 & 0 & 0 & S_{66} \end{pmatrix} \begin{pmatrix} \sigma_{11} \\ \sigma_{22} \\ \sigma_{33} \\ \sigma_{23} \\ \sigma_{13} \\ \sigma_{12} \end{pmatrix} \tag{2.8}
$$

Here σ_{ij}, the stress tensors, are defined in Fig. 2.2. ε_{ij} are the strain tensors defined in a manner analogous to the stress components; it should be noted that the engineering shear strain $\gamma_{ij} = 2\varepsilon_{ij}$ $(i \neq j)$. S_{ij} denote the components of the compliance matrix. For the case of a transversely isotropic lamina with the x_2–x_3 plane being isotropic, the compliance constants are related to the engineering elastic constants as:

$$S_{11} = \frac{1}{E_{11}}$$

$$S_{22} = S_{33} = \frac{1}{E_{22}}$$

$$S_{12} = S_{13} = -\frac{\nu_{12}}{E_{11}} = -\frac{\nu_{21}}{E_{22}} \qquad (2.9)$$

$$S_{23} = -\frac{\nu_{23}}{E_{22}}$$

$$S_{44} = \frac{1}{G_{23}}$$

$$S_{55} = S_{66} = \frac{1}{G_{12}}$$

Fig. 2.2. Stress tensor components.

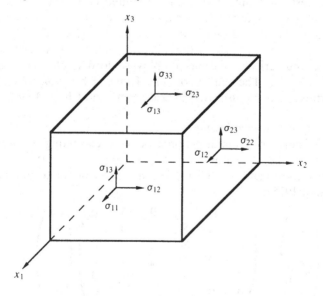

Equation (2.8) can be inverted to obtain the following stress–strain relations

$$
\begin{pmatrix} \sigma_{11} \\ \sigma_{22} \\ \sigma_{33} \\ \sigma_{23} \\ \sigma_{13} \\ \sigma_{12} \end{pmatrix} = \begin{pmatrix} C_{11} & C_{12} & C_{13} & 0 & 0 & 0 \\ C_{12} & C_{22} & C_{23} & 0 & 0 & 0 \\ C_{13} & C_{23} & C_{33} & 0 & 0 & 0 \\ 0 & 0 & 0 & C_{44} & 0 & 0 \\ 0 & 0 & 0 & 0 & C_{55} & 0 \\ 0 & 0 & 0 & 0 & 0 & C_{66} \end{pmatrix} \begin{pmatrix} \varepsilon_{11} \\ \varepsilon_{22} \\ \varepsilon_{33} \\ 2\varepsilon_{23} \\ 2\varepsilon_{13} \\ 2\varepsilon_{12} \end{pmatrix} \tag{2.10}
$$

where C_{ij} are the components of the stiffness matrix. Again, for the case of transverse isotropy in the x_2-x_3 plane, the following relations hold:

$$C_{11} = E_{11}(1 - v_{23}^2)/\Delta$$

$$C_{22} = C_{33} = E_{22}(1 - v_{12}v_{21})/\Delta$$

$$C_{44} = G_{23}$$

$$C_{55} = C_{66} = G_{12} \tag{2.11}$$

$$C_{12} = C_{13} = (v_{21} + v_{21}v_{23})E_{11}/\Delta = (v_{12} + v_{12}v_{23})E_{22}/\Delta$$

$$C_{23} = (v_{23} + v_{12}v_{21})E_{22}/\Delta$$

$$\Delta = 1 - 2v_{12}v_{21} - v_{23}^2 - 2v_{12}v_{21}v_{23}$$

For a unidirectional composite lamina where the thickness is much smaller than the in-plane (x_1-x_2) dimensions, it is sufficient to consider the two-dimensional constitutive relations. Following the convention used in the composites literature, the following contracted notations, σ_i and ε_i, are introduced for the stress and strain components, respectively. Their relations to the tensorial stress and strain components are:

$$\sigma_1 = \sigma_{11}, \quad \sigma_2 = \sigma_{22}, \quad \sigma_3 = \sigma_{33}, \quad \sigma_4 = \sigma_{23}(=\tau_{23}),$$
$$\sigma_5 = \sigma_{13}(=\tau_{13}), \quad \text{and} \quad \sigma_6 = \sigma_{12}(=\tau_{12})$$

$$\varepsilon_1 = \varepsilon_{11}, \quad \varepsilon_2 = \varepsilon_{22}, \quad \varepsilon_3 = \varepsilon_{33}, \quad \varepsilon_4 = 2\varepsilon_{23}(=\gamma_{23}),$$
$$\varepsilon_5 = 2\varepsilon_{13}(=\gamma_{13}), \quad \text{and} \quad \varepsilon_6 = 2\varepsilon_{12}(=\gamma_{12})$$

Under plane stress condition (i.e. $\sigma_{33} = \sigma_{13} = \sigma_{23} = 0$), and using the contracted notations, Eq. (2.8) can be reduced to

$$
\begin{pmatrix} \varepsilon_1 \\ \varepsilon_2 \\ \varepsilon_6 \end{pmatrix} = \begin{pmatrix} S_{11} & S_{12} & 0 \\ S_{12} & S_{22} & 0 \\ 0 & 0 & S_{66} \end{pmatrix} \begin{pmatrix} \sigma_1 \\ \sigma_2 \\ \sigma_6 \end{pmatrix} \tag{2.12}
$$

where the compliance constants S_{ij} are given in Eq. (2.9). Also $\varepsilon_3 = S_{13}\sigma_1 + S_{23}\sigma_2$ and $\varepsilon_4 = \varepsilon_5 = 0$. By inverting Eq. (2.12), the following two-dimensional stress–strain relations are obtained:

$$\begin{pmatrix} \sigma_1 \\ \sigma_2 \\ \sigma_6 \end{pmatrix} = \begin{pmatrix} Q_{11} & Q_{12} & 0 \\ Q_{12} & Q_{22} & 0 \\ 0 & 0 & Q_{66} \end{pmatrix} \begin{pmatrix} \varepsilon_1 \\ \varepsilon_2 \\ \varepsilon_6 \end{pmatrix} \qquad (2.13)$$

Here, the lamina exhibits orthotropic symmetry. The Q_{ij} in Eq. (2.13) are known as the reduced stiffness constants, and are related to the engineering constants as follows:

$$Q_{11} = \frac{E_{11}}{1 - \nu_{12}\nu_{21}}$$

$$Q_{12} = \frac{\nu_{12}E_{22}}{1 - \nu_{12}\nu_{21}} = \frac{\nu_{21}E_{11}}{1 - \nu_{12}\nu_{21}} \qquad (2.14)$$

$$Q_{22} = \frac{E_{22}}{1 - \nu_{12}\nu_{21}}$$

$$Q_{66} = G_{12}$$

It should be noted that the Q_{ij} so obtained by assuming the plane stress condition of the unidirectional lamina are not identical to the C_{ij} given in Eq. (2.11). In fact, the difference between C_{ij} and Q_{ij} increases as the lamina becomes more isotropic. The inter-relations

Table 2.1. *Inter-relations among the different forms of elastic constants. After Chou (1989b)*

	E_{11}	E_{22}	ν_{12}	ν_{21}	G_{12}
Engineering constant					
Compliance	$1/S_{11}$	$1/S_{22}$	$-S_{12}/S_{11}$	$-S_{12}/S_{22}$	$1/S_{66}$
Reduced stiffness	$(Q_{11}Q_{22} - Q_{12}^2)/Q_{22}$	$(Q_{11}Q_{22} - Q_{12}^2)/Q_{11}$	Q_{12}/Q_{22}	Q_{12}/Q_{11}	Q_{66}
	S_{11}	S_{22}	S_{12}		S_{66}
Compliance					
Reduced stiffness	$Q_{22}/(Q_{11}Q_{22} - Q_{12}^2)$	$Q_{11}/(Q_{11}Q_{22} - Q_{12}^2)$	$Q_{12}/(Q_{11}Q_{22} - Q_{12}^2)$		$1/Q_{66}$
Engineering constant	$1/E_{11}$	$1/E_{22}$	$-\nu_{12}/E_{11}$		$1/G_{12}$
	Q_{11}	Q_{22}	Q_{12}		Q_{66}
Reduced stiffness					
Engineering constant	$E_{11}/(1 - \nu_{12}\nu_{21})$	$E_{22}(1 - \nu_{12}\nu_{21})$	$\nu_{12}E_{22}/(1 - \nu_{12}\nu_{21})$		G_{12}
Compliance	$S_{22}/(S_{11}S_{22} - S_{12}^2)$	$S_{11}/(S_{11}S_{22} - S_{12}^2)$	$-S_{12}/(S_{11}S_{22} - S_{12}^2)$		$1/S_{66}$

among the engineering constants, compliance constants and reduced stiffness constants are summarized in Table 2.1. For a unidirectional lamina oriented at an angle θ with respect to the reference axes x–y (Fig. 2.3), the stress–strain relations in the x–y coordinates are

$$\begin{pmatrix} \sigma_{xx} \\ \sigma_{yy} \\ \tau_{xy} \end{pmatrix} = \begin{pmatrix} \bar{Q}_{11} & \bar{Q}_{12} & \bar{Q}_{16} \\ \bar{Q}_{12} & \bar{Q}_{22} & \bar{Q}_{26} \\ \bar{Q}_{16} & \bar{Q}_{26} & \bar{Q}_{66} \end{pmatrix} \begin{pmatrix} \varepsilon_{xx} \\ \varepsilon_{yy} \\ \gamma_{xy} \end{pmatrix} \tag{2.15}$$

where $\bar{Q}_{ij}$, the transformed reduced stiffness, are given by

$$\bar{Q}_{11} = Q_{11} \cos^4 \theta + 2(Q_{12} + 2Q_{66}) \sin^2 \theta \cos^2 \theta + Q_{22} \sin^4 \theta$$

$$\bar{Q}_{12} = (Q_{11} + Q_{22} - 4Q_{66}) \sin^2 \theta \cos^2 \theta$$
$$+ Q_{12}(\sin^4 \theta + \cos^4 \theta)$$

$$\bar{Q}_{22} = Q_{11} \sin^4 \theta + 2(Q_{12} + 2Q_{66}) \sin^2 \theta \cos^2 \theta + Q_{22} \cos^4 \theta$$

$$\bar{Q}_{16} = (Q_{11} - Q_{12} - 2Q_{66}) \sin \theta \cos^3 \theta$$
$$+ (Q_{12} - Q_{22} + 2Q_{66}) \sin^3 \theta \cos \theta \tag{2.16}$$

$$\bar{Q}_{26} = (Q_{11} - Q_{12} - 2Q_{66}) \sin^3 \theta \cos \theta$$
$$+ (Q_{12} - Q_{22} + 2Q_{66}) \sin \theta \cos^3 \theta$$

$$\bar{Q}_{66} = (Q_{11} + Q_{22} - 2Q_{12} - 2Q_{66}) \sin^2 \theta \cos^2 \theta$$
$$+ Q_{66}(\sin^4 \theta + \cos^4 \theta)$$

Fig. 2.3. Fiber axis at an angle θ from the lamina reference axis x.

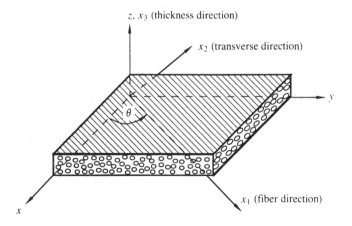

z, x_3 (thickness direction)

x_2 (transverse direction)

y

θ

x_1 (fiber direction)

x

Note that in the $x-y$ coordinate system the notations of τ_{xy} and γ_{xy} are introduced for the shear stress and strain, respectively. The unidirectional lamina referred to the $x-y$ axes is termed generally orthotropic.

Equation (2.15) can be inverted to obtain the strain–stress relations in the following general form:

$$
\begin{pmatrix} \varepsilon_{xx} \\ \varepsilon_{yy} \\ \gamma_{xy} \end{pmatrix} = \begin{pmatrix} \bar{S}_{11} & \bar{S}_{12} & \bar{S}_{16} \\ \bar{S}_{12} & \bar{S}_{22} & \bar{S}_{26} \\ \bar{S}_{16} & \bar{S}_{26} & \bar{S}_{66} \end{pmatrix} = \begin{pmatrix} \sigma_{xx} \\ \sigma_{yy} \\ \tau_{xy} \end{pmatrix}
\tag{2.17}
$$

in which the $\bar{S}_{ij}$ are the transformed compliance constants and their relations to S_{ij} and θ are

$$
\begin{aligned}
\bar{S}_{11} &= S_{11} \cos^4 \theta + (2S_{12} + S_{66}) \sin^2 \theta \cos^2 \theta + S_{22} \sin^4 \theta \\
\bar{S}_{12} &= S_{12}(\sin^4 \theta + \cos^4 \theta) + (S_{11} + S_{22} - S_{66}) \sin^2 \theta \cos^2 \theta \\
\bar{S}_{22} &= S_{11} \sin^4 \theta + (2S_{12} + S_{66}) \sin^2 \theta \cos^2 \theta + S_{22} \cos^4 \theta \\
\bar{S}_{16} &= (2S_{11} - 2S_{12} - S_{66}) \sin \theta \cos^3 \theta \\
&\quad - (2S_{22} - 2S_{12} - S_{66}) \sin^3 \theta \cos \theta \\
\bar{S}_{26} &= (2S_{11} - 2S_{12} - S_{66}) \sin^3 \theta \cos \theta \\
&\quad - (2S_{22} - 2S_{12} - S_{66}) \sin \theta \cos^3 \theta \\
\bar{S}_{66} &= 2(2S_{11} + 2S_{22} - 4S_{12} - S_{66}) \sin^2 \theta \cos^2 \theta \\
&\quad + S_{66}(\sin^4 \theta + \cos^4 \theta)
\end{aligned}
\tag{2.18}
$$

The engineering constants of the unidirectional lamina referring to the $x-y$ axes, which are not aligned with the material principal directions, can be expressed as functions of the off-axis angle, θ, by using Eqs. (2.9) and (2.18)

$$
\frac{1}{E_{xx}} = \frac{1}{E_{11}} \cos^4 \theta + \left(\frac{1}{G_{12}} - \frac{2v_{12}}{E_{11}} \right) \sin^2 \theta \cos^2 \theta + \frac{1}{E_{22}} \sin^4 \theta
$$

$$
v_{xy} = E_{xx} \left(\frac{v_{12}}{E_{11}} (\sin^4 \theta + \cos^4 \theta) \right.
$$

$$
\left. - \left(\frac{1}{E_{11}} + \frac{1}{E_{22}} - \frac{1}{G_{12}} \right) \sin^2 \theta \cos^2 \theta \right)
$$

$$
\frac{1}{E_{yy}} = \frac{1}{E_{11}} \sin^4 \theta + \left(\frac{1}{G_{12}} - \frac{2v_{12}}{E_{11}} \right) \sin^2 \theta \cos^2 \theta + \frac{1}{E_{22}} \cos^4 \theta
$$

$$
\frac{1}{G_{xy}} = 2 \left(\frac{2}{E_{11}} + \frac{2}{E_{22}} + \frac{4v_{12}}{E_{11}} - \frac{1}{G_{12}} \right) \sin^2 \theta \cos^2 \theta
$$

$$
+ \frac{1}{G_{12}} (\sin^4 \theta + \cos^4 \theta)
$$

$$
\tag{2.19}
$$

The variations of E_{xx}, G_{xy}, and v_{xy}, with fiber orientation angle, θ, for a Kevlar-49/epoxy composite are shown in Fig. 2.4.

Jones (1975) discussed the extremum (largest or smallest) values of composite elastic properties, which do not necessarily occur in the principal material directions. It can be shown that E_{xx} is greater than both E_{11} and E_{22} for some values of θ if

$$G_{12} > \frac{E_{11}}{2(1 + v_{12})} \tag{2.20}$$

and that E_{xx} is less than both E_{11} and E_{22} for some values of θ if

$$G_{12} > \frac{E_{11}}{2(E_{11}/E_{22} + v_{12})} \tag{2.21}$$

2.3   Elastic behavior of a composite laminate

2.3.1   *Classical composite lamination theory*

Based upon the constitutive relations for a lamina composed of a generally orthotropic material, Eq. (2.15), the constitutive relations for a laminate formed by bonding several laminae

Fig. 2.4. Variations of engineering elastic constants with fiber orientation angle, θ, for a Kevlar-49/epoxy composite with $V_f = 0.6$, $E_{11} = 76\,\text{GPa}$, $E_{22} = 5.5\,\text{GPa}$, $G_{12} = 2.3\,\text{GPa}$ and $v_{12} = 0.34$.

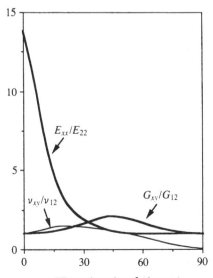

Fiber orientation, θ (degrees)

together is presented in this section. The orientation and material system of each lamina are general. Figure 2.5 depicts the geometry of an n-layered laminate of thickness h; the x–y plane coincides with the laminate geometric middle plane. Following the approach of the classical, linear, thin plate theory, the following assumptions are made (see Vinson and Chou 1975).

(1) A lineal element of the plate extending through the plate thickness, normal to the middle surface (x–y plane) in the unstressed state, upon the application of load: (a) undergoes at most a translation and a rotation with respect to the original coordinate system, and (b) remains normal to the deformed middle surface.

This assumption implies that the lineal element does not elongate or contract, and remains straight upon load applications.

(2) The plate resists lateral and in-plane loads by bending, transverse shear stress, and in-plane action, not through block-like compression or tension in the plate in the thickness direction.

Based upon the foregoing assumptions, also known as the Kirchhoff hypothesis for plates, the strain components can be derived

$$\begin{pmatrix} \varepsilon_{xx} \\ \varepsilon_{yy} \\ \gamma_{xy} \end{pmatrix} = \begin{pmatrix} \varepsilon_{xx}^{o} \\ \varepsilon_{yy}^{o} \\ \gamma_{xy}^{o} \end{pmatrix} + z \begin{pmatrix} \kappa_{xx} \\ \kappa_{yy} \\ 2\kappa_{xy} \end{pmatrix} \tag{2.22}$$

Fig. 2.5. An n-layered laminate.

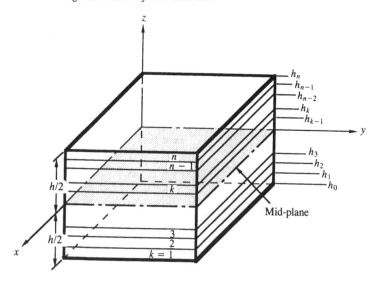

Here, ε_{xx}^{o}, ε_{yy}^{o} and γ_{xy}^{o} are the laminate mid-plane strain, which are expressed in terms of the mid-plane displacements u^{o} and v^{o} in the x and y directions, respectively:

$$\varepsilon_{xx}^{o} = \frac{\partial u^{o}}{\partial x}, \qquad \varepsilon_{yy}^{o} = \frac{\partial v^{o}}{\partial y}, \qquad \gamma_{xy}^{o} = \frac{\partial u^{o}}{\partial y} + \frac{\partial v^{o}}{\partial x} \tag{2.23}$$

The mid-plane curvatures are related to the z direction mid-plane displacement w^{o}

$$\kappa_{xx} = -\frac{\partial^{2} w^{o}}{\partial x^{2}}, \qquad \kappa_{yy} = -\frac{\partial^{2} w^{o}}{\partial y^{2}}, \qquad \kappa_{xy} = -\frac{\partial^{2} w^{o}}{\partial x \, \partial y} \tag{2.24}$$

Note that κ_{xy} represents the twist curvature of the mid-plane. Figure 2.6 depicts the deformation associated with a typical cross-sectional element in a thin plate.

Also, following the approach of the classical plate theory, the resultant forces and moments, instead of the stresses, are utilized in the constitutive relations. Referring to Figs. 2.7 (a) and (b), the force and moment resultants of the laminate are obtained by integrating the stresses of each lamina, through the laminate thickness, h:

$$(N_x, N_y, N_{xy}) = \int_{-h/2}^{h/2} (\sigma_{xx}, \sigma_{yy}, \tau_{xy}) \, dz \tag{2.25}$$

$$(M_x, M_y, M_{xy}) = \int_{-h/2}^{h/2} (\sigma_{xx}, \sigma_{yy}, \tau_{xy}) z \, dz \tag{2.26}$$

Fig. 2.6. Deformation of a typical cross-sectional element in a thin laminated plate.

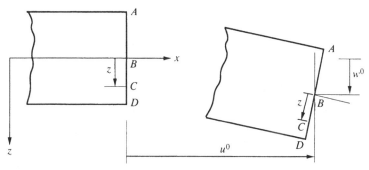

Undeformed cross-section   Deformed cross-section

Substitution of Eqs. (2.15) and (2.16) into Eqs. (2.25) and (2.26) results in the following:

$$
\begin{pmatrix} N_x \\ N_y \\ N_{xy} \end{pmatrix} = \begin{pmatrix} A_{11} & A_{12} & A_{16} \\ A_{12} & A_{22} & A_{26} \\ A_{16} & A_{26} & A_{66} \end{pmatrix} \begin{pmatrix} \varepsilon_{xx}^o \\ \varepsilon_{yy}^o \\ \gamma_{xy}^o \end{pmatrix}
$$

$$
+ \begin{pmatrix} B_{11} & B_{12} & B_{16} \\ B_{12} & B_{22} & B_{26} \\ B_{16} & B_{26} & B_{66} \end{pmatrix} \begin{pmatrix} \kappa_{xx} \\ \kappa_{yy} \\ 2\kappa_{xy} \end{pmatrix} \tag{2.27}
$$

Fig. 2.7. (a) In-plane force resultants. (b) In-plane moment resultants.

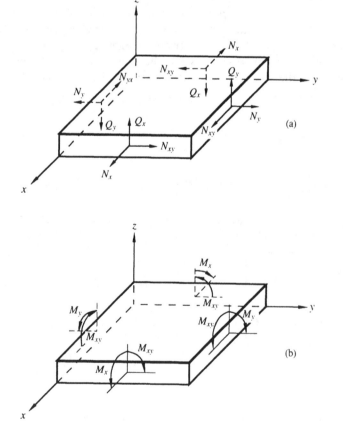

$$
\begin{pmatrix} M_x \\ M_y \\ M_{xy} \end{pmatrix} = \begin{pmatrix} B_{11} & B_{12} & B_{16} \\ B_{12} & B_{22} & B_{26} \\ B_{16} & B_{26} & B_{66} \end{pmatrix} \begin{pmatrix} \varepsilon_{xx}^o \\ \varepsilon_{yy}^o \\ \gamma_{xy}^o \end{pmatrix}
$$

$$
+ \begin{pmatrix} D_{11} & D_{12} & D_{16} \\ D_{12} & D_{22} & D_{26} \\ D_{16} & D_{26} & D_{66} \end{pmatrix} \begin{pmatrix} \kappa_{xx} \\ \kappa_{yy} \\ 2\kappa_{xy} \end{pmatrix} \tag{2.28}
$$

where

$$
A_{ij} = \sum_{k=1}^{n} (\bar{Q}_{ij})_k (h_k - h_{k-1})
$$

$$
B_{ij} = \tfrac{1}{2} \sum_{k=1}^{n} (\bar{Q}_{ij})_k (h_k^2 - h_{k-1}^2) \tag{2.29}
$$

$$
D_{ij} = \tfrac{1}{3} \sum_{k=1}^{n} (\bar{Q}_{ij})_k (h_k^3 - h_{k-1}^3)
$$

In Eqs. (2.27)–(2.29), A_{ij}, B_{ij}, and D_{ij} are called extensional stiffness, extension-bending coupling stiffness, and bending stiffness, respectively. The summation in Eqs. (2.29) is carried out over all the laminae; $(\bar{Q}_{ij})_k$ refers to the reduced stiffness of the kth layer. Eqs. (2.27) and (2.28) are often expressed in the condensed form as

$$
\left(\frac{N}{M} \right) = \left(\frac{A \mid B}{B \mid D} \right) \left(\frac{\varepsilon^o}{\kappa} \right) \tag{2.30}
$$

where $[\kappa]$ is composed of κ_{xx}, κ_{yy} and $2\kappa_{xy}$.

The constitutive relations of Eqs. (2.27) and (2.28) can be rearranged into other useful forms by partially or totally inverting them. The totally inverted forms of Eqs. (2.27) and (2.28) are given in the following condensed matrix expressions:

$$
[\varepsilon^o] = [A'][N] + [B'][M]
$$

$$
[\kappa] = [B'][N] + [D'][M] \tag{2.31}
$$

where

$$
[A'] = [A^*] - [B^*][D^{*-1}][C^*]
$$

$$
[B'] = [B^*][D^{*-1}] = -[D^{*-1}][C^*]
$$

$$
[D'] = [D^{*-1}]
$$

and   (2.32)

$$[A^*] = [A^{-1}]$$
$$[B^*] = -[A^{-1}][B]$$
$$[C^*] = [B][A^{-1}]$$
$$[D^*] = [D] - [B][A^{-1}][B]$$

An application of Eqs. (2.31) is found, for instance, when the stress and moment resultants acting on a laminated plate are specified. Then, with the knowledge of the elastic constants, the mid-plane strain and curvature of the laminate can be determined. The strain components of a specific lamina in terms of the plate reference axes can be derived from Eq. (2.22) and the corresponding stresses from Eq. (2.15). The existing criteria for laminar failure, due to combined in-plane stresses or strains, require the knowledge of stresses and strains along the fiber as well as the transverse directions. This information can be readily obtained by transformation of the stress and strain components to the principal material directions. Thus, the correlation between external loading on the laminated plate and the failure of an individual lamina can be established.

2.3.2   *Geometrical arrangements of laminae*

It has been established in Eqs. (2.29) that the elastic behavior of a composite laminate composed of unidirectional laminae is determined by the constituent material properties as well as the orientation and location of the individual laminae. These geometric aspects of the laminae are indicated by following the convention of the composites literature. For example, $[0°/45_2°/-45_4°/45_2°/0°]$ indicates the stacking sequence of a laminate with one layer at $0°$, two layers at $45°$, four layers at $-45°$, two layers at $45°$, and one layer at $0°$. Because of the mid-plane symmetry, this stacking sequence can also be expressed as $[0°/45_2°/-45_2°]_s$. Following this convention, the basic arrangements of laminae can be expressed as $[0°]$ for unidirectional, $[0°/90°]$ for cross-ply, and $[+\theta/-\theta]$ for angle-ply. The implications of the laminar geometrical arrangements on the laminar elastic behavior, namely, the $[A]$, $[B]$, and $[D]$ matrices, are discussed below.

The $[A]$ matrix relates the stress resultants with the mid-plane strains. The couplings between normal stress resultants and mid-plane shear strains, as well as shear stress resultants and mid-plane normal strains, are due to the components A_{16} and A_{26}. There is also the coupling between mid-plane stress resultants and the bending and twisting of the laminate through the $[B]$ matrix. In particular, the components B_{16} and B_{26} relate normal stress resultants with the twisting of the laminate. The $[B]$ matrix also plays a role in the coupling between the moment resultants and in-plane strains. Finally, the D_{16} and D_{26} terms are responsible for the interaction between the bending moment and twisting.

The various coupling effects in laminated composites can be minimized or eliminated through suitable choices of the laminae stacking sequence. As can be seen from Eqs. (2.29), the B_{ij} terms involve the squares of the z coordinates of the top and bottom faces of each lamina. Each term of B_{ij} vanishes if for every lamina above the mid-plane there is a lamina, identical in properties and orientation, located at the same distance below the mid-plane. Such mid-plane symmetry arrangements eliminate the bending–stretching coupling. The terms A_{16} and A_{26} both vanish under either of the following conditions: (a) all of the laminae assume 0°, 90° or cross-ply [0°/90°] configuration; (b) for every lamina of $+\theta$ orientation there is another lamina of the same property and thickness with a $-\theta$ orientation. The terms D_{16} and D_{26} are zero for the cases: (a) all of the lamina assume 0°, 90° or cross-ply configuration; and (b) for every lamina oriented at $+\theta$ at a given distance above the mid-plane there is an identical layer at the same distance below the mid-plane oriented at $-\theta$. It is obvious that the D_{16} and D_{26} terms are not zero for any mid-plane symmetric laminate, except for the cases of all 0°, all 90° and cross-ply. However, the magnitude of these terms can be made small by increasing the number of layers in the angle-ply configuration. Table 2.2 shows the effect of stacking sequence on the $[A]$, $[B]$ and

Table 2.2. *Effect of stacking sequence on* $[A]$, $[B]$ *and* $[D]$ *matrices. After Chou (1989b)*

	$\theta = 0°, 90°$	$0°/90°$	$\ldots + \theta_2/-\theta_1/$ $+\theta_1/-\theta_2\ldots$ (anti-symmetry)	$\ldots +\theta_2/-\theta_1$ $-\theta_1/+\theta_2\ldots$ (symmetry)	Same number of $+\theta$ and $-\theta$ layers
A_{16}, A_{26}	zero	zero	zero	–	zero
$B_{11}, B_{22}, B_{12}, B_{66}$	zero	–	zero	zero	–
B_{16}, B_{26}	zero	zero	–	zero	–
D_{16}, D_{26}	zero	zero	zero	–	–

[*D*] matrices. The optimization of laminate design for strength has been discussed by Fukunaga and Chou (1988a and b).

2.4   Thick laminates

The term 'thick laminates' here is used to describe composite plates of which the thickness direction properties significantly contribute to the response of the material. Exact elasticity solutions of thick plates have demonstrated that the classical lamination theory of Section 2.3 is not applicable to the thick laminates. Experimental results (for example, Whitney 1972, and Stein and Jegley 1987) have shown significant departure from lamination theory predictions, for such properties as maximum deflections and natural frequencies, when (a) the plate thickness-to-width ratio and (b) the in-plane Young's modulus to interlaminar shear modulus ratio become high.

One reason for the departure of thick plate behavior from classical thin plate theory prediction is the presence of transverse shear deformation. The effect of transverse shear deformation is pronounced in anisotropic materials with high ratios of in-plane Young's moduli to interlaminar shear moduli; this is typical in laminated composites. Other assumptions of the classical plate theory (see Section 2.3) such as negligible transverse normal strains ($\varepsilon_z = 0$), and the linear in-plane strain variation with the z coordinate all contribute to the limitations of the theory. Furthermore, the strong interlaminar shear existing in thick laminates is responsible for delamination, particularly near the free edges. Thus, it is imperative to determine the magnitude and distribution of interlaminar shear in thick laminates.

In the following, the three-dimensional constitutive relations of a thick composite lamina are introduced first. Then, the classical and higher order theory for thick laminated composites is discussed.

2.4.1   *Three-dimensional constitutive relations of a composite lamina*

The three-dimensional constitutive equations of a composite lamina referring to the principal material coordinate system x_1–x_2–x_3 (Fig. 2.1) have been introduced in Eqs. (2.8) and (2.10), for the case of orthotropic symmetry. The relations between the stiffness constants and engineering elastic constants are:

$$C_{11} = E_{11}(1 - v_{23}v_{32})/\Delta$$
$$C_{22} = E_{22}(1 - v_{13}v_{31})/\Delta$$
$$C_{33} = E_{33}(1 - v_{12}v_{21})/\Delta$$
$$C_{44} = G_{23}$$
$$C_{55} = G_{13} \tag{2.33}$$
$$C_{66} = G_{12}$$
$$C_{12} = (v_{21} + v_{23}v_{31})E_{11}/\Delta = (v_{12} + v_{13}v_{32})E_{22}/\Delta$$
$$C_{13} = (v_{31} + v_{21}v_{32})E_{11}/\Delta = (v_{13} + v_{12}v_{23})E_{33}/\Delta$$
$$C_{23} = (v_{32} + v_{12}v_{31})E_{11}/\Delta = (v_{23} + v_{13}v_{21})E_{33}/\Delta$$
$$\Delta = 1 - v_{12}v_{21} - v_{23}v_{32} - v_{13}v_{31} - 2v_{13}v_{21}v_{32}$$

The general three-dimensional constitutive relation of a composite lamina referring to the reference coordinate $x-y-z$ (Fig. 2.3) can be obtained from Eq. (2.10) by tensor transformation:

$$
\begin{pmatrix} \sigma_{xx} \\ \sigma_{yy} \\ \sigma_{zz} \\ \sigma_{yz} \\ \sigma_{xz} \\ \sigma_{xy} \end{pmatrix}
=
\begin{pmatrix}
\bar{C}_{11} & \bar{C}_{12} & \bar{C}_{13} & 0 & 0 & \bar{C}_{16} \\
\bar{C}_{12} & \bar{C}_{22} & \bar{C}_{23} & 0 & 0 & \bar{C}_{26} \\
\bar{C}_{13} & \bar{C}_{23} & \bar{C}_{33} & 0 & 0 & \bar{C}_{36} \\
0 & 0 & 0 & \bar{C}_{44} & \bar{C}_{45} & 0 \\
0 & 0 & 0 & \bar{C}_{45} & \bar{C}_{55} & 0 \\
\bar{C}_{16} & \bar{C}_{26} & \bar{C}_{36} & 0 & 0 & \bar{C}_{66}
\end{pmatrix}
\begin{pmatrix} \varepsilon_{xx} \\ \varepsilon_{yy} \\ \varepsilon_{zz} \\ 2\varepsilon_{yz} \\ 2\varepsilon_{xz} \\ 2\varepsilon_{xy} \end{pmatrix}
\tag{2.34}
$$

Here, the $x-y$ plane coincides with the x_1-x_2 plane and the angle between the x_1 and x axes is θ. The stress and strain tensors in these two coordinate systems are related by

$$
\begin{pmatrix} \sigma_{xx} \\ \sigma_{yy} \\ \sigma_{zz} \\ \sigma_{yz} \\ \sigma_{xz} \\ \sigma_{xy} \end{pmatrix}
= [T]^{-1}
\begin{pmatrix} \sigma_{11} \\ \sigma_{22} \\ \sigma_{33} \\ \sigma_{23} \\ \sigma_{13} \\ \sigma_{12} \end{pmatrix}
\qquad
\begin{pmatrix} \varepsilon_{xx} \\ \varepsilon_{yy} \\ \varepsilon_{zz} \\ \varepsilon_{yz} \\ \varepsilon_{xz} \\ \varepsilon_{xy} \end{pmatrix}
= [T]^{-1}
\begin{pmatrix} \varepsilon_{11} \\ \varepsilon_{22} \\ \varepsilon_{33} \\ \varepsilon_{23} \\ \varepsilon_{13} \\ \varepsilon_{12} \end{pmatrix}
\tag{2.35}
$$

The transformation matrix is

$$
[T] =
\begin{pmatrix}
\cos^2\theta & \sin^2\theta & 0 & 0 & 0 & 2\cos\theta\sin\theta \\
\sin^2\theta & \cos^2\theta & 0 & 0 & 0 & -2\cos\theta\sin\theta \\
0 & 0 & 1 & 0 & 0 & 0 \\
0 & 0 & 0 & \cos\theta & -\sin\theta & 0 \\
0 & 0 & 0 & \sin\theta & \cos\theta & 0 \\
-\cos\theta\sin\theta & \cos\theta\sin\theta & 0 & 0 & 0 & \cos^2\theta - \sin^2\theta
\end{pmatrix}
\tag{2.36}
$$

$[T]^{-1}$ is obtained by changing θ to $-\theta$ in $[T]$. The stiffness matrix is derived from

$$[\bar{C}] = [T]^{-1}[C][T]^{-t} \tag{2.37}$$

with t indicating the matrix transpose and the explicit expressions of $[\bar{C}]$ are

$$\bar{C}_{11} = C_{11}\cos^4\theta + 2(C_{12} + 2C_{66})\sin^2\theta\cos^2\theta + C_{22}\sin^4\theta$$

$$\bar{C}_{12} = (C_{11} + C_{22} - 4C_{66})\sin^2\theta\cos^2\theta + C_{12}(\sin^4\theta + \cos^4\theta)$$

$$\bar{C}_{13} = C_{13}\cos^2\theta + C_{23}\sin^2\theta$$

$$\bar{C}_{16} = (C_{11} - C_{12} - 2C_{66})\sin\theta\cos^3\theta$$
$$\qquad + (C_{12} - C_{22} + 2C_{66})\sin^3\theta\cos\theta$$

$$\bar{C}_{22} = C_{11}\sin^4\theta + 2(C_{12} + 2C_{66})\sin^2\theta\cos^2\theta + C_{22}\cos^4\theta$$

$$\bar{C}_{23} = C_{13}\sin^2\theta + C_{23}\cos^2\theta$$

$$\bar{C}_{26} = (C_{11} - C_{12} - 2C_{66})\sin^3\theta\cos\theta$$
$$\qquad + (C_{12} - C_{22} + 2C_{66})\sin\theta\cos^3\theta \tag{2.38}$$

$$\bar{C}_{33} = C_{33}$$

$$\bar{C}_{36} = (C_{13} - C_{23})\sin\theta\cos\theta$$

$$\bar{C}_{44} = C_{44}\cos^2\theta + C_{55}\sin^2\theta$$

$$\bar{C}_{45} = (C_{55} - C_{44})\sin\theta\cos\theta$$

$$\bar{C}_{55} = C_{55}\cos^2\theta + C_{44}\sin^2\theta$$

$$\bar{C}_{66} = (C_{11} + C_{22} - 2C_{12} - 2C_{66})\sin^2\theta\cos^2\theta$$
$$\qquad + C_{66}(\sin^4\theta + \cos^4\theta)$$

2.4.2   Constitutive relations of thick laminated composites

The classical laminated plate theory does not take into account the effect of transverse shear stress and strain. The inclusion of transverse shear deformation in the classical thin plate theory is achieved by allowing the transverse shear strains, ε_{xz} and ε_{yz}, to be non-zero. This gives rise to definitions of the shear force resultants:

$$(Q_x, Q_y) = \int (\sigma_{xz}, \sigma_{yz})\,dz \tag{2.39}$$

These shear force resultants can be related to the transverse shear strains through the appropriate constitutive relations, Eq. (2.34) (see Vinson and Chou 1975).

Several higher order plate theories have been proposed to account for the transverse shear deformation. This is achieved by retaining higher order terms in the displacement field expansions, which are assumed in the form of power series of the z coordinate. The accuracy of these theories is generally greater for a greater number of terms retained in the series, but the complexity of the governing equations places severe limits on the number of terms for which solutions are realistically attainable.

Among the various proposed displacement field expansions, the simplest one includes the linear term in z; it has been adopted by many workers (for example, Reissner 1945, Whitney and Pagano 1970),

$$u(x, y, z) = u^o(x, y) + z\psi_x(x, y)$$

$$v(x, y, z) = v^o(x, y) + z\psi_y(x, y) \qquad (2.40)$$

$$w(x, y, z) = w^o(x, y)$$

where u, v and w are the displacement components in the x, y and z coordinates (Fig. 2.2), respectively; u^o, v^o and w^o denote the mid-plane displacements of a point (x, y); and ψ_x and ψ_y are the rotations of the normal to the mid-plane about the y and x axes, respectively. It is noted that, unlike the classical plate theory, due to the existence of transverse shear deformation,

$$\psi_x \neq -\frac{\partial w^o}{\partial x}$$

$$\qquad (2.41)$$

$$\psi_y \neq -\frac{\partial w^o}{\partial y}$$

The new curvatures expressions, which are different from Eq. (2.24) are given by

$$\kappa_{xx} = \frac{\partial \psi_x}{\partial x} \neq -\frac{\partial^2 w^o}{\partial x^2}$$

$$\kappa_{yy} = \frac{\partial \psi_y}{\partial y} \neq -\frac{\partial^2 w^o}{\partial y^2} \qquad (2.42)$$

$$\kappa_{xy} = \frac{1}{2}\left(\frac{\partial \psi_x}{\partial y} + \frac{\partial \psi_y}{\partial x}\right) \neq -\frac{\partial^2 w^o}{\partial x \, \partial y}$$

Then, the strain–displacement relations of linear elasticity are

$$\varepsilon_{xx} = \frac{\partial u}{\partial x} = \frac{\partial u^o}{\partial x} + z \frac{\partial \psi_x}{\partial x}$$

$$\varepsilon_{yy} = \frac{\partial v}{\partial y} = \frac{\partial v^o}{\partial y} + z \frac{\partial \psi_y}{\partial y}$$

$$\varepsilon_{zz} = 0$$

$$\varepsilon_{yz} = \frac{1}{2}\left(\frac{\partial v}{\partial z} + \frac{\partial w}{\partial y}\right) = \frac{1}{2}\left(\psi_y + \frac{\partial w^o}{\partial y}\right)$$

$$\varepsilon_{xz} = \frac{1}{2}\left(\frac{\partial u}{\partial z} + \frac{\partial w}{\partial x}\right) = \frac{1}{2}\left(\psi_x + \frac{\partial w^o}{\partial x}\right)$$

$$\varepsilon_{xy} = \frac{1}{2}\left(\frac{\partial u}{\partial y} + \frac{\partial v}{\partial x}\right) = \frac{1}{2}\left[\frac{\partial u^o}{\partial y} + \frac{\partial v^o}{\partial x} + z\left(\frac{\partial \psi_x}{\partial y} + \frac{\partial \psi_y}{\partial x}\right)\right]$$

(2.43)

By substituting Eqs. (2.34) and (2.43) into Eqs. (2.25), (2.26) and (2.39), the constitutive relations of the laminated plate in terms of stress resultants and displacement variables can be obtained as

$$
\begin{pmatrix} N_x \\ N_y \\ N_{xy} \\ M_x \\ M_y \\ M_{xy} \end{pmatrix} =
\begin{pmatrix}
A_{11} & A_{12} & A_{16} & B_{11} & B_{12} & B_{16} \\
A_{12} & A_{22} & A_{26} & B_{12} & B_{22} & B_{26} \\
A_{16} & A_{26} & A_{66} & B_{16} & B_{26} & B_{66} \\
B_{11} & B_{12} & B_{16} & D_{11} & D_{12} & D_{16} \\
B_{12} & B_{22} & B_{26} & D_{12} & D_{22} & D_{26} \\
B_{16} & B_{26} & B_{66} & D_{16} & D_{26} & D_{66}
\end{pmatrix}
\begin{pmatrix}
\dfrac{\partial u^o}{\partial x} \\[2mm]
\dfrac{\partial v^o}{\partial y} \\[2mm]
\dfrac{\partial u^o}{\partial y} + \dfrac{\partial v^o}{\partial x} \\[2mm]
\dfrac{\partial \psi_x}{\partial x} \\[2mm]
\dfrac{\partial \psi_y}{\partial y} \\[2mm]
\dfrac{\partial \psi_x}{\partial y} + \dfrac{\partial \psi_y}{\partial x}
\end{pmatrix}
$$

(2.44)

and

$$\begin{pmatrix} Q_y \\ Q_x \end{pmatrix} = c_k \begin{pmatrix} A_{44} & A_{45} \\ A_{45} & A_{55} \end{pmatrix} \begin{pmatrix} \dfrac{\partial w^o}{\partial y} + \psi_y \\ \dfrac{\partial w^o}{\partial x} + \psi_x \end{pmatrix} \tag{2.45}$$

where

$$A_{ij} = \int \bar{C}_{ij}\, dz \qquad (i, j = 1, 2, 4, 5, 6)$$

$$(B_{ij}, D_{ij}) = \int \bar{C}_{ij}(z, z^2)\, dz \qquad (i, j = 1, 2, 6)$$

and c_k in Eq. (2.45) is a correction factor for the kth lamina which, according to Lo, Christensen and Wu (1977a), is determined by matching the approximated solution with the exact elasticity solution in order to satisfy appropriately the requirements of vanishing transverse shear stress on the top and bottom surfaces of the thick plate.

Having obtained the constitutive relations, the problem of thick laminated plates can be solved by substituting Eqs. (2.44) and (2.45) into the plate equation of motion. Then, a set of partial differential equations in terms of the displacement variables u^o, v^o, w^o, ψ_x and ψ_y can be derived. These unknowns can be solved with the appropriate initial and boundary conditions, which are determined from the total energy of the system (Whitney and Pagano 1970).

The approach outlined above demonstrates an example of the high order laminated plate theories, where only the in-plane displacement terms linear in z are included in Eqs (2.40); and it differs from the classical plate theory only by the terms ψ_x and ψ_y as shown in Eqs (2.41). As pointed out by Lo, Christensen and Wu (1977a&b), despite the increased generality of the shear deformation theory, the related flexural stress distributions show little improvement over the classical laminated plate theory. Thus, it is apparent that higher order terms are needed in the power series expansion of the assumed displacement field to properly model the behavior of thick laminates.

Among the various higher order displacement fields proposed, Lo, Christensen and Wu (1977a&b) suggested the following displacement forms:

$$u = u^o + z\psi_x + z^2\xi_x + z^3\phi_x$$

$$v = v^o + z\psi_y + z^2\xi_y + z^3\phi_y \qquad (2.46)$$

$$w = w^o + z\psi_z + z^2\xi_z$$

where the cubic terms in z for the in-plane displacement field and the square terms in z for the out-of-plane deformations are used; a total of 11 displacement functions (u^o, v^o, w^o, ψ_x, ψ_y, ψ_z, ξ_x, ξ_y, ξ_z, ϕ_x and ϕ_y) are involved. Much improvement over the classical theory predictions is observed; however, the complexity of the analysis has increased tremendously.

The format of solution to higher order systems generally involves the application of the principle of potential energy to derive the pertinent governing equations of equilibrium. Using the strain–displacement relations and the assumed displacement field, in conjunction with the equations of equilibrium, a set of partial differential equations in terms of the displacements used is derived. The number of equations is determined by the number of terms retained in the assumed displacement form. With the appropriate initial and boundary conditions, the solution of these equations describes the elastic behavior of the plate. The details of such approaches can be found, for example, in the work of Whitney and Pagano (1970), Whitney and Sun (1973), Lo, Christensen and Wu (1977a&b), and Reddy (1984).

Although accounting for the higher order plate deformation in thick laminates involves a great deal more complexity than the classical thin plate approach, it is evident that the extra effort to accurately describe their fundamentally different elastic behavior is required. The numerical results of the flexural stress distribution in an infinite [+30, −30]$_s$ laminate of carbon/epoxy composite, subjected to a pressure q, on the top surface ($z = h/2$) of the form

$$q = q_0 \sin\frac{\pi x}{L} \qquad (2.47)$$

are shown in Fig. 2.8 (a) and (b) (see Lo, Christensen and Wu 1977b). Here the length L characterizes the load distribution. The in-plane stress σ_{xx} is normalized as $\bar{\sigma}_{xx} = \sigma_{xx}/q_0 S^2$, $S = L/h$. The results indicate that the higher order theory is necessary for determining the deformation of plates with small L/h ratio.

Sun and Li (1988) and Luo and Sun (1989) have adopted a global–local method for the analysis of thermoelastic fields of thick

Fig. 2.8. (a) Flexural stress distributions for a $[+30, -30]_s$ angle-ply laminate for $L/h = 10$. (b) Flexural stress distributions for a $[+30, -30]_s$ angle-ply laminate for $L/h = 4$. —— exact elastic solution; .... higher order laminated plate theory; –––– classical laminated plate theory. (After Lo, Christensen and Wu 1977b.)

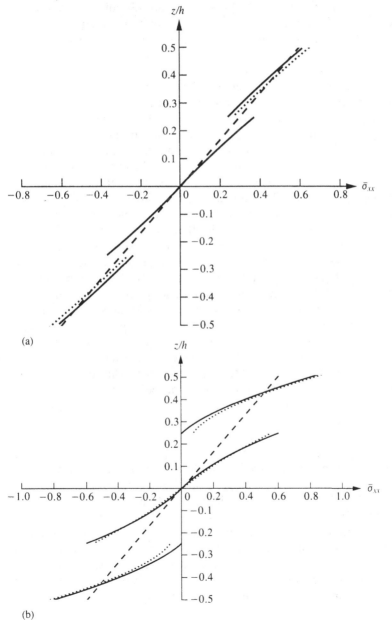

(a)

(b)

laminated composites consisting of a repeating sublaminate (the typical cell). The effective moduli and thermal expansion coefficients are obtained from the sublaminate and used to obtain the global (average) stress and strain solutions. A refining procedure is then introduced in which the global solution is used directly to recover the stresses in the lamina or used as boundary conditions in a sublaminate to perform the exact thermoelastic analysis.

2.5   Thermal and hygroscopic behavior

Besides externally applied load, deformations in laminated composites can also occur due to changes in temperature and absorption of moisture. This is known as the hygrothermal effect. As polymers undergo both dimensional and property changes in a hygrothermal environment, so do composites utilizing polymers as the matrix. Since fibers are fairly insensitive to environmental changes, the environmental susceptibility of composites is mainly through the matrix. Consequently, in a unidirectional composite the temperature–moisture environment has a much greater effect on the transverse and shear properties than the longitudinal properties.

The thermal diffusivity and moisture diffusion coefficient are used as measures of the rates at which the temperature and moisture concentrations change within the material. In general, these parameters depend on the temperature and moisture concentration. However, over the range of temperature and moisture concentration that prevails in typical applications of composites, the thermal diffusivity is usually several orders of magnitude greater than the moisture diffusion coefficient. Consequently, thermal diffusion takes place at a rate much faster than moisture diffusion, and the temperature will reach equilibrium long before the moisture concentration does. This allows one to solve the heat-conduction and moisture-diffusion problems and the resulting elastic fields separately.

The knowledge of anisotropic heat conduction is basic to the solution of thermal stresses in laminated composites. Investigations of such problems have been performed by Poon and Chang (1978), and Chu, Weng and Chen (1983) using transformation theory, by Chang (1977), Huang and Chang (1980), and Nomura and Chou (1986) using Green's function method, by Tauchert and Aköz (1974) using a complex variable method, and by Katayama, Saito and Kobayashi (1974) using a finite difference technique. The solution of the steady-state thermoelastic problem of anisotropic material appears to be initiated by Mossakowska and Nowacki

(1958), Sharma (1958), and Singh (1960). Then Takeuti and Noda (1978), Sugano (1979), and Noda (1983) have examined the transient temperature and thermal stress fields of transversely isotropic elastic medium.

In the category of thermally and elastically orthotropic media, the steady-state temperature and thermal stress field have been investigated for problems of semi-infinite domain (Aköz and Tauchert 1972), a slab bounded by two parallel infinite planes (Tauchert and Aköz 1974) and a rectangular slab (Aköz and Tauchert 1978). The transient thermal stress analysis of thermally and elastically orthotropic laminae has been performed by H. Wang and Chou (1985, 1986), Wang, Pipes and Chou (1986), and Y. Wang and Chou (1988, 1989); their approaches are recapitulated in the following.

In Section 2.5.1, the thermoelastic constitutive equations for a three-dimensional orthotropic material are introduced. These equations are then simplified to the two-dimensional case of unidirectional laminae, and the classical lamination theory is generalized to take into account the thermal and hygroscopic effects. Then, three transient thermal and hygroscopic problems are discussed to illustrate the formulation of the boundary value problems and the solution techniques. The first problem is for the diffusion of moisture through the thickness of a laminated composite (Section 2.5.2). It is assumed that the diffusion equation is one-dimensional (z direction), while the elastic field is two-dimensional ($x-y$ plane). The second problem focuses on the effect of heat conduction on interlaminar thermal stresses (Section 2.5.3). It is assumed, in this case, that heat flows across the width of a laminated plate (one-dimensional heat conduction) and the resulting thermal stress field is three dimensional. Finally, a two-dimensional heat conduction problem is formulated for a rectangular-shaped unidirectional lamina subjected to thermal boundary conditions at its four edges. The two-dimensional thermal elastic field is obtained. In all three problems, the thermal transient effects on stress distribution are demonstrated.

2.5.1 Basic equations

2.5.1.1 Constitutive relations

Deformations of a unidirectional lamina resulting from hygrothermal effects can be described by a modified set of linear constitutive equations: i.e., the total strain minus the non-mechanical strain is linearly related to the stress.

The non-mechanical strain is measured from a stress-free reference state, and the elastic moduli used in the calculation are taken at the final environmental conditions. For example, in the fabrication of polymer matrix composite laminates, the curing of an individual ply results in different deformations along the fiber and transverse directions. The constraint of deformation of a single ply due to the presence of other plies in a multi-directional laminate gives rise to residual stresses. Since most of the cross-linking in the polymer occurs at the highest curing temperature, the polymer matrix can be considered as still viscous enough to allow complete relaxation of the residual stress. Thus, the highest curing temperature can be regarded as the stress-free temperature.

By taking into account the non-mechanical strain in Eq. (2.10) for hygrothermally induced deformation, the laminated plate analysis developed in Section 2.3 can be modified to determine the overall elastic response. The stresses due to moisture absorption and temperature change are identically analogous, in that they are dilatational and self-equilibrating when the whole laminate is considered. In general, the longitudinal properties of polymer matrix composites are far less sensitive to temperature and moisture than the transverse and shear properties of unidirectional composites, because of the excellent retention of mechanical properties by the fibers. The greatest reduction in properties occurs when temperature and moisture are combined, such as in hot and humid environments. However, the combination of temperature and moisture could render a laminate free of residual stresses. This can be understood by considering, for example, a [0°/90°] cross-ply based upon a resin matrix. The thermal stress induced from fabrication is tensile in the transverse direction of a ply, while the residual stresses induced by moisture absorption are compressive. Some details of analysis of such phenomena are developed in the following.

Referring to the principal material coordinate axes of a unidirectional lamina, the three-dimensional orthotropic stress–strain relations of Eq. (2.10) can be written as

$$
\begin{pmatrix} \sigma_{11} \\ \sigma_{22} \\ \sigma_{33} \\ \sigma_{23} \\ \sigma_{13} \\ \sigma_{12} \end{pmatrix} = \begin{pmatrix} C_{11} & C_{12} & C_{13} & 0 & 0 & 0 \\ C_{12} & C_{22} & C_{23} & 0 & 0 & 0 \\ C_{13} & C_{23} & C_{33} & 0 & 0 & 0 \\ 0 & 0 & 0 & C_{44} & 0 & 0 \\ 0 & 0 & 0 & 0 & C_{55} & 0 \\ 0 & 0 & 0 & 0 & 0 & C_{66} \end{pmatrix} \begin{pmatrix} \varepsilon_{11} - \alpha_{11}T - \beta_{11}m \\ \varepsilon_{22} - \alpha_{22}T - \beta_{22}m \\ \varepsilon_{33} - \alpha_{33}T - \beta_{33}m \\ 2\varepsilon_{23} \\ 2\varepsilon_{13} \\ 2\varepsilon_{12} \end{pmatrix}
$$

$$(2.48)$$

where α_{ii} are the coefficients of thermal expansion and β_{ii} are the coefficients of hygroscopic expansion; the subscripts of these coefficients indicate the principal material axes x_i ($i = 1$–3). Also, T denotes a small uniform temperature change from the 'stress-free' temperature; m is the change in moisture concentration referring to a 'moisture-free' environment. Both $\alpha_{ii}T$ and $\beta_{ii}m$ indicate non-mechanical strains.

Referring to the reference axes x–y and following Eq. (2.34), Eq. (2.48) can be rewritten as

$$
\begin{pmatrix} \sigma_{xx} \\ \sigma_{yy} \\ \sigma_{zz} \\ \sigma_{yz} \\ \sigma_{xz} \\ \sigma_{xy} \end{pmatrix} =
\begin{pmatrix}
\bar{C}_{11} & \bar{C}_{12} & \bar{C}_{13} & 0 & 0 & \bar{C}_{16} \\
\bar{C}_{12} & \bar{C}_{22} & \bar{C}_{23} & 0 & 0 & \bar{C}_{26} \\
\bar{C}_{13} & \bar{C}_{23} & \bar{C}_{33} & 0 & 0 & \bar{C}_{36} \\
0 & 0 & 0 & \bar{C}_{44} & \bar{C}_{45} & 0 \\
0 & 0 & 0 & \bar{C}_{45} & \bar{C}_{55} & 0 \\
\bar{C}_{16} & \bar{C}_{26} & \bar{C}_{36} & 0 & 0 & \bar{C}_{66}
\end{pmatrix}
$$

$$
\times \begin{pmatrix}
\varepsilon_{xx} - \alpha_{xx}T - \beta_{xx}m \\
\varepsilon_{yy} - \alpha_{yy}T - \beta_{yy}m \\
\varepsilon_{zz} - \alpha_{zz}T - \beta_{zz}m \\
2\varepsilon_{yz} \\
2\varepsilon_{xz} \\
2\varepsilon_{xy} - \alpha_{xy}T - \beta_{xy}m
\end{pmatrix} \tag{2.49}
$$

where

$$\alpha_{xx} = \alpha_{11}\cos^2\theta + \alpha_{22}\sin^2\theta \qquad \beta_{xx} = \beta_{11}\cos^2\theta + \beta_{22}\sin^2\theta$$

$$\alpha_{yy} = \alpha_{22}\cos^2\theta + \alpha_{11}\sin^2\theta \qquad \beta_{yy} = \beta_{22}\cos^2\theta + \beta_{11}\sin^2\theta$$

$$\alpha_{zz} = \alpha_{33} \qquad\qquad\qquad\qquad \beta_{zz} = \beta_{33}$$

$$\alpha_{xy} = (\alpha_{11} - \alpha_{22})\sin\theta\cos\theta \qquad \beta_{xy} = (\beta_{11} - \beta_{22})\sin\theta\cos\theta$$

$$\tag{2.50}$$

and θ is defined in Fig. 2.3.

The relations given in Eq. (2.49) require that the thermoelastic deformations of the medium are accurately described by linear coefficients of thermal expansion over the range of temperatures of interest, an often used assumption. Similarly, the deformations induced by the hygroscopic nature of the medium are characterized by linear coefficients of hygroscopic expansion, an assumption which follows from existing experimental data.

The elastic constitutive relations for a laminate subjected to both thermal and hygroscopic environments have been formulated by Pipes, Vinson and Chou (1976). For the purpose of laminar analysis, Eq. (2.49) is reduced to

$$
\begin{pmatrix} \sigma_{xx} \\ \sigma_{yy} \\ \tau_{xy} \end{pmatrix} = \begin{pmatrix} \bar{Q}_{11} & \bar{Q}_{12} & \bar{Q}_{16} \\ \bar{Q}_{12} & \bar{Q}_{22} & \bar{Q}_{26} \\ \bar{Q}_{16} & \bar{Q}_{26} & \bar{Q}_{66} \end{pmatrix} \begin{pmatrix} \varepsilon_{xx} - \alpha_{xx}T - \beta_{xx}m \\ \varepsilon_{yy} - \alpha_{yy}T - \beta_{yy}m \\ 2\varepsilon_{xy} - \alpha_{xy}T - \beta_{xy}m \end{pmatrix} \quad (2.51)
$$

Substituting Eq. (2.51) into Eq. (2.25) and following the notation of Eq. (2.30), the constitutive equation is expressed in the following condensed form:

$$
[N] = [A][\varepsilon^{\circ}] + [B][\kappa] - [N]^{T} - [N]^{m} \quad (2.52)
$$

In Eq. (2.52), the effective thermal force resultants, $[N]^{T}$, and effective hygroscopic force resultants $[N]^{m}$ are introduced with the following definitions:

$$
N_x^{T} = \int_{-h/2}^{h/2} (\bar{Q}_{11}\alpha_{xx} + \bar{Q}_{12}\alpha_{yy} + \bar{Q}_{16}\alpha_{xy}) T(z, t)\, dz
$$

$$
N_y^{T} = \int_{-h/2}^{h/2} (\bar{Q}_{12}\alpha_{xx} + \bar{Q}_{22}\alpha_{yy} + \bar{Q}_{26}\alpha_{xy}) T(z, t)\, dz \quad (2.53a)
$$

$$
N_{xy}^{T} = \int_{-h/2}^{h/2} (\bar{Q}_{16}\alpha_{xx} + \bar{Q}_{26}\alpha_{yy} + \bar{Q}_{66}\alpha_{xy}) T(z, t)\, dz
$$

$$
N_x^{m} = \int_{-h/2}^{h/2} (\bar{Q}_{11}\beta_{xx} + \bar{Q}_{12}\beta_{yy} + \bar{Q}_{16}\beta_{xy}) m(z, t)\, dz
$$

$$
N_y^{m} = \int_{-h/2}^{h/2} (\bar{Q}_{12}\beta_{xx} + \bar{Q}_{22}\beta_{yy} + \bar{Q}_{26}\beta_{xy}) m(z, t)\, dz \quad (2.53b)
$$

$$
N_{xy}^{m} = \int_{-h/2}^{h/2} (\bar{Q}_{16}\beta_{xx} + \bar{Q}_{26}\beta_{yy} + \bar{Q}_{66}\beta_{xy}) m(z, t)\, dz
$$

where t denotes time. Consider the kth layer of the laminate; and define $\int^{z} T(\xi, t)\, d\xi \equiv R(z, t)$ and $\int^{z} m(\xi, t)\, d\xi \equiv H(z, t)$. Then,

Eqs. (2.53) are written as summations

$$N_x^{\mathrm{T}} = \sum_{k=1}^{n} (\bar{Q}_{11}\alpha_{xx} + \bar{Q}_{12}\alpha_{yy} + \bar{Q}_{16}\alpha_{xy})_k [R(h_k, t) - R(h_{k-1}, t)]$$

$$N_y^{\mathrm{T}} = \sum_{k=1}^{n} (\bar{Q}_{12}\alpha_{xx} + \bar{Q}_{22}\alpha_{yy} + \bar{Q}_{26}\alpha_{xy})_k [R(h_k, t) - R(h_{k-1}, t)]$$

$$N_{xy}^{\mathrm{T}} = \sum_{k=1}^{n} (\bar{Q}_{16}\alpha_{xx} + \bar{Q}_{26}\alpha_{yy} + \bar{Q}_{66}\alpha_{xy})_k$$
$$\times [R(h_k, t) - R(h_{k-1}, t)]$$

$$\tag{2.54a}$$

$$N_x^{\mathrm{m}} = \sum_{k=1}^{n} (\bar{Q}_{11}\beta_{xx} + \bar{Q}_{12}\beta_{yy} + \bar{Q}_{16}\beta_{xy})_k [H(h_k, t) - H(h_{k-1}, t)]$$

$$N_y^{\mathrm{m}} = \sum_{k=1}^{n} (\bar{Q}_{12}\beta_{xx} + \bar{Q}_{22}\beta_{yy} + \bar{Q}_{26}\beta_{xy})_k [H(h_k, t) - H(h_{k-1}, t)]$$

$$N_{xy}^{\mathrm{m}} = \sum_{k=1}^{n} (\bar{Q}_{16}\beta_{xx} + \bar{Q}_{26}\beta_{yy} + \bar{Q}_{66}\beta_{xy})_k$$
$$\times [(H(h_k, t) - H(h_{k-1}, t)]$$

$$\tag{2.54b}$$

Parallel to the treatment of in-plane response, the flexural response of the laminate is obtained by substituting Eq. (2.51) into Eq. (2.26)

$$[M] = [B][\varepsilon^{\circ}] + [D][\kappa] - [M]^{\mathrm{T}} - [M]^{\mathrm{m}} \tag{2.55}$$

Here, the effective thermal moment resultant, $[M]^{\mathrm{T}}$, and effective hygroscopic moment resultant, $[M]^{\mathrm{m}}$, are defined as

$$M_x^{\mathrm{T}} = \int_{-h/2}^{h/2} (\bar{Q}_{11}\alpha_{xx} + \bar{Q}_{12}\alpha_{yy} + \bar{Q}_{16}\alpha_{xy})T(z, t)z \, \mathrm{d}z$$

$$M_y^{\mathrm{T}} = \int_{-h/2}^{h/2} (\bar{Q}_{12}\alpha_{xx} + \bar{Q}_{22}\alpha_{yy} + \bar{Q}_{26}\alpha_{xy})T(z, t)z \, \mathrm{d}z \tag{2.56a}$$

$$M_{xy}^{\mathrm{T}} = \int_{-h/2}^{h/2} (\bar{Q}_{16}\alpha_{xx} + \bar{Q}_{26}\alpha_{yy} + \bar{Q}_{66}\alpha_{xy})T(z, t)z \, \mathrm{d}z$$

$$M_x^{\mathrm{m}} = \int_{-h/2}^{h/2} (\bar{Q}_{11}\beta_{xx} + \bar{Q}_{12}\beta_{yy} + \bar{Q}_{16}\beta_{xy})m(z, t)z \, \mathrm{d}z$$

$$M_y^{\mathrm{m}} = \int_{-h/2}^{h/2} (\bar{Q}_{12}\beta_{xx} + \bar{Q}_{22}\beta_{yy} + \bar{Q}_{26}\beta_{xy})m(z, t)z \, \mathrm{d}z \tag{2.56b}$$

$$M_{xy}^{\mathrm{m}} = \int_{-h/2}^{h/2} (\bar{Q}_{16}\beta_{xx} + \bar{Q}_{26}\beta_{yy} + \bar{Q}_{66}\beta_{xy})m(z, t)z \, \mathrm{d}z$$

By introducing the integrals of $R(z, t)$ (i.e., $S(z, t) = \int^z R(\xi, t)\,d\xi$), and $H(z, t)$ (i.e., $J(z, t) = \int^z H(\xi, t)\,d\xi$), Eqs. (2.56) are also expressed as summations:

$$M_x^{\mathrm{T}} = \sum_{k=1}^{n} (\bar{Q}_{11}\alpha_{xx} + \bar{Q}_{12}\alpha_{yy} + \bar{Q}_{16}\alpha_{xy})_k [h_k R(h_k, t)$$
$$- h_{k-1}R(h_{k-1}, t) - S(h_k, t) + S(h_{k-1}, t)]$$

$$M_y^{\mathrm{T}} = \sum_{k=1}^{n} (\bar{Q}_{12}\alpha_{xx} + \bar{Q}_{22}\alpha_{yy} + \bar{Q}_{26}\alpha_{xy})_k [h_k R(h_k, t)$$
$$- h_{k-1}R(h_{k-1}, t) - S(h_k, t) + S(h_{k-1}, t)] \tag{2.57a}$$

$$M_{xy}^{\mathrm{T}} = \sum_{k=1}^{n} (\bar{Q}_{16}\alpha_{xx} + \bar{Q}_{26}\alpha_{yy} + \bar{Q}_{66}\alpha_{xy})_k [h_k R(h_k, t)$$
$$- h_{k-1}R(h_{k-1}, t) - S(h_k, t) + S(h_{k-1}, t)]$$

$$M_x^{\mathrm{m}} = \sum_{k=1}^{n} (\bar{Q}_{11}\beta_{xx} + \bar{Q}_{12}\beta_{yy} + \bar{Q}_{16}\beta_{xy})_k [h_k H(h_k, t)$$
$$- h_{k-1}H(h_{k-1}, t) - J(h_k, t) + J(h_{k-1}, t)]$$

$$M_y^{\mathrm{m}} = \sum_{k=1}^{n} (\bar{Q}_{12}\beta_{xx} + \bar{Q}_{22}\beta_{yy} + \bar{Q}_{26}\beta_{xy})_k [h_k H(h_k, t)$$
$$- h_{k-1}H(h_{k-1}, t) - J(h_k, t) + J(h_{k-1}, t)] \tag{2.57b}$$

$$M_{xy}^{\mathrm{m}} = \sum_{k=1}^{n} (\bar{Q}_{16}\beta_{xx} + \bar{Q}_{26}\beta_{yy} + \bar{Q}_{66}\beta_{xy})_k [h_k H(h_k, t)$$
$$- h_{k-1}H(h_{k-1}, t) - J(h_k, t) + J(h_{k-1}, t)]$$

Finally, Eqs. (2.52) and (2.55) are combined as

$$
\begin{pmatrix} N_x \\ N_y \\ N_{xy} \\ M_x \\ M_y \\ M_{xy} \end{pmatrix}
+
\begin{pmatrix} N_x^{\mathrm{T}} \\ N_y^{\mathrm{T}} \\ N_{xy}^{\mathrm{T}} \\ M_x^{\mathrm{T}} \\ M_y^{\mathrm{T}} \\ M_{xy}^{\mathrm{T}} \end{pmatrix}
+
\begin{pmatrix} N_x^{\mathrm{m}} \\ N_y^{\mathrm{m}} \\ N_{xy}^{\mathrm{m}} \\ M_x^{\mathrm{m}} \\ M_y^{\mathrm{m}} \\ M_{xy}^{\mathrm{m}} \end{pmatrix}
$$

$$
=
\begin{pmatrix}
A_{11} & A_{12} & A_{16} & B_{11} & B_{12} & B_{16} \\
A_{12} & A_{22} & A_{26} & B_{12} & B_{22} & B_{26} \\
A_{16} & A_{26} & A_{66} & B_{16} & B_{26} & B_{66} \\
B_{11} & B_{12} & B_{16} & D_{11} & D_{12} & D_{16} \\
B_{12} & B_{22} & B_{26} & D_{12} & D_{22} & D_{26} \\
B_{16} & B_{26} & B_{66} & D_{16} & D_{26} & D_{66}
\end{pmatrix}
\begin{pmatrix}
\varepsilon_{xx}^{\mathrm{o}} \\
\varepsilon_{yy}^{\mathrm{o}} \\
\gamma_{xy}^{\mathrm{o}} \\
\kappa_{xx} \\
\kappa_{yy} \\
2\kappa_{xy}
\end{pmatrix}
\tag{2.58}
$$

The response of a laminate subjected to known mechanical force and moment resultants, and both thermal and hygroscopic effects, can be determined by calculating the effective thermal and hygroscopic resultants and inverting Eq. (2.58). The inversion would yield laminate mid-plane strains, ε°, and curvatures, κ. The strains of the laminae could then be calculated by Eq. (2.22). Given the strains, the stresses within each lamina could be determined according to Eq. (2.51).

2.5.1.2 Thermal and moisture diffusion equations

The three-dimensional heat conduction equation for a general anisotropic solid of constant conductivity coefficients is given by (Ozisik 1980)

$$K_{xx}^{T}\frac{\partial^2 T}{\partial x^2} + K_{yy}^{T}\frac{\partial^2 T}{\partial y^2} + K_{zz}^{T}\frac{\partial^2 T}{\partial z^2} + 2K_{xy}^{T}\frac{\partial^2 T}{\partial x\,\partial y} + 2K_{xz}^{T}\frac{\partial^2 T}{\partial x\,\partial z}$$

$$+ 2K_{yz}^{T}\frac{\partial^2 T}{\partial y\,\partial z} = \rho C_{\mathrm{p}}\frac{\partial T}{\partial t} \quad (2.59)$$

Here, K_{ij}^{T} denote the coefficients of heat conduction, ρ is mass density, and C_{p} is the specific heat. The temperature of the elastic medium, T, is a function of location (x, y, z) and the time, t. It is understood that there is no internal heat generation of the elastic body.

For a thermally orthotropic material, with respect to the reference axes x–y–z, Eq. (2.59) is simplified as

$$K_{xx}^{T}\frac{\partial^2 T}{\partial x^2} + K_{yy}^{T}\frac{\partial^2 T}{\partial y^2} + K_{zz}^{T}\frac{\partial^2 T}{\partial z^2} = \rho C_{\mathrm{p}}\frac{\partial T}{\partial t} \quad (2.60)$$

Here, the thermal conductivities K_{xx}^{T}, K_{yy}^{T} and K_{zz}^{T} are related to the conductivities along the material principal direction, i.e. K_{11}^{T}, K_{22}^{T} and K_{33}^{T} using transformation equations identical in form to those given in Eqs. (2.50).

An equation identical in form to Eq. (2.59) can be written for moisture diffusion. Consider, for instance, the diffusion of moisture along the laminate thickness (z) direction, the governing equation is reduced to

$$K_{zz}^{m}\frac{\partial^2 m}{\partial z^2} = \frac{\partial m}{\partial t} \quad (2.61)$$

where K_{zz}^m is the moisture diffusion coefficient and $m = m(z, t)$ denotes the moisture concentration distribution. Equation (2.61) is further discussed in Section 2.5.2.

In Section 2.5.3, the transient interlaminar stress induced by heat conduction through the laminate width (y) direction is discussed. Then, $T = T(y, t)$, and Eq. (2.59) for each lamina is reduced to

$$K_{yy}^T \frac{\partial^2 T}{\partial y^2} = \rho C_p \frac{\partial T}{\partial t} \tag{2.62}$$

In Section 2.5.4, heat conduction in the plane of a unidirectional composite is considered. The governing equation for heat conduction becomes

$$K_{xx}^T \frac{\partial^2 T}{\partial x^2} + K_{yy}^T \frac{\partial^2 T}{\partial y^2} = \rho C_p \frac{\partial T}{\partial t} \tag{2.63}$$

2.5.2   Hygroscopic behavior

2.5.2.1   Moisture concentration functions

Pipes, Vinson and Chou (1976) assume that the classical diffusion equation (see Jost 1960) governs the absorption and desorption of moisture by a hygroscopic material as given in Eq. (2.61). Consider first the case of moisture absorption. If the laminate is assumed to be initially moisture free, while its surfaces $z = \pm h/2$ are exposed to a moisture concentration M_o, then moisture concentration in the laminate at position z and time t is

$$m(z, t) = M_o \left[1 - \sum_{n=0}^{\infty} m_n \cos(a_n z) \right] \tag{2.64}$$

where

$$a_n = \frac{(2n + 1)\pi}{h}$$

and

$$m_n = \frac{4}{\pi} \left\{ \frac{(-1)^n}{2n + 1} \exp[-a_n^2 K_{zz}^m t] \right\}$$

From Eq. (2.64), the effective hygroscopic force resultant can be readily determined by combining Eqs. (2.54b) and (2.64):

$$N_x^m = \sum_{k=1}^{n} (\bar{Q}_{11}\beta_{xx} + \bar{Q}_{12}\beta_{yy} + \bar{Q}_{16}\beta_{xy})_k M_o$$

$$\times \left[h_k - h_{k-1} - \sum_{n=0}^{\infty} \frac{m_n}{a_n} (\sin(a_n h_k) - \sin(a_n h_{k-1})) \right]$$

$$N_y^m = \sum_{k=1}^{n} (\bar{Q}_{12}\beta_{xx} + \bar{Q}_{22}\beta_{yy} + \bar{Q}_{26}\beta_{xy})_k M_o$$

$$\times \left[h_k - h_{k-1} - \sum_{n=0}^{\infty} \frac{m_n}{a_n} (\sin(a_n h_k) - \sin(a_n h_{k-1})) \right] \quad (2.65)$$

$$N_{xy}^m = \sum_{k=1}^{n} (\bar{Q}_{16}\beta_{xx} + \bar{Q}_{26}\beta_{yy} + \bar{Q}_{66}\beta_{xy})_k M_o$$

$$\times \left[h_k - h_{k-1} - \sum_{n=0}^{\infty} \frac{m_n}{a_n} (\sin(a_n h_k) - \sin(a_n h_{k-1})) \right]$$

The effective hygroscopic moment resultant is then determined from Eqs. (2.57b)

$$M_x^m = \sum_{k=1}^{n} (\bar{Q}_{11}\beta_{xx} + \bar{Q}_{12}\beta_{yy} + \bar{Q}_{16}\beta_{xy})_k M_o$$

$$\times \left[\tfrac{1}{2}(h_k^2 - h_{k-1}^2) - \sum_{n=0}^{\infty} \frac{m_n}{a_n} (h_k \sin(a_n h_k) \right.$$

$$- h_{k-1} \sin(a_n h_{k-1}))$$

$$\left. - \sum_{n=0}^{\infty} \frac{m_n}{a_n^2} (\cos(a_n h_k) - \cos(a_n h_{k-1})) \right]$$

$$M_y^m = \sum_{k=1}^{n} (\bar{Q}_{12}\beta_{xx} + \bar{Q}_{22}\beta_{yy} + \bar{Q}_{26}\beta_{xy})_k M_o$$

$$\times \left[\tfrac{1}{2}(h_k^2 - h_{k-1}^2) - \sum_{n=0}^{\infty} \frac{m_n}{a_n} (h_k \sin(a_n h_k) \right.$$

$$- h_{k-1} \sin(a_n h_{k-1})) \quad (2.66)$$

$$\left. - \sum_{n=0}^{\infty} \frac{m_n}{a_n^2} (\cos(a_n h_k) - \cos(a_n h_{k-1})) \right]$$

$$M_{xy}^m = \sum_{k=1}^{n} (\bar{Q}_{16}\beta_{xx} + \bar{Q}_{26}\beta_{yy} + \bar{Q}_{66}\beta_{xy})_k M_o$$

$$\times \left[\tfrac{1}{2}(h_k^2 - h_{k-1}^2) - \sum_{n=0}^{\infty} \frac{m_n}{a_n} (h_k \sin(a_n h_k) \right.$$

$$- h_{k-1} \sin(a_n h_{k-1}))$$

$$\left. - \sum_{n=0}^{\infty} \frac{m_n}{a_n^2} (\cos(a_n h_k) - \cos(a_n h_{k-1})) \right]$$

Next, consider the desorption of moisture. The laminate containing a uniformly distributed moisture concentration, M_o, is exposed to a moisture-free environment on its surfaces $z = \pm h/2$. The solution of the diffusion Eq. (2.61) corresponding to these boundary conditions is

$$m(z, t) = M_o \sum_{n=0}^{\infty} m_n \cos(a_n z) \tag{2.67}$$

The corresponding effective hygroscopic force and moment resultants are

$$N_x^m = \sum_{k=1}^{n} (\bar{Q}_{11}\beta_{xx} + \bar{Q}_{12}\beta_{yy} + \bar{Q}_{16}\beta_{xy})_k M_o \sum_{n=0}^{\infty} \frac{m_n}{a_n}$$

$$\times (\sin(a_n h_k) - \sin(a_n h_{k-1}))$$

$$N_y^m = \sum_{k=1}^{n} (\bar{Q}_{12}\beta_{xx} + \bar{Q}_{22}\beta_{yy} + \bar{Q}_{26}\beta_{xy})_k M_o \sum_{n=0}^{\infty} \frac{m_n}{a_n}$$

$$\times (\sin(a_n h_k) - \sin(a_n h_{k-1})) \tag{2.68}$$

$$N_{xy}^m = \sum_{k=1}^{n} (\bar{Q}_{16}\beta_{xx} + \bar{Q}_{26}\beta_{yy} + \bar{Q}_{66}\beta_{xy})_k M_o \sum_{n=0}^{\infty} \frac{m_n}{a_n}$$

$$\times (\sin(a_n h_k) - \sin(a_n h_{k-1)}))$$

$$M_x^m = \sum_{k=1}^{n} (\bar{Q}_{11}\beta_{xx} + \bar{Q}_{12}\beta_{yy} + \bar{Q}_{16}\beta_{xy})_k M_o$$

$$\times \left[\sum_{n=0}^{\infty} \frac{m_n}{a_n} (h_k \sin(a_n h_k) - h_{k-1} \sin(a_n h_{k-1})) \right.$$

$$\left. + \sum_{n=0}^{\infty} \frac{m_n}{a_n^2} (\cos(a_n h_k) - \cos(a_n h_{k-1})) \right]$$

$$M_y^m = \sum_{k=1}^{n} (\bar{Q}_{12}\beta_{xx} + \bar{Q}_{22}\beta_{yy} + \bar{Q}_{26}\beta_{xy})_k M_o$$

$$\times \left[\sum_{n=0}^{\infty} \frac{m_n}{a_n} (h_k \sin(a_n h_k) - h_{k-1} \sin(a_n h_{k-1})) \right. \tag{2.69}$$

$$\left. + \sum_{n=0}^{\infty} \frac{m_n}{a_n^2} (\cos(a_n h_k) - \cos(a_n h_{k-1})) \right]$$

$$M_{xy}^m = \sum_{k=1}^{n} (\bar{Q}_{16}\beta_{xx} + \bar{Q}_{26}\beta_{yy} + \bar{Q}_{66}\beta_{xy})_k M_o$$

$$\times \left[\sum_{n=0}^{\infty} \frac{m_n}{a_n} (h_k \sin(a_n h_k) - h_{k-1} \sin(a_n h_{k-1})) \right.$$

$$\left. + \sum_{n=0}^{\infty} \frac{m_n}{a_n^2} (\cos(a_n h_k) - \cos(a_n h_{k-1})) \right]$$

2.5.2.2 *Hygroscopic stress field*

Pipes, Vinson and Chou (1976) have illustrated the hygroscopic effects on a carbon–epoxy system (T300/5208) comprising a six-ply laminate of $[0°/+45°/-45]_s$, where each lamina is of the thickness $\bar{h}$. It is assumed that the diffusion coefficient, K_{zz}^m, and the coefficients of expansion, α and β, are constant over the ranges of

Fig. 2.9. Moisture distribution profiles during absorption. (After Pipes, Vinson and Chou 1976.)

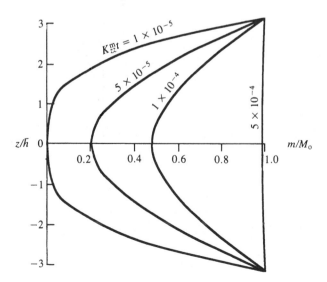

temperature and moisture concentration of interest. The material properties are $E_{11} = 143$ GPa, $E_{22} = 10.1$ GPa, $\nu_{12} = 0.31$, $G_{12} = 4.14$ GPa, $\beta_{11} = 0$, $\beta_{22} = 6.67 \times 10^{-3}/\mathrm{wt}\%$ and $\bar{h} = 0.1397$ mm.

Figure 2.9 illustrates the moisture profiles across the laminate which is moisture free at time $t = 0$, and then exposed to an environment on both surfaces of moisture concentration M_0. The range of $K_{zz}^{m}t$ values is between 1×10^{-5} and 5×10^{-4}. It is seen that by $K_{zz}^{m}t = 5 \times 10^{-5}$ the moisture concentration at the mid-surface is 20% of that at the surface.

Figure 2.10 shows the profiles of σ_{xx}, which is compressive in the outer, $0°$, laminae, because of the expansion caused by the moisture gradients of Fig. 2.9, and the inner four laminae at $\pm45°$ are all in tension. Stress values are maximum at the outer surfaces. The profiles of σ_{yy} follow the same trend as σ_{xx}, and $\sigma_{yy} > \sigma_{xx}$ at each time. In both cases the steady state is achieved at $K_{zz}^{m}t > 5 \times 10^{-4}$.

Figure 2.11 shows that $\tau_{xy} = 0$ in the outer two layers because they are at the orientation of $\theta = 0°$; the same would occur for any

Fig. 2.10. Distribution of stress, σ_{xx} during moisture absorption. (After Pipes, Vinson and Chou 1976.)

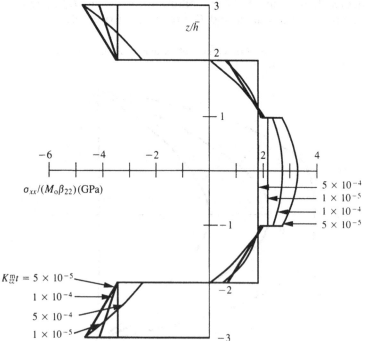

layers at $\theta = 90°$ in balanced laminates. The in-plane shear stresses increase with time, because of the increasing strains caused by increased moisture content; by the time of steady state ($K_{zz}^m t > 5 \times 10^{-4}$) the shear stresses are much larger than either the σ_{xx} or σ_{yy} stress. These large shear stresses imply large interlaminar shear stresses, σ_{zx} and σ_{zy}, near laminate discontinuities.

2.5.3 Transient interlaminar thermal stresses

2.5.3.1 Transient temperature field

Consider an x direction infinite laminated plate subjected to a temperature field $T = T_o$ on two edges ($y = \pm b$) at time $t = 0^+$ (Fig. 2.12). By assuming that the temperature field in each layer is independent of the thickness direction, i.e. $T = T(y, t)$, the heat conduction equation for each lamina follows Eq. (2.62).

Fig. 2.11. Distribution of stress, τ_{xy}, during moisture absorption. (After Pipes, Vinson and Chou 1976.)

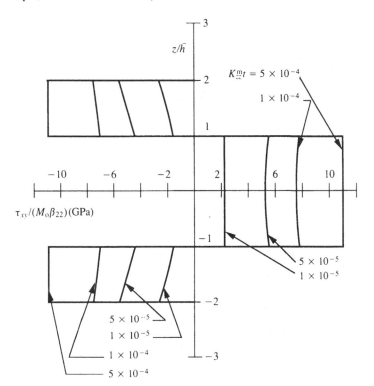

The boundary and initial conditions are

$$T(\pm b, t) = T_o \qquad T(y, 0) = 0 \tag{2.70}$$

The solution of the governing equation Eq. (2.62) by the method of separation of variables is

$$T = T_0 \left(1 + \sum_{n=0}^{\infty} a_n \cos(b_n Y) e^{-c_n t} \right) \tag{2.71}$$

where

$$Y = \frac{y}{b}$$

$$a_n = \frac{(-1)^n 4}{(2n - 1)\pi}$$

$$b_n = (n - \tfrac{1}{2})\pi$$

$$c_n = \left[(n - \tfrac{1}{2}) \frac{\pi g}{b} \right]^2$$

$$g^2 = \frac{K_{yy}}{\rho C_p}$$

2.5.3.2 *Thermal stress field*

Y. Wang and Chou (1989) have considered the transient thermal stresses in an orthotropic composite laminate. Since the

Fig. 2.12. Geometry of an angle-ply laminate for analytical modeling.

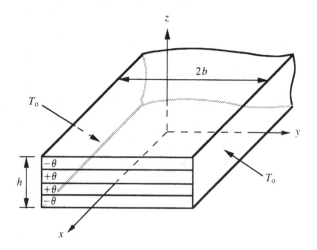

thermal boundary conditions are uniform along the surfaces $y = \pm b$, the displacements are independent of the x axis and expressed as:

$$u = u(y, z, t)$$
$$v = v(y, z, t) \quad (2.72)$$
$$w = w(y, z, t)$$

The stress–strain relations for such an orthotropic laminate follow Eq. (2.49):

$$\sigma_{xx} = \bar{C}_{11}\varepsilon_{xx} + \bar{C}_{12}\varepsilon_{yy} + \bar{C}_{13}\varepsilon_{zz} + 2\bar{C}_{16}\varepsilon_{xy} - \bar{\alpha}_1 T$$
$$\sigma_{yy} = \bar{C}_{12}\varepsilon_{xx} + \bar{C}_{22}\varepsilon_{yy} + \bar{C}_{23}\varepsilon_{zz} + 2\bar{C}_{26}\varepsilon_{xy} - \bar{\alpha}_2 T$$
$$\sigma_{zz} = \bar{C}_{13}\varepsilon_{xx} + \bar{C}_{23}\varepsilon_{yy} + \bar{C}_{33}\varepsilon_{zz} + 2\bar{C}_{36}\varepsilon_{xy} - \bar{\alpha}_3 T$$
$$\sigma_{yz} = 2\bar{C}_{44}\varepsilon_{yz} \quad (2.73)$$
$$\sigma_{xz} = 2\bar{C}_{55}\varepsilon_{xz}$$
$$\sigma_{xy} = \bar{C}_{16}\varepsilon_{xx} + \bar{C}_{26}\varepsilon_{yy} + \bar{C}_{36}\varepsilon_{zz} + 2\bar{C}_{66}\varepsilon_{xy} - \bar{\alpha}_6 T$$

where

$$\bar{\alpha} = \alpha_{xx}\bar{C}_{11} + \alpha_{yy}\bar{C}_{12} + \alpha_{zz}\bar{C}_{13} + \alpha_{xy}\bar{C}_{16}$$
$$\bar{\alpha}_2 = \alpha_{xx}\bar{C}_{12} + \alpha_{yy}\bar{C}_{22} + \alpha_{zz}\bar{C}_{23} + \alpha_{xy}\bar{C}_{26}$$
$$\bar{\alpha}_3 = \alpha_{xx}\bar{C}_{13} + \alpha_{yy}\bar{C}_{23} + \alpha_{zz}\bar{C}_{33} + \alpha_{xy}\bar{C}_{36} \quad (2.74)$$
$$\bar{\alpha}_6 = \alpha_{xx}\bar{C}_{16} + \alpha_{yy}\bar{C}_{26} + \alpha_{zz}\bar{C}_{36} + \alpha_{xy}\bar{C}_{66}$$

The equilibrium equations can be written in terms of the displacements:

$$\bar{C}_{66}\frac{\partial^2 u}{\partial y^2} + \bar{C}_{55}\frac{\partial^2 u}{\partial z^2} + \bar{C}_{26}\frac{\partial^2 v}{\partial y^2} + \bar{C}_{36}\frac{\partial^2 w}{\partial y\,\partial z} = \bar{\alpha}_6\frac{\partial T}{\partial y}$$

$$\bar{C}_{26}\frac{\partial^2 u}{\partial y^2} + \bar{C}_{22}\frac{\partial^2 v}{\partial y^2} + \bar{C}_{44}\frac{\partial^2 v}{\partial z^2} + (\bar{C}_{23} + \bar{C}_{44})\frac{\partial^2 w}{\partial y\,\partial z} = \bar{\alpha}_2\frac{\partial T}{\partial y}$$

$$\bar{C}_{36}\frac{\partial^2 u}{\partial y\,\partial z} + (\bar{C}_{44} + \bar{C}_{23})\frac{\partial^2 v}{\partial y\,\partial z} + \bar{C}_{44}\frac{\partial^2 w}{\partial y^2} + \bar{C}_{33}\frac{\partial^2 w}{\partial z^2} = 0$$

$$(2.75)$$

The equilibrium equations can be solved by a singular perturbation technique (Van Dyke 1975). It is assumed by Y. Wang and Chou (1988, 1989) that, for h/b sufficiently small, i.e. $<10\%$ (see Fig. 2.12), the linear and higher order terms of h/b can be neglected and a zeroth order perturbation approach (Hsu and Herakovich 1976, 1977) is applied. The solution of Eqs. (2.75) for the kth layer in the interior region ($Y = y/b < 1$) of a laminate is

$$U^{(k)} = \frac{2u}{h} = B(Y, t)$$

$$V^{(k)} = \frac{2v}{h} = D(Y, t) \tag{2.76}$$

$$W^{(k)} = \frac{2w}{h} = E(Y, t)Z$$

where

$$Z = \frac{2z}{h}$$

$$\left(\frac{h}{b}\right) B(Y, t) = \frac{q_2 Q_1(Y, t) - q_1 Q_2(Y, t)}{q_1 q_3 - q_2^2}$$

$$\left(\frac{h}{b}\right) D(Y, t) = \frac{q_3 Q_1(Y, t) - q_2 Q_2(Y, t)}{q_2^2 - q_1 q_3}$$

$$E(Y, t) = \frac{\bar{\alpha}_3}{\bar{C}_{33}} T(Y, t)$$

$$Q_1(Y, t) = \sum_{k=1}^{n} \left(\frac{\bar{C}_{23}}{\bar{C}_{33}} \bar{\alpha}_3 - \bar{\alpha}_2\right)_k \bar{T}_k(Y, t) h^{(k)} \tag{2.77}$$

$$Q_2(Y, t) = \sum_{k=1}^{n} \left(\frac{\bar{C}_{36}}{\bar{C}_{33}} \bar{\alpha}_3 - \bar{\alpha}_6\right)_k \bar{T}_k(Y, t) h^{(k)}$$

$$q_1 = \sum_{k=1}^{n} (\bar{C}_{22})_k h^{(k)} \qquad q_2 = \sum_{k=1}^{n} (\bar{C}_{26})_k h^{(k)}$$

$$q_3 = \sum_{k=1}^{n} (\bar{C}_{66})_k h^{(k)} \qquad \bar{T}_k(Y, t) = \int_0^Y T_k(Y, t)\, dY$$

$h^{(k)} = k$th layer thickness

The subscript k indicates the kth lamina of the n-layer laminate.

Following Hsu and Herakovich (1976, 1977), a stretching transformation parameter is introduced to obtain the solution for the boundary layer region ($Y \approx 1$):

$$\eta = (1 - Y)/(h/b) \tag{2.78}$$

Then the equilibrium equations (2.75) become

$$\bar{C}_{66}\frac{\partial^2 U}{\partial\eta^2} + \bar{C}_{55}\frac{\partial^2 U}{\partial Z^2} + \bar{C}_{26}\frac{\partial^2 V}{\partial\eta^2} - \bar{C}_{36}\frac{\partial^2 W}{\partial\eta\,\partial Z} = \left(\frac{h}{b}\right)\left(\frac{\bar{\alpha}_6}{\bar{C}_{\max}}\right)\frac{\partial T}{\partial Y}$$

$$\bar{C}_{26}\frac{\partial^2 U}{\partial\eta^2} + \bar{C}_{22}\frac{\partial^2 V}{\partial\eta^2} + \bar{C}_{44}\frac{\partial^2 V}{\partial Z^2} - (\bar{C}_{23} + \bar{C}_{44})\frac{\partial^2 W}{\partial\eta\,\partial Z}$$

$$= \left(\frac{h}{b}\right)\left(\frac{\bar{\alpha}_2}{\bar{C}_{\max}}\right)\frac{\partial T}{\partial Y} - \bar{C}_{36}\frac{\partial^2 U}{\partial\eta\,\partial Z} - (\bar{C}_{23} + \bar{C}_{44})\frac{\partial^2 V}{\partial\eta\,\partial Z} \quad (2.79)$$

$$+ \bar{C}_{44}\frac{\partial^2 W}{\partial\eta^2} + \bar{C}_{33}\frac{\partial^2 W}{\partial Z^2} = 0$$

where $\bar{C}_{ij} = \bar{C}_{ij}/\bar{C}_{\max}$, and $\bar{C}_{\max}$ is the largest among all the $\bar{C}_{ij}$ values. The following expressions of the displacement field are assumed for matching the solutions in both interior and boundary layer regions, based upon Prandtl's matching principle:

$$U^{(k)} = B(Y, t) + Pe^{\lambda\eta}\cos(\delta Z)$$

$$V^{(k)} = D(Y, t) + Re^{\lambda\eta}\cos(\delta Z) \quad (2.80)$$

$$W^{(k)} = E(Y, t)Z + Se^{\lambda\eta}\sin(\delta Z)$$

Here, $B(Y, t)$, $D(Y, t)$ and $E(Y, t)$ are the interior region solutions (Eqs. 2.77); P, R and S are coefficients to be determined for the correction terms; δ is an undetermined positive constant; λ is the negative characteristic of Eqs. (2.79). It is seen from Eqs. (2.80) that away from the boundary layer region ($\eta \gg 1$), the correction terms have no influence on the displacement field; their effects become significant in and near the boundary layer region.

Substituting the $U^{(k)}$, $V^{(k)}$ and $W^{(k)}$ expressions into the equilibrium equations (2.79), the six roots of λ for non-trivial solutions of P, R and S are obtained:

$$\lambda_{1,2} = \pm a_k\delta$$

$$\lambda_{3,4} = \pm b_k\delta \quad (2.81)$$

$$\lambda_{5,6} = \pm c_k\delta$$

where a_k, b_k and c_k are three positive constants. The positive roots of λ are dropped to avoid divergence in the displacement field. Thus, the displacements for both the interior and boundary layer regions can be written as follows:

$$U^{(k)} = B(Y, t) + (P_1 e^{-a_k \delta \eta} + P_2 e^{-b_k \delta \eta} + P_3 e^{-c_k \delta \eta}) \cos(\delta Z)$$

$$V^{(k)} = D(Y, t) + (R_1 e^{-a_k \delta \eta} + R_2 e^{-b_k \delta \eta} + R_3 e^{-c_k \delta \eta}) \cos(\delta Z)$$

$$W^{(k)} = E(Y, t)Z + (S_1 e^{-a_k \delta \eta} + S_2 e^{-b_k \delta \eta} + S_3 e^{-c_k \delta \eta}) \sin(\delta Z)$$

$$(2.82)$$

There are ten unknowns for the displacement solution of the kth layer (P_1, P_2, P_3, R_1, R_2, R_3, S_1, S_2, S_3 and δ).

The available equations for the solution of these constants are: (i) three stress boundary conditions, $\sigma_{yy}(b, z) = \sigma_{xy}(b, z) = \sigma_{yz}(b, z) = 0$; (ii) six equilibrium equations (2.79); and (iii) the integrated equilibrium condition

$$\int_0^{1/2} \sigma_{xy}(0, Z) \frac{h}{2} \, dZ = \int_0^1 \sigma_{xz}\left(Y, \frac{1}{2}\right) b \, dY \qquad (2.83)$$

A four-layer angle-ply composite is taken as a numerical example. Each layer is 5 mm in thickness $h^{(k)}$, 200 mm in width (b). The SiC/borosilicate glass laminate is used as a baseline composite system for demonstration of the results. The transient interlaminar normal stress distribution of a $[-45°/45°]_s$ SiC/borosilicate glass laminate, which is subjected to a sudden edge heating of the magnitude $T_o = 1°C$ at $t = 0^+$, is demonstrated in Fig. 2.13. No stress singularity is found as a consequence of the assumed displacement

Fig. 2.13. Transient interlaminar thermal stress of a SiC/borosilicate glass $[-45°/45°]_s$ laminate for $V_f = 30\%$ and $T_o = 1°C$. (After Y. Wang and Chou 1989.)

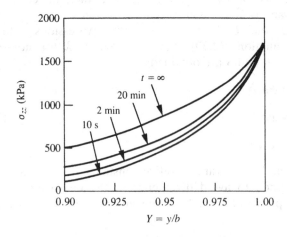

field, but it is apparent that the interlaminar normal stress concentration increases very significantly as approaching to the free edge of the plate ($Y = 1$). The stress at $Y = 1.0$ is about three to twenty times higher than that at $Y = 0.90$ for $t = 10$ s to ∞. As the heating proceeds, the overall interlaminar normal stress increases smoothly, while the stress which is very close to the boundary remains almost constant. Also, the interlaminar normal stress tends to zero away from the free edge of the laminate due to the adoption of the classical lamination theory in the interior region.

Figure 2.14 shows the results of a parametric study of the stress solution sensitivity to the composite elastic and thermal properties. Here the $[-45°/45°]_s$ SiC/borosilicate glass laminate is taken as the baseline system, and Δ indicates an increment. The Young's modulus (E_{33}) and thermal expansion coefficient (α_{33}) along the plate thickness direction have a more significant effect on the stress σ_{zz} than the thermal conductivity (K_{33}) and specific heat (C_p). The transient thermal stress analysis can be applied for the characterization of thermal shock resistance capability of composite materials. (See, for example, Cheng 1951; Kingery 1955; Y.Wang and Chou 1991.)

2.5.4 Transient in-plane thermal stress

Having discussed the thermoelastic field due to one-dimensional heat and moisture diffusion in Sections 2.5.2 and 2.5.3,

Fig. 2.14. Parametric studies of stress solution sensitivity to composite elastic and thermal properties. The base material is a SiC/borosilicate glass $[-45°/45°]_s$ laminate. Calculations of $|\Delta\sigma_{zz}|/\sigma_{zz}$ are based upon $t = 2$ min and $Y = 0.99$. (After Wang and Chou 1989.)

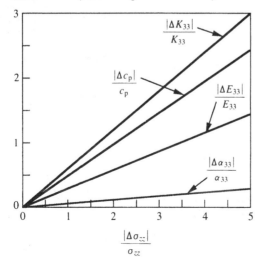

a two-dimensional transient heat conduction problem is examined in the following. The model material considered is a unidirectional lamina. The interaction of thermal stresses among the layers of a laminate is thus not included in order to clearly demonstrate the effect of transient heat conduction.

2.5.4.1 *Transient temperature field*

Consider the two-dimensional problem of an orthotropic slab with a rectangular region $(0 \leq x \leq l_1, \; 0 \leq y \leq l_2)$ as shown in Fig. 2.15. The slab is initially held at a uniform temperature and then the edge $y = l_2$ is suddenly subjected to an arbitrary temperature distribution or heat flux $f(x)$. The two-dimensional temperature distribution, $T(x, y; t)$ in the rectangular region is assumed to satisfy the heat conduction equation (2.63).

The initial condition is

$$T(x, y; 0) = 0 \qquad \text{for } t = 0 \tag{2.84}$$

The boundary conditions of the rectangle assume the following

Fig. 2.15. Thermal stress variations with time for $K = 0.1$ at the cross-section $x = 0$. (After H. Wang and Chou 1985.)

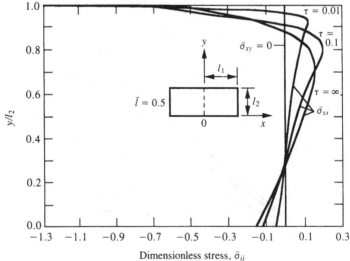

general forms:

$$-a_1 \frac{\partial T}{\partial x} + b_1 T = 0 \qquad \text{for } x = 0 \tag{2.85a}$$

$$a_2 \frac{\partial T}{\partial x} + b_2 T = 0 \qquad \text{for } x = l_1 \tag{2.85b}$$

$$-a_3 \frac{\partial T}{\partial y} + b_3 T = 0 \qquad \text{for } y = 0 \tag{2.85c}$$

$$a_4 \frac{\partial T}{\partial y} + b_4 T = f(x) \qquad \text{for } y = l_2 \tag{2.85d}$$

Here, a_i ($i = 1, 2, 3, 4$) are conductivities for the respective directions, and b_i are the coefficients of surface heat transfer. The various types of boundary conditions can be obtained through the proper selections of the constant ratio b_i/a_i (see, for example, Carslaw and Jaeger 1959). Equations (2.85a)–(2.85c) correspond to zero surface temperature or heat flow, whereas the nonhomogeneous boundary condition of Eq. (2.85d) is for an arbitrary variation of surface thermal condition.

Equation (2.85d) suggests the use of the principle of superposition. The problem has a steady-state solution as $t \to \infty$. It is assumed that

$$T(x, y; t) = \phi(x, y) + \psi(x, y; t) \tag{2.86}$$

such that $\phi(x, y)$ and $\psi(x, y; t)$ satisfy

$$K_{xx}^T \frac{\partial^2 \phi}{\partial x^2} + K_{yy}^T \frac{\partial^2 \phi}{\partial y^2} = 0 \tag{2.87}$$

$$-a_1 \frac{\partial \phi}{\partial x} + b_1 \phi = 0 \qquad \text{for } x = 0 \tag{2.88a}$$

$$a_2 \frac{\partial \phi}{\partial x} + b_2 \phi = 0 \qquad \text{for } x = l_1 \tag{2.88b}$$

$$-a_3 \frac{\partial \phi}{\partial y} + b_3 \phi = 0 \qquad \text{for } y = 0 \tag{2.88c}$$

$$a_4 \frac{\partial \phi}{\partial y} + b_4 \phi = f(x) \qquad \text{for } y = l_2 \tag{2.88d}$$

and

$$K_{xx}^T \frac{\partial^2 \psi}{\partial x^2} + K_{yy}^T \frac{\partial^2 \psi}{\partial y^2} = \rho C_p \frac{\partial \psi}{\partial t} \qquad (2.89)$$

$$\psi(x, y; 0) = -\phi(x, y) \qquad \text{for } t = 0 \qquad (2.90)$$

$$-a_1 \frac{\partial \psi}{\partial x} + b_1 \psi = 0 \qquad \text{for } x = 0 \qquad (2.91a)$$

$$a_2 \frac{\partial \psi}{\partial x} + b_2 \psi = 0 \qquad \text{for } x = l_1 \qquad (2.91b)$$

$$-a_3 \frac{\partial \psi}{\partial y} + b_3 \psi = 0 \qquad \text{for } y = 0 \qquad (2.91c)$$

$$a_4 \frac{\partial \psi}{\partial y} + b_4 \psi = 0 \qquad \text{for } y = l_2 \qquad (2.91d)$$

H. Wang and Chou (1986) have obtained the general solution of ϕ and ψ with the unknown constants in the infinite series expressions to be determined by the boundary conditions of Eqs. (2.88) and (2.91) and initial condition of Eq. (2.90).

An example of this solution technique is given by H. Wang and Chou (1985) for a slab initially held at a constant temperature and suddenly subjected to an arbitrary temperature variation along one of its edges. The constants in Eqs (2.85) are $a_1 = -1$, $a_2 = a_3 = b_1 = 0$, and $b_2 = b_3 = 1$. The temperature field solution is

$$T(x, y; t) = \sum_{n=1}^{\infty} \left\{ I_n \cos \delta_n x \sinh \frac{\delta_n}{K} y \right.$$
$$\left. + \sum_{m=1}^{\infty} I_{nm} \cos \delta_n x \sin \frac{\mu_m}{K} y \exp[-d(\delta_n^2 + \mu_m^2)t] \right\} \qquad (2.92)$$

where

$$d = K_{xx}/\rho C_p, \quad K^2 = K_{yy}/K_{xx}$$

$$I_{nm}(\delta_n, \mu_m) = \frac{2}{l_2^2} I_n(\delta_n) \frac{(-1)^m \frac{\mu_m}{K} l_2}{\left(\frac{\delta_n}{K}\right)^2 + \left(\frac{\mu_m}{K}\right)^2} \sinh \frac{\delta_n}{K} l_2$$

$$I_n(\delta_n) = \frac{2}{a_4 \frac{\delta_n}{K} \cosh \frac{\delta_n l_2}{K} + b_4 \sinh \frac{\delta_n l_2}{K}} \frac{1}{l_1} \int_0^{l_1} f(x) \cos \delta_n x \, dx \qquad (2.93)$$

$$\delta_n = \frac{2n-1}{2l_1} \pi \qquad n = 1, 2, 3, \ldots, \infty$$

Also, μ_m/K are the positive roots of

$$\left(\frac{\mu}{K}l_2\right)\cot\left(\frac{\mu}{K}l_2\right) + \frac{b_4}{a_4}l_2 = 0$$

If $a_4 = 0$ and $b_4 = 1$,

$$\frac{\mu_m}{K} = \frac{m}{l_2}\pi = \delta_m \qquad m = 1, 2, 3, \ldots, \infty \tag{2.94}$$

H. Wang and Chou (1986) have tabulated the solution of temperature field from the various combinations of a_1, a_2, a_3, b_1, b_2 and b_3 values of Eqs. (2.85).

2.5.4.2 *Thermal stress field*

Consider a unidirectional fiber composite; let the principal material directions x_1 and x_2 coincide with the reference axes x and y, respectively. The stress–strain relations follow Eq. (2.49) with the $\bar{C}_{ij}$ replaced by C_{ij}. Depending upon the thickness of the elastic medium in the z direction, the thermoelastic problem is in the state of either plane strain or plane stress. In the case of plane strain, the stress components in the x–y plane are related to the in-plane displacements, $u(x, y; t)$ and $v(x, y; t)$, and the temperature, $T(x, y; t)$, by substituting the strain–displacement relations into the stress–strain relations of Eqs. (2.73). The results are

$$\sigma_{xx} = C_{11}\frac{\partial u}{\partial x} + C_{12}\frac{\partial v}{\partial y} - \bar{\alpha}_1 T(x, y; t)$$

$$\sigma_{yy} = C_{12}\frac{\partial u}{\partial x} + C_{22}\frac{\partial v}{\partial y} - \bar{\alpha}_2 T(x, y; t)$$

$$\sigma_{zz} = C_{13}\frac{\partial u}{\partial x} + C_{23}\frac{\partial v}{\partial y} - \bar{\alpha}_3 T(x, y; t) \tag{2.95}$$

$$\sigma_{xy} = C_{66}\left(\frac{\partial u}{\partial y} + \frac{\partial v}{\partial x}\right)$$

where

$$\bar{\alpha}_1 = C_{11}\alpha_{11} + C_{12}\alpha_{22} + C_{13}\alpha_{33}$$

$$\bar{\alpha}_2 = C_{12}\alpha_{11} + C_{22}\alpha_{22} + C_{23}\alpha_{33} \tag{2.96}$$

$$\bar{\alpha}_3 = C_{13}\alpha_{11} + C_{23}\alpha_{22} + C_{33}\alpha_{33}$$

The relations corresponding to Eqs. (2.95) for plane stress condition are obtained by replacing C_{ij} and $\bar{\alpha}_i$ by $C_{ij} - C_{3j}/C_{33}$ and $\bar{\alpha}_i - \bar{\alpha}_3 C_{3i}/C_{33}$, respectively.

The displacement equations of equilibrium governing the plane strain conditions are

$$C_{11} \frac{\partial^2 u}{\partial x^2} + C_{66} \frac{\partial^2 u}{\partial y^2} + (C_{12} + C_{66}) \frac{\partial^2 v}{\partial x \, \partial y} = \bar{\alpha}_1 \frac{\partial T}{\partial x}$$

$$C_{66} \frac{\partial^2 v}{\partial x^2} + C_{22} \frac{\partial^2 v}{\partial y^2} + (C_{12} + C_{66}) \frac{\partial^2 u}{\partial x \, \partial y} = \bar{\alpha}_2 \frac{\partial T}{\partial y}$$

(2.97)

The equilibrium equations are solved by introducing the displacement potentials ψ_1, ψ_2 and ϕ defined by

$$u(x, y; t) = \frac{\partial \psi_1}{\partial x} + \frac{\partial \psi_2}{\partial x} + \frac{\partial \phi}{\partial x}$$

$$v(x, y; t) = v_1 \frac{\partial \psi_1}{\partial y} + v_2 \frac{\partial \psi_2}{\partial y} + \lambda \frac{\partial \phi}{\partial y}$$

(2.98)

where v_1, v_2 and λ are unknown constants. Also, ϕ is the homogeneous solution and ψ_1 and ψ_2 are particular solutions of Eqs. (2.97).

An example of the transient thermal stress solution is given by H. Wang and Chou (1985) for a rectangular slab $(-l_1 \le x \le l_1$ and $0 \le y \le l_2)$ with fibers oriented in the x direction. The initial temperature of the slab is $T = 0$. Then the following form of temperature rise at the upper edge $(y = l_2)$ is adopted:

$$T = f(x) = T_0 \cos \frac{\pi}{2l_1} x \qquad \text{for } t > 0 \qquad (2.99)$$

while the temperature over the remainder of the boundary is maintained at the initial value. All edges of the rectangle are assumed to be traction free:

$$\sigma_{xx} = \sigma_{xy} = 0 \qquad \text{for } x = \pm l_1$$

$$\sigma_{yy} = \sigma_{xy} = 0 \qquad \text{for } y = 0, \, l_2$$

(2.100)

The thermal and elastic properties as given by Aköz and Tauchert (1978) simulating a boron/epoxy composite are adopted for the numerical calculations. Owing to the symmetry of the assumed temperature rise, only one half of the rectangle $(0 \le x \le l_1)$ needs to be considered. Thus, the boundary condition Eq. (2.85d) is reduced to $a_4 = 0$ and $b_4 = 1$. For the convenience of presenting the numerical results, the following dimensionless quantities are introduced for temperature stress, time, and lamina dimension, respectively: $\bar{T} = T(x, y; t)/T_o$; $\bar{\sigma}_{ij} = \sigma_{ij}(x, y; t)/\bar{\alpha}_2 T_o$, $\tau = dt/l_1^2$, and $\bar{l} = l_2/l_1$.

Figure 2.15 shows the y direction variation of thermal stresses at the cross-section $x = 0$ for the various dimensionless time intervals. It is clear that large longitudinal stresses σ_{xx} occur in the vicinity of the heated boundary, where the relatively large temperature gradient, $\partial T/\partial y$, exists. On the other hand, the transverse stresses, σ_{yy}, and the shear stresses σ_{xy} are fairly small. Also, for σ_{xx}, the maximum transient tensile stress is 25% higher than that in the steady state; the maximum transient compressive stress near the upper edge ($y = l_2$) is 78% higher than the corresponding steady-state stress. An examination of the plots of σ_{xx} and σ_{xy} at a given time interval indicates that each stress is in self-equilibrium when the slab is free to deform, i.e. no boundary constraints. This is consistent with the nature of thermal residual stresses.

3 Strength of continuous-fiber composites

3.1 Introduction

Fiber-reinforced composites are a valuable class of engineering materials because they can exhibit both high stiffness and strength simultaneously, in contrast to more homogeneous materials which are generally brittle and defect sensitive. In fiber composites, the inherent lack of toughness of the reinforcing fiber, or its sensitivity to microstructural defects, is overcome by the local redundancy of the composite structure, so that its strength may be utilized effectively. Individual fibers are relatively weakly coupled by the matrix so that failure of one fiber does not generally precipitate immediate failure of the composite as a whole, allowing high strength and stiffness to be achieved in the fiber direction.

The tensile failure of a fiber-reinforced material is a complex process which involves an accumulation of microstructural damage. Unlike homogeneous brittle materials, fiber composites do not contain a population of observable pre-existing defects, one of which ultimately precipitates failure. Instead, an accumulation of fiber or matrix fractures develops as the material is loaded and this constitutes a 'critical defect' in a macroscopic view of the fracture. Fracture mechanics may successfully account for the strength of single fibers, but it is inadequate to extend its application to unidirectional fiber composites when the overall behavior is dominated by the probability of defects in fibers propagating under the stress concentrations surrounding previous fiber fractures as well as the probability of defects in the matrix which are responsible for the multiplication of transverse cracks. Consequently, the statistical process of damage development in composites needs to be emphasized (Manders, Bader and Chou 1982).

The development of a rigorous analysis of fracture, considering all the sequences of fiber and matrix fractures which result in fracture of the composite, is a formidable task, and for this reason the strength of composites with realistic dimensions is much less well understood than their elastic properties.

This chapter treats the strength of continuous fiber composites with a combination of statistical and fracture mechanics approaches. The statistical analysis of unidirectional composites is better developed than that for cross-ply laminates. No comprehensive statistical methodology is available at this time for treating the strength and failure of composites from the fiber and bundle level up to composite laminates. Thus, the fracture mechanics approach to laminate failure is necessary.

In this chapter, the classical approximation of the rule-of-mixtures is adopted as a starting point for composite axial strength. This approximation is substantially altered due to stress concentrations induced at fiber breakages. The statistical variations of fiber and bundle strengths are then discussed. The knowledge of the stress redistribution at fiber breaks is then incorporated into the statistical strength analysis of unidirectional fiber composites. Next, the strength analysis is extended to the case of cross-ply laminates which serve as model systems for laminate composites. Finally, an attempt is made to shed some light on the failure of laminated composites in general where both inter- and intralaminar failures play key roles in the failure modes. A method of analysis based upon the fracture mechanics approach is introduced. Section 3.4.6.2 is contributed by S. L. Phoenix, and Sections 3.4.7.4 and 3.4.8 are contributed by A. S. D. Wang.

Another approach to the strength and damage of fiber composites is based upon the overall properties degradation. The strength behavior can be modeled by regarding the composite with damage as a continuum with changing microstructure. A phenomenological theory of constitutive behavior then provides relationships between the severity of damage and the overall stiffness properties of a composite (Reifsnider, Henneke, Stinchcomb and Duke 1983; Talreja, 1985, 1986, 1987, 1989).

Strength theories dealing with short-fiber and hybrid composites are discussed in Chapters 4 and 5, respectively.

3.2   Rule-of-mixtures

The classical approximation of unidirectional continuous-fiber composite strength takes the form of the rule-of-mixtures. By assuming equal strain in the fiber and matrix phases, the stress in the composite under uniaxial loading can be expressed as (see Kelly and Nicholson 1971 and Vinson and Chou 1975)

$$\sigma_c = \sigma_f V_f + \sigma_m (1 - V_f) \tag{3.1a}$$

where σ and V_f denote, respectively, stress and fiber volume fraction. The subscripts c, f and m are for composite, fiber and matrix, respectively. Then, the ultimate composite strength is

$$\sigma_{cu} = \sigma_{fu}V_f + \sigma_{mu}(1 - V_f) \tag{3.1b}$$

Here, the subscript u denotes ultimate strength. Equation (3.1b) is valid provided that both the fiber and matrix have the same ultimate strain.

Equation (3.1b) is not sufficient in determining the strength of continuous-fiber composites. Aveston, Cooper and Kelly (1971) have discussed the strength of composites based upon the transfer of load at the fiber/matrix interface and the mode of failure. For the case of brittle fiber-reinforced ductile matrix, the matrix ultimate strain is often higher than that of the fiber (Fig. 3.1a); then *single fractures* of the composite occur when

$$\sigma_{fu}V_f + \sigma'_{mu}(1 - V_f) > \sigma_{mu}(1 - V_f) \tag{3.1c}$$

Fig. 3.1. (a) Stress–strain relation of a brittle fiber/ductile matrix composite. (b) Composite strength vs. fiber volume fraction for brittle fiber/ductile matrix composites. (c) Stress–strain relation of a ductile fiber/brittle matrix composite. (d) Composite strength vs. fiber volume fraction for ductile fiber/brittle matrix composites.

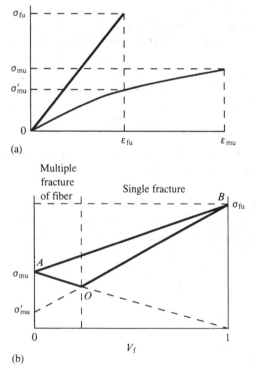

where σ'_{mu} is the stress in the matrix when the fibers fail. The matrix is unable to withstand the additional load transferred to it due to the fiber fracture, and thus single fracture prevails at sufficiently high fiber volume fractions. At low fiber fractions,

$$\sigma_{fu} V_f + \sigma'_{mu}(1 - V_f) < \sigma_{mu}(1 - V_f) \tag{3.1d}$$

and the load is essentially born by the matrix material. The failure of the composite is characterized by *multiple fractures* of the fibers into shorter and shorter segments as the strain on the matrix increases (Fig. 3.1b). Experimental data on the ultimate strength of unidirectional fiber composites usually fall within the triangular region of Fig. 3.1b specified by the solid line segments.

Provided the failure strain of the matrix is sufficiently large, the fibers are fractured into lengths between x and $2x$. Assuming a constant fiber–matrix interfacial shear stress τ, the fiber fracture

Fig. 3.1. (*cont.*).

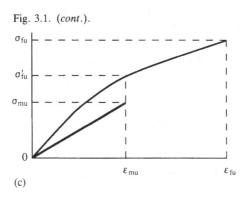

(c)

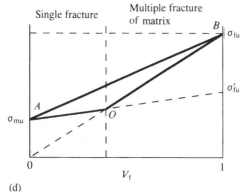

(d)

spacing is determined from a simple force balance

$$x = \frac{\sigma_{fu} r}{2\tau} \qquad (3.2)$$

where r denotes fiber radius.

The above analysis does not fully account for the fact that the strength of a fiber is a statistical quantity which results from flaws being randomly distributed along the length, as is discussed later. One result is that the strength depends on the fiber length, and thus is not really a fixed quantity σ_{fu}. Using the accepted Poisson/Weibull model, Henstenburg and Phoenix (1989) have developed a modified version of Eq. (3.2) which includes a factor connected to the variability in fiber strength. The revised formula typically produces values which are 15 to 20% larger. Also, these authors have delved further into the nature of the statistical distribution for fragment length, and experimental examples can be found in Netravali, Henstenburg, Phoenix and Schwartz (1989).

For the case of ductile fiber-reinforced brittle matrix composites, multiple fracture of the matrix occurs when the fiber ultimate strain, ε_{fu}, is higher than that of the matrix, ε_{mu} (Fig. 3.1c). The condition of multiple fracture, according to Aveston, Cooper and Kelly (1971), is

$$\sigma'_{fu} V_f + \sigma_{mu}(1 - V_f) < \sigma_{fu} V_f \qquad (3.3)$$

Here, σ'_{fu} is the stress in the fiber at the failure strain of the matrix. A single fracture of the composite occurs if the fibers cannot withstand the increase in loading due to the matrix failure (Fig. 3.1d).

The spacing between two adjacent matrix cracks can again be determined from a simple force balance, and the separation distance is between x' and $2x'$

$$x' = \frac{1 - V_f}{V_f} \frac{\sigma_{mu} r}{2\tau} \qquad (3.4)$$

In deriving Eq. (3.4), it is understood that the number of fibers per unit area transverse to the fiber direction is given by $V_f / \pi r^2$.

Composites containing ductile fibers in a ductile matrix have shown work-hardening behavior. Mileiko (1969) has theorized that the instability or necking of the matrix can be suppressed due to the constraint of the matrix, and the ultimate strain of the composite, in this case, is shown to lie in between the ultimate strains of the fiber and matrix materials.

3.3 Stress concentrations due to fiber breakages

Fiber breakages in a continuous-fiber composite can occur at fabrication or during the early stage of loading. Stress redistribution takes place in the vicinity of a fiber breakage because load can no longer be transferred along the fiber in a continuous manner. The resulting stress concentrations in the neighboring fibers are detrimental to the strength of continuous-fiber composites. In the following, the shear-lag analysis is introduced to examine both the static and dynamic stress concentrations in unidirectional continuous-fiber composites.

3.3.1 Static case

The problem of static stress concentration in composites has been treated by the shear-lag method (see Hedgepeth 1961; Hedgepeth and Van Dyke 1967; Fichter 1969, 1970; Van Dyke and Hedgepeth 1969; Zweben 1974; Fukuda and Kawata 1976a, 1980; Goree and Gross 1979, 1980; Hikami and Chou 1990), elasticity theory (see Burgel, Perry and Scheider 1970; Takao, Taya and Chou 1981), and numerical methods (see Carrara and McGarry 1968; Chen 1971).

Among these approaches, the shear-lag method, which is based upon simplified assumptions, often provides good physical insights of rather complex problems. The shear-lag method was first adopted by Hedgepeth (1961) to treat multi-filament failure problems of unidirectional composites. The technique also has been extended to include the effects of plasticity of the matrix (Hedgepeth and Van Dyke 1967; Goree and Gross 1979; Hikami and Chou 1984a), and the condition of interfacial debonding (Van Dyke and Hedgepeth 1969). The major assumptions of this method are that: (1) the fibers sustain only the axial loads, and (2) matrix between fibers transmits only the shear force.

In the following the single filament failure model of Fukuda and Kawata (1976a) is reproduced first to demonstrate the fundamentals of this method, and the nature of stress redistribution in unidirectional composites. Next, the work of Hikami and Chou (1990) is introduced for the explicit solutions of multi-filament failure problems.

3.3.1.1 Single filament failure

Figure 3.2 shows the model of analysis by Fukuda and Kawata (1976a) which contains three parallel fibers with the middle one being broken. This model can also be considered as the two-dimensional representation of a laminate with a broken middle layer. Because of symmetry, only half of the model needs to be considered and the fibers are denoted as $n = 1$ and 2. The equilibrium of forces in the fibers in the free-body diagram of Fig. 3.3 gives

$$\frac{1}{2}\frac{dP_1}{dx} + \tau_1 = 0 \tag{3.5}$$

$$\frac{dP_2}{dx} - \tau_1 = 0 \tag{3.6}$$

Fig. 3.2. A three-fiber composite model for shear-lag analysis.

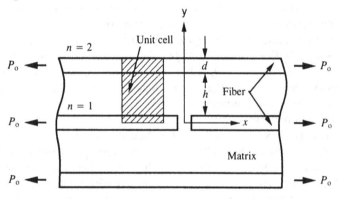

Fig. 3.3. Free-body diagrams for the 'unit cell' of the composite shown in Fig. 3.2.

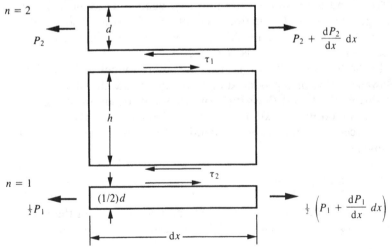

where P_1 and P_2 denote fiber axial force per unit thickness, and τ_1 is the matrix shear stress. Let the displacement of the nth fiber be denoted as u_n. Then,

$$P_n(x) = Ed\frac{\mathrm{d}u_n(x)}{\mathrm{d}x} \qquad n = 1, 2 \tag{3.7}$$

$$\tau_1(x) = \frac{G}{h}(u_2(x) - u_1(x)) \tag{3.8}$$

where E is the Young's modulus of the fibers; G is the effective shear modulus of the matrix; h is the effective fiber spacing; d is the fiber width; and the lamina is of unit thickness.

Using Eqs. (3.7) and (3.8), and the following non-dimensional parameters

$$\xi = x/d \tag{3.9}$$

$$\alpha = Eh/Gd \tag{3.10}$$

Equations (3.5) and (3.6) become

$$\tfrac{1}{2}\alpha\frac{\mathrm{d}^2u_1}{\mathrm{d}\xi^2} + u_2 - u_1 = 0 \tag{3.11}$$

$$\alpha\frac{\mathrm{d}^2u_2}{\mathrm{d}\xi^2} + u_1 - u_2 = 0 \tag{3.12}$$

From Eq. (3.11), u_2 can be expressed by u_1 and its derivatives as follows:

$$u_2 = u_1 - \tfrac{1}{2}\alpha\frac{\mathrm{d}^2u_1}{\mathrm{d}\xi^2} \tag{3.13}$$

Substitution of Eq. (3.13) into Eq. (3.12) yields

$$\frac{\mathrm{d}^4u_1}{\mathrm{d}\xi^4} - \frac{3}{\alpha}\frac{\mathrm{d}^2u_1}{\mathrm{d}\xi^2} = 0 \tag{3.14}$$

The general solution of Eq. (3.14) is

$$u_1 = A + B\xi + Ce^{\lambda\xi} + De^{-\lambda\xi} \tag{3.15}$$

where $\lambda = \sqrt{(3/\alpha)}$ and A, B, C and D are integration constants. Substituting Eq. (3.15) into Eq. (3.13), the general solution of u_2 is obtained as

$$u_2 = A + B\xi - \tfrac{1}{2}Ce^{\lambda\xi} - \tfrac{1}{2}De^{-\lambda\xi} \tag{3.16}$$

A, B, C and *D* in Eqs. (3.15) and (3.16) can be determined from Eq. (3.7) and the following boundary conditions:

$$(u_2)_{\xi=0} = 0, \qquad (P_1)_{\xi=0} = 0, \qquad (P_1)_{\xi=\infty} = P_o \qquad (3.17)$$

Finally, the fiber displacements and axial loads are obtained

$$u_1 = \frac{P_o}{E}\left(\frac{1}{2\lambda} + \xi + \frac{1}{\lambda}e^{-\lambda\xi}\right)$$

$$u_2 = \frac{P_o}{E}\left(\frac{1}{2\lambda} + \xi - \frac{1}{2\lambda}e^{-\lambda\xi}\right) \qquad (3.18)$$

$$P_1 = P_o(1 - e^{-\lambda\xi})$$

$$P_2 = P_o(1 + \tfrac{1}{2}e^{-\lambda\xi}) \qquad (3.19)$$

Values of P_1 and P_2 in Eq. (3.19) are shown in Fig. 3.4. The stress concentration factor of this model, $(P_2/P_o)_{\xi=0}$, is 1.5. According to Eq. (3.19), the distributions of fiber displacements and axial loads are functions of the material constant λ. However, the stress concentration factor is independent of λ. The above treatment has been extended to composites containing a finite number of fibers with any number of adjacent fiber breakages on the same transverse plane.

3.3.1.2 *Multi-filament failure*

Hikami and Chou (1984b, 1990) have examined the two-dimensional multi-filament failure problem of unidirectional fiber

Fig. 3.4. Variations of fiber axial forces.

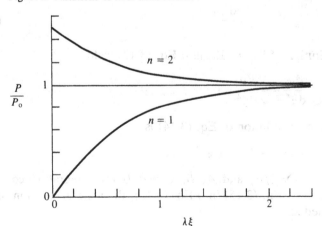

composites, focussing specifically on the stress concentration factors of fibers adjacent to the cracks. The physical problems are analyzed by the two-dimensional shear-lag method under two loading conditions: (A) uniform tensile force applied to all fibers at infinity (Fig. 3.5), and (B) concentrated force dipole applied at a particular fiber, $n = b - a$, on the crack plane (Fig. 3.7).

These analyses are unique in that the general solution of the governing equations of the elastic field has been obtained in explicit forms in terms of the Legendre polynomials for the loading condition (A). Based upon this solution, closed form expressions of stress concentration factors in all fibers have been derived. These analyses also provide rigorous proofs of both Hedgepeth and Van Dyke's inspection (1967) on the general form of the tensile stress concentration factor at the tip of a crack and Fichter's inspection (1969) on the general form of the shear stress concentration factor for the loading condition (A). Since there exists a reciprocal relation between the influence function matrices for the loading conditions (A) and (B), the solution for the condition (B) can be readily derived from the solution for the condition (A).

The analyis considers a two-dimensional unidirectional continuous-fiber composite containing a slit notch in the transverse direction, as shown in Fig. 3.5. The fiber direction is taken along the x axis. The broken fibers are denoted as $n = 1, 2, 3, \ldots, b$, starting from the left tip of the notch with b being the total number of fibers in the notch.

Under the assumption of shear-lag analysis, the matrix material transfers only shear force, $\bar{\tau}_n(x)$, per unit fiber length between two adjacent fibers. Thus $\bar{\tau}_n(x)$ is related to the difference of displacements $u_n(x)$ in the fiber direction as

$$\bar{\tau}_n(x) = \frac{G}{h} \{u_{n+1}(x) - u_n(x)\} \tag{3.20}$$

where G is the effective shear modulus of the matrix, and h is the effective fiber spacing. The tensile force $P_n(x)$ per unit thickness in the nth fiber is related to the displacement by

$$P_n(x) = Ed \frac{du_n(x)}{dx} \tag{3.21}$$

where d is the width of the fiber. The equilibrium of forces in the x direction gives

$$\frac{dP_n}{dx} + \bar{\tau}_n - \bar{\tau}_{n-1} = 0 \tag{3.22}$$

The non-dimensionalized axial force, displacement and coordinate are given, respectively, by:

$$F_n(\xi) = P_n(x)/P_0$$

$$U_n(\xi) = u_n(x)\sqrt{(EdG/hP_0^2)} \qquad (3.23)$$

$$\xi = \sqrt{(G/Edh)}x$$

Then, the equilibrium equation (3.22) can be written as

$$\frac{d^2U_n(\xi)}{d\xi^2} = 2U_n(\xi) - U_{n+1}(\xi) - U_{n-1}(\xi) \qquad (3.24)$$

The boundary conditions are:

$$F_n(0) = 0 \qquad (1 \le n \le b)$$

$$U_n(0) = 0 \qquad (n \le 0, n \ge b+1) \qquad (3.25)$$

$$F_n(\pm\infty) = 1 \qquad \text{(all } n\text{)}$$

Fig. 3.5. Model of a multi-filament crack in a unidirectional composite under uniform force at infinity (After Hikami and Chou 1990.)

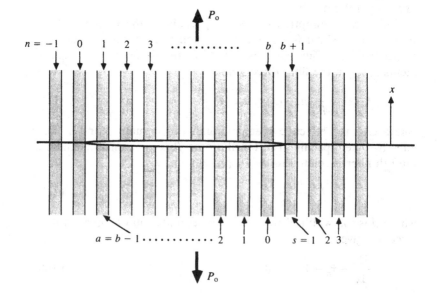

for loading condition (A), and

$$F_n(0) = 0 \qquad (n = 1, 2, \ldots, b - a - 1, b - a + 1, \ldots, b)$$
$$U_n(0) = 0 \qquad (n \leq 0, n \geq b + 1)$$
$$F_n(\pm\infty) = 0 \qquad (\text{all } n) \tag{3.26}$$
$$F_{b-a}(0) = -1$$

for loading condition (B).

The general solutions of the multi-filament failure problem have been obtained explicitly by Hikami and Chou (1990) using the Legendre polynomials and Fourier transformation. The stress concentration factors in all fibers on the crack plane are given in closed forms. First, for the loading condition (A), the stress concentration factor of the sth fiber ahead of the tip of a crack containing b broken fibers is given by

$$K_b^s = (b + 2s - 1)$$

$$\times \frac{2s \cdot (2s + 2) \cdot (2s + 4) \cdots (2s + 2b - 2)}{(2s - 1) \cdot (2s + 1) \cdot (2s + 3) \cdots (2s + 2b - 3) \cdot (2s + 2b - 1)} \tag{3.27}$$

As a special case of Eq. (3.27), the stress concentration factor in the first intact fiber ($s = 1$) adjacent to b broken fibers is

$$K_b^1 = \frac{4 \cdot 6 \cdot 8 \cdots (2b + 2)}{3 \cdot 5 \cdot 7 \cdots (2b + 1)} \tag{3.28}$$

Hedgepeth (1961) deduced Eq. (3.28) by inspecting the numerical results of the cases $b = 1, 2, \ldots, 6$. This inspection on the general form of the stress concentration factor has been rigorously proven by Hikami and Chou. Figure 3.6 depicts the numerical results for K_b^s.

Furthermore, the maximum shear stress takes place in the matrix at the tip of the crack. Thus, the dimensionless displacement at the crack tip $U_b(0)$ is termed the maximum shear stress concentration factor, S_{max}. Hikami and Chou (1990) have obtained

$$S_{max} = \frac{\pi(2b - 1)!}{2^{2b}[(b - 1)!]^2} \tag{3.29}$$

Fichter (1969) deduced the above result by calculating the cases of $b = 1, 2, \ldots, 6$. The axial stress in fibers away from the crack plane has also been obtained.

In the case of loading condition (B), Fig. 3.7, Hikami and Chou

Fig. 3.6. Stresses concentration factor K_b^s in the $(b+s)$th fiber. b denotes the number of broken fibers; $s=1$ corresponds to the special case of Hedgepeth (1961). (After Hikami and Chou 1990.)

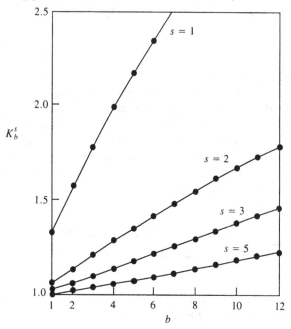

Fig. 3.7. Model of a multi-filament crack in a unidirectional composite under concentrated force dipole in the $(b-a)$th fiber on the crack plane. (After Hikami and Chou 1990.)

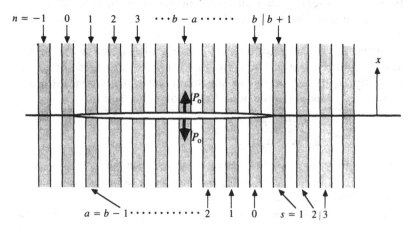

(1990) assume that a unit force dipole is applied on the $(b - a)$th fiber. Then the more general cases with multiple dipoles can be obtained by the linear combination of the solutions of the simple problem.

The closed form solution of the stress concentration factor at the sth fiber in front of the tip of a crack containing b fibers and a unit force dipole at the n $(=b - a)$th fiber is given as

$$K_b^{s,a} = \frac{1}{2} \frac{(2a + 1)!! \, (2b - 2a - 1)!! \, (2s - 3)!! \, (2s + 2b - 2)!!}{(2a)!! \, (2b - 2a - 2)!! \, (2s - 2)!! \, (2s + 2b - 1)!!} \frac{1}{(s + a)}$$

(3.30)

where !! denotes double factorial (i.e. $n!! = (n!)!$). The highest fiber stress concentration takes place at the edge of the crack ($s = 1$)

$$K_b^{1,a} = \frac{(2a + 1)!! \, (2b - 2a - 1)!! \, (2b)!!}{(2a + 2)!! \, (2b - 2a - 2)!! \, (2b + 1)!!}$$

(3.31)

Figure 3.8 depicts the numerical results for $K_b^{1,a}$.

For a semi-infinite crack the stress concentration factor at the sth fiber from the crack tip due to the unit applied force dipole at the

Fig. 3.8. Stress concentration $K_b^{1,a}$ in the $(b + 1)$th fiber when the unit load is applied at the $(b - a)$th fiber. b denotes the number of broken fibers. (After Hikami and Chou 1990.)

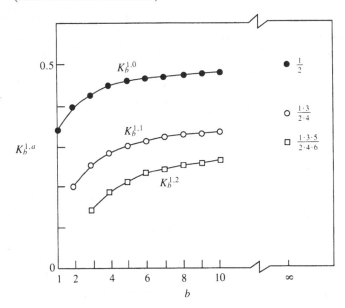

*a*th fiber has the following value:

$$\lim_{b \to \infty} K_b^{s,a} = \frac{1}{2(s+a)} \cdot \frac{(2s+1)!!}{(2a)!!} \cdot \frac{(2s-3)!!}{(2s-2)!!} \tag{3.32}$$

The axial fiber stress distributions away from the fracture plane for both loading conditions (A) and (B) also have been obtained by Hikami and Chou (1990). Also Fukuda and Kawata (1980) have shown in their analysis of a finite number of fibers that the stress concentration factor tends to that of Hedgepeth as the total fiber number increases.

The static stress concentration factors in a layer of unidirectional composites containing dacron fibers imbedded in a polyurethane elastomer have been measured by Zender and Deaton (1963). The number of fiber breakages in this experiment is controlled by partially slitting the specimens in the transverse direction. The slit length determines the number of broken fibers. The results of the experiments show reasonably close agreement with the theoretical analysis. It should be noted that although the broken fibers induce the adjacent fibers to fail in the vicinity of the cut, the chances are that such a location is not the weakest location of the fiber. This has to do with the statistical nature of fiber strength distribution and will be discussed in Section 3.4.

The problem of static stress concentration factors in a three-dimensional fiber array has been examined by Van Dyke and Hedgepeth (1969). They consider square and hexagonal arrays where a specified number of fibers are broken. Other stress concentration problems including the effects of finite length of fibers (Fichter 1970), relative locations of fiber breaks (Chen 1973), holes (Kulkarni, Rosen and Zweben 1973) and notches (Zweben 1974) also have been treated.

3.3.2   *Dynamic case*

When fibers are suddenly broken in a composite under stress, the load in the broken fibers must be transferred through the matrix to the adjacent fibers in order to restore equilibrium. Of interest is not only the resulting static stress, but also the dynamic overshoot which occurs during the transient phase. Hedgepeth (1961) examined the dynamic aspect of stress concentration for the two-dimensional fiber array as shown in Fig. 3.5. The analytical model is also based upon the assumptions of the shear-lag analysis;

that is, it is composed of tension-carrying elements connected by purely shear-carrying material.

The formulation of boundary value problem for the evaluation of dynamic stress concentration is outlined below. The fibers are separated by a constant distance and are numbered from $n = -\infty$ to $n = \infty$ (Fig. 3.5). The coordinate along the fiber is denoted by x and the displacement of the nth fiber at the location x and time t is given by $u_n(x, t)$. Similarly, the force per unit thickness in the nth fiber is denoted by $P_n(x, t)$ and is given in terms of u_n by

$$P_n = Ed\frac{\partial u_n}{\partial x} \tag{3.33}$$

where E and d are, respectively, the fiber Young's modulus and width. The equilibrium of an element of the nth filament then requires

$$Ed\frac{\partial^2 u_n}{\partial x^2} + \frac{G}{h}(u_{n+1} - 2u_n + u_{n-1}) = m\frac{\partial^2 u_n}{\partial t^2} \tag{3.34}$$

Here, G and h denote matrix shear modulus and width, respectively; m is the mass per unit area of the nth filament.

In general, for b broken filaments, let $1 \le n \le b$ denote the broken filaments. The boundary conditions are:

$$\begin{aligned} P_n(0, t) &= 0 \qquad (1 \le n \le b) \\ u_n(0, t) &= 0 \qquad (n \le 0 \quad \text{or} \quad n \ge b + 1) \end{aligned} \tag{3.35}$$

For large x, of course, the force in each filament approaches the uniform applied force per unit thickness, P_o. Thus

$$P_n(\pm\infty, t) = P_o \tag{3.36}$$

For the time-dependent problem, the following initial conditions are required:

$$\begin{aligned} P_n(x, 0) &= P_o \\ \frac{\partial u_n}{\partial t}(x, 0) &= 0 \end{aligned} \tag{3.37}$$

Using a Laplace transform of the time-dependent differential equation and boundary conditions, the resulting equations are similar in form to those of the static problem discussed in Section 3.3.1. The variation of stress concentration factor with time is shown in Fig. 3.9 for one, two and three broken fibers. As can be seen from Fig. 3.9, the stress concentration factor, K_b^1, varies with the dimensionless time $\bar{t}$ ($=t/\sqrt{(md/G)}$), and approaches the steady-state value. In all cases, the first peak is the largest one and the value of the stress at this peak determines the dynamic overshoot.

Hedgepeth (1961) defines the dynamic-response factor as the ratio between the maximum stress and the static stress. Values for one, two and three broken fibers are, respectively, 1.15, 1.19 and 1.20. It can be shown that the dynamic-response factor approaches 1.27 as the number of broken fibers tends to infinity. Further discussions of dynamic stress concentration factors are given in Section 3.4.9.

Following the approach of Hedgepeth (1961), Ji, Liu and Chou (1985) have investigated the variation of dynamic stress concentration along the length of a fiber next to a broken fiber. Define the dimensionless parameter in fiber axial location as

$$\xi = \frac{x}{\sqrt{\left(\dfrac{E}{G}\right) \cdot d}} \tag{3.38}$$

The asymptotic expressions of the stress concentration factor

Fig. 3.9. The variation of dynamic stress concentration factor K_b^1 with dimensionless time $\bar{t}$ for $b = 1, 2$ and 3 (After Hedgepeth 1961.)

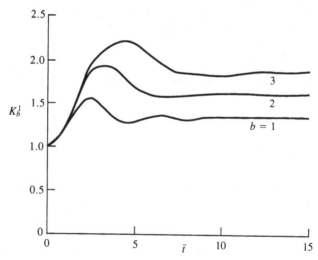

$K_1^1(\xi, \bar{t})$ for the fiber $s = 1$ at $x = 0$ due to the fracture of the fiber $n = b = 1$ (see Fig. 3.5) has been obtained. The results are depicted in Fig. 3.10, and the following observations can be made: (a) the fiber axial stress is always tensile at $\xi = 0$. For $\xi \neq 0$, the initial stress induced by fiber fracture is compressive, and the magnitude of this initial compressive stress increases with ξ; (b) the dynamic stress concentration factor, which is defined by the maximum initial tensile stress, decreases as ξ increases, i.e. away from the plane of fiber fracture; (c) the dynamic stress concentration factor is appreciable (say, $K_1^1 (\xi, \bar{t}) > 1.1$) within the range of $0 \leq \xi \leq 1$. When $\bar{t} > 10$, the dynamic stress concentration factor results for $\xi < 1$ approach the static stress concentration values. The change of stress concentration factor with the location on a fiber needs to be taken into account when there is a scattering in fiber strength and variation of fiber strength with fiber length. The results of Ji, Liu and Chou indicate that the variation of stress concentration is significant for $\xi \leq 1$, namely x is of the order of fiber diameter times

Fig. 3.10. Dynamic stress concentration factor K_1^1 with dimensionless time $\bar{t}$ for $0 \leq \xi \leq 1$. (After Ji, Liu and Chou 1985.)

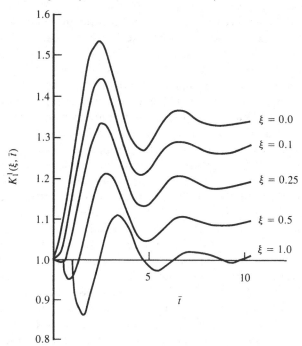

$\sqrt{(E/G)}$. For $\xi \geq 1.0\text{--}2.0$, the dynamic response diminishes with increasing $\bar{i}$ value, and the static stress concentration factor approaches 1.0; there is virtually no static stress concentration. On the other hand, dynamic response in fiber stress concentration exists at small $\bar{i}$ even for $\xi \geq 2$; this factor needs to be taken into account in the statistical composite strength models.

The variation of stress concentration along the length of a fiber has implications on the dynamic failure characteristics of fiber composites. For instance, in the experimental observation of Ji (1982), carbon composite specimens often fracture at locations near specimen end-tabs. The reflection and hence magnification of the stress waves at specimen ends could cause fiber fractures at locations away from the plane of the existing fiber breakages.

3.4   Statistical tensile strength theories

3.4.1   *Preliminary*

Statistics is concerned with scientific methods for collecting and analyzing data, as well as drawing valid conclusions and making reasonable decisions on the basis of such analysis. Spiegel (1961) and Kirkpatrick (1974) provide introductions to the basics of statistics. Statistical treatment of composite strength has emerged as an important analytical tool for the obvious reason that the strengths of brittle fibers and yarns are *statistical* in nature, and not deterministic such as in metals. A concise outline of the fundamentals in statistics based upon Spiegel (1961) is given below.

In collecting data concerning characteristics of a group of objects, it is often impractical to observe the entire group or *population* if it is large. A small part of the group examined is known as a *sample*. Valid conclusions can often be inferred from analysis of the sample. Because such inference cannot be absolutely certain, the language of *probability* is often used in stating conclusions.

When summarizing large masses of raw data, it is often useful to distribute the data into *classes* or *categories*. The number of individuals belonging to each class is called the *class frequency*. Figure 3.11 gives a graphical representation of the *frequency distribution* of the measured strength of carbon fibers (M. G. Bader and B. Gul-Mohammed, private communication, 1990; see also Dhingra 1980). The *relative frequency* of a class is the frequency of the class divided by the total frequency of all classes and is generally expressed as a percentage. A histogram can be approximated by a continuous frequency distribution curve as shown schematically in

Fig. 3.12. Also shown in Fig. 3.12 is the *cumulative frequency*, which, for a particular class or strength level, is the total frequency of all classes observed at equal to and less than this particular class. Cumulative frequency can also be presented on a relative or percentage basis.

Several types of averages can be defined for a given frequency distribution. The most commonly used ones may include the arithmetic mean, geometric mean, quadratic mean (root mean square), median and mode. The degree to which numerical data tend to spread about an average value is called the *variation* or *dispersion* of the data. The *standard deviation* is often used to measure dispersion, and is defined as the root mean square of the

Fig. 3.11. Distributions of carbon fiber tensile strength in air at gauge-lengths of 5, 12, 30 and 75 mm. (After Bader and Gul-Mohammed 1990.)

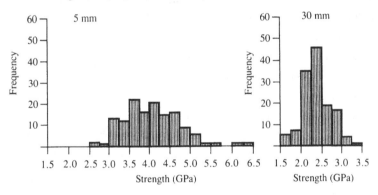

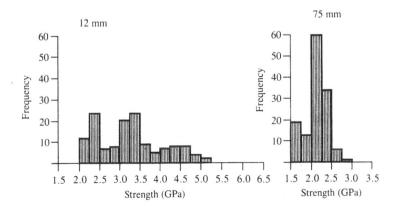

deviations from the mean. Furthermore, the *variance* is defined as the square of the standard deviation, and the *coefficient of variation* is the ratio of the standard deviation to the mean. The coefficient of variation is independent of units used and it fails to be useful when the mean is close to zero.

The probability of occurrence of an event e is denoted by

$$Pr = P\{e\} \tag{3.39}$$

The probability of non-occurrence of the event is denoted by

$$1 - Pr = P\{\text{not } e\} = 1 - P\{e\} \tag{3.40}$$

Some basic relations of probabilities of events are summarized below. Consider two events e_1 and e_2. The probability that e_2 occurs given that e_1 has occurred is the conditional probability of e_2 relative to e_1; it is denoted by $P\{e_2 \mid e_1\}$. If e_1 and e_2 are *independent* events and hence the occurrence or non-occurrence of e_2 is not affected by e_1, then

$$P\{e_2 \mid e_1\} = P\{e_2\} \tag{3.41}$$

Otherwise, they are *dependent* events. The probability that *both e_1 and e_2 occur* is denoted by

$$P\{e_1 e_2\} = P\{e_1\} P\{e_2 \mid e_1\} \tag{3.42}$$

Fig. 3.12. Relative frequency and cumulative frequency vs. fiber tensile strength.

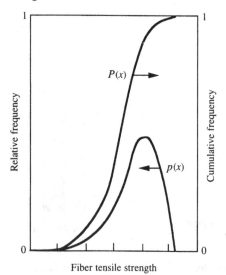

For independent events, the above equation is simplified to

$$P\{e_1e_2\} = P\{e_1\}P\{e_2\} \tag{3.43}$$

In the case of three events e_1, e_2 and e_3, Eq. (3.42) is modified to become

$$P\{e_1e_2e_3\} = P\{e_1\}P\{e_2 \mid e_1\}P\{e_3 \mid e_1e_2\} \tag{3.44}$$

If e_1 and e_2 are *mutually exclusive* events, namely the occurrence of one excludes the occurrence of the other, Eq. (3.42) becomes

$$P\{e_1e_2\} = 0 \tag{3.45}$$

Finally, the event that *either e_1 or e_2 or both occur* is given by

$$P\{e_1 + e_2\} = P\{e_1\} + P\{e_2\} - P\{e_1e_2\} \tag{3.46}$$

For the special case of n mutually exclusive events $e_1, e_2, \ldots, e_n$, the probability of occurrence of either e_1 or e_2 or $\cdots e_n$ is then

$$P\{e_1 + e_2 + \cdots + e_n\} = P\{e_1\} + P\{e_2\} + \cdots + P\{e_n\} \tag{3.47}$$

The applications of these relations to the probabilities of various events in composite failure are given in this chapter as well as in Chapters 4 and 5.

The function representing the frequency distribution in Fig. 3.12 is also known as the *probability density function*. The knowledge of the probability density function is fundamental to any analysis based upon a statistical approach. One of the well-known probability density functions is the normal distribution given by

$$p(x) = \frac{1}{s\sqrt{(2\pi)}} \exp\left(-\frac{1}{2}\left(\frac{x - \bar{x}}{s}\right)^2\right) \qquad s > 0 \tag{3.48}$$

where $\bar{x}$ and s are the mean and standard deviation, respectively. It can be shown, for normal distribution, that 68.27% of the cases are included between $(\bar{x} - s)$ and $(\bar{x} + s)$, and 99.73% of the cases are between $(\bar{x} - 3s)$ and $(\bar{x} + 3s)$. Given a continuous probability density function $p(x)$ the *cumulative distribution function* is defined by

$$P(x) = \int_{-\infty}^{x} p(x)\,\mathrm{d}x \tag{3.49}$$

Other commonly used distribution functions may include the binomial distribution, Bernoulli distribution, and Poisson distribution. However, the Weibull distribution (Weibull 1939a&b, 1951) is probably best known in composite strength theories. Weibull proposed a cumulative distribution function in the general form of

$$P(x) = \begin{cases} 0 & x \le x_u \\ 1 - \exp\left(-\dfrac{(x - x_u)^m}{x_0}\right) & x > x_u \end{cases} \quad (3.50)$$

where m is a shape parameter and x_0 is a scale parameter. The function $(x - x_u)^m / x_0$ has the characteristics of being positive, non-decreasing and vanishing at constant value of x_u, which is not necessarily equal to zero.

3.4.2 Strength of individual fibers

Coleman (1958) examined the strength of long fibers from a common source (say, from the same spool) for the case that their tensile strengths are independent of the rate of loading. To obtain a form for the cumulative strength distribution function $P(\sigma_f)$, Coleman observed that (a) when a fiber is tested it breaks at its weakest cross-section, (b) the strength of a fiber must be positive regardless of the fiber length, and (c) $P(\sigma_f)$ must be a monotonically increasing function of σ_f. Coleman postulated that a fiber may be regarded as composed of a set of N non-interacting unit lengths (or *links*). It is further assumed that all the links in a fiber have the same cumulative strength distribution function $P(\sigma_f)$.

The probability that a link has a strength greater than σ_f is $1 - P(\sigma_f)$, and the probability that all *links* do not fail at σ_f is $[1 - P(\sigma_f)]^N$ (Eq. (3.43)). It follows then the probability that at least one link breaks at σ_f is

$$P_f(\sigma_f) = 1 - [1 - P(\sigma_f)]^N \quad (3.51)$$

$P_f(\sigma_f)$ can be regarded as the cumulative distribution function of the strength of fibers.

Coleman has shown that $P(\sigma_f)$ has the form of a Weibull distribution. For long fibers ($N \to \infty$), Eq. (3.51) gives the cumulative probability of failure

$$P_f(\sigma_f) = 1 - \exp\left[-L\left(\frac{\sigma_f}{\sigma_o}\right)^\beta\right] \quad (3.52)$$

$P_f(\sigma_f)$ is the probability of failure of a fiber at a stress level equal to or less than σ_f. Here, L is the length ratio with respect to a reference length, σ_o is the scale parameter for unit fiber length ratio (i.e. $L = 1$), and β is the shape parameter. Equation (3.52) implies a probability density function of

$$p_f(\sigma_f) = L\sigma_o^{-\beta}\beta\sigma_f^{\beta-1}\exp\left[-L\left(\frac{\sigma_f}{\sigma_o}\right)^{\beta}\right] \qquad (3.53)$$

Following Coleman, the mean fiber strength, $\bar{\sigma}_f$, and standard deviation, s, are given by

$$\bar{\sigma}_f = \sigma_o L^{-1/\beta}\Gamma\left(1+\frac{1}{\beta}\right) \qquad (3.54a)$$

$$s = \sigma_o L^{-1/\beta}\sqrt{\left[\Gamma\left(1+\frac{2}{\beta}\right)-\Gamma^2\left(1+\frac{1}{\beta}\right)\right]} \qquad (3.54b)$$

where Γ denotes the gamma function. An important feature of Eq. (3.54a) is that the fiber strength depends upon the fiber length. The coefficient of variation, which is a function of β only, is

$$\frac{s}{\bar{\sigma}_f} = \sqrt{\left[\frac{\Gamma\left(1+\frac{2}{\beta}\right)}{\Gamma^2\left(1+\frac{1}{\beta}\right)}-1\right]} \qquad (3.55)$$

Over the range of practical interest, β is approximately equal to 1.2/(coefficient of variation). Thus, β is an inverse measure of the dispersion of material strength. For values of β between 20 and 2, the coefficient of variation can be expressed approximately as $\beta^{-0.92}$. Values of β between 2 and 4 correspond to brittle fibers, whereas a value of 20 is appropriate for a ductile metal. β is about 4 for carbon fibers, between 2.7 and 5.8 for boron fibers and about 11 for glass fibers. The factor $\sigma_o L^{-1/\beta}$ in Eqs. (3.54) is often referred to as a characteristic strength level of the fibers (Kelly 1973, Rosen 1964).

Manders and Chou (1983a) have shown that the scale and shape parameters of the Weibull distribution function for fiber strength can be estimated from experimental measurements in a number of ways. First, by taking logarithms of Eq. (3.54a), it is seen that a graph of $\ln(\bar{\sigma}_f)$ against $\ln(L)$ is linear and has gradient $-1/\beta$. The shape parameter can be obtained in this way by testing single fibers of a range of gauge-lengths. The second procedure is to plot the cumulative distribution on appropriate logarithmic axes as follows. The cumulative probability of survival is simply

$$P_s = 1 - P_f \qquad (3.56)$$

and Eq. (3.52) can be rewritten after taking logarithms as

$$\ln(P_s) = -L\left(\frac{\sigma_f}{\sigma_o}\right)^{\beta} \tag{3.57}$$

Taking logarithms a second time with a change of sign

$$\ln(-\ln(P_s)) = \ln(L) + \beta \ln(\sigma_f) - \beta \ln(\sigma_o) \tag{3.58}$$

shows that a graph of $\ln(-\ln(P_s))$ against $\ln(\sigma_f)$ is linear with gradient β (at fixed gauge-length).

The procedures outlined above rely on testing many separate fibers. If a single fiber could be uniformly stressed along its length it would fracture into a series of unequal fragments of which the average length would decrease with higher applied stress. The distribution of lengths between fractures at any given stress should be exponential following Eq. (3.57), and plotting $\ln(P_s)$ against L should give a straight line passing through the origin with gradient $-(\sigma_f/\sigma_o)^{\beta}$. Taking logarithms with a change of sign gives

$$\ln(-\text{gradient}) = \beta \ln(\sigma_f) - \beta \ln(\sigma_o) \tag{3.59}$$

so that a graph of $\ln(-\text{gradient})$ against $\ln(\sigma_f)$ is linear with gradient β.

3.4.3 Strength of fiber bundles

Having examined the strength of single fibers, the strength theory of fiber bundles can be developed (see Daniels 1945, Epstein 1948, Coleman 1958, Kelly 1973, Phoenix 1974). Following the treatment of Coleman (1958), a bundle composed of a very large number, M, of fibers of equal length is considered. The fibers are further assumed to have the same cross-sectional area and the same shape of stress–strain curves, but differ in their values of the elongation at break. It can be shown that the probability density function of bundle tensile strength σ_b (breaking load for the bundle/total fiber cross-sectional area) tends for large M toward a normal distribution (Eq. 3.48)

$$p_b(\sigma_b) = \frac{1}{s_b\sqrt{(2\pi)}} \cdot \exp\left[\frac{-(\sigma_b - \bar{\sigma}_b)^2}{2s_b^2}\right] \tag{3.60}$$

with a mean bundle strength

$$\bar{\sigma}_b = \sigma_{fm}[1 - P_f(\sigma_{fm})] \tag{3.61}$$

and standard deviation

$$s_b = \sigma_{fm}\sqrt{\{P_f(\sigma_{fm})[1 - P_f(\sigma_{fm})]\}}M^{-1/2} \tag{3.62}$$

Here, $P_f(\sigma_{fm})$ is the cumulative fiber strength distribution function and σ_{fm} is the value of fiber stress σ_f which gives $\sigma_f[1 - P_f(\sigma_f)]$ its maximum value, namely

$$\frac{d}{d\sigma_f}\{\sigma_f[1 - P_f(\sigma_f)]\}_{\sigma_f=\sigma_{fm}} = 0 \tag{3.63}$$

Equation (3.63) implies that the maximum fiber stress σ_{fm} is found from the condition that at failure the load borne by the bundle is a maximum.

Assuming $P_f(\sigma_f)$ follows the Weibull distribution of Eq. (3.52) for fiber length L, Eqs. (3.61) and (3.63) give, respectively,

$$\sigma_{fm} = \sigma_o(L\beta)^{-1/\beta} \tag{3.64}$$

and

$$\bar{\sigma}_b = \sigma_o(L\beta e)^{-1/\beta} \tag{3.65}$$

where $e = 2.71828\cdots$. Equation (3.65) implies that the proportion of surviving fibers is $\exp(-1/\beta)$. The strength of loose bundles is lower than the mean strength of single fibers of the same length by the ratio of Eq. (3.65) to Eq. (3.54a), which is termed the 'Coleman factor'

$$\frac{\bar{\sigma}_b}{\bar{\sigma}_f} = \left[\beta^{1/\beta}\exp(\beta^{-1})\,\Gamma\left(1 + \frac{1}{\beta}\right)\right]^{-1} \tag{3.66}$$

It is noticed that when there is no dispersion in the strength of the component fibers of a bundle $\bar{\sigma}_b = \bar{\sigma}_f$. As the coefficient of variation of the fibers increases above zero, however, the bundle strength efficiency decreases monotonically and approaches zero in the limit of infinite dispersion. $\bar{\sigma}_b/\bar{\sigma}_f \approx 70\%$ for the coefficient of variation about 17%.

The ratio given in Eq. (3.66) is independent of the length of the fibers so that the strength of loose bundles decreases with length in the same way as the mean strength of single fibers. The Weibull parameters can therefore be obtained by plotting ln(strength of loose bundle) against ln(length) as described above for single fibers. The above analysis is concerned with single bundles, whereas some situations are better modeled as a chain-of-bundles, such as a moderately twisted yarn where the link length is a frictional load transfer length among fibers. A review of this problem is given by Smith and Phoenix (1981).

3.4.4   *Correlations between single fiber and fiber bundle strengths*

Equation (3.54a) indicates that the Weibull shape parameter of single fiber strength can be determined from the measurement of strength at several fiber gauge-lengths. There are shortcomings in such measurements. First, it is rather tedious to extract individual fibers from a bundle and to perform numerous tests on fibers with very small diameters. Second, the extraction of fibers from a bundle inevitably has 'selected' the stronger ones, since the weaker fibers are prone to damage and fracture in the process. Third, experiments based upon laser diffraction fringes have shown that the measured fiber diameters vary along the fiber length due to fiber twist and the non-circular fiber cross-section.

In this section, following the approach of Chi, Chou and Shen (1984), a theoretical expression of the load–strain relationship for a bundle of fibers under tension is derived first. Then, two methods for determining the two parameters of Weibull distribution for single fiber strength are developed. This is done by analyzing the characteristics of the load–strain curves. The open circles in Fig. 3.13 show the experimental results of a displacement-controlled test for a loose bundle of carbon fibers.

3.4.4.1 *Analysis*

The correlation between single fiber and fiber bundle strengths is established based upon the following assumptions: (1) the single fiber strength under tension obeys the cumulative Weibull distribution function, $P_f(\sigma_f)$, of Eq. (3.52); (2) the relationship between stress, σ_f, and strain ε_f for a single fiber obeys Hooke's law up to fracture:

$$\sigma_f = E_f \varepsilon_f \tag{3.67}$$

where E_f is the fiber Young's modulus; (3) the applied load is distributed uniformly among the surviving fibers at any instant during a bundle tensile test.

To establish the tensile load–strain $(F-\varepsilon)$ relation, Eq. (3.52) is rewritten in terms of fiber strain:

$$P_f(\varepsilon_f) = 1 - \exp\left[-L\left(\frac{\varepsilon_f}{\varepsilon_o}\right)^{\beta}\right] \tag{3.68}$$

Here, ε_o is the scale parameter for unit fiber length ratio (i.e. $L = 1$) and is given by

$$\varepsilon_o = \sigma_o/E_f \tag{3.69}$$

Assume iso-strain conditions for the fibers in a bundle. At an applied strain, ε_f, the number of surviving fibers in a bundle, which consists of N_o fibers, is

$$N = N_o[1 - P_f(\varepsilon_f)] = N_o \exp[-L(\varepsilon_f/\varepsilon_o)^{\beta}] \tag{3.70}$$

N can be related to the applied tensile force, F, on the bundle by

$$F = \sigma_f AN = AE_f\varepsilon_f N_o \exp[-L(\varepsilon_f/\varepsilon_o)^{\beta}] \tag{3.71}$$

Fig. 3.13. Comparison of a theoretical $F-\varepsilon_f$ curve (solid line) with experimental data (open circles) for carbon fiber, $E_f = 225$ GPa, $d_f = 7\ \mu$m, $N_o = 1000$, $\beta = 4.5$ and $\varepsilon_o = 0.026$. (After Chi, Chou and Shen 1984.)

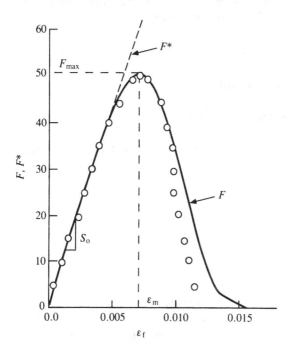

Equation (3.71) is the relationship of $F-\varepsilon_f$ for a bundle of fibers under tension, where A is the cross-sectional area of a single fiber. If A, N_o, L, E_f, ε_0 and β are known, the $F-\varepsilon_f$ curve for a bundle of fibers can be drawn. The solid line in Fig. 3.13 shows the result of the theoretical prediction.

According to Eq. (3.71), the $F-\varepsilon_f$ curve is continuous and smooth. After reaching the point of maximum load, F_{max}, the tensile force on the bundle decreases gradually to zero. The slope of the curve, S_0, at $\varepsilon_f = 0$ is

$$S_o = AE_fN_o \tag{3.72}$$

and the tensile load defined by the tangent line of the $F-\varepsilon_f$ curve at $\varepsilon_f \approx 0$ is

$$F^* = AE_fN_o\varepsilon_f \tag{3.73}$$

Based upon the $F-\varepsilon_f$ relation, the survivability of single fibers in the bundle can be determined from Eqs. (3.71) and (3.73)

$$\frac{F}{F^*} = 1 - P_f(\varepsilon_f) = P_s \tag{3.74}$$

Next, the strain corresponding to the maximum load on the $F-\varepsilon_f$ curve, ε_m, is obtained from $dF/d\varepsilon_f = 0$

$$\varepsilon_m = \varepsilon_o\left(\frac{1}{L\beta}\right)^{1/\beta} \tag{3.75}$$

Thus, the maximum load is

$$F_{max} = AN_oE_f\varepsilon_o\left(\frac{1}{L\beta e}\right)^{1/\beta} \tag{3.76}$$

From Eqs. (3.72), (3.75) and (3.76), the slope of the straight line connecting the origin and the point (F_{max}, ε_m) in Fig. 3.13 is

$$S = F_{max}/\varepsilon_m = S_o\left(\frac{1}{e}\right)^{1/\beta} \tag{3.77}$$

As a result,

$$\beta = 1/\ln\left(\frac{\varepsilon_mS_o}{F_{max}}\right) \tag{3.78}$$

3.4.4.2 Single fiber strength distribution

Based upon the analysis of the fiber and bundle strength relations, Chi, Chou and Shen (1984) proposed the following two methods for determining single fiber strength distribution (shape parameter β and scale parameter ε_o) from measurements on fiber bundles, and constructing the theoretical $F-\varepsilon_f$ curve.

Method (A)

The method is based upon Eqs. (3.68) and (3.74) and the experimental $F-\varepsilon_f$ curve. The procedure is outlined below:

(1) Calculate S_o from Eq. (3.72) and the data of A, E_f and N_o of the fiber bundle.
(2) Calculate F^* from Eq. (3.73), $F^* = S_o E_f$. Measure F from the $F-\varepsilon_f$ curve. Then determine from Eq. (3.74) the fiber survivability as a function of strain, $P_s(\varepsilon_f) = F/F^*$.
(3) The shape parameter, β, can be obtained from the gradient of the graph of $\ln(-\ln(P_s))$ vs. $\ln(\varepsilon_f)$, based upon the relation

$$\ln(-\ln(P_s)) = \ln(L) + \beta \ln(\varepsilon_f) - \beta \ln(\varepsilon_o) \quad (3.79a)$$

(4) The scale parameter, ε_o, is determined either from Eq. (3.75) using the measured ε_m value, or from the value of $\ln(L) - \beta \ln(\varepsilon_o)$ measured from the graph of $\ln(-\ln(P_s))$ vs. $\ln(\varepsilon_f)$.

Method (B)

In this method, $F_{\max}$ and ε_m are known from experiments. The calculation steps are:

(1) Determine S_o, β and ε_o from Eqs. (3.72), (3.78) and (3.75), respectively.
(2) From Eqs. (3.71) and (3.72), the $F-\varepsilon_f$ relation can be written as

$$F = S_o \varepsilon_f \exp\left(-L\left(\frac{\varepsilon_f}{\varepsilon_o}\right)^{\beta}\right) \quad (3.79b)$$

3.4.5 *Experimental measurements of Weibull shape parameter*

It is understood that the shape parameter β gives a measurement of the scattering of the strength data. On a $p_f(\sigma_f)$ vs. σ_f plot, the range of strength distribution is narrower for higher β values. The discussions of Sections 3.4.2–3.4.4 for the estimation of the Weibull shape parameter are summarized below (see Manders and Chou 1983a; Chi, Chou and Shen 1984).

Single fibers

(i)   Variation of mean strength with length: The method requires tests at different gauge-lengths. Fiber diameters are measured to obtain true stress.

(ii)   Distribution of strength at fixed gauge-length: Diameters are measured to obtain true stress. The method measures both inherent variability, and also artificial scatter introduced by experimental techniques.

(iii) Distribution of lengths between multiple fractures of a single fiber: Estimate is based on strain, not stress. The method requires correction for non-uniformity of strain near fractures.

Loose bundles

(iv)   Variation of mean strength with length: The method assumes identical fiber diameters and stiffness.

(v)   Proportion of surviving fibers is obtained from the load–strain curve. Estimate is based on strain not stress. The method assumes fibers are identical.

(vi)   Determination of the initial slope of the load/strain curve and the strain corresponding to the maximum load on the bundle.

Manders and Chou (1983a) have established the Weibull shape parameter based upon the methods (i)–(v) by performing tests on a single batch of PAN-based carbon fiber (Hercules AS-4, 12 000 filament unsized tow) while the loose bundle tests (iv) and (v) are carried out with the E-glass fiber (St. Gobain, vetrotex type DCN56 filament, unsized tow). Chi and Chou (1983), and Chi, Chou, and Shen (1984) have examined methods (i), (ii), (v) and (vi) using Thornel-300 carbon fibers and bundles containing 1000 fibers.

3.4.5.1 *Single fiber tests*

In order to obtain the strengths of single filaments and their distributions, it is necessary to measure the diameters and ultimate tensile load of the filaments. For the measurement of filament diameters, Chi and Chou (1983) used a helium–neon laser, and the diameters were determined from the laser diffraction fringes (see Lipson and Lipson 1981) The results indicate an average filament diameter of 7.12 μm with the standard deviation of 0.2 μm.

Fiber strength measurements are performed for fiber gauge-lengths of 10, 30 and 60 mm, and the number of measurements are 80, 81 and 64, respectively. The results are presented on the Weibull probability paper as shown in Fig. 3.14. Here σ_f denotes fiber ultimate strength; $P_f(\sigma_f)$ is the fiber cumulative probability of failure at stresses equal to or less than σ_f and $\ln\{-\ln[1 - P_f(\sigma_f)]\}$ is a representation of failure probability. The variations of fiber failure probability with strength can be approximated as linear with the exceptions of the low strength range for 60 mm length fibers, and the high strength range for 10 mm and 30 mm length fibers.

The Weibull shape parameter, β, can be obtained by following method (i), by plotting the mean fiber strength $\ln(\bar{\sigma}_f)$ vs. fiber length $\ln(L)$ (see Eq. (3.54a)) as shown in Fig. 3.15. A measurement of the slope of the straight line gives the value of $\beta = 6.2$. It is worth noting that because of the high scatter in strengths a large number of tests needs to be performed to determine with high accuracy whether the Weibull distribution is an accurate description of strength, and this is where the loose bundle approach is

Fig. 3.14. Strength distributions of single filaments on Weibull probability paper. (After Chi and Chou 1983.)

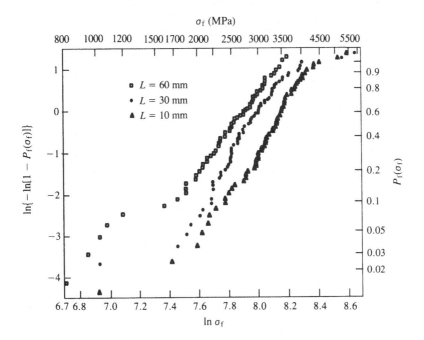

advantageous. If method (ii) is followed, then the slope is measured from the linear portion of the data of Fig. 3.14 for $\ln\{\ln[1 - P_f(\sigma_f)]\}$ vs. $\ln(\sigma_f)$ (see Eq. (3.58)). The distributions of the three sets of data are reasonably linear and parallel, and an average of the approximate gradients is taken to obtain the shape parameter of 5.3.

Method (iii) requires multiple fracture tests of a single fiber. In the experiments of Manders and Chou (1983a), single carbon fibers are bonded to the surface of a 2 mm thick filled PVC carrier sheet using a film of polystyrene adhesive approximately 50 μm thick. The fiber is strained to successively higher levels by bending the carrier strip around mandrels of decreasing radii. The strain in the fiber is virtually uniform because the ratio of the carrier thickness to fiber diameter is $\sim$200. The combination of adhesive and carrier is found to be quite resistant to repeated straining, and facilitates visual location of the fiber fractures. At each strain level the lengths between fiber fractures are measured by travelling microscope and are ranked and plotted as the cumulative distribution on the logarithmic axes. According to Eq. (3.57), the distributions should be linear and pass through the origin, but, while they are relatively straight, they intersect the fracture spacing axis at some positive intercept. The minimum crack spacing given by the intercept represents the effective 'unstressed' length of fiber over which the

Fig. 3.15. Relationship between filament average strength and gauge-length. (After Chi and Chou 1983.)

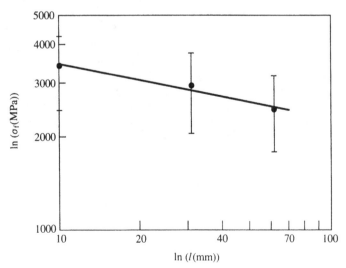

load builds up. The logarithm of the gradient has been plotted against the logarithm of strain and at low strains, the curve is relatively linear with a gradient corresponding to $\beta = 6.4$. At high strain the curve becomes horizontal because the fiber debonds from the adhesive film and no new fractures occur. Despite this short-coming the technique is able to measure the shape parameter for shorter fibers than the other techniques.

Henstenburg and Phoenix (1989) have considered the problem of measuring the Weibull parameters for fiber strength using data from a multiple fracture test of a single fiber. Using a Monte-Carlo approach they arrived at a simple method which applies to fibers of length equal to the mean fragmentation length.

3.4.5.2 *Loose bundle tests*

In the loose bundle tests of Manders and Chou (1983a), based on method (iv), tows of different lengths are cemented into grooved end-tabs while particular care is taken to ensure that none of the fibers are slack. Manders and Chou obtained between five and ten results for each gauge-length, and they are plotted in the same way as for the single fiber tests in Fig. 3.16. The cross-sectional area of the tow is calculated from the manufacturer's value of its density and weight per length. Because each failure of a loose bundle involves the independent fracture of many fibers, there is much less scatter than for the single fibers. According to Manders and Chou, the mean strength ratios of loose bundles and single carbon fibers of the same length range from 0.67 to 0.85 for fibers with lengths between 10 and 200 mm, and this compares quite well with the theoretical Coleman factor which ranges from 0.65 to 0.76 for fibers with β equal to 5 and 10, respectively (Coleman 1958). The discrepancy may be due to the fact that the strengths are not perfectly Weibull distributed, and that the optical technique for measuring fiber diameter overestimates the cross-sectional area of non-circular crenelated fibers. Also, fiber breaks may be pre-existing in the bundle, becoming more noticeable at longer bundle lengths.

It has been noticed that both single fibers and loose bundles show an increase in strength variability at longer gauge-lengths. This could be interpreted as the influence of a relatively small population of severe and broadly distributed flaws. The majority of short gauge-lengths would not contain one of these severe flaws and the population would have little influence on the mean strength, but longer fibers would be more likely to contain one or more such flaws and their mean strength would be significantly lowered. The observation of similar behavior in glass fiber suggests that a 'double' Weibull distribution with two shape and scale parameters may be more appropriate (Metcalfe and Schmitz 1964; Harlow and Phoenix 1981a & b). It is also noticed, in the case of loose bundles, that the recoil and entanglement of failed fiber causes neighboring fibers to fail, thereby weakening the bundle.

In the loose bundle tests of Chi, Chou and Shen (1984), the shape parameter and scale parameter were determined based upon methods (v) and (vi), which correspond to methods (A) and (B) of Section 3.4.4.2. The relevant data are: $N_o = 1000$, fiber diameter = $7 \, \mu m$, $E_f = 255 \, GPa$ and gauge-length = 60 mm. The shape parameters obtained from methods (v) and (vi) are 4.6 and 4.5, respectively. The scale parameter, ε_o, corresponding to a fiber of unit length (1 mm in this case), is 0.026 for both methods. The experimental data points indicating the load–strain ($F-\varepsilon_f$) relationship are shown in Fig. 3.13. The consistency between the theory and experiment is rather satisfactory in the range of bundle strain not much greater than ε_m.

Fig. 3.16. Variation of mean strength with length for loose bundles of carbon and E-glass fibers. (After Manders and Chou 1983a.)

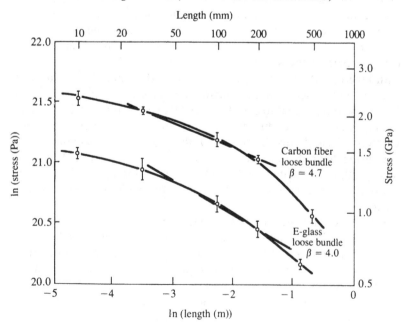

3.4.6 *Strength of unidirectional fiber composites*

This section deals with statistical strength theories of unidirectional fiber composites. Upon the fracture of a fiber, the load originally carried by the fiber needs to be transferred to its neighboring fibers. A simple approximation of the load redistribution is to assume that the load is shared equally by all the unbroken fibers. A more precise treatment takes into account the local concentration of load on neighboring fibers. A Monte-Carlo simulation is also presented to illustrate the statistical nature of composite failure.

3.4.6.1 *Equal load sharing*

In general, the high-strength high-stiffness fibers used in composites are brittle and their tensile strength should be characterized statistically. Parratt (1960) notes that the tensile failure of composites reinforced with brittle fibers occurs when the fibers have been broken up into lengths so short that any increase in applied load cannot be transmitted to the fibers because the limit of interface or matrix shear has been reached. Rosen (1964), following Gucer and Gurland (1962), considers fibers as having a statistical distribution of flaws or imperfections that result in individual fiber breaks at various stress levels. The fracture initiated in a fiber is contained by the matrix material. Composite failure occurs when the remaining unbroken fibers, at the weakest cross-section, are unable to resist the applied load. Then composite failure results from tensile fracture of the fibers. In Rosen's failure model, the composite is assumed to be strained uniformly and the load in a broken fiber is distributed equally among the remaining unbroken fibers in a cross-section. Harlow and Phoenix (1978a) have labelled such a model as *equal load sharing*. Scop and Argon (1967) also have dealt with the problem of equal load sharing in their treatment of the strength of laminated composites.

Figure 3.17 depicts Rosen's failure model. In the vicinity of an internal fiber end in such a composite, the axial load carried by the fiber is transmitted by shear through the matrix to adjacent fibers (see Section 3.3.1). A portion of the fiber at each end is, therefore, not fully effective in resisting the applied stress. At some distance from an internal break, the fiber stress will reach a given fraction of the undisturbed fiber stress. Rosen considers that the fiber length δ, measured from the fiber end, over which the stress is less than a given fraction (i.e. 90%) of the uniform stress that would exist in infinite fibers, as ineffective. δ is thus known as the ineffective length. The model composite in Fig. 3.17 is assumed to be composed of a series of layers of height δ. The segment of a fiber within a layer may be considered as a link in the chain that constitutes the fiber. Each layer is then a bundle of such links and the composite is a series of such bundles.

The treatment of a fiber as a chain of links is appropriate to the hypothesis that fracture is a result of local imperfections in the fibers. The links may be considered to have a statistical strength distribution that is equivalent to the statistical flaw distribution along the fibers. Rosen defines the link dimension by a shear-lag analysis of the stress distribution in the vicinity of a fiber end (see Section 3.3.1). The length of the composite specimen is designated by L and the number of links is given by $N = L/\delta$.

The relationship between fiber strength and the strength of links has been briefly discussed in the formulation of Eq. (3.51). Obviously, the probability density function $p_1(\sigma_1)$ for fiber links can be characterized if the experimental data on fiber strength distribution $p_f(\sigma_f)$ are known. Suppose that the fibers are characterized by a strength distribution of the Weibull type (Eq. (3.53)), the link strength density function can be readily written as

$$p_1(\sigma_1) = \delta \sigma_o^{-\beta} \beta \sigma_1^{\beta-1} \exp\left[-\delta \left(\frac{\sigma_1}{\sigma_o} \right)^\beta \right] \qquad (3.80)$$

Fig. 3.17. Chain-of-links model for a unidirectional fiber composite.

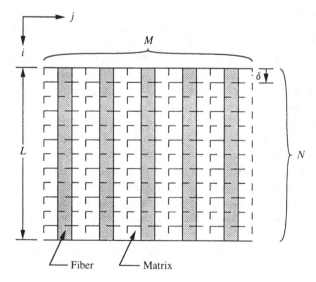

Fiber   Matrix

For a bundle of links and a large number, M, of fibers, the distribution of bundle strength $p_b(\sigma_b)$ and the mean bundle strength $\bar{\sigma}_b$ are given by Eqs. (3.60) and (3.61), respectively.

The bundles may be treated as links in a chain, which now represents the whole composite of Fig. 3.17. The weakest link theorem can again be applied to define the failure of the composite. For N bundles forming a chain (composite) the probability density function $p_c(\sigma_c)$ for the average fiber stress at *composite* failure, σ_c, is given by

$$p_c(\sigma_c) = Np_b(\sigma_c)[1 - P_b(\sigma_c)]^{N-1} \tag{3.81}$$

where

$$P_b(\sigma_c) = \int_0^{\sigma_c} p_b(\sigma)\,d\sigma \tag{3.82}$$

The notations of $p_c(\sigma_c)$, $p_b(\sigma_b)$, $p_f(\sigma_f)$ and $p_l(\sigma_l)$ have been used to denote the strength density functions of the fibers at the level of composite, bundle, fiber and link, respectively. Thus, it is understood that σ_c, σ_b, σ_f and σ_l all refer to stresses in the reinforcements; the contribution of matrix to composite strength is not considered.

The most probable composite failure stress σ_c^* is obtained by setting

$$\frac{d}{d\sigma_c}[p_c(\sigma_c)]_{\sigma_c=\sigma_c^*} = 0 \tag{3.83}$$

Following Rosen (1964), the substitution of Eq. (3.81) into Eq. (3.83) yields

$$\sigma_c^* = \bar{\sigma}_b - s_b\sqrt{(2 \cdot \log N)} + s_b\frac{\log\log N + \log 4\pi}{2\sqrt{(2 \cdot \log N)}} \tag{3.84}$$

It can be seen from Eq. (3.62) that, for composite dimensions large relative to fiber cross-section ($M \gg 1$), $s_b \to 0$ and Eq. (3.84) is reduced to the mean bundle strength expression of Eq. (3.65)

$$\sigma_c^* = \sigma_0(\delta\beta e)^{-1/\beta} \tag{3.85}$$

When the fiber volume content is considered, the tensile strength of the composite is given by $V_f \sigma_c^*$. In Eq. (3.85), the ineffective length δ can be determined from the stress analyses discussed in Section 3.3.1. It is obvious that the composite strength is enhanced due to a reduction in fiber ineffective length and fiber strength dispersion. The statistical nature of fiber fracture and the resulting weakest link mode of failure have been demonstrated experimentally in a glass/epoxy system by Rosen (1964). This experiment also points out the very significant phenomenon in brittle fiber composites: fiber breakages may exist in a composite of continuous fibers at stress levels well below the maximum load.

If the composite strength (Eq. (3.85)) is compared with the mean strength of the tested fibers of length L (Eq. (3.54)), some interesting conclusions can be drawn (Rosen 1970). Figure 3.18 shows that for reference fibers of ineffective length δ, the strength of the composite is less than the mean fiber strength. When the fiber length is greater than δ, the composite strength is larger than the mean fiber strength of a fiber bundle of length $L > 7\delta$. Also for a fiber strength coefficient of variation (s/σ) less than 15% (or the shape parameter $\beta > 8$), the composite strength is close to the mean fiber strength, as shown in Fig. 3.18.

3.4.6.2 Idealized local load sharing

When a fiber breaks in a composite there is inevitably a redistribution of load in the vicinity of the fiber breakage. Thus, local load sharing takes place (see Zweben 1968; Scop and Argon 1969; Zweben and Rosen 1970; Fukuda and Kawata 1976b; Harlow and Phoenix, 1978a&b; Harlow 1979; Phoenix 1979). The localized nature of stress redistribution around a random fiber break has been discussed in Section 3.3. Zweben (1968) first considered the

Fig. 3.18. Composite strength/mean fiber strength vs. β at various L/δ values.

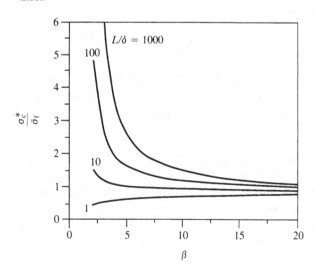

micromechanical stress transfer process and the probabilistic aspects of the generation of clusters of breaks to form catastrophic breaks. Fukuda and Kawata (1976b) generalized the original concept of Zweben and derived the cumulative strength distribution for the composite. In the following, an analysis is presented under the Weibull distribution for fiber strength, and somewhat simplified assumptions on local fiber load sharing but with the advantage that various quantities can be worked out either exactly or asymptotically. The result is that insight can be gained on the *approximate* Weibull behavior for composite strength where the Weibull parameters for the composite will be connected to various fiber and matrix properties, and in particular to the composite volume. The size effect law for the composite will also be discussed. Most of the features have been experimentally observed but have been difficult to explain. The ideas for this section are taken from Harlow and Phoenix (1978a&b, 1979, 1981a&b); Smith (1980, 1982); Phoenix and Smith (1983); Smith *et al.* (1983); and Phoenix, Schwartz and Robinson (1988).

The model considered is the planar, chain-of-bundles model of Fig. 3.17 where M is the number of fibers and N is the number of bundles each with fiber elements of length δ, which might better be termed 'the effective load transfer length'. Following the notation of Phoenix, the cumulative distribution function for the failure of a single fiber element of length δ is taken as the Weibull distribution and expressed as

$$F(\sigma) = 1 - \exp\{-(\sigma/\sigma_\delta)^\beta\} \qquad \sigma \geq 0 \qquad (3.86)$$

where σ is the fiber stress, and β and σ_δ are the Weibull shape and scale parameters, respectively. (At this point it should be mentioned that δ should take into account certain statistical aspects of fiber strength which modify its magnitude somewhat as described by Harlow and Phoenix (1979), and Phoenix, Schwartz and Robinson (1988). Roughly, δ varies inversely as the shape parameter β.) According to principles discussed earlier, the strength of a fiber element of length δ can be expressed in terms of those for a longer reference length L (used, say, for tension tests) according to

$$\sigma_\delta = \sigma_L \left(\frac{\delta}{L}\right)^{-1/\beta} \qquad (3.87)$$

where σ_L is the Weibull scale parameter for fiber strength at the reference length. Often σ_δ will be about double σ_L in magnitude.

The local load-sharing rule is 'idealized' as follows: In a bundle, if the stress is nominally σ (ignoring the matrix), a surviving fiber element carries load $K_r\sigma$, where

$$K_r = 1 + r/2, \qquad r = 0, 1, 2, 3, \ldots \tag{3.88}$$

and r is the number of consecutive failed elements immediately adjacent to the surviving element (counting on both sides). At the same time a failed fiber element carries no stress over length δ. Essentially the load of a failed fiber is shifted equally onto its two nearest surviving neighbors, one on each side. This rule is more severe than the true situation where the stress redistribution is somewhat more diffuse, as described say by Hedgepeth (1961), but it captures the essential features and has the advantages of simplicity and being fully described for all configurations.

Before proceeding with an approximate analysis of this model, it is useful to review an extensive numerical analysis performed by Harlow and Phoenix (1978a&b), where the basic insight into its behavior was uncovered. To eliminate boundary effects, they considered circular bundles (composite tubes), and studied the behavior of the cumulative strength distribution as the bundle size M increases. They defined $G_M(\sigma)$ as the cumulative distribution function for failure of a bundle with M fibers under the stress σ, and worked out exact formulas for $G_M(\sigma)$ for M up to 5 by considering all configurations of failed and surviving fibers and all ways that failure could proceed through these configurations and then summing all probabilities for these ways. For example, for $M = 2$,

$$G_2(\sigma) = F(\sigma)^2 + 2F(\sigma)[F(2\sigma) - F(\sigma)]$$
$$= 2F(\sigma)F(2\sigma) - F(\sigma)^2 \tag{3.89}$$

where in the intermediate step the first term represents direct failure under the applied stress of both fiber elements, and the second term represents the two ways one element can fail under the direct stress and the other under the overstress, which is naturally taken as 2σ in this situation (rather than $3\sigma/2$). For $M = 4$, they obtained by a tedious calculation

$$\begin{aligned} G_4(\sigma) = {}& 16F(4\sigma)F(2\sigma)F(3\sigma/2)F(\sigma) - 4F(4\sigma)F(2\sigma)F(\sigma)^2 \\ & - 4F(4\sigma)F(3\sigma/2)^2F(\sigma) + 4F(4\sigma)F(\sigma)^3 \\ & - 8F(2\sigma)^2F(3\sigma/2)F(\sigma) \\ & + 2F(2\sigma)^2F(\sigma)^2 - 8F(4\sigma)F(3\sigma/2)F(\sigma)^2 \\ & + 4F(3\sigma/2)^2F(\sigma)^2 - F(\sigma)^4 \end{aligned} \tag{3.90}$$

Generally no simple pattern emerged except that each term involved a product of M quantities in F. The evaluation procedure was automated on a computer, but results were only obtained at that time for M up to 9 because of the tremendous increase in computational complexity resulting from the increasing number of ways the bundle can fail as the bundle size increases. (Even with present supercomputer capability the limit is still about $M = 14$.) At the same time we desire results for M orders of magnitude larger.

Suspecting an eventual weakest-link type relationship, Harlow and Phoenix (1978b) considered plotting the 'renormalization'

$$W_M(\sigma) = 1 - [1 - G_M(\sigma)]^{1/M} \tag{3.91}$$

since in reverse this yields the weakest-link relation

$$G_M(\sigma) = 1 - [1 - W_M(\sigma)]^M \tag{3.92}$$

They discovered an extremely rapid numerical convergence

$$W_M(\sigma) \to W(\sigma) \qquad \text{as } M \to \infty \tag{3.93}$$

where $W(\sigma)$ was called the *characteristic distribution function* for failure. This convergence is shown in Fig. 3.19 for the Weibull shape parameter $\beta = 5$, which is typical of brittle fibers. The coordinates are Weibull coordinates ($\ln\{-\ln(1 - W)\}$ vs. $\ln(\sigma/\sigma_\delta)$) wherein a Weibull distribution always plots as a straight line. For each value of σ the convergence is abrupt at some value of M, which increases slowly with decreasing values of σ. Also the convergence becomes complete far into the lower tail of $W(\sigma)$ (probabilities below 10^{-10}) for $M = 9$. In an extremely complex calculation, Harlow and Phoenix (1981a&b) uncovered the analytical character of $W(\sigma)$ in terms of the largest eigenvalue of a Markov recursion matrix. It suffices to say here that $W(\sigma)$ has no simple analytical form, though shortly we will develop an approximation which will give us considerable insight.

The importance of $W(\sigma)$ is that, from Eq. (3.92), the distribution function for bundle failure can be given extremely accurately by the approximation

$$G_M(\sigma) \approx 1 - [1 - W(\sigma)]^M \tag{3.94}$$

and this works for M many orders of magnitude larger than the values used in the calculation of $W(\sigma)$ on the computer. Perhaps one should note that any boundary effects, which may come into play for small bundles, are being ignored.

Because the composite is seen as a weakest-link arrangement of its N bundles (Fig. 3.17), and the bundles are treated as statistically independent, the cumulative distribution function for the failure of the composite, denoted as $H_{M,N}(\sigma)$, is given as

$$H_{M,N}(\sigma) = 1 - [1 - G_M(\sigma)]^N \tag{3.95}$$

Combining Eqs. (3.94) and (3.95) and writing $V = MN$ yields the accurate approximation

$$H_{M,N}(\sigma) \approx 1 - [1 - W(\sigma)]^V \tag{3.96}$$

which surprisingly, perhaps, is a result which is symmetric in M and

Fig. 3.19. Convergence of the renormalized distribution functions $W_M(\sigma)$ to the characteristic distribution function $W(\sigma)$ as M increases. (After Harlow and Phoenix 1978b.)

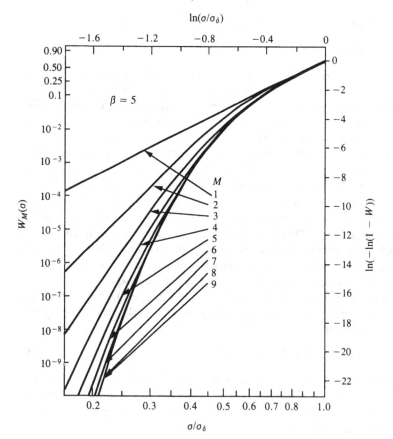

N. Note that by the binomial expansion $H_{M,N}(\sigma) \approx VW(\sigma)$. Thus if V is large, say 10^6 elements, it is necessary to know $W(\sigma)$ where its value is much less than 10^{-6}. As mentioned, this is provided for in Fig. 3.19. Note that despite the fact that Eq. (3.96) is a 'weakest-link' relation, in terms of $V = MN$ elements, there is no identifiable and independent material element to which one can attach $W(\sigma)$. At best, $W(\sigma)$ characterizes the effects of local failure events which are actually statistically dependent.

Figure 3.20 displays $W(\sigma)$ for values of β from 3 to 50. Now Fig. 3.20 can be used to construct a figure for $H_{M,N}(\sigma)$ upon noting that $\ln\{-\ln(1 - H)\} = \ln\{-\ln(1 - W)\} + \ln V$, which on Weibull prob-

Fig. 3.20. Characteristic distribution function $W(\sigma)$ for various values of the Weibull shape parameter β for fiber strength. (After Harlow and Phoenix 1978b.)

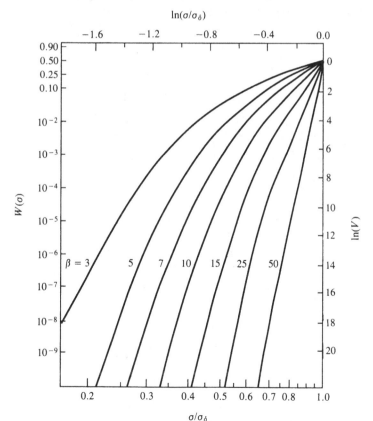

ability paper amounts to a simple translation of each curve upward (or the left-hand scale downward) the amount $\ln(V)$ on the right-hand scale provided for this purpose. Figure 3.21 shows the result of such a translation for $V = 10^6$ elements, which amounts to a display of the original region on Fig. 3.20 below 10^{-5}. This yields plots of the cumulative distribution function of composite failure, $H_{M,N}(\sigma)$, for various β for a relatively small composite specimen.

Several features of Fig. 3.21 warrant discussion. First, all the lines are approximately straight over a very wide probability range, which suggests that the strength of a composite approximately (but not exactly) follows a Weibull distribution. In fact, an empirical plot to cover the probability range shown would require testing about

Fig. 3.21 Cumulative distribution function $H_{M,N}(\sigma)$ for composite strength for volume $MN = 10^6$ and various values of the fiber shape parameter β. (After Harlow and Phoenix 1978b.)

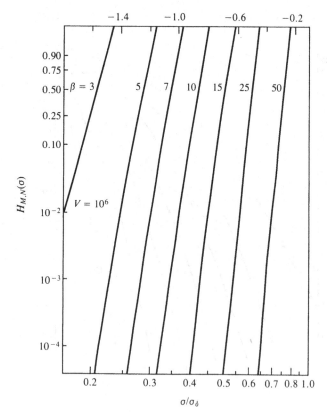

20 000 specimens, and using standard statistical techniques it is probable that a Monte-Carlo simulation would *not* lead to rejection of the hypothesis that the Weibull distribution is actually the correct distribution! Second, the lines show only a modest change in slope, by a factor of less than three, as the original Weibull shape parameter for the fiber β decreases from 50 to 3, which is a factor of more than ten. Since the slope is directly proportional to the Weibull shape parameter, this indicates that the *effective* Weibull shape parameter for the composite decreases modestly, from about 50 to 20 as that for the fiber decreases drastically, from about 50 to 3. On the other hand, the horizontal location of the plots is quite strongly influenced by the value of β, which suggests that an increase in variability in fiber strength substantially decreases composite strength. It is seen, for example, that the median strength drops from about $0.75\sigma_\delta$, to about $0.2\sigma_\delta$ as β drops from 50 to 3. Note also that the median strength of the composite is much less than that for a fiber element of length δ, being only about $\frac{1}{3}$ for the typical case $\beta = 7$. On the other hand, standard tension tests on fibers are performed at gauge-lengths L about two orders of magnitude larger than δ, and by Eq. (3.87) their strengths are about one-half of σ_δ. Fortuitously then, the strength of the composite will be little different from the strength of the fiber from typical laboratory tension tests as is often observed. Finally, the method of constructing Fig. 3.21 indicates that there is a mild size effect in composite strength and a mild shift in the effective Weibull shape parameter for the composite. Had a larger volume $V = 10^9$ been chosen rather than 10^6, the curved nature of the graphs on Fig. 3.20, from which Fig. 3.21 was derived, would produce a slightly lower strength and a slightly higher effective shape parameter for the composite depending on β.

Attention is now turned toward a simple but approximate theoretical explanation based on some key ideas motivated by the above numerical analysis and results. First, the range for the composite failure stress lies $\ll \sigma_\delta$, as we saw from Fig. 3.21. (Note that both the median and the stress at 0.99 probability of failure lie well below σ_δ for typical values of β below 15.) Second, the 'initial' failures, that is fiber elements which fail directly under the applied stress σ, are viewed as 'seeds' for the growth of failure clusters, which are lateral strings of adjacent fiber breaks contained within bundles. Third, the *number* of such seeds is easily seen to follow the binomial distribution with parameters MN and $F(\sigma)$ (the number depends, of course, on σ) with the mean number being $MNF(\sigma)$.

Fourth, cluster growth from a seed is viewed for calculation purposes in terms of the sequential failure of adjacent fibers in a bundle, with growth in either direction to form a string. Fifth, instability occurs when a string of k breaks occurs such that $F(K_{k-1}\sigma) < \frac{1}{2}$, say, but $F(K_k\sigma) \approx 1$; thus, subsequent fiber failures become almost certain leading to catastrophic growth of a transverse 'crack' and failure of the composite. This value of k, which depends on the stress level σ, is called the *critical crack size*, and in view of Eq. (3.88) is better defined as the k value for which

$$K_{k-1}\sigma \leq \sigma_\delta < K_k\sigma \tag{3.97}$$

Sixth, the following analysis is based on the Weibull shape parameter β for fiber strength being 'large', but fortunately the results work quite well for β down to about 4.

Proceeding with the analysis, it is first important to realize that the initial breaks or 'seeds' are actually quite far apart. For example, from Fig. 3.21 we recall that the median composite strength was about $0.27\sigma_\delta$ for $\beta = 5$, and $F(0.27\sigma_\delta) = 0.0014$. This means that the average spacing of seeds along a fiber is the inverse of this value times δ, or about 700δ, and laterally in a bundle is about 700 fiber diameters. Moreover this spacing grows larger as the composite volume increases due to the size effect. To see why, we note that the size effect means that the median strength will decrease as the volume increases. As an example, repeat the process used to develop Fig. 3.21 from Fig. 3.20 but for a volume $MN = 10^9$ instead of 10^6. One can see that the median strength will now be only $0.22\sigma_\delta$ instead of $0.27\sigma_\delta$ and since $F(0.22\sigma_\delta) = 0.00052$, the average spacing is almost 2000δ. Note that although the seeds are now farther apart (fewer per unit volume), there are more of them in the composite because the volume grew by a factor of 10^3. (It may come as a surprise to the reader that a small composite will show lots of single breaks per unit volume just before failure, but a large composite will show relatively few!) Thus, as a first approximation we can ignore the possible interactions of two clusters growing near each other since the critical k will turn out to be quite small.

The probability of a given fiber element becoming a seed *and* its immediate neighbors developing further into a failure string of size k is approximately

$$P\{\text{seed } and \text{ string}\} \approx F(\sigma)2F(K_1\sigma)2F(K_2\sigma) \cdots 2F(K_{k-1}\sigma) \tag{3.98}$$

where the factors '2' appear because, at each step of the growth beyond the seed, there are two choices for the next failure (one on each side) which approximately doubles the probability for that step. Thus, such a string can stretch out variously to the left, or to the right, or be centered relative to the original break. Clearly Eq. (3.98) ignores considerable detail about the events of cluster growth, as discussed more fully in Phoenix and Smith (1983), but it works mainly because $F(K_j\sigma) \gg F(K_{j-1}\sigma)$ when β is large. (The nature of the simplification can be appreciated upon studying Eqs. (3.89) and (3.90) for small bundles where in each case the first term will dominate all the others when β is large.) Using a Taylor series expansion in $(\sigma/\sigma_\delta)^\beta$ it can be seen that

$$F(\sigma) \approx (\sigma/\sigma_\delta)^\beta \qquad (3.99)$$

This is especially true when $\sigma \ll \sigma_\delta$, but it turns out that for present purposes we can take this as a good approximation for $0 \le \sigma \le \sigma_\delta$, particularly in Eq. (3.98). Substituting Eq. (3.99) in Eq. (3.98), we have

$$P\{\text{seed } and \text{ string}\} \approx 2^{k-1}(\sigma/\sigma_\delta)^\beta(K_1\sigma/\sigma_\delta)^\beta \cdots (K_{k-1}\sigma/\sigma_\delta)^\beta$$
$$= 2^{k-1}(K_1K_2 \cdots K_{k-1})^\beta(\sigma/\sigma_\delta)^{k\beta} \qquad (3.100)$$

This factorization and collapse of terms, to yield an exponent of $k\beta$ instead of β, is an important feature which follows from the use of the Weibull distribution. It is the point at which the effect of micromechanical 'redundancy' in the composite emerges as a reduction in variability.

In the composite there are MN potential seed fibers, each of which may produce a string, and the composite will fail if at least one such event occurs. Treating the MN seed *and* string events as statistically independent (which works because of the wide spacing mentioned above), we actually have a weakest-link situation so that the probability of composite failure is

$$H_{M,N}(\sigma) \approx 1 - [1 - P\{\text{seed } and \text{ string}\}]^{MN}$$
$$\approx 1 - [1 - 2^{k-1}(K_1K_2 \cdots K_{k-1})^\beta(\sigma/\sigma_\delta)^{k\beta}]^{MN} \qquad (3.101)$$

From the calculus, $(1 - a\sigma^b)^n \to \exp\{-na\sigma^b\}$ as $n \to \infty$ so that

$$H_{M,N}(\sigma) \approx 1 - \exp\{-MN2^{k-1}(K_1K_2 \cdots K_{k-1})^\beta(\sigma/\sigma_\delta)^{k\beta}\} \qquad (3.102)$$

which is of the Weibull form, though k depends on the stress σ following Eq. (3.97).

Before discussing several important features of Eq. (3.102), it is useful to develop a connection to the characteristic distribution function $W(\sigma)$. For $k = 1, 2, 3, \ldots$, let

$$\mathbf{F}^{[k]}(\sigma) = 1 - \exp\{-d_k(\sigma/\sigma_\delta)^{k\beta}\} \tag{3.103}$$

where

$$d_k = 2^{k-1}(K_1 K_2 \cdots K_{k-1})^\beta \tag{3.104}$$

Equation (3.103) gives us a family of Weibull distributions with increasing shape parameter $k\beta$ in k. Furthermore, following Eq. (3.97) we can partition the important stress range $0 \le \sigma \le \sigma_\delta$ into the segments

$$\sigma_\delta/K_k < \sigma \le \sigma_\delta/K_{k-1} \qquad k = 1, 2, 3, \ldots \tag{3.105}$$

and for each k restrict the corresponding distribution to its appropriate stress range. Then Eq. (3.102) becomes

$$H_{M,N}(\sigma) \approx 1 - [1 - \mathbf{F}^{[k]}(\sigma)]^{MN} \tag{3.106}$$

where k and σ are chosen to follow Eq. (3.105). An approximation to $W(\sigma)$ then follows from a comparison of Eqs. (3.96) and (3.106) yielding

$$W(\sigma) \approx \mathbf{F}^{[k]}(\sigma) \tag{3.107}$$

where again k and σ satisfy Eq. (3.105).

Figure 3.22 shows a plot of $W(\sigma)$ for $\beta = 5$ together with the family of Weibull distributions $\mathbf{F}^{[k]}(\sigma)$ for $k = 1, 2, 3, \ldots$, where each is extended over the whole stress range $0 \le \sigma \le \sigma_\delta$. For each stress level σ one of these Weibull distributions comes very close to $W(\sigma)$, and indeed it is normally the one whose k value satisfies Eq. (3.105). Unfortunately, Eq. (3.107) has a jagged appearance when plotted because of small 'jumps' occurring as k changes at the transition stresses of the boundaries of Eq. (3.105). A graphically pleasant 'repair' with a smooth appearance is to work with the inner 'envelope' of the family of Weibull distributions, that is

$$W(\sigma) \approx \min\{\mathbf{F}^{[1]}(\sigma), \mathbf{F}^{[2]}(\sigma), \mathbf{F}^{[3]}(\sigma), \ldots\} \tag{3.108}$$

Figure 3.22 indicates that this approximation works extremely well.

In principle we could develop similar graphs to Fig. 3.22 for the other cases $\beta = 3, 7, 10, \ldots, 50$ in Fig. 3.20. In developing Fig. 3.21 from Fig. 3.20 for a given volume V, it is quickly seen that one of the Weibull cases, that is one value of k, would 'dominate' for each value of β, which is why each line in Fig. 3.21 is approximately straight. For each plot, the appropriate k and Weibull shape parameter $k\beta$ would be determined through Eq. (3.105) from the relevant stress range in Fig. 3.21, especially near the median. For example, for $\beta = 10$, the case $k = 3$ is appropriate in developing Fig. 3.21, as the *effective* Weibull shape parameter for composite strength is about $3 \times 10 = 30$ (as determined from the slope of the $\beta = 10$ line in Fig. 3.21).

It is now possible to determine the appropriate Weibull distribution for each plot in Fig. 3.21. Substituting the appropriate Weibull distribution $\mathbf{F}^{[k]}(\sigma)$ into Eq. (3.106) (which actually returns us to Eq. (3.102)) yields the following Weibull approximation for composite strength:

$$H_{M,N}(\sigma) \approx 1 - \exp\{-(\sigma/\sigma_{k,MN})^{k\beta}\} \tag{3.109}$$

where

$$\sigma_{k,MN} = \sigma_\delta (MNd_k)^{-1/(k\beta)} \tag{3.110}$$

For each value of β, this Weibull approximation closely fits the plot on Fig. 3.21, provided k is chosen by the above graphical scheme.

Fig. 3.22. Envelope construction from Weibull family $\mathbf{F}^{[k]}(\sigma)$ to approximate the characteristic distribution function $W(\sigma)$ for composite strength. Reprinted with permission from *International Journal of Solids and Structures*, **19**, Phoenix and Smith, Copyright © (1983), Pergamon Press, plc.

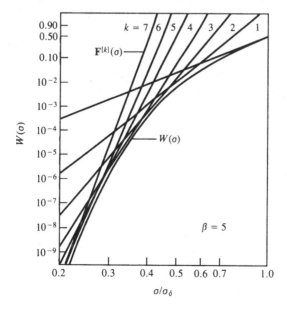

Of course k will change if the volume $V = MN$ is changed significantly.

At this stage it is important to recall the interpretation of k as the 'critical crack size'. It is now appreciated that given the composite volume MN and the Weibull shape parameter β for the fiber strength, a special value of k emerges which is the size of the longest crack or string of fiber breaks when such a composite fractures. This value of k also determines the effective Weibull shape parameter for composite strength, $k\beta$. Thus far, the calculation of the appropriate k value has been performed graphically, but it is possible to estimate k explicitly. The method is given in Phoenix and Smith (1983), and begins by the study of

$$\sigma_\delta / K_k < \sigma_{k,MN} < \sigma_\delta / K_{k-1} \qquad (3.111)$$

For large MN, this leads to the appropriate k being the value which satisfies

$$\gamma(k) > \ln(MN)/\beta > \gamma(k-1) \qquad (3.112)$$

where

$$\gamma(r) = r \ln(K_r) - \{\ln(K_1) + \ln(K_2) + \cdots + \ln(K_{r-1})\} \qquad (3.113)$$

for $r = 1, 2, 3, \ldots$ and $\gamma(0) = 0$. For $K_j = 1 + j/2$, we obtain the values given in Table 3.1. According to Eq. (3.112) the critical value of k depends on the ratio $\ln(MN)/\beta$, and thus it increases slowly as the composite volume is increased but decreases more rapidly as the variability in fiber strength is decreased (β is increased).

As an example, for the case $\beta = 5$ on Fig. 3.21, the graphical procedure puts the stress range near $0.27\sigma_\delta$ which by Fig. 3.22 or Eq. (3.105) puts $k = 5$. On the other hand, $\ln(10^6)/5 = 2.76$, and by Eqs. (3.112) and (3.113) and Table 3.1 one also obtains $k = 5$. Thus the effective Weibull shape parameter for the composite being represented is $k\beta = 25$.

Table 3.1.

r	$\gamma(r)$	r	$\gamma(r)$
0	0	5	3.15
1	0.405	6	3.95
2	0.981	7	4.78
3	1.65	8	5.62
4	2.38	9	6.48

Finally, it is interesting to consider the ultimate size effect for the composite. In the case of a Weibull distribution, we recall that the strength decreases as the volume V in proportion to $V^{-1/\beta}$. On the other hand, the curvatures of the lines on Figs. 3.20 and 3.21, together with our finding that k slowly increases as the volume $V = MN$ increases suggest that the strength of the composite will not ultimately have a Weibull size effect, but one which is increasingly milder as V increases. Smith (1980, 1982) considered this question and concluded that

$$\text{composite strength} \approx \beta 2^{1-1/\beta} \sigma_\delta / \ln(V) \tag{3.114}$$

which indicates that the strength decreases as the inverse of the log of the volume. It turns out that Eq. (3.114) tends to be an overestimate and a composite must be astronomically huge ($V > 10^{20}$) for this result to be accurate.

In conclusion, a few extensions and limitations of the above analysis should be mentioned. As stated earlier, the results given are based on β being 'large'. This allowed us to write the approximation Eq. (3.98), which led us to Eq. (3.100) and then to the definition of d_k in Eq. (3.104). As mentioned earlier, the calculation of the event implied in Eq. (3.98) is more complex if 'double counting' of certain failure possibilities is to be avoided. For example, for $k = 2$, a more accurate rendition is

$$\begin{aligned} P\{\text{seed } and \text{ string}\} &\approx 2F(\sigma)[F(K_1\sigma) - F(\sigma)] + F(\sigma)^2 \\ &= 2F(\sigma)F(K_1\sigma) - F(\sigma)^2 \\ &\approx [2(K_1)^\beta - 1](\sigma/\sigma_\delta)^{2\beta} \end{aligned} \tag{3.115}$$

so d_2 should be $[2(K_1)^\beta - 1]$ rather than just $2(K_1)^\beta$. The same sort of analysis shows that d_3 should actually be $4(K_1K_2)^\beta - (K_1)^{2\beta} - (K_2)^\beta - 2(K_1)^\beta + 1$ and so on for higher k. But it turns out that these refinements make very little difference, especially when calculating the scale parameter values $\sigma_{k,MN}$ in Eq. (3.110) where the error is typically one or two per cent.

The above results were developed for the idealized case of local load sharing defined by Eq. (3.88), but appear also to work for more realistic cases provided one chooses K_r to be the largest load sharing constant at the edge of a failure configuration. Generally such values of K_r tend to be smaller than $1 + r/2$ (see, for example, Hedgepeth 1961). Following through the above analysis, the main effects are not only to increase the scale parameters for strength, thus increasing the composite strength itself, but also to increase the critical k values thus reducing the composite variability. Second, an analysis has been carried out by Smith *et al.* (1983), for three-dimensional composites, with the parallel fibers forming a two-dimensional hexagonal array. Here the clusters of broken fibers can take on many different geometric configurations other than a linear string, but for large β one still comes up with a form for d_k that is similar in structure to Eq. (3.104) except that 2^{k-1} is replaced by a much more complex configurational constant. Many of the ideas carry through except that one no longer finds quite the same simple relationship between the critical cluster size k and the effective Weibull shape parameter for composite strength. The strength of such a three-dimensional composite is typically larger than in the two-dimensional planar case described above. The reason is that while there are many more failure configurations, the load sharing occurs over many more fibers at the boundary of a failure cluster so that the reduction in the K_r values more than compensates for the increased number of failure possibilities, especially for larger β.

Finally, experimental data to illustrate the above features have been presented by Phoenix, Schwartz and Robinson (1988), who also extend the ideas, through viscoelasticity of the matrix, to explain creep rupture phenomena under constant stress.

3.4.6.3 *Monte-Carlo simulation*

The Monte-Carlo method is a numerical technique suitable for simulating complicated stochastic processes, and it has been employed to analyze a wide range of physical processes of a statistical nature (Oh 1979). The Monte-Carlo simulation of composite strength can be regarded as testing the composite materials 'analytically' in an automated fashion. In each Monte-Carlo experiment, random numbers are generated and assigned to the underlying random variables and the outcome of the process of interest can be observed. When the number of such independent experiments is sufficiently high, the observations will yield a good assessment of the statistical characteristics of the process. In dealing with the strength of fibers as well as composites, the Monte-Carlo experiment involves the partitioning of fiber or a composite into elements, then random numbers are assigned to the strength of the elements. For a given applied load, the stress in the elements of a fiber or a composite can be determined as described in Section 3.3. From the assigned strength value and the arrangement of breaks of elements the failure load is then obtained. In the following, fractures of fibers as well as composites based upon the Monte-Carlo simulation

(Fukuda and Kawata 1977; Oh 1979; Manders, Bader and Chou 1982) are considered. Several common procedures for generating the normal random numbers are available.

Fukuda and Kawata studied the fracture of a two-dimensional fiber composite based upon the Monte-Carlo method by choosing a mean strength of 100 and a standard deviation of 10. A simulation of the fracture process is shown in Fig. 3.23 for $E_f/E_m = 20$, and $M = N = 20$, where M and N are defined in Fig. 3.17. The elements or links in the partitioned composite specimen are specified by the position (i, j). Here, 0 indicates that the link is not broken and the other numerals indicate the sequence of link breakage. As the initial condition, each link (i, j) is assumed to have a stochastic strength, $STR(i, j)$, which is the normal random number with a specific value of mean and standard deviation. Both the Weibull distribution and normal distribution have been used for expressing the link strength distributions. Stress concentration factors of all links, $SCF(i, j)$, are initially assigned as 1. A link with the least value of $STR(i, j)/(SCF(i, j)$ is sought, and let this link be (i_o, j_o). The link breaks first at the tensile stress of $STR(i_o, j_o)$. When this link breaks, stress concentration occurs in the two adjacent links $(i_o, j_o \pm 1)$. The values of $STR(i, j)/SCF(i, j)$ are again calculated for the remaining $M \times N - 1$ links. A link which has the least of this value breaks second. This procedure is repeated until all the links in a plane transverse to the loading direction $(j = 1, 2, \ldots, M)$ are broken.

Fig. 3.23. Monte-Carlo simulation of fiber link fractures. (After Fukuda and Kawata 1977.)

```
 0  2  0  0  0  0  0  0  0  0  0  0  0  0  0  0  0  0  0  0
 0  0  0 33  0  0  0  0  0  0  0  0  0  0  0  0  0  0  0  0
 0  0  0  0 39  0  0  0  0  0  0  0  0  0  0  0  0  0  0  0
 0  0  0  0  0  1  0  0  0  0  0  0  0  0  0  0  0 38  0  0
 0  0  0 40  0  0  0  0  0  0  0  0  0  0  0  0  0  0  0  0
 0  0  0  0  0  0  0  0  0  0  0  0  0  0  0  0  0  0  0  0
 0  0  7  0  0  0  0  0  0  0  0  0  0 43  0  0  0  0  0  0
 0  0  0  0  0  0  0  0  0  0  0  0  0 41  0  0  0  0  0  0
18 44 45 31 46 47  0  0  0  0 11  0  0  0  0  0  0  0  0  0
42  0  0  0  0  0  0  0 10  0  0  0  0  0  0  0  0  0  0  0
 0  0  0  0  0  0  0  0  0  0  0  0  0  0 30  0  0  0  0  0
 0  0  0  0  0  0  5  0  0  0  0  0  0  0  0  0  0  0  0  0
 0  0  0  0  9  0  0  0  0  0  0  3  0  0  0  0  0  0  0  0
62 59 58 57 56 55 54 53 52 51 50 49 48 60 61 37 16 36 35  4
 0  0  0  0  0  0  0  0  0  0  0  0  0  0  0  0  0  0  0  0
 0 19  0  0  0  0  0 20  0  0  0  0 32 12  0  0  0  0  0  0
 0  0  0  0  0  0  0  0  0  0  0  0  0  0  0  0  0  0  0  0
 0  0  0  0  0  0  0  0  0  0  0  0  0 21  0 13  0  0  0  0
 0  0  0  0  0  0  0  0 34 29 28 14 27 26 25  6 15 24 22  8
 0 17  0  0  0  0  0  0 23  0  0  0  0  0  0  0  0  0  0  0
```

The result given in Fig. 3.17 resembles the sequence of fiber failure observed in the experimental work of Rosen (1964). The predictions of composite strength are shown in Fig. 3.24. It should be noted that the Monte-Carlo approaches are generally limited to $MN < 50\,000$ under current supercomputer power which may not be enough for a realistic composite. Also the Monte-Carlo approach is inherently poor at handling the lower tails of the distributions.

3.4.7   *Strength of cross-ply composites*
Cross-ply construction is the simplest form of lamination of unidirectional laminae. This simple geometric configuration facilitates the understanding of the fundamental problems concerning laminate strength. It provides a model system for investigating the matrix cracking of laminates under tensile loading. This section analyzes the problem from both deterministic and statistical viewpoints. The treatment of Aveston, Cooper and Kelly (1971) of multiple fracture, although it deals with unidirectional composites, is basic to matrix cracking of laminated composites in general. Hence, it is outlined first.

3.4.7.1   *Energy absorption during multiple fracture*
Section 3.2 discusses the mode of fracture of unidirectional composites as affected by the ultimate failure strains of the fiber and

Fig. 3.24. Numerical results of Monte-Carlo simulation. (After Fukuda and Kawata 1976b.)

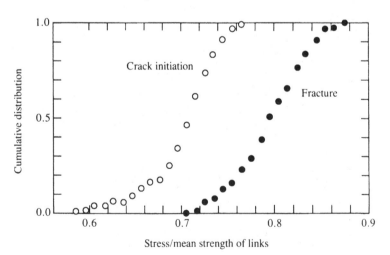

Stress/mean strength of links

matrix materials as well as the fiber volume fraction. The energy absorption of composites during the failure process was first investigated by Aveston, Cooper and Kelly (1971). Contributions to the fracture surface energy during single fracture may be derived from deformation of the fiber or matrix, the work done in fracturing the fiber–matrix interfacial bond, and work done in pulling the fibers out of the matrix against frictional forces. It is found that the work of fracture increases with increasing fiber diameter and decreasing fiber–matrix interfacial strength.

Multiple fracture of fibers occurs in ductile matrix composites at low fiber volume fraction. Multiple fracture of matrix, on the other hand, takes place in brittle matrix composites at high fiber volume fraction, as a result of applied tensile loads or thermal stresses induced by cooling from the stress-free temperature. The energy consideration for the development of multiple matrix cracking in a unidirectional lamina subject to axial tensile loading is introduced below (see Aveston, Copper and Kelly 1971; Aveston and Kelly 1973, 1980; Kelly 1976).

Consider the formation of a single matrix crack normal to the fiber direction, at the strain ε_{mu} under conditions of fixed load. It is assumed that the stress in the matrix is equal to the matrix fracture stress and there is a decrease in the combined energy of the specimen and the loading system. The energy changes due to the formation of a crack at a fixed load include ΔW = the work done by the applied load per unit area of the composite, γ_{db} = energy absorbed per unit area of debonded fiber, U_s = the work done per unit area of the composite against the frictional force between the fiber and matrix, ΔU_m = the elastic strain energy lost due to the relaxation of the strain in the matrix, and ΔU_f = the increase in strain-energy of the fibers per unit area of the composite. If the surface energy in forming a matrix crack is γ_m, a crack will occur provided

$$2\gamma_m(1 - V_f) + \gamma_{db} + U_s + \Delta U_f \leq \Delta W + \Delta U_m \qquad (3.116)$$

The terms in Eq. (3.116) have been evaluated by Aveston, Cooper and Kelly under the assumption that the changes in stress (strain) in the matrix and fiber due to the formation of the crack vary linearly with distance from the crack surface. By further assuming purely frictional bond between the fiber and matrix, Eq. (3.116) yields the following expression for the failure strain of the matrix:

$$\varepsilon_{mu} = \left\{ \frac{12\tau\gamma_m E_f V_f^2}{E_c E_m^2 r(1 - V_f)} \right\}^{1/3} \tag{3.117}$$

where τ = fiber–matrix interfacial shear strength (See Eq. (3.2)), r = fiber radius, and E = Young's modulus with the subscripts f, m and c indicating fiber, matrix and composite, respectively. Equation (3.117) indicates that the composite strain at the formation of the transverse matrix crack can be enhanced by suitable control of the elastic moduli of the fiber and matrix, fiber volume fraction and diameter, matrix surface energy, and the fiber–matrix interfacial strength.

Budiansky, Hutchinson and Evans (1986) have generalized the results of Aveston, Cooper and Kelly for unbonded, frictionally constrained slipping fibers initially held in the matrix by thermal or other strain mismatches. The other case considered by Budiansky *et al.* for the onset of matrix cracking involves fibers that initially are weakly bonded to the matrix, but may be debonded by the stresses near the tip of an advancing matrix crack. McCartney (1987) has used an energy-balance calculation for a continuum model of brittle matrix cracking in a uniaxially fiber-reinforced composite and confirmed that the Griffith fracture criterion is valid for matrix cracking.

3.4.7.2 *Transverse cracking of cross-ply laminates*

Multiple transverse cracks in the matrix of unidirectional fiber composites have been observed in a number of systems, for example, glass-reinforced cement, and gypsum reinforced with polyvinyl chloride or glass, where the failure strains of the fibers are greater than those of the matrices. Transverse cracking also occurs in the 90° plies of cross-ply laminates. Experimental observations and analytical modeling of this behavior have been made by Bailey, Garrett, Parvizi, Bader and Curtis (see Garrett and Bailey 1977a&b; Parvizi and Bailey 1978; Parvizi, Garrett and Bailey 1978; Bader, Bailey, Curtis and Parvizi 1979; Bailey, Curtis and Parvizi 1979; Parvizi 1979; Bailey and Parvizi 1981 who followed Aveston and Kelly's shear-lag approach and interpreted this pheno-menon by the concept of constrained cracking). Manders, Chou, Jones and Rock (1983) proposed a statistical treatment of multiple cracks. Wang, Crossman, Warren and Law (see Wang and Crossman 1980; Crossman, Warren, Wang and Law 1980; Crossman and Wang 1982; Wang 1984), on the other hand, theorized it based

based upon the strain-energy release rate of crack extension. The theory of Bailey *et al.* is introduced in this section. The work of Manders *et al.* is discussed in Section 3.4.7.3 and that of Wang *et al.* is introduced in Section 3.4.7.4.

(A)   Cross-ply laminate
        The cross-ply construction of [0°/90°/0°] is shown in Fig. 3.25. For the cases of glass/epoxy and carbon/epoxy systems, the mechanical properties of unidirectional laminates are shown in Table 3.2. The glass/epoxy 0° test curves are essentially linearly elastic to fracture but the 90° specimens show a pronounced *knee* at a strain of about 0.3%, after which a whitening effect can be observed. The 0° carbon/epoxy test curves are elastic to failure but they are not linear, there being an increase in the modulus with increasing strain. The 90° carbon/epoxy is linear to failure with no knee or acoustic emission prior to failure. The failure strains of the 90° specimens in both systems are characteristically low due to strain concentrations in the matrix (see Kies 1962).

Fig. 3.25. Illustration of a [0°/90°/0°] specimen.

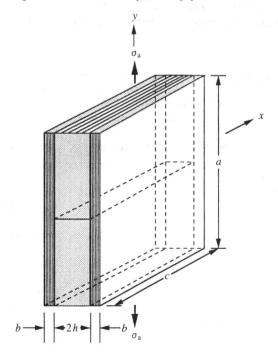

When extended in tension, initial failure of the cross-ply laminate is usually in the central 90° ply, which cracks in a direction normal to the applied tension and parallel to the fibers in that layer (Fig. 3.26). The failure sequence in both laminates follows a similar pattern. Two knees appear on the stress–strain curve of glass/epoxy, first at 0.3% strain, associated with the visual whitening effect and at 0.5% strain due to transverse cracking, but this is not apparent in the carbon/epoxy laminate. On further extension, more cracks are formed until the whole gauge portion of the test-piece is filled with a regular array of cracks. The strain at which the first crack occurs increases as the thickness ($2h$) of the 90° layer is reduced and at the same time the crack spacing tends to become smaller. In the case of the thinnest transverse layers, transverse cracking is not observed at all before the final catastrophic failure of the test-piece. Microscopy has shown that the earliest indications of failure are debonds at or near the fiber/matrix interface. These occur at strains even lower than those at which the whitening is observed in the glass/epoxy systems. The next stage is a coalescence of a number of debonds to form a microcrack, which grows rapidly when it reaches a critical size, about three to four fiber diameters.

Longitudinal splitting is observed to occur in the 0° plies of the cross-ply laminate at strains intermediate between the transverse

Table 3.2. *Mechanical properties of unidirectional laminates (after Bader et al. 1979), Reprinted with permission from Mechanical Behaviour of Materials-Copyright © 1979, Pergamon Press, plc.).*

Property	0° CFRP*	0° GRP**	90° CFRP	90° GRP	Units
Low-strain Young's modulus	127	42	8.3	14	GPA
Fracture stress	1.7	0.92	0.039	0.056	GPa
Fracture strain	1.2	2.2	0.48	0.50	%
Poisson's ratio	0.29	0.27	0.02	0.09	–

* CFRP: carbon fiber-reinforced plastic
**GRP: glass fiber-reinforced plastic

cracking strain for the 90° plies and final failure (Fig. 3.27). Longitudinal splitting is due to mismatches in the Poisson's ratios and the coefficients of thermal expansion of the 0° and 90° plies. The strain to initiate splitting increases as the thickness of the longitudinal plies is reduced. Splitting has not been observed in the carbon/epoxy cross-ply laminates.

(B)   Transverse crack spacing
        The low strain failure behavior was first explained by Kies (1962), who predicts the magnification of strain in the matrix when a unidirectional composite is stressed in the transverse direction. In the limit when the fibers are almost touching one another, the strain magnification factor approaches the value E_f/E_m. It should be noted that even at comparatively low fiber volume fractions there are invariably regions in the lamina where fibers almost touch one another. The glass fibers are nearly

Fig. 3.26. Transverse-ply crack in a [0°/90°/0°] carbon fiber-reinforced cross-ply laminate with an inner-ply thickness of $2h = 0.125$ mm. (After Bailey, Curtis and Parvizi 1979.)

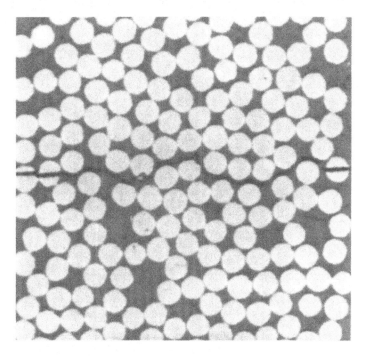

isotropic, but the transverse Young's modulus of carbon is much lower than its longitudinal modulus and it is this modulus which should be used for calculating the strain magnification factor. The first matrix crack usually forms between fibers which are touching or nearly touching along a direction perpendicular to the loading axis.

The crack density, and hence the crack spacing, is related to the geometry of the laminate. These can be explained by the cross-ply laminate shown in Fig. 3.25. When the strain has reached the fracture strain, ε_{tu}, of the 90° ply, the first crack occurs in the transverse ply, and an additional stress $\Delta\sigma$ is placed on the longitudinal plies. From a shear-lag analysis similar to that given in Section 3.3.1,

$$\Delta\sigma = \Delta\sigma_o \exp(-\sqrt{(\phi)}y) \tag{3.118}$$

where

$$\phi = \frac{E_c G_{12}}{E_{11} E_{22}} \left(\frac{b+h}{bh^2}\right)$$

Fig. 3.27. Longitudinal-ply splitting in a [0°/90°/0°] glass fiber-reinforced cross-ply specimen. (After Bailey, Curtis and Parvizi 1979.)

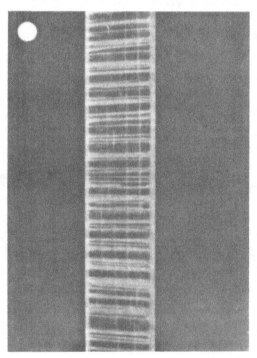

E_c is the laminate Young's modulus in the y direction, E_{11} and E_{22} are the Young's moduli of a unidirectional ply in the fiber and transverse directions, respectively, and G_{12} is the shear modulus of a unidirectional ply. This additional stress has its maximum value $\Delta\sigma_o$ in the plane of the crack ($y = 0$) and decays with distance y from the crack plane as some load is transferred back into the transverse ply through interlaminar shear stress

$$\tau_i = -b\frac{d\Delta\sigma}{dy} \tag{3.119}$$

The tensile load in the transverse ply is zero at the crack plane but builds up by shear transfer from the longitudinal plies. At a given distance y from the crack, the load F in the inner ply is given by

$$F = \int_0^y 2c\tau_1\,dy \tag{3.120a}$$

where c is defined in Fig. 3.25. The first crack in the transverse ply occurs when the load carried by it is equal to $2ch\sigma_{tu}$ where σ_{tu} denotes the ultimate tensile strength of the 90° ply in the cross-ply laminate, which may be different from the transverse tensile strength of a unidirectional ply. This load is then transferred onto the longitudinal plies. Another crack can only occur when the transverse ply is again loaded to $2ch\sigma_{tu}$. The transverse ply will not be loaded to this value except at infinity and $\Delta\sigma_o = \sigma_{tu}h/b$, if the applied stress on the laminate is maintained at $\sigma_a = E_c\varepsilon_{tu}$ after the first cracking. For another crack to occur, σ_a and hence $\Delta\sigma_o$ must be increased to such a value that $F = 2ch\sigma_{tu}$.

If the first crack is assumed to take place in the middle of the specimen ($y = 0$) of length a, the following cracking sequence will occur:

(1) Initial crack at $\sigma_a = E_c\varepsilon_{tu}$, and

$$F = 2bc\,\Delta\sigma_o[1 - \exp(-\sqrt{(\phi)}y)] \tag{3.120b}$$

(2) Second and third cracks occur simultaneously at the ends of the specimen when the applied load increases to such a value that

$$\Delta\sigma_o = \sigma_{tu}\frac{h}{b}[1 - \exp(-\sqrt{(\phi)}a/2)]^{-1} \tag{3.121}$$

The crack spacing is $a/2$.

(3) The next series of cracks will occur midway between the present cracks. The total shear stress between two existing cracks is

$$\tau_i = b \, \Delta\sigma_o \sqrt{(\phi)}\{\exp(-\sqrt{(\phi)}y)$$
$$- \exp[\sqrt{(\phi)}(y - a/2)]\} \qquad (3.122)$$

and from Eq. (3.120a)

$$F = 2bc \, \Delta\sigma_o[1 + \exp(-\sqrt{(\phi)}a/2)$$
$$- 2 \exp(-\sqrt{(\phi)}a/4)] \qquad (3.123)$$

The value of $\Delta\sigma_o$ when the cracks occur now at intervals of $a/4$ is

$$\Delta\sigma_o = \sigma_{tu} \frac{h}{b}[1 + \exp(-\sqrt{(\phi)}a/2)$$
$$- 2 \exp(-\sqrt{(\phi)}a/4)]^{-1} \qquad (3.124)$$

(4) For crack spacing of $a/8$

$$\Delta\sigma_o = \sigma_{tu} \frac{h}{b}[1 + \exp(-\sqrt{(\phi)}a/4)$$
$$- 2 \exp(-\sqrt{(\phi)}a/8)]^{-1} \qquad (3.125)$$

This crack sequence will continue until the strength of the longitudinal plies is exceeded or the spacing between neighboring cracks is so small that the normal stress in the 90° ply cannot be built up to σ_{tu}.

(C)   Transverse cracking constraint
    The strain required to initiate transverse cracks is greater when the transverse lamina is thinner, and in some cases cracking is constrained completely up to the strain at which the longitudinal laminae fail catastrophically. This phenomenon of constrained cracking is attributed to the fact that in order for a crack to form it must be both mechanistically possible and energetically favorable. The former requirement is satisfied for cross-ply laminates from the viewpoint of strain magnification as discussed in (B). The effect of lamina thickness on the transverse failure strain can be understood from the viewpoint of energetics.
    For a specimen under constant load, a crack initiates if the following condition is satisfied:

$$\Delta W > \Delta U + U_D + 2\gamma A \qquad (3.126)$$

where ΔW is the work done by the applied stress per unit area of the specimen, ΔU is the increase in stored energy per unit area of the specimen, U_{D} is the energy loss per unit area due to any dissipative processes present (e.g. sliding friction between debonded fiber and matrix), γ is the fracture surface energy per unit fracture surface area, and A denotes the fracture surface area. It has been found that for practical ply thicknesses the interface between the longitudinal and transverse plies remains bonded during the cracking of the transverse ply and the laminate behaves in a fully elastic manner, thus Eq. (3.126) becomes

$$\Delta W > \Delta U + 2 \frac{h}{h + b} \gamma_{\mathrm{t}} \tag{3.127}$$

Here, γ_{t} is the fracture surface energy of the transverse ply in a direction parallel to the fibers. Since half of the work done by the applied stress is stored as elastic energy of the specimen, it follows that

$$\tfrac{1}{2} \Delta W > 2\gamma_{\mathrm{t}} \frac{h}{h + b} \tag{3.128}$$

When the first crack occurs in the transverse ply at a strain of $\varepsilon_{\mathrm{tu}}$ an additional stress $\Delta\sigma$, Eq. (3.118), is thrown onto the outer plies and the laminate increases in length by δa, given by

$$\delta a = 2 \int_{\mathrm{o}}^{a/2} \frac{\Delta\sigma}{E_{11}} \, \mathrm{d}y \tag{3.129}$$

For $a/h \gg 1$, Eq. (3.129) becomes

$$\delta a = \frac{2hE_{22}\varepsilon_{\mathrm{tu}}}{bE_{11}\sqrt{(\phi)}} \tag{3.130}$$

The work done by the applied stress σ_{a} at the strain of first transverse failure is

$$\Delta W = \delta a \sigma_{\mathrm{a}} \tag{3.131}$$

Hence

$$\Delta W = \frac{2hE_{\mathrm{c}}E_{22}\varepsilon_{\mathrm{tu}}^2}{bE_{11}\sqrt{(\phi)}} \tag{3.132}$$

The substitution of Eq. (3.132) into Eq. (3.128) yields the minimum value of the transverse failure strain

$$(\varepsilon_{\mathrm{tu}})_{\min} = (\varepsilon_{\mathrm{c}})_{\min} = \sqrt{\left[\frac{2bE_{11}\gamma_{\mathrm{t}}\sqrt{(\phi)}}{(h + b)E_{22}E_{\mathrm{c}}}\right]} \tag{3.133}$$

The theoretical values of the minimum cracking strain have been calculated from Eq. (3.133) as a function of h and are compared with the experimental results in Fig. 3.28 for glass/epoxy laminates. Close agreement is observed between theory and experiment in the region where $h < 0.25$ mm, indicating an energy controlled crack propagation. For the thicker laminates, however, this theory does not apply and cracking occurs at a constant strain of 0.5% which is close to the cracking strain of the unidirectional 90° lamina.

According to Bader *et al.* (1979), microscopic cracks usually develop in glass- and carbon-reinforced plastic laminates in regions where fibers lie normal to the principal tension axis, at strains which are, at the most, only 30% of the final failure strain. Thus designers are faced with a dilemma: whether to base the design on strains below the cracking threshold (typically 0.5% for glass-reinforced plastics) or the ultimate failure strain, which might be 1.5% or more. Microcracks which do not appear to be detrimental to the short-term mechanical properties of laminates may act as nuclei for further local damage leading to ultimate failure under cyclic loading and a hostile environment. Experimental evidence suggests that the formation of transverse cracks and longitudinal splitting can be constrained or inhibited by constructing the laminate from thinner individual plies.

Fig. 3.28. Plot of the theoretical and experimental transverse cracking strain, $(\varepsilon_c)_{min}$, as a function of the inner-ply thickness, $2h$, for glass-reinforced sandwich laminates. The outer ply thickness is 0.5 mm. — Eq. (3.133); --- cracking strain of the unidirectional 90° lamina; ⦶ experiment. Reprinted with permission from Bader *et al.* in *Mechanical Behaviour of Materials*, Copyright © (1979), Pergamon Press plc.

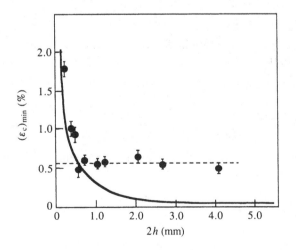

3.4.7.3 *Statistical analysis*

The deterministic multiple cracking theory of Garrett, Bailey and Parvizi attempts to account for the measured distribution of crack spacing in [0°/90°/0°] glass fiber/resin matrix laminates. Manders *et al.* (1983) have proposed a statistical model which fits the experimental data and predicts a dependence of strength on size. The origins and implications of this variability of strength are discussed below after descriptions of the experimental observations.

The three-ply [0°/90°/0°] laminates of Manders *et al.* are composed of Silenka E-glass fibers in an Epikote epoxy resin. The central 90° ply is 1.1 mm thick and is sandwiched between two 0.55 mm plies. A close match between the refractive indices of the fiber and matrix makes the laminate virtually transparent so that cracking and microscopic damage in the 90° can be closely observed (Fig. 3.29).

Fig. 3.29. Photographs of specimens at the indicated strain levels (%) under bright-field ((a) to (i)) and dark-field ((j) to (r)) illumination, showing multiple transverse cracks in the 90° ply, stress 'whitening' and longitudinal splitting in the 0° plies. (After Manders *et al.* 1983.)

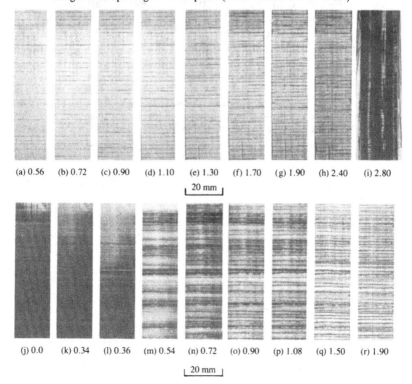

(a) 0.56   (b) 0.72   (c) 0.90   (d) 1.10   (e) 1.30   (f) 1.70   (g) 1.90   (h) 2.40   (i) 2.80

⊢ 20 mm ⊣

(j) 0.0   (k) 0.34   (l) 0.36   (m) 0.54   (n) 0.72   (o) 0.90   (p) 1.08   (q) 1.50   (r) 1.90

⊢ 20 mm ⊣

The pattern of cracks is photographed at regular intervals of applied load using either bright- or dark-field illumination. The dark-field illumination shows fiber–matrix debonding ('stress whitening' which scatters light) with good contrast, whereas bright-field illumination gives better definition of the cracks, although in this case the fiber–matrix debonding appears dark with relatively poor contrast. The thermal residual tensile strain of the 90° ply is estimated to be about 0.22% due to cooling from the postcure temperature of 150° to ambient.

As the specimens are loaded the initial whitening progressively increases, most noticeably at about 0.34% strain (Fig. 3.29k). A knee is visible in the stress–strain curve of Fig. 3.30 at about 0.1% which is attributed to the onset of fiber–matrix debonding. Cracks appear instantaneously at about 0.4% strain, often in the bands of more pronounced whitening (Fig. 3.29l and m). It is concluded from this observation that a crack forms by the joining up of the fiber–matrix debonds. The beginning of multiple cracking is reflected on the stress–strain curve by a second knee. The rate of crack formation with applied strain decreases throughout the loading. At higher strains the crack spacing becomes more uniform. At a strain of about 0.7% stress whitening appears in the longitudinal 0° ply (Fig. 3.29n–r); this is seen as darkening in Figs.

Fig. 3.30. Low-strain portion of a stress-strain curve. Changes of gradient are associated with a rapid increase in stress whitening and with the beginning of multiple cracking. (After Manders *et al.* 1983.)

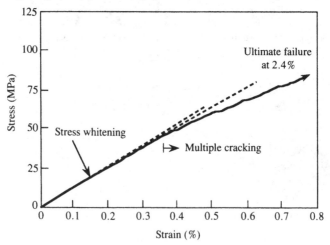

3.29(b)–(i), and it develops into longitudinal cracks at about 1.8% strain.

Manders *et al.* have measured the positions of every crack in a photograph by traveling microscope and calculated the spacings between cracks and their cumulative distribution functions for each load. These distributions illustrate the overall trend towards closer spacing at higher strains. In their study of the variation of crack spacing with stress, Manders *et al.* assume that the 90° ply is an ideal homogeneous brittle material with an inherent distribution of strength which is described by a cumulative distribution function termed S_o for failure of a unit volume. It is also expected that the strength of the 90° ply will be statistically the same throughout its volume; i.e. the constituent volumes which are substantially larger than the microstructure should have strengths which are independent of each other and which are identically distributed.

Thus, the cumulative distribution function of strength S_V for a volume V can be written as

$$1 - S_V = (1 - S_o)^V \tag{3.134}$$

Then the 'risk of rupture', R_V, proposed by Weibull (1939a and b) is given by

$$\ln(1 - S_V) = V \ln(1 - S_o) = -R_V \tag{3.135}$$

Let

$$\ln(1 - S_o) = -\phi(\sigma) \tag{3.136}$$

then the risk of rupture dR for a volume element dV is

$$dR = -\ln(1 - S_o)\,dV = \phi(\sigma)\,dV \tag{3.137}$$

For a non-uniform state of stress

$$R_V = \int_V \phi(\sigma)\,dV \tag{3.138}$$

and

$$S_V = 1 - \exp(-R_V) = 1 - \exp\left[-\int_V \phi(\sigma)\,dV\right] \tag{3.139}$$

Assuming that the stress is uniform in the cross-sectional area, A, the volume integral may be replaced by an integration over the length L. Then Eq. (3.135) becomes

$$\ln(1 - S_V) \cong -A\phi(\sigma)L \tag{3.140}$$

The quantity $A\phi$ is found from the gradient when $\ln(1 - S_V)$ is plotted against L.

Manders *et al.* adopted a two-parameter Weibull distribution for the strength of the 90° ply in which

$$A\phi = A\left(\frac{\sigma}{\sigma_o}\right)^\beta = A\left(\frac{\varepsilon}{\varepsilon_o}\right)^\beta \qquad (3.141)$$

The constants σ_o and ε_o are the scale parameters in terms of stress and strain, respectively, and β is the shape parameter. Taking logarithms of Eq. (3.141) gives

$$\ln(A\phi) = \beta \ln \varepsilon - \beta \ln \varepsilon_o + \ln A \qquad (3.142)$$

It is seen from Eq. (3.142) that a graph of the gradients obtained from $\ln(1 - S_V)$ vs. L and applied strain is linear with gradient β if the Weibull distribution is valid. This is demonstrated in Fig. 3.31, which shows two linear regions intersecting at a strain of about 0.4% (corrected for thermal residual strain), or 0.6% of applied strain. The values of β are about 8.5 and 1.0, respectively, for low

Fig. 3.31. Variation of gradients with 90° ply strain, corrected for residual thermal strain. Solid and open circles correspond to two nominally identical specimens. (After Manders *et al.* 1983.)

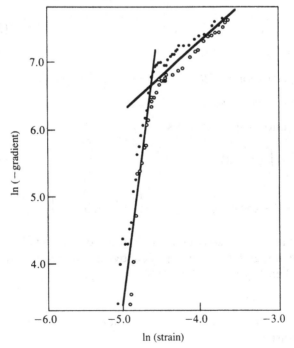

ln (strain)

strain and high strain. The two intercepts ($\ln A - \beta \ln \varepsilon_o$) for the two linear segments are 47 and 11.

Finally, Eq. (3.140) can be evaluated after substitution of Eq. (3.142) using the fitted values of $\beta = 8.5$ and intercept = 47 to obtain median crack spacings ($S_V = 0.5$) as a function of strain. The results of the theoretical correlations are shown by the solid curve in Fig. 3.32.

It is suggested by Manders *et al.* that the deterministic model of Garrett, Bailey and Parvizi and the probabilistic models are complementary. At low strains, the crack spacing is large and the length necessary to build up stress in the 90° ply on either side of a crack is relatively small. Therefore, most of the region between cracks is fairly uniformly stressed and the positions of new cracks are determined by the distribution of flaws in the matrix; a new crack rarely forms exactly midway between two existing cracks. Consequently, the distribution of crack spacings covers a wider range than the factor of two predicted by Garrett, Bailey and Parvizi. At high strains the opposite is true. The region between cracks is non-uniformly stressed. Since the highest stress is found midway between two existing cracks, this is where the new crack forms as described by the deterministic model. When the crack

Fig. 3.32. Crack spacing vs. strain. Solid curve is based upon the statistical model predictions. (After Manders *et al.* 1983.)

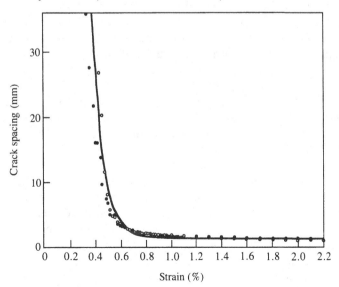

Strain (%)

spacing is significantly higher than the 'unstressed length' (approximately equal to the 90° ply thickness) the probabilistic model is appropriate, and when it is of similar magnitude the deterministic model is more appropriate.

Further analytical treatments of the statistical strength of cross-ply laminates can be found in the work of Fukunaga, Peters, Schulte and Chou (1984) and Peters and Chou (1987).

3.4.7.4 *Transverse cracking and Monte-Carlo simulation*

The occurrence of transverse cracks in cross-plied laminates under ascending tension can be regarded as a kind of stochastic process due to the presence of randomly distributed microflaws. As discussed in Section 3.4.6.3, a stochastic process can be simulated by the Monte-Carlo procedure. In this case, it is postulated that 'intralaminar flaws' exist randomly in the unidirectional ply, which lie in the ply thickness direction and align with the fibers, Fig. 3.33(a). When the transverse ply in the cross-plied laminate is subjected to tension, these flaws effect the observed transverse cracking. For purpose of simulation, the identity of the intralaminar flaws is represented by randomly generated 'effective flaws'. The effective flaws are not, of course, the real flaws. However, if chosen properly, they represent an inherent property of the ply system and effect the essential characteristics of the transverse cracking process in the simulation model.

Wang and Crossman (1980) first conducted an energy analysis to predict the onset of a single transverse crack based on the classical fracture mechanics concept, in conjunction with the effective flaw postulation. Their analysis was validated by a series of experiments (see Crossman, Warren, Wang and Law 1980; Crossman and Wang 1982). Later, Wang, Chou and Lei (1984) and Wang (1984, 1987) incorporated the energy method into a Monte-Carlo procedure to simulate the stochastic nature of multiple cracking. In this section, the work of Wang *et al.* is discussed in some detail.

(A)   Ply-elasticity and three-dimensional stress states

At the outset, it is useful to describe briefly the basis of the energy method. The method is simply derived within the confines of ply elasticity and the classical theory of fracture mechanics. The theory of ply elasticity regards each unidirectional ply as a three-dimensional, elastic, homogeneous and anisotropic solid; and the laminate is modeled as a three-dimensional layered medium containing flaws. An individual effective flaw is handled as a small crack; hence the elastic stress field surrounding the flaw is almost always three-dimensional. Under certain simplifying assumptions, however, some three-dimensional fields may be reduced to generalized plane-strain fields. Even then, numerical techniques are usually required for solutions (see Pipes and Pagano 1970; Wang and Crossman 1977).

(B)   Effective flaw distribution
        The exact mechanism of transverse cracking is rather complicated when viewed at the fiber–matrix scale. It is usually postulated that the crack is caused initially by the coalescence of material microflaws which lie aligned with the fibers in the transverse ply. When viewed at the ply scale, however, a transverse

Fig. 3.33. Schematic view of (a) effective intralaminar flaws, and (b) effective interlaminar flaws. (After Wang 1987.)

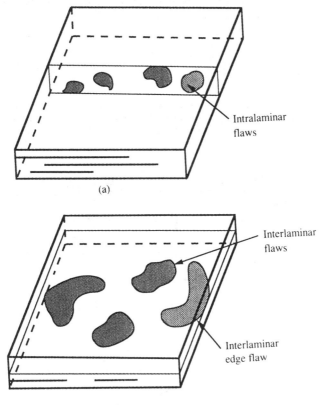

crack represents a separation of the transverse ply along the fiber–matrix interface (see Fig. 3.29). To facilitate a mathematical description of the event at the ply level, the concept of effective flaws is now introduced. Assume that in each unidirectional ply there exists a characteristic probability density distribution of effective flaw sizes as shown in Fig. 3.34. The linear size of the flaws is denoted by $2a$ and the location by x. Then, the discrete random variables $\{a_i, i = 1, 2, \ldots, M\}$ and $\{x_i, i = 1, 2, \ldots, M\}$ characterize the size and the location distributions of the flaws. When two or more plies are grouped together, such as in the $[0°/90°_n/0°]$ laminate (with $n > 1$), the flaw size distribution in the grouped 90° plies is represented by the volumetric rule (see Lei 1986):

$$a_{i,n} = a_i(n)^{2/\lambda} \tag{3.143}$$

where $i = 1, 2, \ldots, M$ and λ is a constant related to the distributional characteristics of $\{a_i\}$.

For simplicity, the flaw location distribution in the grouped 90° plies is assumed to be independent of n

$$x_{i,n} = x_i \qquad i = 1, 2, \ldots, M \tag{3.144}$$

Fig. 3.34. (a) The size ($2a_i$) and location (x_i) of an intralaminar flaw. (b) The probability density distribution of effective intralaminar flaw size in transverse plies. (After Wang 1987.)

(a)

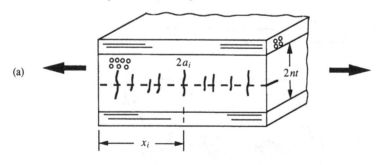

(b)

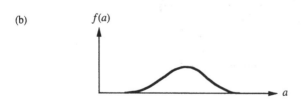

(C)   Onset of the first transverse crack

The $[0°/90°_n/0°]$ laminate shown in Fig. 3.34a is now used to illustrate the energy method. Consider that the laminate is under both the applied tensile strain ε_{xx} and the temperature change ΔT (ΔT is positive for a temperature drop). Let the distribution of the flaws be characterized by Eqs. (3.143) and (3.144), Fig. 3.34b. With the size and the location of a particular flaw known, an elastic stress analysis can be performed; and by treating the flaw as a small crack, one can also calculate the crack-tip strain-energy release rate $G(a_{i,n}, \varepsilon_{xx}, \Delta T)$ (see Wang 1987). The condition governing the propagation of the small crack into a full transverse crack is then given by

$$G(a_{i,n}, \varepsilon_{xx}, \Delta T) = G_{Ic} \qquad (3.145)$$

where G_{Ic} is the material fracture toughness for mode I matrix crack propagation.

Now, for the first crack to form, it is assumed that the crack is caused by the largest of $\{a_{i,n}\}$, denoted by a_{max}. The critical laminate strain $(\varepsilon_{xx})_{cr}$ for the onset of the first crack is then determined from Eq. (3.145) by setting $a_{i,n} = a_{max}$. Now, this first crack is physically detectable.

(D)   Shear-lag effect

When the first transverse crack is formed, the local tensile stress σ_{xx} formerly existing in the unbroken 90° plies is now zero. If the 0°/90° interface bonding is strong, a localized interlaminar shear stress τ_{xz} is then developed in the vicinity of the transverse crack, as shown in Fig. 3.35. This interlaminar shear stress decays exponentially a small distance away from the transverse crack; while within the same distance, the tensile stress σ_{xx} in the 90° plies regains its original magnitude. This local stress-transfer zone, or the shear-lag zone, is proportional to the thickness of the grouped 90° plies, $2nt$.

When there is an effective flaw located near a transverse crack, Fig. 3.36, the flaw may be under the shear-lag zone of the transverse crack. The degree of the shielding effect depends on the relative spacing, s/nt. Specifically, if the size of this flaw is $2a$ and the associated strain-energy release rate at the flaw tip is $G(a, \varepsilon_{xx}, \Delta T, s)$, then the shear-lag effect on the strain-energy release rate can be expressed by the factor, $R(s)$, defined by

$$R(s) = G(a, \varepsilon_{xx}, \Delta T, s)/G(a, \varepsilon_{xx}, \Delta T) \qquad (3.146)$$

where $G(a, \varepsilon_{xx}, \Delta T)$ is calculated without the influence of shear-lag. It may be noted that the range of the retention factor $R(s)$ is between zero and unity over the range of the shear-lag zone, as shown in Fig. 3.36, for a carbon/epoxy composite.

When a flaw is situated between two consecutive transverse cracks, then it is under the shear-lag effect from both cracks. The associated strain-energy release rate, G^*, is given by

$$G^*(a, \varepsilon_{xx}, \Delta T) = R(s_L)G(a, \varepsilon_{xx}, \Delta T)R(s_R) \qquad (3.147)$$

where s_L and s_R are the distances from the flaw to the left crack and to the right crack, respectively.

(E)   Multiple cracks as a function of loading
       After the formation of the first crack from the largest flaw in $\{a_{i,n}\}$, subsequent cracks can form from the remaining flaws at laminate strains appropriately higher than $(\varepsilon_{xx})_{cr}$. A search is then

Fig. 3.35. (a) A transverse crack in a cross-ply laminate. (b) Local stress transfer caused by transverse cracking and the shear-lag zone. (After Wang 1987.)

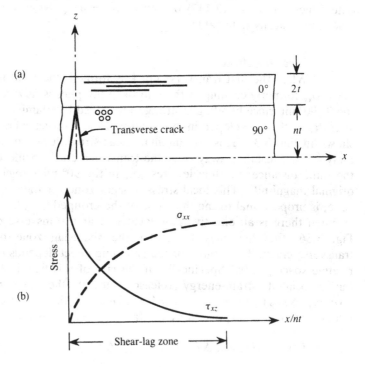

commenced to determine the next flaw that yields the highest strain-energy release rate G^* (with due regard to the shear-lag effect cast by the existing cracks). The applied laminate strain corresponding to the next crack, which should be higher than $(\varepsilon_{xx})_{cr}$, is determined by using G^* in Eq. (3.145).

Successive searches for the next most energetic flaw follow, and the entire load sequence of transverse cracks is simulated until it is no longer energetically possible to produce any more transverse cracks, or until some other failure modes (e.g. delamination, fiber break, etc.) set in during the loading process.

(F)   Determining the effective flaw distribution

One difficulty in the above simulation procedure lies in the fact that the effective flaws are hypothetical quantities, and that they must be chosen properly to yield the essential features of transverse cracking. Appropriate experiments are required to determine the effective flaw distribution.

In the work of Lei (1986), the effective intralaminar flaw distribution in the AS4-3501-06 carbon–epoxy unidirectional ply was determined by testing $[0°_2/90°_2]_s$ tensile coupons. In the test, transverse cracks were detected by X-radiography and were recorded as a function of the laminate tensile stress. The shaded band

Fig. 3.36. The energy retention factor, $R(s)$, vs. s/nt due to the shear-lag effect (after Wang 1987.)

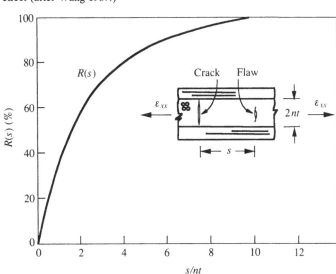

in Fig. 3.37 is formed by plotting data obtained from four specimens, in terms of crack density (cracks per unit length of specimen) versus the applied laminate stress. This band, representing a cumulative formation of the transverse cracks during loading, resembles a form of the output from a certain stochastic process.

It is noted that the experimental band possesses a certain position on the stress scale, a certain characteristic curvature in the coordinate plane and an asymptotic value on the crack density scale. These features will now be used to determine the effective flaw distribution in the $[90°_4]$ layer. To do so, a random number generator is used to form a set of M random values in the interval of $(0, 1)$. These M values are assigned to be $\{x_i\}$, the locations of M flaws along the unit length of the $[90°_4]$ layer. The sizes of the M flaws $\{a_{i,4}\}$ are assumed to fit a Weibull cumulative function,

$$F(a) = 1 - \exp[-(a/\alpha)^\beta] \tag{3.148}$$

Fig. 3.37. Cumulative crack density (number of cracks per millimeter specimen length) vs. applied laminate stress for $[0°_2/90°_2]_s$ laminates. The shaded data band indicates experimental range of four specimens. The dots represent results of Monte-Carlo simulations. (After Wang 1987.)

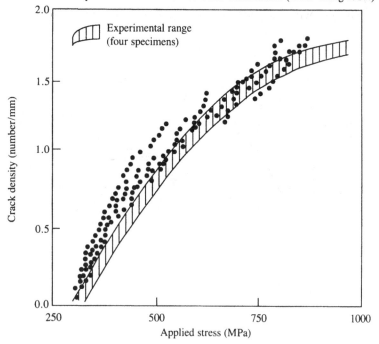

At this point, the parameters M, α and β are assumed known. And a new set of M random values is again generated in the interval $(0, 1)$. These values are assigned to $\{F_i\}$, corresponding to the values of $F(a)$ at $a = a_{i,4}$. The flaw size $\{a_{i,4}\}$ is then determined using Eq. (3.148).

With the assumed values of α, β and M, a simulation of the transverse cracking process as described earlier can now be performed. An appropriate choice of α, β and M is one that simulates closely the experimental data band shown in Fig. 3.37. Generally, α affects primarily the curvature of the band, β shifts the band along the stress scale, and M determines the asymptotic value of the band on the crack density scale (see Lei 1986). Figure 3.37 shows also the simulated crack density vs. laminate stress data from five simulation specimens. Properly selected values of α, β and M can fit the experimental data band very well.

Once the values α, β and M are chosen, the effective flaw size distribution in any number of grouped 90° plies can be found using Eq. (3.143); and then the transverse cracking in the grouped 90° plies in laminates can be simulated. Figure 3.38 shows the simulated results for four $[0°_2/90°]_s$ coupons along with the experimental data band from four test specimens. Figure 3.39 shows a similar comparison between experiment and simulation for four $[0°_2/90°_4]_s$ coupons. In both Figs. 3.38 and 3.39, the simulated data were based on the flaw distribution found from the $[0°_2/90°_2]_s$ coupons in conjunction with Eq. (3.143).

As was mentioned in Section 3.4.6.3, the Monte-Carlo method depends on the nature of the input random variables; and in this case, the input is the distribution of the assumed effective flaws. In the examples discussed above, the values of α, β and M determined by fitting the experiment could not be proved unique. Nevertheless, the simulation, which is performed in conjunction with fracture mechanics analysis, provides not only a quantitative description of the mechanisms but also an assessment of the statistical characteristics of the transverse cracking process.

3.4.8　*Delamination in laminates of multi-directional plies*

Delamination is another mode of failure in multi-directional laminated plates and shells. At the ply level, delamination may be viewed as a plane crack propagating in the interface between two adjacent plies, Fig. 3.40. Cracking of this kind is peculiar because the crack plane is parallel rather than perpendicular to the applied tension; the driving force stems from the interlaminar stresses. As most laminates are designed to carry in-plane loading, interlaminar stresses are generally absent throughout the laminate except near free edges, cut-outs, large defects and other such locations where local interactions from mismatched ply properties cause stress concentrations. Again, these local stress fields are almost always three-dimensional in character.

The three-dimensional stress analysis model and the energy method discussed in Section 3.4.7.4 can be applied to describe the initiation and propagation of delamination. Crossman *et al.* (1980) and Wang and Crossman (1980) followed this approach and investigated free-edge delamination in laminates loaded in uniaxial tension; Wang, Slomiana and Bucinell (1985) considered free-edge delamination in compressively loaded laminates; and Wang, Kishore and Li (1985) examined delamination near interacting laminate defects. In all cases, experimental correlation was performed to validate the analysis.

Fig. 3.38. Cumulative crack density (number of cracks per millimeter specimen length) vs. applied laminate stress for $[0°_2/90°_2]_s$ laminates. The shaded data band indicates experimental range of four specimens. The dots represent results of Monte-Carlo simulations. (After Wang 1987.)

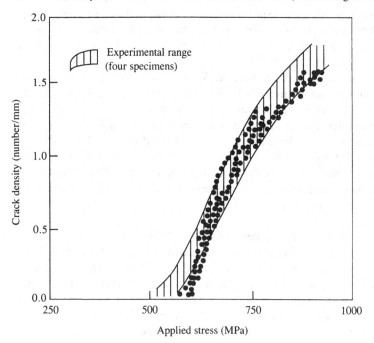

Fig. 3.39. Cumulative crack density (number of cracks per millimeter specimen length) vs. applied laminate stress for $[0°_2/90°_4]_s$ laminates. The shaded data band indicates experimental range of four specimens. The dots represent results of Monte-Carlo simulations. (After Wang 1987.)

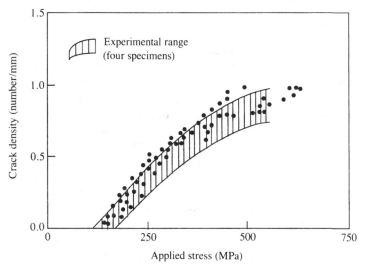

Fig. 3.40. Inter-ply cracking (edge delamination) in a multi-ply laminate. (After Wang 1987.)

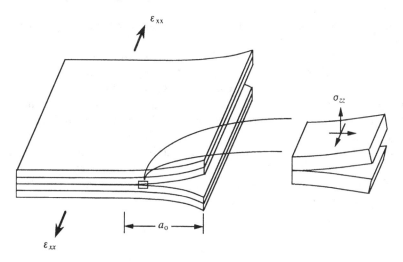

For conciseness, the problem of free-edge delamination in laminates loaded in uniaxial tension is discussed in this section, and only the logic underlying the formulation of the analytical method is presented.

3.4.8.1 *Free-edge delamination*

The free-edge delamination problem has attracted considerable interest for both its scientific challenge and engineering importance. Early laboratory tests have shown that laminate tensile strength can be greatly reduced if free-edge delamination occurs during the course of loading (Pagano and Pipes 1971; Bjeletich, Crossman and Warren 1979). A similar effect on laminate compressive strength has also been confirmed (Wang, Slomiana and Bucinell 1985). Further analyses of the delamination mechanisms have established that the physical behavior of delamination is profoundly influenced by ply stacking sequence, ply fiber orientation, individual ply thickness and laminate width to thickness ratio (Crossman and Wang 1982).

While there have been many predictive models describing delamination growth, the energy method developed by Wang and Crossman (1980) accounts for all these intrinsic and extrinsic factors operating in a severely concentrated three-dimensional stress field near the free edges.

To illustrate this method, the symmetric laminate having straight edges shown in Fig. 3.40 is considered as an example. Assume that the laminate under the applied laminate tensile strain ε_{xx} is such that free-edge delamination is induced in one of its ply interfaces. The problem is then to determine which interface is most likely to delaminate and at what load.

For long, symmetrically stacked and finite-width laminates, it may be assumed that the laminate stress field is independent of the loading axis, x. Hence, it can be described by ply elasticity formulation under the generalized plane strain condition (Pipes and Pagano 1970). The induced free-edge delamination would then extend uniformly along the length of the laminate and advance from the free edges toward the center of the laminate piece, as shown in Fig. 3.40; and the delamination crack can be considered as a self-similar line crack with a linear size, a, propagating in the preferred ply interface.

(A)   Effective interlaminar flaws and conditions for propagation
        To render a prediction for delamination initiation, the

assumption of effective flaws (Section 3.4.7.4) will again be invoked here. In this case, random interfacial flaws are assumed to exist in each ply interface of the laminate as illustrated in Fig. 3.33(b). In particular, along the laminate free edges there is a dominant interlaminar edge flaw. It is further assumed that this flaw is located in a known interface and has a linear size, a_o, in the sense depicted in Fig. 3.40. This flaw is treated as a starter delamination crack, with its size a_o still a random variable. Thus, one can proceed to calculate the crack-tip strain-energy release rate $G(a_o, \varepsilon_{xx})$ if the elastic constants of the unidirectional plies, the ply stacking sequence, the ply fiber orientations and the ply interface in which a_o is residing are known.

The general character of $G(a_o, \varepsilon_{xx})$ as a function of a_o is shown in Fig. 3.41 (for a unit of the applied laminate strain ε_{xx}). G rises sharply from zero at $a_o = 0$, and reaches an asymptotic value, G_{asy}, as a_o becomes greater than a_m. It should be noted that in Fig. 3.41, G_{asy} can be expressed in terms of ε_{xx}^2. The physical meaning of a_m is that, at this size, the delamination no longer interacts with the free-edge boundary. Generally, this boundary effect extends roughly to a distance of about one-half the thickness of the laminate. Beyond this distance, the delamination problem merely involves the extension of cracks between two anisotropic elastic media and the free-edge effect vanishes.

The calculated strain-energy release rate G may be expressed

Fig. 3.41. Variation of the strain-energy release rate G with delamination crack size a_o for a given ε_{xx} value. (After Wang 1984). Copyright ASTM, reprinted with permission.)

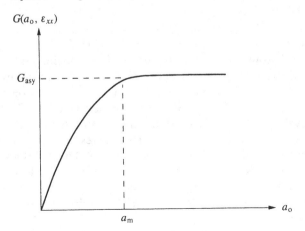

explicitly in terms of the applied laminate strain ε_{xx}:

$$G(a_o, \varepsilon_{xx}) = C_e(a_o)2t\varepsilon_{xx}^2 \tag{3.149}$$

with $C_e(a_o)$ an exclusive function of delamination size a_o. In the above, $2t$ is this thickness of the ply.

Effects of thermal residual stresses due to cooling in fabrication can be readily included in the calculation of G. If the laminate stress-free temperature is T_0 and the ambient temperature is T, then the laminate is exposed to a temperature drop of $\Delta T = T_o - T$. The calculated strain-energy release rate G can be expressed in explicit terms of ε_{xx} and ΔT as

$$G(a_o, \varepsilon_{xx}, \Delta T) = [\sqrt{(C_e)}\varepsilon_{xx} + \sqrt{(C_T)}\,\Delta T]^2 2t \tag{3.150}$$

where C_T is also an exclusive function of a_o.

From fracture mechanics, the condition governing the onset of delamination is given by:

$$G(a_o, \varepsilon_{xx}, \Delta T) = G_c \tag{3.151}$$

where G_c is the fracture toughness of the laminate under delamination.

Equation (3.151) provides a prediction for the critical laminate strain ε_{xx} at the onset of delamination when the delaminating interface, the values of a_o and G_c are given. These values, however, are not readily available; a further analysis of the problem is still needed.

(B)   The effective edge flaw size

Given the functional character of $G(a_o, \varepsilon_{xx})$ shown in Fig. 3.41, a one-to-one relationship between the critical ε_{xx} and a_o can be obtained from Eq. (3.151) assuming the delaminating interface and the associated G_c are known. If a_o is represented by some probability density function, $f(a_o)$, then there is a corresponding range of ε_{xx} for which Eq. (3.151) is satisfied (see Fig. 3.42). The limiting value of ε_{xx} as a_o becomes equal to or greater than a_m is determined by setting $G_{asy}/G_c = 1$. This serves as the lower-bound of the critical strain, ε_{xx}. Since a_m is about one-half the thickness of the laminate, it is small compared to the observable delamination size in relatively thin laminates. In effect, the lower-bound value for ε_{xx} is usually regarded as the critical strain at the onset of delamination.

(C)   The critical delaminating ply interface

Given a specific laminate, the most probable delaminating interface cannot be presupposed from experience. It requires an analysis in which the values of G_{asy}/G_c on all possible interfaces can be compared. According to Eq. (3.151), delamination shall occur on the interface which yields the largest value of G_{asy}/G_c (for the same ε_{xx}).

While G_{asy} at each interface can be calculated readily, the G_c associated with each interface may differ from one interface to another. To elucidate this fact, consider a specific example: the $[\pm 25°/90°]_s$ laminate made of the AS4-3501-06 carbon–epoxy system. Based on the generalized plane strain model mentioned earlier, the entire laminate stress field is calculated first. Of interest are the interlaminar stresses near the free edges before delamination. Figure 3.43 shows near the free edge, the through thickness distribution of the interlaminar normal stress σ_{zz}. Note that σ_{zz} is tensile and unbounded approaching the $-25°/90°$ interface; and is tensile but bounded on the laminate mid-plane ($90°/90°$ interface). Figure 3.44 shows the interlaminar shear stress τ_{xz} near the free edge. Here, an unbounded τ_{xz} exists on both the $25°/-25°$ and the $-25°/90°$ interfaces. These results suggest only qualitatively that free edge delamination may occur either in the $90°/90°$ interface as a mode I crack, or in the $-25°/90°$ interface as a mixed-mode (mode I and mode III) crack.

Further energy analysis provides $(G_I)_{asy}$ for mode I cracks in the

Fig. 3.42. Relation between applied strain ε_{xx} and flaw size a_0. Flaw size distribution $f(a_0)$ is shown schematically. (After Wang 1987.)

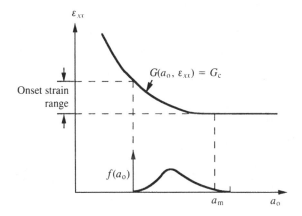

mid-plane, and $(G_I + G_{III})_{asy}$ for the mixed-mode crack in the $-25°/90°$ interface. In the latter, the mixed-mode ratio for G_{III}/G_I is also obtained.

Fracture toughness for mode I delamination may actually be different from that for mixed-mode delamination. Indeed, interfacial fracture of various mixed modes often manifest themselves

Fig. 3.43. The distribution of normal stress σ_{zz} through the laminate thickness. (After Wang 1987.)

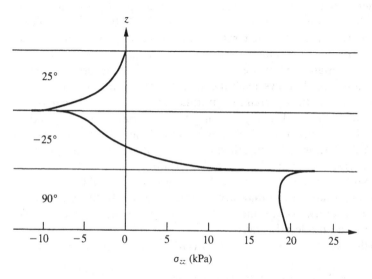

Fig. 3.44. The distribution of shear stress τ_{xz} through the laminate thickness. (After Wang 1987.)

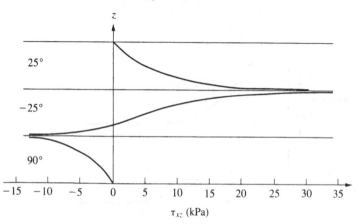

through differences in the fractured surface morphology, which in turn implies differences in the G_c value measured on the ply scale (Bradley and Cohen 1985). Laboratory tests using carbon–epoxy specimens have shown that G_c for mode I delamination is generally lower than G_c for mixed-mode delamination. And, the latter often increases with the amount of the shearing mode. The cause for variable G_c in mixed-mode delamination is complex; several recent studies cited local crack-tip matrix yielding and fiber bridging across the crack surfaces possibly due to shear deformation (see Russell and Street 1985). To use Eq. (3.151) for mixed-mode delamination, G_c must be first obtained as a function of mixed-mode ratio.

For the example problem, as it turned out, mid-plane delamination was predicted because it yielded a larger G_{asy}/G_c than the $-25°/90°$ interface. The prediction agreed with the experiment (see Wang, Slomiana and Bucinell 1985). It should be noted that besides Eq. (3.151) many other fracture criteria for mixed-mode cracks have been suggested in the literature.

3.4.8.2 *General delamination problems*

The free-edge delamination problem discussed above serves to illustrate the basic rationale in the formulation of the energy method. The assumption of effective interfacial flaws allows a fracture analysis from which the onset of delamination could be determined. The assumption may seem awkward at first glance; but it is no more inconvenient than to assume the existence of a stress-based interlaminar strength that is used to determine delamination onset in the highly concentrated free-edge stress fields.

It should also be remarked that delamination problems encountered in practice are very complicated. Frequently, the delamination plane has a two-dimensional contour. To describe the growth of a contoured delamination may require a criterion which is directionally dependent, due to different material characteristics along the contoured crack front. In addition, delamination growth in practical laminates is almost always accompanied and/or preceded by other types of damages such as transverse cracks. Interactions amongst the various local cracks with delamination can be both deterministic and probabilistic in nature. The energy method discussed in this section appears to have sufficient generality for application to the more complex delamination problems. Generic extension of the method could conceivably be developed which can provide quantitiative, if approximate, predictions for a wide class of delamination problems.

3.4.9   *Enhancement of composite strength through fiber prestressing*

The scattering in fiber strength has been attributed to the existence of surface and bulk defects (See Section 3.4.2). Owing to the statistical strength distribution of fibers, it is necessary to design fiber composite structural components based upon a high level of survivability. The enhancement of composite strength can be achieved by eliminating some of the weak spots or defects in the fibers. One way of attaining this goal is to stress the fibers and to induce fracture at the defect sites before they are incorporated into the matrix.

Mills and Dauksys (1973) were the first to adopt the concept of fiber prestressing. In their work, carbon fiber prepregs are pre-stressed at temperatures as low as −18°C. The prestress of prepregs by bending induces non-uniform tensile stress which reaches maximum values at the outer surfaces with fibers near the center of the prepreg stressed the least.

Manders and Chou (1983b) provide a theoretical analysis of enhancement of strength in composites reinforced with previously stressed fibers. The basis of their reasoning is as follows. The failure of a fiber in an aligned composite causes a stress wave to propagate outwards placing a dynamic overstress on the neighboring fibers (see Section 3.3.2). The resulting dynamic stress concentration is generally greater than the static stress concentration which prevails after the system has settled, and increases the probability that adjacent fibers also fail, weakening the composite. This analysis shows how weak fibers may be prefractured to eliminate the dynamic overstress, thereby increasing the strength of the composite. Manders and Chou discussed this strength enhancement with reference to the level of prestress, fiber variability, stress concentrations, and size of the composite.

Chi and Chou (1983) have measured in a systematic fashion the effect of fiber prestressing on the mean strength of composites as well as the dispersion of composite strength. Thornel-300 carbon fibers are used as the reinforcement materials for composites. A loose bundle contains 1000 fibers with a fiber diameter of 7 μm. In order to obtain consistent results in composite strength enhancement, it is essential that all the defect sites of the fibers with strength less than a certain value should be broken when they are subject to prestressing at a given level. It would be most ideal if a uniform tensile stress could be applied uniformly to each small segment of a fiber with length comparable to the ineffective length of the fiber. However, this is impractical in real experiments, where the gauge-length for fiber testing is much larger than the ineffective length. Thus, a fiber already broken at its weakest site can no longer be stressed under tensile loading.

The prestressing of carbon fibers is achieved by pulling the bundle through a pair of circular bars of the same diameter at a tensile force of 30 g. The relationship among the maximum prestress in fibers, σ_p, the bar diameter, D, and the fiber diameter, d, is

$$\sigma_p = E_f d / D \tag{3.152}$$

where E_f denotes the fiber axial Young's modulus. The stress in the fiber caused by the applied tensile force is much smaller than σ_p and hence it is neglected. Composite specimens are fabricated by impregnating prestressed and non-prestressed fiber bundles in

Fig. 3.45. Negative strength enhancement in composites reinforced with prestressed loose carbon fiber bundle. (After Chi and Chou 1983.)

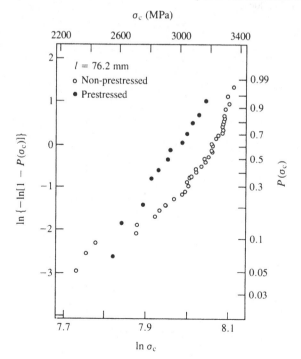

epoxy resin. The strength data obtained for prestressed fiber composites with gauge-length of 76.2 mm are shown in Figs. 3.45 and 3.46, using Weibull probability paper. Here, σ_c denotes composite strength, $P(\sigma_c)$ is the cumulative strength distribution and $\ln\{-\ln[1 - P(\sigma_c)]\}$ indicates the failure probability. The D values for specimens presented in Figs. 3.45 and 3.46 are 0.711 mm and 1.168 mm, respectively; the resulting σ_p values are 2.21 GPa and 1.35 GPa. The mean strength of the composites with non-prestressed fiber bundles is 3.01 GPa. The strength data of Fig. 3.45 show negative enhancement while significant strength enhancement can be seen in Fig. 3.46. It is noted that the strength data of prestressed composites can be fitted approximately by straight lines. Chi and Chou (1983) have concluded that the composite strength for high survivability (low failure probability) is low. These low strength tails can be eliminated by stressing the loose fiber bundles. Enhancement in strength as high as 25% for survivability of 99.9% has been achieved.

Fig. 3.46. Positive strength enhancement in composites reinforced with prestressed loose carbon fiber bundle. (After Chi and Chou 1983.)

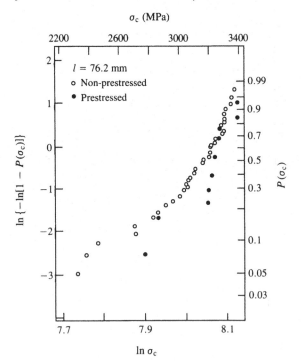

4   Short-fiber composites

4.1   Introduction

Composites reinforced with discontinuous fibers are categorized here as short-fiber composites. The fiber aspect ratio (length/diameter $= l/d$) is often used as a measurement of fiber relative length. Depending upon the dispersion of fibers in the matrix, the relevant d values may include those of the filaments, strands, rovings, as well as other forms of fiber bundles. Although discontinuous fibers such as whiskers have been used to reinforce metals and ceramics, the majority of short-fiber composites are based upon polymeric matrices. Discontinuous fiber-reinforced plastics are attractive in their versatility in properties and relatively low fabrication costs. The concern of the rapid depletion of world resources in metals and the search for energy-efficient materials has contributed to the increasing interest in composite materials. Discontinuous fiber-reinforced plastics will constitute a major portion of the demand of composites in automotive, marine and aeronautic applications.

A discontinuous fiber composite usually consists of relatively short, variable length, and imperfectly aligned fibers distributed in a continuous-phase matrix. In polymeric composites the fibers are mostly glass, although carbon and aramid are also used; non-fibrous fillers are often added. The orientation of the fibers depends upon the processing conditions employed and may vary from random in-plane and partially aligned to approximately uniaxial.

The understanding of the behavior of short-fiber composites is complicated by the non-uniformity in fiber length and orientation as well as the interaction between the fiber and matrix at fiber ends (Chou and Kelly 1976, 1980). These factors are examined in the following discussions on the physical and mechanical properties of short-fiber composites.

4.2   Load transfer

Various attempts have been made to evaluate the stress transfer from the matrix to the fiber in a short-fiber composite. Analyses based upon the shear-lag theory, elasticity theory, and finite element method have been performed. Considerations regarding fiber aspect ratio (Fukuda and Kawata 1974), the effects of bonded ends and loose ends as well as the geometric shapes of fiber ends (Burgel, Perry and Schneider 1970), and the distribution of radial and circumferential stresses near the interface at fiber ends (Haener and Ashbaugh 1967; Carrara and McGarry 1968) have been made. Experimental measurements of interfacial strength have been made using a single fiber pull out test (Favre and Perrin 1972), a fiber fragmentation test (Wadsworth and Spilling 1968), a microtension test (Miller, Muri and Rebenfeld 1987) and a microcompression test (Mandell, Grande, Tsiang and McGarry 1986). (Also see Piggott 1987 and Piggott and Dai 1988.)

Although the shear-lag approach is not as rigorous as the other methods, it does provide a simplistic analysis for gaining some insights into a complex problem and it will be employed in the following. The fiber axial and interfacial stresses are discussed with or without the consideration of interactions among neighboring short fibers.

4.2.1   A single short fiber

Cox (1952) first dealt with the problem of a single short fiber embedded in an infinite matrix material. In this essentially one-dimensional approach the load on the fiber is considered to be built up entirely due to the generation of shear stress in the matrix. Under the assumptions of shear-lag analysis no tensile stress is permitted to transmit across a fiber end.

Consider a long cylindrical composite of radius R which contains a fiber of radius r_o and length l along the cylinder axis. The composite as a whole is subjected to a normal strain ε in the direction of the fiber. The assumption of the shear-lag analysis leads to the following relation:

$$\frac{\mathrm{d}P}{\mathrm{d}x} = H(u - v) \tag{4.1}$$

where $u(x)$ is the displacement of the fiber at the point x; $v(x)$ is the matrix displacement; H is a constant; and P is the fiber axial force. The force–displacement relation of the fiber is

$$P = E_f A_f \frac{\mathrm{d}u}{\mathrm{d}x} \tag{4.2}$$

where E_f and A_f denote the fiber axial Young's modulus and cross-sectional area, respectively. Substituting Eq. (4.2) and $\mathrm{d}v/\mathrm{d}x = \text{constant} = \varepsilon$ into Eq. (4.1), and applying the boundary conditions

$P = 0$ at $x = 0$ and l, the fiber axial stress, σ_f, and interfacial shear stress, τ, are obtained:

$$\sigma_f = \frac{P}{A_f} = E_f \varepsilon \left[1 - \frac{\cosh \beta(l/2 - x)}{\cosh (\beta l/2)} \right]$$

$$\tau = E_f \varepsilon \sqrt{\left[\frac{G_m}{2E_f \ln(R/r_o)} \right]} \frac{\sinh \beta(l/2 - x)}{\cosh(\beta l/2)}$$

(4.3)

where

$$\beta = \sqrt{[H/E_f A_f]}$$

$$H = \frac{2\pi G_m}{\ln(R/r_o)}$$

Here, G_m and E_f are the matrix shear modulus and fiber Young's modulus, respectively. Figure 4.1 shows schematically the variation of σ_f and τ along the length of the fiber. The largest axial stress in the fiber occurs at the center and it reaches $E_f \varepsilon$ for a very long fiber. The magnitude of τ reaches its maximum at the fiber ends, i.e., at $x = 0$ and l, and it vanishes at the middle point of the fiber.

4.2.2 Fiber–fiber interactions

The interactions among fibers in a short-fiber composite are more complex than those in a continuous fiber composite. This is because the axial load carried by a short fiber has to be transferred to the neighboring fibers at locations near its ends. To illustrate the load transfer in a short-fiber composite, the work of Fukuda and Chou (1981a) is recapitulated in the following. This approach, based upon the shear-lag model, introduces axial load into the

Fig. 4.1. The variation of σ_f and τ along a short fiber.

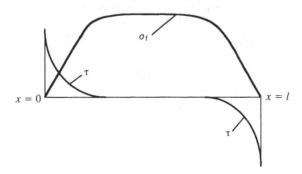

matrix, and the fiber ends are assumed to be bonded to the matrix. These assumptions of the modified shear-lag analysis are valid if the bonding between the fiber and matrix at the fiber end is perfect such as the cases often observed in metal matrix composites and in polymeric matrix composites under compression.

The two-dimensional model for analysis is given in Fig. 4.2, where the hatched parts of the matrix sustain axial load and behave as if they are fibers with a Young's modulus different from the actual fibers. The fiber diameter and matrix layer width are denoted by d and h, respectively. A representative region in Fig. 4.2 containing fiber ends is divided into n parts along the fiber direction x. Fibers in this region are numbered from $i = 1$ to $i = m$. Figure 4.3 shows a free body diagram of a fiber and the adjacent matrix. The equilibrium of forces in the x direction gives

$$\frac{dP_{1j}}{dx} + \tau_{1j} = 0$$

$$\frac{dP_{ij}}{dx} + \tau_{ij} - \tau_{i-1j} = 0 \qquad (i = 2, \ldots, m-1) \tag{4.4}$$

$$\frac{dP_{mj}}{dx} - \tau_{m-1j} = 0$$

Fig. 4.2. The general model of analyses, $\xi = x/d$. (After Fukuda and Chou 1981a).

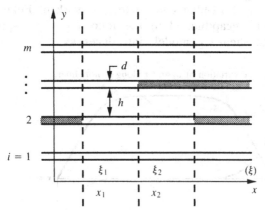

where P_{ij} and τ_{ij} are, respectively, the axial force of the *i*th fiber and the interfacial shear stress in the *j*th region. The condition of linear elastic deformation leads to the following stress–strain relations:

$$P_{ij} = E_{ij} d \frac{\mathrm{d}u_{ij}}{\mathrm{d}x}$$

(4.5)

$$\tau_{ij} = \frac{G}{h}(u_{i+1j} - u_{ij}).$$

where *E, G* and *u* denote the Young's modulus of the fiber, shear modulus of the matrix, and displacement of the fiber, respectively. The subscripts *i* and *j* indicate the *i*th fiber and *j*th region as shown in Fig. 4.2. Thus, E_{ij} is either E_f (Young's modulus of the fiber) or E_m (Young's modulus of the matrix).

The above general formulation is now applied to a model composite shown in Fig. 4.4. This model is composed of a row of short fibers of equal length and two surrounding long fibers. This simple model is adopted for demonstrating the load transfer of short

Fig. 4.3. Free body diagram of the *i*th fiber. (After Fukuda and Chou 1981a.)

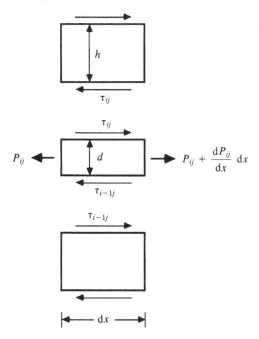

fibers. Given i (=1 and 2) and j (=1 and 2) as shown in Fig. 4.4, the following general solutions of u_{ij} and P_{ij} are obtained from Eqs. (4.4) and (4.5):

$$u_{11} = A_1 + B_1\xi + C_1 e^{\lambda_1 \xi} + D_1 e^{-\lambda_1 \xi}$$

$$u_{21} = A_1 + B_1\xi - \tfrac{1}{2}(C_1 e^{\lambda_1 \xi} + D_1 e^{-\lambda_1 \xi})$$

$$P_{11} = E_f\{B_1 + \lambda_1(C_1 e^{\lambda_1 \xi} - D_1 e^{-\lambda_1 \xi})\}$$

$$P_{21} = E_f\{B_1 - \tfrac{1}{2}\lambda_1(C_1 e^{\lambda_1 \xi} - D_1 e^{\lambda_1 \xi})\}$$

$$u_{12} = A_2 + B_2\xi + C_2 e^{\lambda_2 \xi} + D_2 e^{-\lambda_2 \xi}$$

$$u_{22} = A_2 + B_2\xi - \frac{k}{2}(C_2 e^{\lambda_2 \xi} + D_2 e^{-\lambda_2 \xi})$$

$$P_{12} = E_m\{B_2 + \lambda_2(C_2 e^{\lambda_2 \xi} - D_2 e^{-\lambda_2 \xi})\}$$

$$P_{22} = E_f\left\{B_2 - \frac{k}{2}\lambda_2(C_2 e^{\lambda_2 \xi} - D_2 e^{-\lambda_2 \xi})\right\}$$

(4.6)

where

$$\xi = x/d, \quad \alpha_{ij} = E_{ij}h/Gd$$

$$k = E_m/E_f$$

$$\lambda_j = \sqrt{\left(\frac{\alpha_{1j} + 2\alpha_{2j}}{\alpha_{1j}\alpha_{2j}}\right)}$$

and A_1, B_1, C_1, D_1, A_2, B_2, C_2 and D_2 are unknown constants.

The axial force and boundary conditions of the model of Fig. 4.4 are

(i) *symmetry conditions*

$$(u_{11})_{\xi=0} = 0, \quad (u_{21})_{\xi=0} = 0, \quad (u_{12})_{\xi=\xi_1} = (u_{22})_{\xi=\xi_1}$$

(4.7)

Fig. 4.4. An example of a three-row fiber model. (After Fukuda and Chou 1981a.)

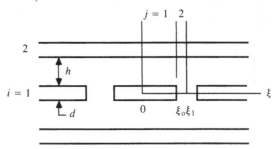

(ii) *continuity conditions*

$$(u_{11})_{\xi=\xi_o} = (u_{12})_{\xi=\xi_o}, \quad (u_{21})_{\xi=\xi_o} = (u_{22})_{\xi=\xi_o}$$
$$(P_{11})_{\xi=\xi_o} = (P_{12})_{\xi=\xi_o}, \quad (P_{21})_{\xi=\xi_o} = (P_{22})_{\xi=\xi_o}$$
(4.8)

(ii) *equilibrium of force*

$$P_{11} + 2P_{21} = 3P_o, \qquad P_{12} + 2P_{22} = 3P_o \tag{4.9}$$

where $3P_o$ denotes the total applied load, and ξ_o and ξ_1 are given in Fig. 4.4. The above conditions provide nine equations, of which eight are independent, to determine the eight integral constants of Eqs. (4.6). Finally, the axial load distribution becomes:

$$P_{11}/P_o = 1 - \frac{2\lambda_1(1-k)}{3F} \sinh \lambda_2(\xi_1 - \xi_o) \cosh \lambda_1 \xi$$

$$P_{21}/P_o = 1 + \frac{\lambda_1(1-k)}{3F} \sinh \lambda_2(\xi_1 - \xi_o) \cosh \lambda_1 \xi$$

(4.10)

$$P_{12}/P_o = \frac{3k}{2+k} \left\{ 1 + \frac{2\lambda_2(1-k)}{3F} \sinh \lambda_1 \xi_o \cosh \lambda_2(\xi - \xi_1) \right\}$$

$$P_{22}/P_o = \frac{3}{2+k} \left\{ 1 - \frac{k\lambda_2(1-k)}{3F} \sinh \lambda_1 \xi_o \cosh \lambda_2(\xi - \xi_1) \right\}$$

where

$$F = k\lambda_2 \cosh \lambda_2(\xi_1 - \xi_o) \sinh \lambda_1 \xi_o$$
$$+ \frac{2+k}{3} \lambda_1 \sinh \lambda_2(\xi_1 - \xi_o) \cosh \lambda_1 \xi_o$$

The displacement field can also be obtained with the given boundary conditions.

Limiting cases such as a single short fiber, a semi-infinite fiber and two semi-infinite fibers separated by a gap can be deduced from Fig. 4.4. Furthermore, the solution for the case where no load is transferred at fiber ends can be obtained by setting $k = 0$ in Eqs. (4.10). Figure 4.5 shows the axial load distributions, for several E_f/E_m values, in the continuous and discontinuous fibers. Fukuda and Chou (1981a) also concluded that the axial load distributions near the fiber ends are essentially the same for fibers of different length for given h/d and E_f/E_m values. This is demonstrated in Fig. 4.6 for the fiber configuration of Fig. 4.4. This finding is consistent with Rosen's (1964) definition of ineffective length (Section 3.4.6.1) which is independent of the actual fiber length (Chen 1971; Fukuda and Kawata 1977). Fukuda and Chou (1981b) have also considered the effects of load transfer at fiber ends and plastic deformation in the matrix.

4.3   Elastic properties

The elastic behavior of short-fiber composites has been extensively studied. It is convenient to subdivide short-fiber composites into three categories, according to their fiber orientations,

Fig. 4.5. Effect of E_f/E_m on fiber axial load distribution, for $h/d = 1$, $\xi_o/d = 100$, $\xi_1/d = 120$. ξ_o and ξ_1 are defined in Fig. 4.4. (a) Continuous fiber, and (b) discontinuous fiber. (After Fukuda and Chou 1981a.)

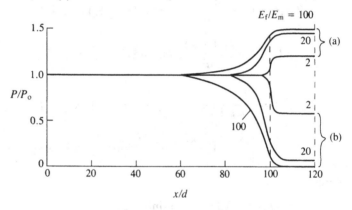

Fig. 4.6. Axial load distribution for fiber 1, and definition of fiber ineffective length. $h/d = 1$, $E_f/E_m = 20$, and $\delta =$ ineffective length. (After Fukuda and Chou 1981a.)

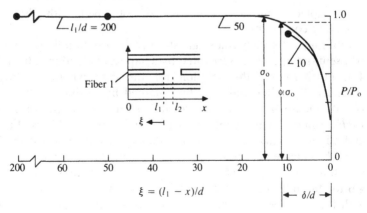

for the purpose of stiffness discussions: (1) unidirectionally aligned short fibers, (2) partially aligned short fibers, and (3) random short fibers.

4.3.1 *Unidirectionally aligned short-fiber composites*

For unidirectionally aligned short-fiber composites, the focus is on the effect of fiber length. Two major approaches are presented for the prediction of elastic moduli of aligned short-fiber composites. The first one is based upon a self-consistent model and the second one gives the upper and lower bounds of elastic moduli. The validity of some semi-empirical and numerical solutions is also examined.

4.3.1.1 *Shear-lag analysis*

Using Cox's fiber stress expression of Eqs. (4.3), the average fiber stress is

$$\bar{\sigma}_f = \frac{1}{l} \int_0^l \sigma_f \, dx$$

$$= E_f \varepsilon \left(1 - \frac{\tanh(\beta l/2)}{\beta l/2} \right) \tag{4.11}$$

Based upon Eq. (4.11), the effective axial Young's modulus of the short-fiber composite is approximated by

$$E_{11} = E_f V_f f(l) + E_m (1 - V_f) \tag{4.12}$$

where

$$f(l) = 1 - \frac{\tanh(\beta l/2)}{\beta l/2} \tag{4.13}$$

and it represents a reduction of the composite elastic modulus due to the finite length of the fiber.

4.3.1.2 *Self-consistent method*

The self-consistent method is a rigorous approach based upon the assumptions that the fiber and matrix materials are isotropic, homogeneous and linearly elastic, the fiber–matrix interfacial bonding is perfect, and the composite with aligned fibers is macroscopically homogeneous and transversely isotropic. As reviewed by Chamis and Sendeckyj (1968), there exist two basic variants of the self-consistent approach, namely the method by Hill (1965a&b) and that used by Kilchinskii (1965, 1966) and Hermans

(1967). Hill followed the method proposed by Kröner (1958) for aggregates of crystals and modeled the composite as a single long fiber embedded in an unbounded homogeneous medium which is macroscopically indistinguishable from the composite. The model of Kilchinskii and Hermans, on the other hand, consists of three concentric cylinders: the innermost cylinder has the elastic properties of the fiber, the middle one simulates the pure matrix material, and the outer one is unbounded and has the properties of the composite. Hill has shown that the prediction of the self-consistent method is more reliable at low and intermediate fiber contents.

The approach of Hill has been adopted by Chou, Nomura and Taya (1980) to treat the stiffness of short-fiber composites. In their work, a single inclusion is assumed to be embedded in a continuous and homogeneous medium (see Hill 1952; Eshelby 1957; Hashin and Rosen 1964; Mura 1982). The inclusion has the elastic properties of a short fiber while the surrounding material possesses the properties of the composite. It is the unknown elastic property of the composite that needs to be found. The work of Chou *et al.* does not restrict the number of component phases in the composite and is hence applicable to hybrid composites (Chapter 5). Numerical examples of this self-consistent approach are given for the special case of a binary system of one kind of fiber in a matrix. Figure 4.7 shows the variation of longitudinal modulus E_{11} of a glass/epoxy system with inclusion volume fraction V_f at three different inclusion aspect ratios (l/d). For $l/d = 100$ the self-consistent theory predicts that the inclusions behave like continuous fibers and the rule-of-mixtures is valid. Also shown in Fig. 4.7 are the predictions of the semi-empirical relation of Halpin and Tsai (see Halpin 1984). The discrepancy between the self-consistent theory and the Halpin–Tsai equation is most pronounced at intermediate values of the aspect ratio. Comparisons of the self-consistent approach with experiments are given in Section 4.3.2, where the effect of fiber misorientation is taken into account.

The predictions of elastic stiffness for particulate-filled composites have been performed by a number of investigators, including Kerner (1956), van der Poel (1958), Hashin and Shtrikman (1963) and Budiansky (1965). The self-consistent theory reduces to Budiansky's solution for the special case of $l/d = 1$.

4.3.1.3 *Bound approach*

Nomura and Chou (1984) also adopted an alternate approach to short-fiber composite effective moduli by deriving their upper and lower bounds. This approach was motivated by the work of Eshekby (1961), Hashin (1965a), Kröner (1967, 1972, 1977), Dederichs and Zeller (1973), Zeller and Dederichs (1973), Wu and McCullough (1977), and Christensen (1979). Nomura and Chou adopted a perturbation expansion of the composite local strain based upon the elastic Green's function. The effective elastic constants can be expressed in infinite series form. When the series are written in terms of the stiffness constants, the first term is the well known Voigt average (1889). The first term of the series represents the Reuss average (1929) when the expression is written in terms of the compliance constants. Based upon the assumptions that the short fibers are modeled as aligned ellipsoidal inclusions and distributed in the matrix material in a statistically homogeneous manner, Nomura and Chou have evaluated the series expressions of the elastic constants up to the third-order term. A variational treatment has been utilized to derive the bounds of the effective elastic moduli of the unidirectional short-fiber composite.

Fig. 4.7. The variation of E_{11}/E_m with V_f at various l/d values for $E_f/E_m = 20$, $v_f = 0.3$ and $v_m = 0.35$. —— self-consistent approach; – – – – Halpin–Tsai equation. (After Chou, Nomura and Taya 1980).

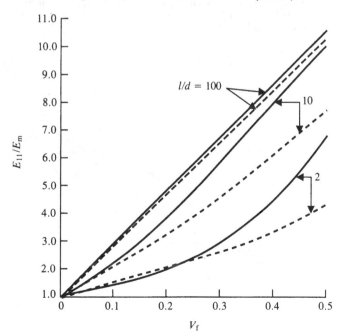

Figure 4.8 illustrates the variation of the axial Young's modulus E_{11} (normalized by the matrix modulus E_m) with fiber volume fraction V_f at fiber aspect ratios $l/d = 1$, 5 and ∞, for glass/epoxy composites. The solid lines indicate the upper and lower bound predictions; the predictions of the self-consistent model of Chou, Nomura and Taya (1980) are indicated by broken lines. The self-consistent model prediction is close to the lower bound at low fiber volume fraction and approaches the upper bound at high fiber volume fraction. The gap between the bounds at a fixed fiber volume fraction narrows as the fiber aspect ratio increases. For long continuous fibers, the bound approach and the self-consistent model all predict the rule-of-mixtures relation. Although fiber volume fraction in the full range of 0 to 1 is used in Fig. 4.8, it is understood that the maximum attainable fiber volume fraction in a composite is determined by the fiber geometric packing and fiber cross-sectional shape.

Figure 4.9 shows the comparison of the bound approach with Hashin's (1965a) results for the effective axial shear modulus G_{12} of continuous fiber composites. The theory of Nomura and Chou (1984) predicts tighter bounds than those of Hashin. This is due to

Fig. 4.8. The variation of E_{11}/E_m with V_f for $E_f/E_m = 20$, $\nu_m = 0.4$ and $\nu_f = 0.3$ —— bound approach; – – – – self-consistent model. (After Nomura and Chou 1984).

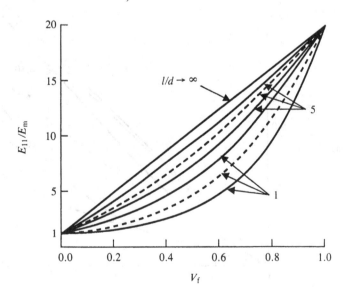

the fact that Hashin's result is equivalent to the evaluation of the series expression of the elastic constants up to the second-order term. The bounds of effective elastic moduli of multi-phase systems such as hybrid composites can also be examined by this approach.

4.3.1.4 *Halpin–Tsai equation*

The Halpin–Tsai equation (see Halpin 1984) was obtained by reducing Hermans' solution (1967) to a simpler analytical form while the filament geometries are taken into account through the use of some empirical factors. The pertinent relations are

$$\frac{\bar{P}}{P_m} = \frac{1 + \zeta \eta V_f}{1 - \eta V_f}$$

$$v_{12} \cong v_f V_f + v_m V_m \tag{4.14}$$

Fig. 4.9. The variation of G_{12}/G_m with V_f for $E_f/E_m = 20$, $v_m = 0.4$, $v_f = 0.3$ and $l/d \to \infty$. —— bound approach; – – – – self-consistent model; —·—·— bounds of Hashin and Shtrikman. (After Nomura and Chou 1984).

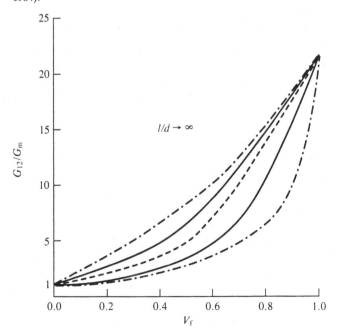

where

$$\eta = (P_f/P_m - 1)(P_f/P_m + \zeta)$$

$$\zeta(E_{11}) = 2\left(\frac{l}{d}\right) + 40V_f^{10}$$

$$\zeta(E_{22}) = 2 + 40V_f^{10}$$

$$\zeta(G_{12}) = 1 + 40V_f^{10}$$

$$\zeta(G_{23}) \cong 1/(4 - 3v_m)$$

$$\bar{P} = E_{11}, \ E_{22}, \ G_{12} \ \text{or} \ G_{23}$$

$$P_f = E_f \ (\text{for } E_{11} \text{ and } E_{22}) \ \text{or} \ G_f \ (\text{for } G_{12} \text{ and } G_{23})$$

$$P_m = E_m \ (\text{for } E_{11} \text{ and } E_{22}) \ \text{or} \ G_m \ (\text{for } G_{12} \text{ and } G_{23})$$

Other solutions of effective elastic constants can be found from, for instance, the numerical work of Conway and Chang (1971), and Chang, Conway and Weaver (1972). Experimental data on short-fiber composite elastic properties have been reported by Lees (1968), and Blumentritt, Vu and Cooper (1974).

4.3.2   *Partially aligned short-fiber composites*

It is usually desirable to orient the fibers for enhanced stiffness and strength properties. However, perfect alignment of short fibers in a composite is normally very difficult to achieve. Partial fiber alignment is typical in, for example, injection molded composites. Several different approaches have been adopted by researchers to predict the stiffness of short-fiber composites with biassed fiber orientation. The following discussions of these approaches begin with a brief summary of the original treatments on misaligned continuous fibers.

The first attempt in examining the effect of fiber orientation is attributed to the work of Cox (1952), who studied the elastic properties of paper and other fibrous materials. Cox's model is concerned with continuous fibers of negligible thickness with orientations either random or defined by some distribution rules. The contribution of matrix to stiffness is ignored. It is also assumed in this model that under load the fibers do not slide across each other at the points of intersection (see Cook 1968).

Cook (1968) provides the elastic properties of continuous fiber composites in three dimensions. The systems of misorientation examined by Cook include the axially symmetric type, a fan shaped array and systems of crossed fibers. Fiber orientation distribution functions are generated analytically to describe the characteristics of these systems. A case of most practical significance is the axially symmetric fiber distribution, which is also termed the *witch's broom* by Cook. The degree of fiber scatter from perfect alignment is described by the root-mean-square deviation of orientation from the symmetry axis

$$s^2 = \int_0^{\pi/2} \eta(\theta)\theta^2 \sin\theta \, d\theta \qquad (4.15)$$

where $\eta(\theta)$ is the fiber orientation distribution function. According to Cook, for a composite such as glass fibers in a polymer resin $(V_f E_f / V_m E_m \sim 20)$ the orientation effect can be minimized if the fibers are sufficiently long and, hence, a high degree of orientation can be achieved. On the other hand, for whisker reinforced composites the reduction in stiffness may be significant if the fibers are short and alignment becomes a difficult technical problem. Cook reported that for a silicon nitride whisker reinforced epoxy resin composite examined, stiffness reduction of 4–19% could occur for the root-mean-square scatter between 4.5° and 10°.

Fukuda and Kawata (1974) considered the Young's modulus of short-fiber composites and took into consideration variations in both fiber length and orientation. The analysis is based upon the plane stress elasticity solution of load transfer between the fiber and matrix in a single short-fiber model, and the assumption of negligible interactions between neighboring fibers. The prediction of the composite Young's modulus is given in the general form

$$E_c = C_l C_\theta E_f V_f + E_m (1 - V_f) \qquad (4.16)$$

The factors C_l and C_θ reflect the effects of fiber length and orientation distributions, respectively. Both C_l and C_θ are unity in the case of aligned continuous fibers.

Figure 4.10 shows the variations of C_l with the factor $(l/d)(E_f/E_m)$ where l/d denotes the fiber aspect ratio. The open and solid circles in Fig. 4.10 are experimental values of Anderson and Lavengood (1968) for glass/epoxy and boron/epoxy, respectively. The solid line is obtained from a two-dimensional analysis and the broken line is the modified result when the fiber circular cross-sectional shape is taken into account. Figure 4.11 shows the variations of C_θ with $\eta(\theta)$, which is the probability density of fibers at the orientation angle θ. Fukuda and Kawata assume that $\int_0^{\pi/2} \eta(\theta) \, d\theta = 1$. Three forms of $\eta(\theta)$ are assumed: rectangular, sinusoidal and triangular. θ_o defines the range of fiber angular distribution. Comparisons of the predictions of Eq. (4.16) with the measured modulus of an α-SiC whisker/aluminum composite (Schierding and Deex 1969) are favorable.

The above discussions have centered upon either continuous fibers or short fibers in planar arrangement. A treatment of the three-dimensional fiber orientation effect has been developed by Chou and Nomura (1981). They considered an axially symmetrical fiber orientation distribution. Referring to Fig. 4.12, the general orientation of a short fiber can be considered as derived from the

Fig. 4.10. Relation between C_l and $(l/d)/(E_f/E_m)$. (After Fukuda and Kawata 1974).

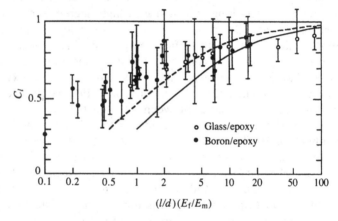

$(l/d)(E_f/E_m)$

Fig. 4.11. Relation between C_θ and θ_o. (After Fukuda and Kawata 1974).

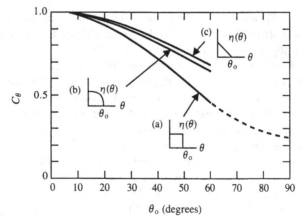

θ_o (degrees)

original position along the z axis by two rotations. The corresponding rotational angles are φ and θ, as indicated in Fig. 4.12. The transformation matrix, from the original coordinate system to the current system, is defined as

$$[T] = \begin{bmatrix} \sin\theta\cos\varphi & \sin\theta\sin\varphi & \cos\theta \\ \cos\theta\cos\varphi & \cos\theta\sin\varphi & -\sin\theta \\ -\sin\varphi & \cos\varphi & 0 \end{bmatrix} \qquad (4.17)$$

Let the bold-faced letter indicate a tensor. The transformation of an elastic stiffness tensor **C** (or compliance tensor **S**) of a unidirectionally aligned short-fiber composite can be performed through the application of the **T** matrix and the resulting tensor is denoted by **C′** (or **S′**). The effective elastic tensor of a misaligned short-fiber composite is then given by

$$\mathbf{C}'' = \int \mathbf{C}'(\theta, \varphi)\eta(\theta, \varphi)\, \mathrm{d}A$$

$$= \int_0^{2\pi} \mathrm{d}\varphi \int_0^{\pi} \mathbf{C}'(\theta, \varphi)\eta(\theta, \varphi)\sin\theta\, \mathrm{d}\theta \qquad (4.18)$$

Here, $\eta(\theta, \varphi)$ in the above equation is the probability density function of fiber orientation determined from experiments. The integration is carried out over the surface area of a unit sphere to include all the fibers in the composite.

Fig. 4.12. Reference coordinate axes.

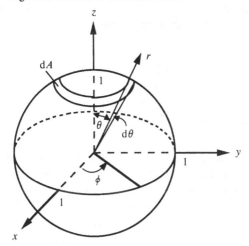

Two cases of fiber orientation distribution are of practical importance. In the case of injection molded objects, fiber orientation distribution is independent of the angle φ if the direction of flow is along the z axis, and $\eta = \eta(\theta)$. The composite in this case is isotropic in the plane transverse to the z axis, and $\mathbf{C}''$ is independent of φ. In sheet molding compounds, it is reasonable to assume that the short fibers all lie on the xz plane and the problem is two-dimensional. The transformation matrix is

$$T = \begin{bmatrix} \sin \theta & 0 & \cos \theta \\ \cos \theta & 0 & -\sin \theta \\ 0 & 1 & 0 \end{bmatrix} \tag{4.19}$$

Equation (4.18) is then reduced to

$$\mathbf{C}'' = 2\pi \int_0^\pi \mathbf{C}'(\theta)\eta(\theta) \sin \theta \, d\theta \tag{4.20}$$

It has been pointed out in the variational treatment of Section 4.3.1 that the first term in the series expression of composite stiffness constant or compliance constant gives the well known Voigt's upper bound or Reuss' lower bound. The averaging principles of Voigt and Reuss were first used to predict the elastic properties of a polycrystalline aggregate in terms of the basic properties of a single crystal and its orientation in the aggregate. The Voigt and Reuss averages are equivalent to assuming that the single crystals are arrayed in parallel and in series, respectively. These concepts of Voigt and Reuss averages are also useful in dealing with misaligned composites. They can be expressed in the general forms for the stiffness constant $\mathbf{C}$ and compliance constant $\mathbf{S}$ as

$$\langle C \rangle = \int_V \mathbf{C}(r, \theta, \varphi) \, dV \Big/ \int_V dV$$

$$\tag{4.21}$$

$$\langle S \rangle = \int_V \mathbf{S}(r, \theta, \varphi) \, dV \Big/ \int_V dV$$

where, in general, $\mathbf{C}$ and $\mathbf{S}$ are functions of position (r, θ, φ) as shown in Fig. 4.12. Furthermore, it can be shown that in the Voigt and Reuss averaging processes for small fiber misalignment there is negligible difference between the model involving a distribution of fiber orientations and the model in which all the fibers are aligned along the direction of the root-mean-square average angle (see

Knibbs and Morris 1974). The treatment of effective elastic moduli for partially aligned short-fiber composites can also be achieved through the laminated plate analogue, which is discussed in Section 4.3.3.

4.3.3 *Random short-fiber composites*

The treatment of Cox (1952) discussed in Section 4.3.2 deals with the stiffness of *continuous* fibers distributed in a plane. For completely random fiber distribution, Cox's results are reduced to the simple forms

$$E_c = E_f V_f / 3$$
$$G_c = E_f V_f / 8 \qquad (4.22)$$
$$v_c = \tfrac{1}{3}$$

where E_c, G_c and v_c are, respectively, the Young's modulus, shear modulus and Poisson's ratio of the composite. The random distribution of fibers imparts isotropic properties of the composite at the macroscopic scale. Hence, E_c, G_c and v_c satisfy the relationship for isotropic materials:

$$G_c = E_c / 2(1 + v_c) \qquad (4.23)$$

The contribution of matrix material is neglected in this treatment but has been taken into account in the works of Arridge (1963), and Pakdemirli and Williams (1969), who also derived approximate expressions for E_c and G_c.

Nielsen and Chen (1968) proposed that the in-plane Young's modulus of a random composite with *continuous* fibers can be approximated by an averaging process. Basic to this process is the knowledge of the elastic moduli of a unidirectional fiber composite measured at an angle θ from the fiber direction (see Eqs. 2.19). The effective in-plane Young's modulus of a random composite, for example, is then given by

$$E_c = \frac{2}{\pi} \int_0^{\pi/2} E(\theta) \, d\theta \qquad (4.24)$$

In applying Eq. (4.24), the fiber volume fraction of the composite used for calculating $E(\theta)$ should be the same as that in the random composite. It should also be noted that $E(\theta)$ is not a component of a tensor. Hence, the averaging process defined in Eq. (4.24) bears no relation to the Voigt and Reuss averages discussed in Section 4.3.2.

The elastic moduli of a composite where the short fibers exhibit in-plane random orientation can also be examined by the method of a laminate analogue (Halpin 1969; Halpin and Pagano 1969; Halpin, Jerine and Whitney 1971). The following discussions are based upon reviews by Kardos (1973) and Nicolais (1975). In the laminate analogue the mechanical response of the composite is simulated by that of a laminate composed of unidirectional short fibers (Kardos 1973). The laminate is symmetric about the mid-

Fig. 4.13. (a) Laminate analogue of a composite with random in-plane orientation of short fibers. The quasi-isotropic laminate has the [+45°/−45°/90°/0°]$_s$ configuration. (b) Dependence of tensile modulus on volume fraction of 3.2 mm E-glass/polycarbonate composites for random in-plane (dashed curve) and biassed (solid curves) fiber orientations. − − − − is quasi-isotropic calculation; —— weighted distribution calculations; ○, ● experimental data. Fiber aspect ratio is about 313. (After Halpin, Jerine and Whitney 1971).

Quasi-isotropic laminate

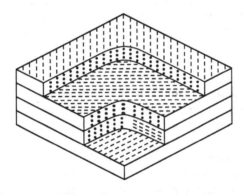

Random in-plane orientation

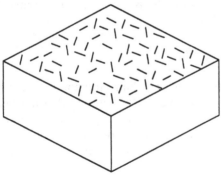

(a)

plane and has the same number of $+\theta$ and $-\theta$ orientation plies (Fig. 4.13a).

The concept of the laminate analogue is outlined in the following. First, the four independent elastic moduli E_{11}, E_{22}, ν_{12} and G_{12} of a unidirectional short-fiber lamina can be derived from the fiber and matrix properties based upon the self-consistent model, the variational method, or the Halpin–Tsai equation. The stiffness matrix components Q_{ij} are given by Eqs. (2.14). The effective engineering stiffness constants E_{11} E_{22}, ν_{21} and G_{12} for the aligned short-fiber lamina can be expressed in terms of the Q_{ij}'s as given in Table 2.1.

The stiffness matrix components $\bar{Q}_{ij}$ for a unidirectional lamina oriented at an angle θ with respect to the x axis are given in Eqs. (2.16). They can also be written in the following alternate forms:

$$\bar{Q}_{11} = U_1 + U_2 \cos 2\theta + U_3 \cos 4\theta$$
$$\bar{Q}_{22} = U_1 - U_2 \cos 2\theta + U_3 \cos 4\theta$$
$$\bar{Q}_{12} = U_4 - U_3 \cos 4\theta$$
$$\bar{Q}_{66} = U_5 - U_3 \cos 4\theta \tag{4.25}$$
$$\bar{Q}_{16} = \tfrac{1}{2}U_1 \sin 2\theta + U_3 \sin 4\theta$$
$$\bar{Q}_{26} = \tfrac{1}{2}U_2 \sin 2\theta - U_3 \sin 4\theta$$

Fig. 4.13. (*cont.*).

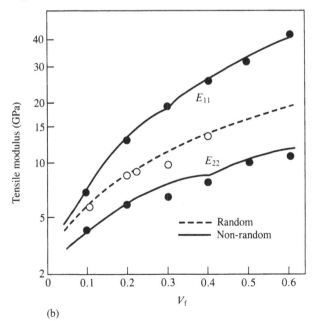

(b)

where the U_i are defined as

$$U_1 = \tfrac{1}{8}(3Q_{11} + 3Q_{22} + 2Q_{12} + 4Q_{66})$$

$$U_2 = \tfrac{1}{2}(Q_{11} - Q_{22})$$

$$U_3 = \tfrac{1}{8}(Q_{11} + Q_{22} - 2Q_{12} - 4Q_{66})$$

$$U_4 = \tfrac{1}{8}(Q_{11} + Q_{22} + 6Q_{12} - 4Q_{66})$$

$$U_5 = \tfrac{1}{8}(Q_{11} + Q_{22} - 2Q_{12} + 4Q_{66})$$

When the plies are stacked together to form a laminate, the in-plane stretching stiffness A_{ij} is given by Eqs. (2.29). For the case of a balanced angle-ply ($\pm\theta$) composite with mid-plane symmetry, the bending stiffness B_{ij} (Eqs. 2.29) and the coupling terms A_{16} and A_{26} vanish, and the A_{ij} components can be represented by

$$A_{11} = [U_1 + U_2 \cos 2\theta + U_3 \cos 4\theta]h$$

$$A_{22} = [U_1 - U_2 \cos 2\theta + U_3 \cos 4\theta]h$$

$$A_{12} = [U_4 - U_3 \cos 4\theta]h \tag{4.26}$$

$$A_{66} = [U_5 - U_3 \cos 4\theta]h$$

Here, h denotes the total laminate thickness. Following the same reasoning for the derivation of Eq. (2.15), the effective engineering constants of the laminate are given by

$$E_{11} = \frac{A_{11}A_{22} - A_{12}^2}{A_{22} \cdot h}$$

$$E_{22} = \frac{A_{11}A_{22} - A_{12}^2}{A_{11} \cdot h}$$

$$\nu_{12} = \frac{A_{12}}{A_{22}} \tag{4.27}$$

$$G_{12} = \frac{A_{66}}{h}$$

If a random short-fiber composite assumes the form of a thin sheet while the sheet thickness is less than the average fiber length, the composite can be modeled as a 'quasi-isotropic laminate'. In principle, the laminate can be constructed by stacking up unidirectional laminae in all orientations to achieve a balanced and symmetric arrangement. Because the fiber orientation covers all the values between $0°$ and $180°$, the angular dependent terms in the A_{ij}

components of Eqs. (2.16) cancel one another. Consequently, the effective engineering constants can be simplified as

$$E_c = 4U_5 \frac{U_1 - U_5}{U_1}$$

$$v_c = \frac{U_1 - 2U_5}{U_1} \tag{4.28}$$

$$G_c = U_5$$

It is obvious that the above elastic constants satisfy the necessary relation for in-plane isotropy. Expressions for random fiber composite elastic constants equivalent to Eqs. (4.28) also have been obtained by Akasaka (1974). Halpin, Jerine and Whitney (1971) have demonstrated the validity of the laminate analogue by comparing the analytical predictions with the measurement of effective tensile modulus of E-glass/polycarbonate with random fiber orientation as shown in Fig. 4.13(b).

The laminate analogue can also be applied to quasi-isotropic short-fiber composites using lay-ups such as $0°/\pm 60°$ and $0°/\pm 45°/90°$ (Warren and Norris 1953). Other works dealing with the elastic stiffness of random fiber composites can be found from Tsai and Pagano (1968); Manera (1971); Christensen and Waals (1972); Wilczynki (1978); and Hahn (1978). As pointed out by Bert (1979), the accuracy of these approximations is affected by the fiber volume fraction and the ratio E_f/E_m. The laminate analogue can also be used for examining the elastic properties of short-fiber composites with layered microstructures. Figure 4.14 shows the scanning electron micrograph of the cross-section of an injection molded polyethylene terephthalate with short glass fibers. This type of layered structure has been found in many types of short-fiber reinforced thermoplastics.

Attempts also have been made to predict the stiffness of composites with random fibers in three-dimensional distribution. Rosen and Shu (1971) and Christensen and Waals (1972) examined the case of continuous fibers. Halpin, Jerine and Whitney (1971) treated the case of layers of plain woven fabric in which the unit weave cell is pierced by a straight yarn perpendicular to the fabric plane. The problem of random short fiber orientation in three dimensions has been treated by Chou and Nomura (1981). By taking $\eta = 1/2\pi$ in Eq. (4.18), elastic moduli for completely random orientation can be obtained. Figure 4.15 illustrates the theoretical variations of E_c/E_m with V_f for a random glass/epoxy system and the experimental data of Manera (1971).

The laminated plate analogue developed above can also be applied to consider in-plane partially aligned short fibers (Halpin, Jerine and Whitney 1971; Kardos 1973) discussed in Section 4.3.2. In this case the angular fiber distribution function $\eta(\theta)$ needs to be measured from the composite specimen. The laminate simulating the composite is treated as composed of weighted groups of angle plies $(\pm\theta)$ with fixed fiber volume fraction. The percentage of materials oriented at the angles $\pm\theta$ is obtained from $\eta(\theta)$. The contributions to the overall response of laminate stiffness from different layers are proportioned to their fractional thickness in the laminate.

Table 4.1 gives an example of the orientation distributions of discontinuous glass fibers in a polymeric matrix. The composite is

Fig. 4.14. SEM micrograph of short glass fiber/polyethylene terephthalate showing layered structure of fiber orientations. (After Friedrich and Karger-Kocsis 1989.)

200 μm
⊢——⊣

compression molded from extrudate. It can be seen that most of the fibers are oriented quite close to the extrusion direction. Whereas previously each $\pm\theta$ ply was weighted equally in summing up the stiffness contributions to the laminate, one must now account for the fact that more of the laminate thickness may be made up of one angle than the other. Define $a(\theta)/h$ as the percentage of the material oriented at the angles $\pm\theta$, and it is obtained from the experimental angular distribution $\eta(\theta)$ where $\int_0^\pi \eta(\theta)\,d\theta = 1$. The stiffness moduli A_{ij} of the laminate is related to the stiffness of the plies $A_{ij}(\theta_k)$, oriented at the angles $\pm\theta_k$, by

$$A_{ij} = \sum_{k=1}^{n} \frac{a(\theta_k)}{h} A_{ij}(\theta_k) \qquad (4.29)$$

where n is the total number of plies.

In summary, the calculation of the effective engineering stiffness of short-fiber composites with biassed fiber orientations should first follow the procedure outlined in Section 2.3 to obtain the A_{ij} components for each fiber angle. These are then summed according to their fiber angular distributions such as that given in Table 4.1 and Eq. (4.29) to obtain the A_{ij} terms. The engineering constants are then obtained from Eqs. (4.27). The solid lines in Fig. 4.13(b)

Fig. 4.15. The comparison of E_c/E_m (——— bound approach; – – – – self-consistent model) with experimental data for $E_f/E_m = 32.4$, $v_m = 0.4$, $v_f = 0.25$ and $l/d \to \infty$. (After Chow and Nomura 1981.)

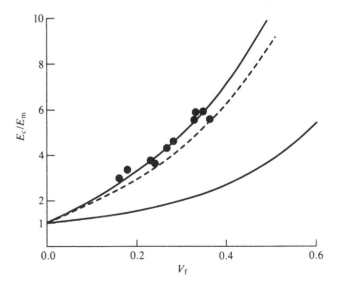

are theoretical predictions of the tensile moduli based upon this procedure. The bumps in the curves are attributed to the fact that the angular distribution functions are not smooth functions of fiber volume fraction.

4.4   Physical properties

The physical properties described below include thermal conductivity and thermal expansion coefficients. These properties are essential to the study of the thermomechanical behavior of short-fiber composites.

4.4.1   *Thermal conductivity*

The important transport properties of composites include dielectric constant, heat conduction, electrical conduction, magnetic

Table 4.1. *Fiber orientation distributions in composites compression molded from rod extrudate. Short glass fiber aspect ratio ≈ 313. After Halpin et al.* (1971)

Orientation θ (degrees)	Percent fibers having θ orientation			
2.5	23.4	25.4	25.0	36.5
7.5	17.9	18.1	23.8	23.9
12.5	12.0	12.3	16.4	14.2
17.5	16.0	7.7	10.0	5.7
22.5	6.2	6.4	6.8	3.0
27.5	5.9	5.6	4.8	2.7
32.5	4.4	4.6	3.1	1.8
37.5	4.6	3.1	2.4	2.0
42.5	2.6	3.4	1.6	1.0
47.5	1.7	1.9	1.3	0.4
52.5	0.4	1.3	0.8	0.7
57.5	0.7	0.7	1.1	0.8
62.5	1.0	1.4	0.9	0.5
67.5	0.7	1.1	0.7	0.7
72.5	0.1	2.1	0.4	0.5
77.5	0.9	0.9	0.6	0.8
82.5	0.5	2.3	0.3	0.9
87.5	1.0	1.4	0.1	0.8
Fiber volume fraction	20	30	40	50

permeability and diffusion coefficients. Since all these properties are second rank tensors, only the bounds of thermal conductivity are demonstrated. The linear relation between the heat flux $\mathbf{q}$ and gradient of temperature T is given by

$$\mathbf{q} = \mathbf{k}(-\nabla T) \tag{4.30}$$

where $\mathbf{k}$ denotes thermal conductivity and is assumed to be a function of position only. It is understood that $\mathbf{k}$ is a symmetric tensor quantity. The governing equation for a steady-state heat conduction is

$$\nabla \cdot \mathbf{q} = 0 \tag{4.31}$$

Several approaches to this subject have been employed by researchers. These include the statistical method by Beran (1965), Beran and Molyneux (1966), and Hori and Yonezawa (1975) as well as the self-consistent and variational approaches of Hashin and Shtrikman (1962), Hashin (1968) and Willis (1977).

Nomura and Chou (1980), following their development of bounds of elastic moduli (1984), derived bounds of effective thermal conductivity of unidirectional short-fiber composites. The short fibers are again modeled as ellipsoidal inclusions of the same length and are distributed in a statistically homogeneous manner in the matrix material. The composite exhibits transverse isotropy. This approach is also valid for composites containing more than one type of fiber. Consider the case of a binary system and denote the thermal conductivity and volume fraction of the fiber and matrix phases by k_f, V_f and k_m, V_m, respectively. The bounds of the effective composite conductivity k_{11} along the fiber directions are

$$\left\{ \frac{V_f}{k_f} + \frac{V_m}{k_m} - \frac{V_f V_m \left(\frac{1}{k_f} - \frac{1}{k_m} \right)^2 h(t)}{(V_m - V_f)\left(\frac{1}{k_f} - \frac{1}{k_m} \right)h(t) + \frac{V_f}{k_f} + \frac{V_m}{k_m}} \right\}^{-1} \leq k_{11}$$

$$\leq V_f k_f + V_m k_m - \frac{V_f V_m (k_f - k_m)^2 (1 - h(t))}{(V_m - V_f)(k_f - k_m)(1 - h(t)) + V_f k_f + V_m k_m} \tag{4.32}$$

The bounds of the conductivity k_{22} and k_{33} in the transverse direction are

$$\left\{ \frac{V_f}{k_f} + \frac{V_m}{k_m} - \frac{V_f V_m \left(\frac{1}{k_f} - \frac{1}{k_m}\right)^2 (1 - h(t)/2)}{(V_m - V_f)\left(\frac{1}{k_f} - \frac{1}{k_m}\right)(1 - h(t)/2) + \frac{V_f}{k_f} + \frac{V_m}{k_m}} \right\}^{-1} \leq k_{22} = k_{33}$$

$$\leq V_f k_f + V_m k_m - \frac{V_f V_m (k_f - k_m)^2 h(t)}{(V_m - V_f)(k_f - k_m)h(t) + 2(V_f k_f + V_m k_m)} \qquad (4.33)$$

where

$$h(t) = \frac{t^2}{t^2 - 1} \left\{ 1 - \frac{1}{2} \left[\sqrt{\left(\frac{t^2}{t^2 - 1}\right)} \right.\right.$$
$$\left.\left. - \sqrt{\left(\frac{t^2 - 1}{t^2}\right)} \ln\left(\frac{t + \sqrt{(t^2 - 1)}}{t - \sqrt{(t^2 - 1)}}\right) \right] \right\} \qquad (4.34)$$

and t denotes the aspect ratio l/d of the short fiber.

For the special case of spherical inclusions ($h(t) = \frac{2}{3}$), the composite is isotropic and Eqs. (4.32) and (4.33) are simplified as

$$\left[\frac{V_f}{k_f} + \frac{V_m}{k_m} - \frac{2V_f V_m \left(\frac{1}{k_f} - \frac{1}{k_m}\right)^2}{2(V_m - V_f)\left(\frac{1}{k_f} - \frac{1}{k_m}\right) + 3\left(\frac{V_f}{k_f} + \frac{V_m}{k_m}\right)} \right]^{-1} \leq k \leq V_f k_f + V_m k_m$$

$$- \frac{V_f V_m (k_f - k_m)^2}{(V_f - V_m)(k_f - k_m) + 3(V_f k_f + V_m k_m)} \qquad (4.35)$$

In the case of continuous fibers, $h(t) = 1$ and Eqs. (4.32) and (4.33) become

$$k_{11} = V_f k_f + V_m k_m \qquad (4.36)$$

$$\frac{(k_m + k_f)k_m k_f}{(V_f k_m + V_m k_f)^2 + k_m k_f} \leq k_{22}(= k_{33}) \leq \frac{(V_m k_m + V_f k_f)^2 + k_m k_f}{k_m + k_f}$$

$$(4.37)$$

Figure 4.16 illustrates the variations of k_{11}/k_m with fiber volume fraction of an E-glass/epoxy system for the limiting cases of $l/d \to \infty$ and $l/d = 1$. The bounds of k_{11} converge to a single line for continuous fibers as indicated by Eq. (4.36).

For axially symmetrical fiber arrangement at an angle θ with respect to the x_1 axis, the fiber orientation effect can be investigated as in Section 4.3.2. By transforming the effective thermal conductivity tensor k_{ij} based upon the $[T]$ matrix of Eq. (4.17) and subsequently integrating the tensor components over the 2π range of φ, the resulting components are transversely isotropic with respect to the x_2–x_3 plane:

$$k_{11}^* = k_{11} \cos^2 \theta + k_{22} \sin^2 \theta$$

$$k_{22}^* = k_{33}^* = \tfrac{1}{2} k_{11} \sin^2 \theta + k_{22}\left(\frac{1 + \cos^2 \theta}{2}\right) \tag{4.38}$$

By substituting the bounds of k_{ij} (Eqs. (4.32) and (4.33)) into the above expressions, the bounds of thermal conductivity can be expressed as functions of fiber orientation θ. Again, Eq. (4.18) can be used to find the effective thermal conductivity of a composite with a given $\eta(\theta)$.

For completely random fiber orientation, the result can be simplified to

$$k_{11}^* = k_{22}^* = k_{33}^* = \frac{k_{11}}{3} + \frac{2k_{22}}{3} \tag{4.39}$$

Fig. 4.16. The variation of the upper and lower bounds of k_{11}/k_m with V_f for an E-glass/epoxy system. (After Nomura and Chou 1980.)

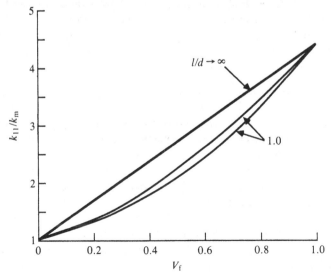

4.4.2   *Thermoelastic constants*

Knowledge of the thermoelastic constants, including thermal expansion coefficients and thermal stress coefficients, is basic to the understanding of the hygrothermal effects in composites. So far as it is assumed that these quantities obey the linear constitutive equation, their solutions can be obtained in a manner similar to the determination of effective elastic moduli or thermal conductivities. The problem of effective thermoelastic constants for non-homogeneous materials has been investigated by several researchers. The works of Kerner (1956), Levin (1967), Schapery (1968) and Budiansky (1970) are mainly concerned with composites reinforced with spherical inclusions. Rosen and Hashin (1970) extended Levin's model of a binary composite to general anisotropic composites by adopting a variational approach. Laws (1973) studied the thermoelastic behavior of anisotropic composites based upon Hill's self-consistent approximation.

By focussing attention on thermostatics and considering composites at uniform temperature, heat conduction can be excluded and the problem is uncoupled with that given in Section 4.4.1. Consider a composite subjected to a stress field, σ, and a uniform temperature rise, ΔT. The total strain of the elastic medium is given as

$$\varepsilon = \mathbf{S}\sigma + \alpha \Delta T \tag{4.40}$$

where $\mathbf{S}$ denotes the elastic compliance tensor, and α is the thermal expansion coefficient. The constitutive relation of the thermal elastic field can also be expressed in the following general form:

$$\sigma = \mathbf{C}(\varepsilon - \alpha \Delta T) \tag{4.41}$$

where $\mathbf{C}$ is the elastic stiffness tensor.

Nomura and Chou (1981) have shown that for composites reinforced with ellipsoidal inclusions and exhibiting statistical homogeneity, the effective thermoelastic constants can be evaluated following the technique for deriving elastic moduli. Figure 4.17 shows the variation of α_{ij} (normalized by the fiber thermal expansion coefficient α_f) with V_f and fiber aspect ratio l/d for a glass/epoxy system, assuming $E_f = 72.3$ GPa, $E_m = 2.76$ GPa, $\nu_m = 0.35$, $\nu_f = 0.2$, $\alpha_m = 36 \times 10^{-6}/°C$ and $\alpha_f = 5.04 \times 10^{-6}/°C$. At a given fiber volume fraction, the thermal expansion coefficient along the fiber direction (α_{11}) is smaller than that transverse to the fiber direction (α_{22}). Figure 4.18 shows a comparison of the theoretical

Fig. 4.17. The variation of α_{ij}/α_f with V_f and l/d for an E-glass/epoxy system. —— $l/d = 1$; —·— $l/d = 5$; ---- $l/d = \infty$. (After Nomura and Chou 1981.)

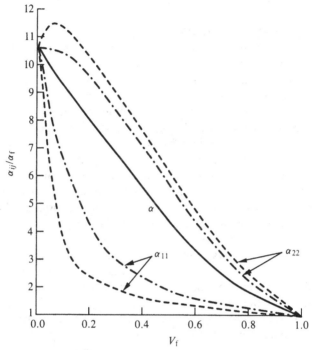

Fig. 4.18. Comparison of the predicted α_{22}/α_f with experimental data of Yates *et al.* (1978).

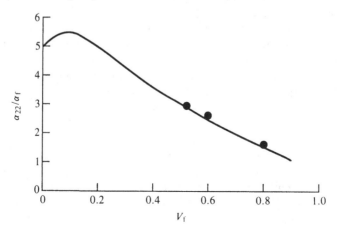

prediction of Nomura and Chou with the experimental results of Yates *et al.* (1978) on a carbon/epoxy system where $E_f/E_m = 53.4$, $v_m = v_f = 0.34$, $\alpha_m = 5 \times 10^{-5}/°C$ and $\alpha_f = 0.5-1.9 \times 10^{-5}/°C$.

4.5   Viscoelastic properties

The viscoelastic properties of composite materials were first examined by Hashin (1965b, 1969, 1972), who dealt with matrices reinforced with spherical inclusions and continuous fibers. Hashin showed that viscoelastic problems in composite materials can be solved by considering the corresponding problems in elasticity. Although application of the elastic–viscoelastic correspondence principle (see, for example, Christensen 1971) is well known, there are practical difficulties. This is due to the fact that very often the creep compliances or relaxation moduli of the constituents of a multi-component system are not known, and, even if they are given, the inverse transformation process would be formidable. Approximate methods for inverting the Laplace transform have been proposed by Schapery (1967, 1974).

The work of Laws and McLaughlin (1978) on viscoelastic composite materials adopted a self-consistent approximation. They derived the creep compliance, and numerical calculations were performed for the limiting cases of composites containing spherical inclusions and continuous fibers. Eimer (1971) derived formal effective relaxation moduli expressions of multi-phase media by considering the many point correlation functions.

Chou and Nomura (1980) and Nomura and Chou (1985) obtained the effective relaxation moduli of short-fiber composites based upon their work on effective elastic properties. Explicit expressions of composite relaxation moduli are given in terms of the elastic and viscoelastic properties of the constituent phases, fiber volume fraction, and fiber aspect ratio. Numerical calculations for a typical glass/epoxy composite system based upon the collocation approximation method as well as the self-consistent model have been performed by Nomura and Chou. It is assumed that the fiber is elastic while the matrix phase is viscoelastic. Figure 4.19 shows the time dependence of the effective axial Young's modulus of relaxation (normalized by the initial value of the matrix Young's modulus) for the fiber volume fraction of $V_f = 0.2$ and fiber aspect ratios $l/d = 5$ and ∞. The matrix behavior is shown by the lowermost curve in Fig. 4.19. The effective axial Young's modulus of relaxation at each fiber aspect ratio is calculated from the effective relaxation moduli (upper curve), the self-consistent model (middle curve), and the effective creep compliances (lower curve). The self-consistent approximation always lies in between the predictions of the two other approaches. The results also indicate that the increase in fiber length or aspect ratio makes the effective axial Young's modulus of relaxation less sensitive to the time effect. The fiber length effect also has been examined by Nomura and Chou for other effective moduli, i.e. the transverse Young's modulus of relaxation and the shear relaxation modulus, and they found no such sensitivity for these effective relaxation moduli, as in the elastic case.

4.6 Strength

Unlike continuous-fiber composites the mechanical behavior of short-fiber composites is often dominated by complex stress distributions due to fiber discontinuities. In particular, the local stress concentration at fiber ends plays a critical role in affecting the performance of short-fiber composites, and it often reduces the strength of a short-fiber composite to a level far less than that of a continuous-fiber composite with the same fiber volume content. Several theories (see Vinson and Chou 1975) have been proposed to predict the strength of discontinuous-fiber com-

Fig. 4.19. Time dependence of effective axial Young's modulus E_L/E_m for $l/d = 5$ and ∞ and $V_f = 0.2$. The viscoelastic material properties are $E_m(t) = E_m(0) = 3.2$ GPa, $E_m(\infty) = 0.04$ GPa, $\nu_m(0) = 0.365$, $\nu_m(\infty) = 0.485$, $E_f = 71.5$ GPa and $\nu_f = 0.2$. t denotes time. For each l/d value, the upper, middle and lower curves are obtained from the effective relaxation moduli, self-consistent model and effective creep compliances, respectively. (After Nomura and Chou 1985.)

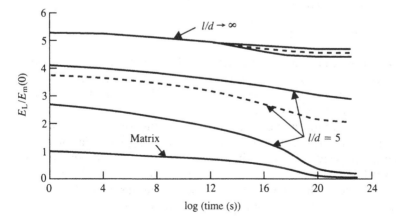

posites. One type of theory is based on a modification of the 'rule-of-mixtures', which was originally developed for continuous-fiber composites. Since the axial stress distribution in a short fiber is not uniform, the rule-of-mixtures has been modified by researchers to take into account the effect of fiber length.

Among short-fiber composites, aligned-fiber composites have many attractive properties (see Edward and Evans 1980; Richter 1980; Manders and Chou 1982). When complicated shapes and double curvatures are fabricated by matched-die molding techniques, aligned short-fiber composites have an advantage over their equivalent continuous mats (Kacir and Narkis 1975). The ability of aligned-fiber composites to elongate both parallel and perpendicular to the fiber direction without splitting complements the predominant shear deformation of woven materials. Because of their useful properties, highly aligned short-fiber composites have been commercially produced by the centrifuge (Edward and Evans 1980) and hydrodynamic alignment (Richter 1980) processes.

In the following, the strength of short-fiber composites is discussed first for the case of aligned fibers. Then, the effect of fiber orientation is considered for partially aligned and random fiber arrangements.

4.6.1   *Unidirectionally aligned short-fiber composites*

To examine the strength of short-fiber composites it is necessary to recall the original strength predictions developed by Kelly and co-workers (see Kelly and Davies 1965; Kelly and Tyson 1965a&b; Kelly 1971; Hale and Kelly 1972) for continuous-fiber composites. The ultimate axial tensile strength expression of Kelly *et al.* is (see Section 3.2)

$$\sigma_{cu} = \sigma_{fu}V_f + \sigma'_{mu}(1 - V_f) \tag{4.42}$$

where σ_{cu} and σ_{fu} are the ultimate tensile strengths of the composite and the fiber, respectively. σ_{fu} is identical with the fracture strength of brittle fibers. σ'_{mu} denotes the stress in the matrix at the failure strain of the composite.

Equation (4.42) was derived based upon the assumptions that the tensile strain in the composite is uniform along the axial direction and the applied load is distributed among the fibers and the matrix. When fibers are discontinuous, the iso-strain condition of Eq. (4.42) is no longer valid. The difference of the strains in the fiber and matrix near a fiber end induces shear stresses along the fiber axis.

The shear forces acting near both ends of a fiber stress the fiber in tension or compression. It is through this transferring of stress that applied load can be dispersed among the short fibers.

4.6.1.1 *Fiber length considerations*

Figure 4.20 shows schematically the variation of fiber axial tensile stress with fiber length. The profile of linear stress variation from fiber ends originates from the assumption of constant interfacial shear stress. The fiber critical length l_c is defined as the minimum fiber length necessary to build up the axial stress to σ_{fu}. The ultimate strength of a short fiber can be realized if its length reaches l_c.

Kelly and Tyson (1965a) proposed a linear transfer of stress from the tip of a fiber to a maximum value when the strain in the fiber is equal to that in the matrix. By assuming constant interfacial stress τ, the fiber critical length can be easily derived by considering the balance of tensile and shear stresses:

$$\frac{l_c}{d} = \frac{\sigma_{fu}}{2\tau} \tag{4.43}$$

τ is the shear strength of either the matrix or the interface, whichever is smaller. Experimental measurement techniques for l_c have been discussed by Vinson and Chou (1975).

Using the concept of critical fiber length and replacing σ_{fu} in Eq. (4.42) by the average fiber stress $\bar{\sigma}_f$, Kelly (1973) derived the following expression of composite strength for $l \geq l_c$:

$$\begin{aligned}
\sigma_{cu} &= \bar{\sigma}_f V_f + \sigma'_{mu}(1 - V_f) \\
&= \sigma_{fu}[1 - (1 - \delta)l_c/l] + \sigma'_{mu}(1 - V_f)
\end{aligned} \tag{4.44}$$

where δ is defined as the ratio of the area under the stress distribution curve over the length $l_c/2$ in Fig. 4.20 to the area of

Fig. 4.20. Variations of fiber tensile stress with fiber length.

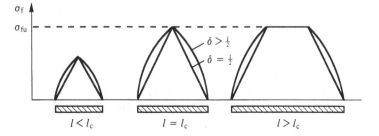

$\sigma_{fu}l_c/2$. For constant interfacial shear strength, $\delta = \frac{1}{2}$ and

$$\sigma_{cu} = \sigma_{fu}(1 - l_c/2l)V_f + \sigma'_{mu}(1 - V_f) \qquad l \geq l_c$$

$$\sigma_{cu} = \sigma_{fu}V_fl/2l_c + \sigma'_{mu}(1 - V_f) \qquad l \leq l_c$$

(4.45)

Equations (4.45) predict that for short fibers with $l/l_c = 10$, $\bar\sigma_f$ reaches 95% of the value for continuous fibers. Equations (4.45) have been shown to be a good approximation for metallic (Kelly and Tyson 1965a&b; Kelly 1973) and polymer matrices (Kelly 1973; Riley and Reddaway 1968; Hancock and Cuthbertson 1970). It should be noted that Eqs. (4.45) do not consider fiber end stress concentration which occurs in short-fiber composites. There exist several variants of Kelly's formulation of short-fiber composite strength. For example, Outwater (1956) has taken into consideration the effect of interfacial friction load due to resin cure shrinkage. However, there lies the difficulty of measuring the friction coefficient and radial shrinkage pressure (Kardos 1973).

For pure elastic deformation of the fiber, $\sigma_{fu} = E_f\varepsilon_{cu}$ where ε_{cu} is the composite ultimate strain. Equation (4.43) can be rewritten as

$$\frac{l_c}{d} = \frac{E_f\varepsilon_{cu}}{2\tau}$$

(4.46)

For composites with variation in fiber length, Bowyer and Bader (1972) pointed out that at any value of composite strain ε_c there is a critical fiber length given by

$$l_\varepsilon = \frac{E_f\varepsilon_c d}{2\tau}$$

(4.47)

Fibers shorter than l_ε will carry the average stress

$$\bar\sigma_f = \frac{l\tau}{d}$$

(4.48)

which is always less than $\frac{1}{2}E_f\varepsilon_c$. Fibers longer than l_ε carry the average stress

$$\bar\sigma_f = E_f\varepsilon_c\left(1 - \frac{E_f\varepsilon_c d}{4l\tau}\right)$$

(4.49)

which is always greater than $\frac{1}{2}E_f\varepsilon_c$.

Following Bowyer and Bader, for a composite containing a spectrum of fibers of different lengths, its strength can be estimated by dividing the length of fibers into sub-fractions at a given composite strain level (Lees 1968). Sub-critical fractions are denoted by l_i and their respective volume fractions V_i while super-critical fractions are denoted by l_j and V_j. Thus the composite stress can be expressed as

$$\sigma_c = \sum_i^{l_i \le l_\epsilon} \frac{\tau l_i V_i}{d} + \sum_j^{l_j \ge l_\epsilon} E_f \varepsilon_c \left(1 - \frac{E_f \varepsilon_c d}{4 l_j \tau}\right) V_j + E_m \varepsilon_c (1 - V_f) \quad (4.50)$$

Equation (4.47) indicates that at low composite strain l_ϵ is small and all fibers will contribute to the reinforcement as given by Eq. (4.49). As the strain is increased, a progressively smaller proportion of the fibers will reinforce according to Eq. (4.49) and an increasing proportion will follow Eq. (4.48). Thus, the load-extension curve for such a material as indicated by Eq. (4.50) is expected to show smaller slope as the strain is increased. The work of Bowyer and Bader on short-fiber-reinforced thermoplastics has further shown that improvements in the fiber–matrix bond strength have led to small improvements in strength. Also the fibers which are too short to be strained coherently with the matrix tend to fail at very low strains preventing the potential of the longer fibers from being realized. Thus the very short fibers should be eliminated if full strengthening potential is to be achieved.

4.6.1.2 *Probabilistic strength theory*

The following discussions of the probabilistic strength theory of short-fiber composites begin with a consideration of fiber length variations and their effect on fiber axial stress distribution. Then, the influence of local stress concentrations due to fiber–fiber interaction is introduced. A probabilistic strength theory is developed to consider the maximum stress concentration induced by the clustering of ends of short fibers.

(A) Modification of the rule-of-mixtures

Consider a unidirectional short-fiber composite material with fibers of uniform length and strength. The mechanisms of failure can be categorized according to fiber length (Fig. 4.21). When fibers are very short, a crack formed at a fiber end can circumvent the neighboring fibers without breaking them (Fig. 4.21a). Final failure of the composite is then attributed to fiber pull-out. On the other hand, if fibers are sufficiently long, fiber end cracks will cause fracture of the neighboring fibers and, hence, failure of the composite (Fig. 4.21b). The strength model of Fukada

Fig. 4.21. Two failure modes in short-fiber composites. (After Fukuda and Chou 1981b.)

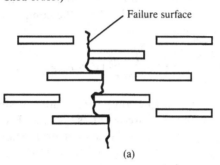

(a)

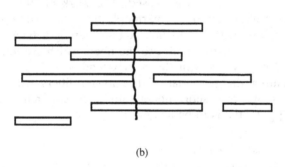

(b)

Fig. 4.22. Stress distribution in a short fiber. (After Fukuda and Chou 1981b).

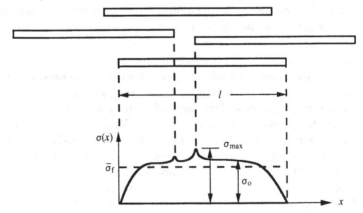

and Chou (1981a&b), and Hikami and Chou (1984a&b) aim at the latter case.

The composite ultimate strength σ_{cu} is defined as the stress level which causes first fiber fracture. Consequently, the maximum stress in a fiber is of primary importance in predicting composite strength. Figure 4.22 shows schematically stress distributions in a short fiber. Here σ_{max} and σ_o are, respectively, the maximum and plateau stress of the profile. The average fiber stress at failure is given by

$$\bar{\sigma}_f = \frac{1}{l} \int_0^l \sigma(x) \, dx \tag{4.51}$$

In the case the composite has a distribution of fiber length, Eq. (4.51) should be replaced by

$$\bar{\sigma}_f = \int_0^\infty f(l) \left\{ \frac{1}{l} \int_0^l \sigma(x) \, dx \right\} dl \tag{4.52}$$

where $f(l)$ is a probability density function of fiber length and has the following characteristics:

$$\int_0^\infty f(l) \, dl = 1 \tag{4.53}$$

$$\int_0^\infty f(l)l \, dl = \bar{l} \tag{4.54}$$

$\bar{l}$ in Eq. (4.54) denotes the average fiber length. Then $\bar{\sigma}_f$ of Eq. (4.52) should be used in the rule-of-mixtures expression of Eq. (4.44). The values of $\bar{\sigma}_f$ and σ_o are not the same. However, the difference diminishes as the fiber length increases. For relatively large fiber aspect ratios it is reasonable to assume $\bar{\sigma}_f \approx \sigma_o$. Furthermore, by defining the stress concentration factor K in the following expression:

$$\sigma_{max} = \sigma_{fu} = K\sigma_o \tag{4.55}$$

Eq. (4.44) can be written as

$$\sigma_{cu} = \frac{\sigma_{fu}}{K} V_f + \sigma'_{mu}(1 - V_f) \tag{4.56}$$

(B)   Critical zone model

A systematic experimental study of short-fiber composite strength has been performed by Curtis, Bader and Bailey (1978) using polyamide thermoplastic reinforced with short glass and carbon fibers. Their experimental findings led Bader, Chou and Quigley (1979) to propose a damage model. The basic concepts are that microcracks are most likely to develop at fiber ends at microscopic strains well below the fiber failure strain, and that failure is finally initiated in a critical cross-section that has been weakened by the accumulation of cracks.

Figure 4.23 depicts a typical volume element in a short-fiber composite used by Bader, Chou and Quigley. The width of a 'critical zone' in the strength model is denoted by βl where $0 < \beta \leq 1$ is a constant parameter and l is the average fiber length. The critical zone width is assumed to be of the same order as the fiber ineffective length (Sections 3.4.6.1 and 4.2.2).

A discontinuous fiber can end in the zone (ending fiber), in which case it bears no load, or it can bridge the zone (bridging fiber) and contribute to the strength of the critical zone. The probabilities of finding an ending fiber and a bridging fiber are β and $1 - \beta$, respectively. All fibers are assumed to have uniform strength σ_{fu}. Within each transverse section of the composite, ending fibers and bridging fibers are distributed randomly. A typical fiber configuration on a transverse section in a two-dimensional fiber array is shown in Fig. 4.24. The ending fibers and bridging fibers are depicted, respectively, by solid circles and open circles. Under the applied stress, the stress in the bridging fibers is enhanced by the

Fig. 4.23. A typical critical zone in a short-fiber composite. (After Bader, Chou and Quigley 1979).

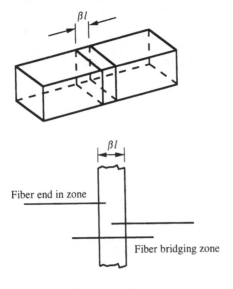

stress transferred from the neighboring ending fibers. For example, the stress in the bridging fiber no. 8 in this figure is enhanced by the ending fibers nos. 1, 5, 6, 7, 9, 12 and 13. In other words, it is enhanced by the neighboring fiber-end-gaps *A, B, C* and *D*.

The strength of the composite is determined by the relative numbers of fibers that bridge the zone vs. those with ends within the zone. These latter will develop matrix cracks when the strain exceeds a critical value. The critical situation arises when the bridging fibers are unable to sustain the load transfer due to matrix cracking and failure occurs. The critical stress and strain values for a wide range of fiber aspect ratio, fiber critical length, fiber–matrix interfacial strength and critical zone width have been evaluated by Bader, Chou and Quigley.

(C)   Stress concentration

The stress concentration factor for the unidirectional fiber arrangement of Fig. 4.25 is difficult to evaluate in a precise manner. The following assumptions are adopted to facilitate the calculation of K: (a) fibers are of the same length, l; (b) they are arranged in rows along the axial direction; (c) the spacing between two neighboring rows is uniform (Fig. 4.25a); and (d) fibers with ends in the critical zone of width βl are assumed to have the ends aligned along the cross-section zz' (Fig. 4.25b). This collection of fiber ends is termed a 'fiber-end-gap' in a two-dimensional array. It is assumed that the fiber length l is much larger than the critical length l_c and, hence, results for stress concentrations due to the fracture of long fibers can be used. Also, in Fig. 4.25(a), the number 1 and number 4 fibers are labeled as 'bridging fibers' and the number 2 and number 3 fibers as 'ending fibers'.

Since the stress concentration factor, K, cannot be readily calculated by considering the enhancement effect from all the fiber-end-gaps, assumptions need to be introduced for the load sharing rule. Hikami and Chou (1984a) have examined the first and

Fig. 4.24. Schematic cross-sectional view of fiber configuration. Solid circles depict ending fibers and open circles indicate bridging fibers. A group of adjacent ending fibers is termed a fiber-end-gap (i.e. *A, B, C* and *D*). (After Hikami and Chou 1984a.)

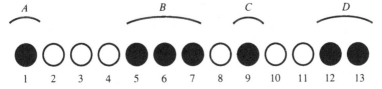

simplest approximation for K by only considering the stress enhancement effects of the first nearest neighboring fiber-end-gap of a bridging fiber. This is known as the *weak local load sharing rule* and the assumption is allowable if the probability of finding the ending fibers is relatively small. Using the shear-lag method, the stress concentration factor due to the presence of n_l and n_r ending fibers (Fig. 4.26) has been obtained by Hikami and Chou (1984a and b, 1990).

It can be shown that the failure of the $(n_l + 1)$th fiber does not cause the composite failure since the stress concentration factor for the $(n_l + 1)$th fiber is larger than that for the zeroth bridging fiber after the failure of the $(n_l + 1)$th bridging fiber. Clearly, the failure of the zeroth bridging fiber causes the total failure of the composite. Thus neglecting the load bearing capacity of the matrix, the strength of the composite is given by

$$\sigma_{cu} = \sigma_{fu}/K_b \qquad (4.57)$$

The explicit expression of elastic stress concentration factor K_b due to b broken fibers is given in Section 3.3.1.2.

Fig. 4.25. (a) Critical zone in a two-dimensional fiber array. (b) A fiber-end-gap. (After Fukuda and Chou 1981b.)

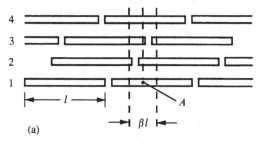

(a)

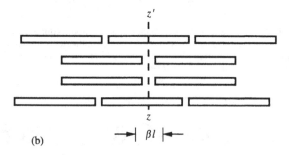

(b)

The explicit expression of stress concentration factor for composites with plastically deformed matrices (Fig. 4.26) has also been obtained by Hikami and Chou (1984a). For the small-scale plastic deformation case, the plastic stress concentration factor, $\tilde{K}_b$, can be expressed in series expansion form in terms of the dimensionless plastic deformation zone length α.

In the large-scale plastic deformation case, $\tilde{K}_b$ at the tip of a fiber-end-gap can be approximated by

$$\tilde{K}_b \cong 1 + \frac{2}{\pi}\left(\frac{T_o}{\sigma_a}\right)[nl(b\sigma_a/T_o) + \gamma'] \tag{4.58}$$

where

$$T_o = \tau_m\sqrt{(hE_f/G_mA_f)} \tag{4.59}$$

Fig. 4.26. Model of stress concentration calculations for a fiber-end-gap in short-fiber composites with matrix plastic deformation zone at the tip of the gap. (After Hikami and Chou 1984.)

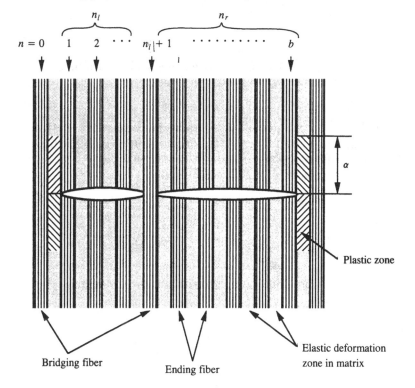

Also, σ_a = applied stress, b = number of fibers in the gap, γ' = Euler's constant (≈ 0.577), τ_m = matrix shear strength, G_m = matrix shear modulus, E_f = fiber axial Young's modulus, h = fiber spacing, and A_f = fiber cross-sectional area. The fibers are of unit thickness.

(D)   Probability distribution of maximum fiber-end-gap
The fiber-end-gap size has been analyzed by Hikami and Chou (1984a) for the case of the two-dimensional array shown in Fig. 4.25(b). Focussing attention on a single fiber end, the probability, P_n, that this fiber end is in the gap consisting of n fiber ends is

$$P_n = n\beta^{n-1}(1 - \beta)^2 \tag{4.60}$$

and

$$\sum_{n=1}^{\infty} P_n = 1 \tag{4.61}$$

The probability that a given fiber end is not in any one of the gaps with more than n fiber ends is

$$Q_n = 1 - \sum_{i=n+1}^{\infty} P_i \tag{4.62}$$

When the above probability is independent for each fiber, the probability that there is no gap larger than size n is

$$\bar{P}(n) = (Q_n)^N \tag{4.63}$$

where N is the total number of fibers in the composite. However, actually Q_n for a given fiber is not independent of the other fibers. When N is sufficiently larger than the average gap size, $\bar{n}$, it is more suitable to express $\bar{P}(n)$ of Eq. (4.63) as

$$\bar{P}(n) = (Q_n)^{N/\bar{n}} \tag{4.64}$$

where

$$\bar{n} = \sum_{n=1}^{\infty} nP_n \tag{4.65}$$

Using Eqs. (4.60) and (4.62), Eq. (4.64) can be rewritten as

$$\bar{P}(n) = \{1 - \beta^n[n(1 - \beta) + 1]\}^{N\beta'} \tag{4.66}$$

and

$$\beta' = \frac{1}{\bar{n}} = \frac{1 - \beta}{1 + \beta} \tag{4.67}$$

$\bar{P}(n)$ can be used to determine the strength of short-fiber composites through the relation between gap size, n, and the corresponding stress concentration. Figure 4.27 demonstrates the variation of $\bar{P}(n)$ with N and β. It can be shown that $\bar{P}(n)$ behaves like a step function and $\bar{P}(n)$ changes from 0 to 1 at $n \cong M$, where M is determined from

$$\beta^n[n(1-\beta)+1]N\beta' = 1 \tag{4.68}$$

M obtained from Eq. (4.68) is termed the 'most probable maximum gap size'. Figure 4.28 shows M as a function of β and N. For actual composites, the values of M do not vary tremendously with β and N. When N is sufficiently large, using the formula $1 - x \cong \exp(-x)$, $\bar{P}(n)$ can be approximated as

$$\bar{P}(n) \cong \exp[-N\beta^n n(1-\beta)^2] \tag{4.69}$$

(E)   Strength predictions
       Based upon the considerations of fiber-end-gap size and stress concentrations, Hikami and Chou (1984a) have proposed a modification of the rule-of-mixtures for composite strength. The composite ultimate strength σ_{cu} is defined as the stress level at which fracture of the composite occurs. Based upon the approxima-

Fig. 4.27. Cumulative probability distribution functions for the maximum fiber-end-gap size. ○: $N = 10^6$, $\beta = 0.2$; ●: $N = 10^6$, $\beta = 0.1$; △: $N = 10^8$, $\beta = 0.2$. (After Hikami and Chou 1984a.)

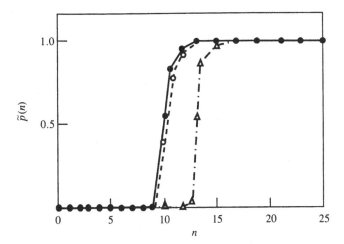

tions discussed above, σ_{cu} is given as

$$\sigma_{cu} = \sigma_a V_f + \sigma'_{mu}(1 - V_f) \qquad (4.70)$$

Here, σ'_{mu} is the matrix stress at the ultimate tensile strain of the fiber. σ_a is the applied fiber stress at the instant when the fiber stress at the site of stress concentration reaches σ_{fu}. Thus, σ_a satisfies the following relation:

$$\sigma_{fu} = K[\sigma_a - \eta \sigma_{my}(1 - V_f)/V_f] \qquad (4.71)$$

for the weak local load sharing rule, where $K = K_b$ or $\bar{K}_b$. σ_{my} is the matrix yield strength. The parameter η in Eq. (4.71) reflects the loading condition of the matrix in the fiber-end-gap. If the matrix is brittle, a crack can propagate in the matrix along the fiber-end-gap prior to the failure of the intact bridging fiber. In this case, the matrix in the fiber-end-gap will bear no load and η is taken to be zero. However, in a ductile matrix composite the matrix in the fiber-end-gap can deform plastically to the yield strength, σ_{my}. Then each fiber in the fiber-end-gap sustains the stress $\sigma_{my}(1 - V_f)/V_f$, thus reducing the applied stress σ_a, and $\eta = 1$. Since the fracture of a composite initiates at the weakest point, the stress concentration factor for the most probable maximum gap size M of Eq. (4.68) should be used.

Fig. 4.28. Most probable maximum gap size, M, vs. critical zone parameter, β. N denotes the total number of fibers. (After Hikami and Chou 1984a.)

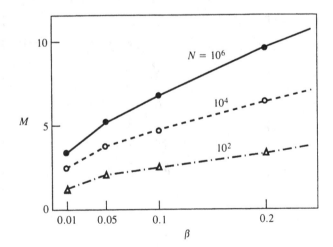

In the case of three-dimensional fiber arrays, the problem is more complicated and there is no rigorous probabilistic treatment available. The shape of the fiber-end-gap cannot be uniquely defined for a given number of fiber ends and it is fairly involved to obtain the highest stress concentration factor in the intact bridging fibers. Furthermore, the fiber failure process here is more complex than that in the two-dimensional case. To circumvent these difficulties, Fukuda and Chou (1981b) took only compact fiber-end-gaps as the first approximation. Following this approximation, Hikami and Chou (1984a) have considered the special type of fiber-end-gap which consists of square-arrayed ending fibers. A typical example of such a fiber-end-gap is shown in Fig. 4.29, where ending fibers are indicated by solid circles and bridging fibers by open circles in the two-dimensional square lattice. Approximations for the most probable maximum gap size and the resulting composite strength have been obtained and the details can be found in the reference.

The relation between the fiber volume fraction, V_f, and composite strength normalized by the matrix stress at failure, σ_{cu}/σ'_{mu}, is shown in Fig. 4.30 for the case of an elastic matrix. The properties of a glass fiber/thermoplastic matrix composite are used; fiber length $(l) = 1\,\text{mm}$; fiber diameter $(d) = 0.01\,\text{mm}$; fiber critical length $(l_c) = 0.1\,\text{mm}$; and critical zone parameter $(\beta) = 0.1$. Also $\sigma_{fu}/\sigma'_{mu} = E_f/E_m = 35.2$.

In Fig. 4.30, line A shows the simple rule-of-mixtures for continuous fibers, while line B depicts the rule-of-mixtures modified for short fibers. Neither case takes the effect of local stress

Fig. 4.29. Schematic cross-sectional view of a three-dimensional fiber array. Solid circles indicate ending fibers and open circles are for bridging fibers. (After Hikami and Chou 1984a.)

Square-packed fiber-end-gap

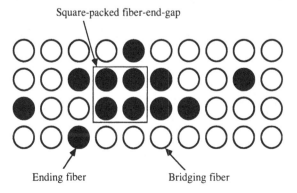

Ending fiber   Bridging fiber

concentrations into consideration. Lines C and E indicate the results of Hikami and Chou (1984a) for a three-dimensional fiber array and a two-dimensional fiber array, respectively, based on the local load sharing rule. The composite strength is expected to lie between these two bounds, which are far less than the values obtained from the rule-of-mixtures because of local stress concentrations.

4.6.2   *Partially oriented short-fiber composites*

Cox (1952) first proposed the idea of orientation factor in the strength equation for continuous fiber composites to account for fiber misalignment. Bowyer and Bader (1972) adopted this concept in their study of short-fiber systems, and Eq. (4.50) was modified by multiplying the fiber dependent terms on the right-hand side of this equation by the orientation factor C_o, $C_o = 1$ for perfectly aligned fibers and C_o assumes values less than unity for partially oriented fibers. Bowyer and Bader concluded that the orientation factor is independent of strain and is the same for all fiber length at least at small strains. The orientation factor can then be calculated from Eq. (4.50) based upon the knowledge of fiber length distribution, interfacial bond strength and composite ultimate tensile strength.

Curtis, Bader and Bailey (1978) investigated the strength of a

Fig. 4.30. Strength of the composite as a function of V_f. A: rule-of-mixtures; B: Kelly and Tyson (1965b); C: three-dimensional fiber array, weak local load sharing; E: two-dimensional fiber array, weak local load sharing. (After Hikami and Chou 1984a.)

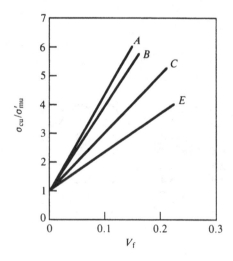

polyamide thermoplastic reinforced with glass and carbon fibers, and calculated the fiber orientation factor from the measured composite modulus and the knowledge of the fiber and matrix properties. Their results indicate that fiber alignment increases with increasing fiber volume fraction, which agrees with the qualitative assessment of optical micrographs.

In general, when there are variations in both fiber length and orientation, the rule-of-mixtures (Eq. (4.42)) can be modified as

$$\sigma_{cu} = \sigma_{fu} V_f F(l_c/\bar{l}) C_o + \sigma'_{mu}(1 - V_f) \qquad (4.72)$$

Here, the factor $F(l_c/\bar{l})$ is a function of fiber average length $\bar{l}$ and critical length l_c. Equations (4.45), for instance, give the forms of $F(l_c/\bar{l})$ for aligned short fibers of uniform length. If the necessary information with respect to fiber orientation is known, C_o can be estimated analytically.

Fukuda and Chou (1982) have used a probabilistic theory to predict the strength of short-fiber composites with variable fiber length and orientation. They introduced two kinds of probability density functions to describe the fiber length and orientation distributions and neglected the effect of stress concentration in this particular treatment. The analytical result of composite strength is given only in the form of an average value. The theory of Fukuda and Chou is introduced below in three parts.

(A)   Geometrical consideration of a single short fiber

First, the geometrical arrangement of a single short fiber is described. Figure 4.31(a) shows an obliquely positioned short fiber

Fig. 4.31. Several notations on short-fiber arrangement. (a) Obliquely oriented fiber. (b) Bridging fiber and ending fiber. (c) Critical angle. (After Fukuda and Chou 1982.)

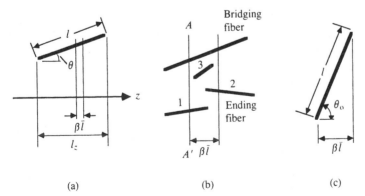

(a)   (b)   (c)

of length l. In accordance with the terminology of Section 4.6.1, a bridging fiber and an ending fiber are defined in Fig. 4.31(b); that is, if a fiber crosses a critical zone (Section 4.6.1.2) of width $\beta\bar{l}$, it is termed a bridging fiber; and if the end of a fiber is within the critical zone, it is defined as an ending fiber. Here, $\bar{l}$ denotes average fiber length. The probability density function of fiber length distribution $h(l)$ satisfies the following condition:

$$\int_0^\infty h(l)\,dl = 1 \tag{4.73}$$

Then, the average fiber length is defined as

$$\bar{l} = \int_0^\infty lh(l)\,dl \tag{4.74}$$

From Fig. 4.31(a),

$$l_z = l\cos\theta \tag{4.75}$$

and from Fig. 4.31(c) the critical angle θ_o within which a fiber of length l is a bridging fiber becomes

$$\theta_o = \cos^{-1}\beta\bar{l}/l \tag{4.76}$$

for $\beta\bar{l} \le l$. If $\beta\bar{l} > l$, θ_o cannot be defined, and a fiber in such a case is inevitably an ending fiber. If the fibers are distributed randomly with respect to the z axis, the probability p_e that a fiber of length l is an ending fiber in the critical zone becomes

$$p_e = \frac{\beta\bar{l}}{l_z} = \begin{cases} \beta\bar{l}/l\cos\theta & (0 \le \theta \le \theta_o \text{ and } \beta\bar{l} \le l) \\ 1 & (\theta_o \le \theta \le \pi/2 \text{ or } \beta\bar{l} \ge l) \end{cases} \tag{4.77}$$

and the probability p_b for finding a bridging fiber is, by definition,

$$p_b = 1 - p_e \tag{4.78}$$

The probability density function with respect to fiber orientation $(g(\theta))$ should satisfy the condition

$$\int_0^{\pi/2} g(\theta)\,d\theta = 1 \tag{4.79}$$

(B)   Load transfer in a short fiber
First, consider a short fiber situated parallel to the applied tensile stress, σ_o, along the z axis. The average fiber stress is

$$\sigma_{fo} = \frac{1}{l}\int_0^l \sigma_f(z)\,dz \tag{4.80}$$

The fiber axial stress $\sigma_f(z)$ has, in general, the profile shown in Fig. 4.1. Consider the simplest form of $\sigma_f(z)$ by assuming a constant interfacial shear stress (Fig. 4.20). Then σ_{fo} becomes

$$\sigma_{fo} = \begin{cases} \sigma_{fu}\left(1 - \dfrac{l_c}{2l}\right) & (l > l_c) \\[2ex] \sigma_{fu}\left(\dfrac{l}{2l_c}\right) & (l < l_c) \end{cases} \tag{4.81}$$

The average force in a fiber of cross-sectional area A_f is $\sigma_{fo}A_f$.

Next, consider a single short fiber situated at an angle θ to the applied stress σ_o. The applied stress can be decomposed into an axial and a shear component, with respect to the fiber axis, as

$$\sigma_o' = \sigma_o \cos^2 \theta \tag{4.82}$$

$$\tau_o' = \sigma_o \sin \theta \cos \theta \tag{4.83}$$

If the effect of τ_o' on the fiber stress distribution can be neglected, the average force of the fiber becomes $A_f\sigma_{fo}\cos^2 \theta$ and the z direction force component is

$$F_z = A_f\sigma_{fo}\cos^3\theta \tag{4.84}$$

(C)   Strength of short-fiber composites
        Based upon the above preparations, the strength of short-fiber composites can be derived. In the following discussion, $h(l)$ and $g(\theta)$ are assumed to be independent of each other. This means that $g(\theta)$ is the same for all the samples with different fiber length distributions. A rectangular-shaped specimen with the lengths of the three mutually perpendicular edges denoted by a, b and c is considered. The c axis is so chosen as to be parallel to the z axis. The volume of the specimen is

$$V = abc \tag{4.85}$$

and from the definition of fiber volume fraction, V_f becomes

$$V_f = NA_f\bar{l}/V \tag{4.86}$$

where N and A_f denote, respectively, the total number of fibers and fiber cross-sectional area.

Recall that Eq. (4.76) gives the length of the projection of a fiber on the z axis. Then the average length of the projection of fibers can be written as

$$\bar{l}_z = \int_0^{\pi/2} \int_0^{\infty} l \cos\theta \, h(l) g(\theta) \, dl \, d\theta$$

$$= \bar{l} \int_0^{\pi/2} g(\theta) \cos\theta \, d\theta \tag{4.87}$$

The value of $N\bar{l}_z$ gives the total length of projection of all fibers on the z axis and if this value is divided by the specimen length c, the average number of fibers which cross an arbitrary section in the specimen normal to the z axis is obtained. That is,

$$N_c = \frac{N\bar{l}_z}{c} = \frac{abV_f}{A_f} \int_0^{\pi/2} g(\theta) \cos\theta \, d\theta \tag{4.88}$$

Equation (4.77) gives the probability of a specific fiber being an ending fiber. Therefore, the average probability of finding an arbitrary fiber being an ending fiber is

$$q_e = \int_0^{\pi/2} \int_0^{\infty} p_e h(l) g(\theta) \, dl \, d\theta \tag{4.89}$$

Similarly, the average probability of finding an arbitrary fiber being a bridging fiber is

$$q_b = \int_0^{\pi/2} \int_0^{\infty} p_b h(l) g(\theta) \, dl \, d\theta$$

$$= 1 - q_e \tag{4.90}$$

Substituting Eqs. (4.77) and (4.78) into Eqs. (4.89) and (4.90),

$$q_e = \int_0^{\theta_o} d\theta \left(\int_0^{\beta\bar{l}} g(\theta) h(l) \, dl + \int_{\beta\bar{l}}^{\infty} \frac{\beta\bar{l}}{l \cos\theta} g(\theta) h(l) \, dl \right)$$

$$+ \int_{\theta_o}^{\pi/2} \int_{\beta\bar{l}}^{\infty} g(\theta) h(l) \, dl \, d\theta \tag{4.91}$$

$$q_b = \int_{\beta\bar{l}}^{\infty} dl \int_0^{\theta_o} \left(1 - \frac{\beta\bar{l}}{l \cos\theta} \right) g(\theta) h(l) \, dl \tag{4.92}$$

Then, the total numbers of ending and bridging fibers in the specimen are

$$N_e = N_c q_e \tag{4.93}$$

$$N_b = N_c q_b \tag{4.94}$$

Strictly speaking, the value of N_e is not precise because only *one* cross-section, for example AA' in Fig. 4.31(b), has been examined. The fibers denoted by 2 and 3 in Fig. 4.31(b) are not considered. However, the objective is to calculate the number of bridging fibers, which is not affected by N_e in the subsequent discussions.

Based upon Eq. (4.84) for the z direction component of the axial load of one specific fiber, the average value among the bridging fibers is

$$\bar{F}_z = \int_0^{\theta_0} \int_{\beta\bar{l}}^{\infty} F_z h(l) g(\theta) \, dl \, d\theta \tag{4.95}$$

Then the total load that all of the bridging fibers can sustain in the zone $\beta\bar{l}$ is

$$F_T = N_b \cdot \bar{F}_z \tag{4.96}$$

and the composite strength becomes

$$\sigma_{cu} = \frac{F_T}{ab} + \sigma'_{mu}(1 - V_f) \tag{4.97}$$

where the matrix is assumed to sustain part of the applied load. Substituting Eqs. (4.81), (4.84), (4.88) and (4.91)–(4.96) into Eq. (4.97), the composite ultimate strength is determined as

$$\sigma_{cu} = \sigma_{fu} V_f \int_0^{\pi/2} g(\theta) \cos\theta \, d\theta \int_0^{\theta_0} g(\theta) \cos^3\theta \, d\theta$$

$$\times \int_{\beta\bar{l}}^{\infty} \int_0^{\theta_0} \left(1 - \frac{\beta\bar{l}}{l\cos\theta}\right) g(\theta) \, d\theta \, h(l) \, dl$$

$$\times \left[\int_{\beta\bar{l}}^{l_c} \frac{l}{2l_c} h(l) \, dl + \int_{l_c}^{\infty} \left(1 - \frac{l_c}{2l}\right) h(l) \, dl\right] + \sigma'_{mu}(1 - V_f) \tag{4.98}$$

Equation (4.98) is a general strength expression of short-fiber composites. In order to conduct further analysis, it is necessary to know the functions $g(\theta)$ and $h(l)$ together with σ_{fu}, σ'_{mu}, V_f and l_c.

Some limiting cases of Eq. (4.98) are discussed in the following. First, consider a unidirectional short-fiber composite with uniform fiber length $\bar{l}$. Equation (4.98) can be reduced to

$$\sigma_c = \sigma_{fu} V_f(1 - \beta)\left(1 - \frac{l_c}{2\bar{l}}\right) + \sigma'_{mu}(1 - V_f) \qquad (\bar{l} > l_c)$$

$$\tag{4.99}$$

$$\sigma_c = \sigma_{fu} V_f(1 - \beta)\frac{\bar{l}}{2l_c} + \sigma'_{mu}(1 - V_f) \qquad (\bar{l} < l_c)$$

Equations (4.99) coincide with the result of the original failure model of Bader, Chou and Quigley (1979).

Secondly, consider the effect of fiber length distribution on the strength of a unidirectional composite. By assuming the limiting case of $\beta \to 0$, namely all fibers are bridging, and the following probability density function of fiber length distribution

$$h(l) = \frac{\pi}{4\bar{l}} \sin\left(\frac{\pi l}{2\bar{l}}\right) \qquad (0 \le l/\bar{l} \le 2) \tag{4.100}$$

Eq. (4.98) is reduced to

$$\sigma_{cu} = \sigma_{fu} V_f \left\{ \frac{1}{2} + \frac{\bar{l}}{2\pi l_c} \sin\left(\frac{\pi l_c}{2\bar{l}}\right) + \frac{1}{4} \cos\left(\frac{\pi l_c}{2\bar{l}}\right) \right.$$
$$\left. - \frac{\pi l_c}{8\bar{l}} \left[\operatorname{Si}(\pi) - \operatorname{Si}\left(\frac{\pi l_c}{2\bar{l}}\right) \right] \right\} + \sigma'_{mu}(1 - V_f) \tag{4.101}$$

where $\operatorname{Si}(x)$ is the integral sine function defined by

$$\operatorname{Si}(x) = \int_0^x \frac{\sin t}{t} \, dt \tag{4.102}$$

The result of $F(l_c/\bar{l})$ from Eq. (4.101) is shown in Fig. 4.32 by the solid line. In the case of constant fiber length, the strength can be obtained from Eqs. (4.45) and the value is also shown in Fig. 4.32 by a broken line. It can be concluded from Fig. 4.32 that the strength of a composite material is reduced if the fiber length is not

Fig. 4.32. $F(l_c/\bar{l})$ vs. $l_c/\bar{l}$ —— fiber length distribution considered; – – – – fiber length assumed to be constant. (After Fukuda and Chou 1982.)

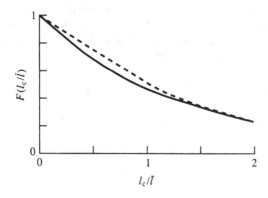

uniform. However, the difference in composite strength between the non-uniform fiber length system (Eq. (4.100)) and the uniform fiber length system is not very significant and, hence, the ordinary theory based upon an average fiber length may be used as a first approximation.

As a third example, the case of uniform fiber length and biassed fiber orientation distribution is considered. The following two types of fiber orientation are examined.

(a) $g(\theta) = 1/\alpha$ for $0 \le \theta \le \alpha$ and $g(\theta) = 0$ for $\theta > \alpha$;
(b) $g(\theta) = (\pi/2\alpha)\cos(\pi\theta/2\alpha)$ for $0 \le \theta \le \alpha$ and $g(\theta) = 0$ for $\theta > \alpha$.

These functions are taken so as to satisfy Eq. (4.79). The shapes of these functions are shown schematically in Fig. 4.33 and θ is defined in the three-dimensional view of Fig. 4.12. Note that $g(\theta)$ does not mean the probability per unit area. The probability per unit area is proportional to $g(\theta)/\sin\theta$. The limit of $\beta \to 0$ is again considered. At this limit, θ_0 tends to $\pi/2$ from Eq. (4.76). Considering this condition, C_0 is calculated from Eq. (4.98) for the two types of $g(\theta)$

Fig. 4.33. Values of C_0 for two types of fiber orientation distribution. (After Fukuda and Chou 1982.)

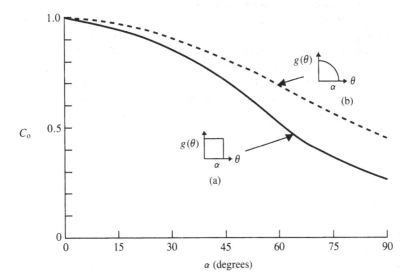

given above:

(a) $\quad \lim_{\beta \to 0} C_o = \frac{\sin \alpha}{\alpha} \frac{1}{\alpha} \left(\frac{1}{12} \sin 3\alpha + \frac{3}{4} \sin \alpha \right)$

(4.103)

(b) $\quad \lim_{\beta \to 0} C_o = \frac{1}{16} \left[\frac{1}{1+q} \sin \frac{\pi}{2}(1+q) + \frac{1}{1-q} \sin \frac{\pi}{2}(1-q) \right]$

$\qquad \times \left[\frac{3}{1+q} \sin \frac{\pi}{2}(1+q) + \frac{3}{1-q} \sin \frac{\pi}{2}(1-q) \right.$

$\qquad \left. + \frac{1}{1+3q} \sin \frac{\pi}{2}(1+3q) + \frac{1}{1-3q} \sin \frac{\pi}{2}(1-3q) \right]$

where $q = 2\alpha/\pi$. These values are shown in Fig. 4.33. Bowyer and Bader (1972) estimated the value of C_o by their experimental data. For laboratory glass/nylon injection molded materials, C_o was 0.66. If a retangular distribution for $g(\theta)$ is used, the value of α corresponding to $C_o = 0.66$ is approximately 45° from Fig. 4.33.

The orientation factor C_o discussed here is slightly different from the factor C_θ discussed in Section 4.3.2. The bridging effect of fibers is considered in the derivation of C_o, while the Poisson's effect of the composite is taken into account in evaluating C_θ. C_o and C_θ are essentially the same for the limiting case of $\beta \to 0$. The effect of β is discussed in Section 4.6.3.

4.6.3   *Random short-fiber composites*

Both Lees (1968a&b) and Chen (1971) attempted an averaging technique to treat the strength of random fiber composites. They adopted the failure mechanisms of Stowell and Liu (1961) and Jackson and Cratchley (1966), namely fiber failure, matrix failure in shear and matrix failure in plane strain. The operative failure mechanism in composites is dictated by the angle between the fiber direction and the direction of applied stress

$$\sigma_c = \begin{cases} \sigma_1 = \sigma_c'/\cos^2 \theta & (0 \le \theta \le \theta_1) \\ \sigma_2 = \tau_m/\sin \theta \cos \theta & (\theta_1 \le \theta \le \theta_2) \\ \sigma_3 = \sigma_m/\sin^2 \theta & (\theta_2 \le \theta \le \pi/2) \end{cases}$$

(4.104)

where σ_c' denotes the strength along the fiber direction of the unidirectional composite given by a rule-of-mixtures type of relationship. τ_m and σ_m are, respectively, the shear and tensile failure stresses of the matrix and the interface. Local stress perturbation due to fiber–fiber interaction can also be included in σ_1 of Eq.

(4.104). The strength for random fiber composites can be obtained by considering the angular strength dependence as a piecewise continuous function integrated over 90°:

$$\sigma_c = \frac{2}{\pi}\left\{\int_0^{\theta_1} \sigma_1 \, d\theta + \int_{\theta_1}^{\theta_2} \sigma_2 \, d\theta + \int_{\theta_2}^{\pi/2} \sigma_3 \, d\theta\right\} \qquad (4.105)$$

The predictions of this approach agree reasonably well with experimental results on glass-reinforced polyethylene and PMMA random mat (Lees 1968a) as well as random Al_2O_3–aluminum–silicon and glass/epoxy composites (Chen 1971).

Treatments of the strength of random short-fiber composites can also be found in the works of Lee (1969), Lavengood (1972), Kardos (1973), McNally (1977) and Blumentritt, Vu and Cooper (1975). The method of laminate analogue discussed for stiffness (Section 4.3.3) can also be applied to prediction of the strength of two-dimensional random fiber composites; the strength behavior of an isotropic laminate can be simulated by unidirectionally oriented plies laid up to approximate random orientation.

The strength prediction method of Fukuda and Chou (1982) can also be applied to determine the orientation factor C_o (Eq. (4.72)) for random fiber composites. By assuming that the fiber length is uniform and is larger than the critical length l_c, Eq. (4.98) becomes

$$\sigma_{cu} = \sigma_{fu} V_f\left(1 - \frac{l_c}{2l}\right)\int_0^{\pi/2} g(\theta) \cos\theta \, d\theta \int_0^{\theta_o} g(\theta) \cos^3\theta \, d\theta$$

$$\times \int_0^{\theta_o}\left(1 - \frac{\beta}{\cos\theta}\right)g(\theta) \, d\theta + \sigma'_{mu}(1 - V_f) \qquad (4.106)$$

By comparing Eqs. (4.72) and (4.106), the following expression for C_o is obtained:

$$C_o = \int_0^{\pi/2} g(\theta) \cos\theta \, d\theta \int_0^{\theta_o} g(\theta)\cos^3\theta \, d\theta$$

$$\times \int_0^{\theta_o}\left(1 - \frac{\beta}{\cos\theta}\right)g(\theta) \, d\theta \qquad (4.107)$$

Now consider both two-dimensional and three-dimensional random fiber arrays. In a two-dimensional random array model, $g(\theta)$ must be constant in the whole region of $0 \le \theta \le \pi/2$, and

$$g(\theta) = 2/\pi \qquad (4.108)$$

from Eq. (4.79). Substituting Eq. (4.108) into Eq. (4.107), the following result is obtained:

$$C_o = \frac{8}{3\pi^3}(2 + \beta^2)\sqrt{(1 - \beta^2)}\left[\cos^{-1}\beta - \tfrac{1}{2}\beta\log\left(\frac{1 + \sqrt{(1 - \beta^2)}}{1 - \sqrt{(1 - \beta^2)}}\right)\right]$$

(4.109)

The solid line of Fig. 4.34 depicts this result. As β increases, the composite contains more ending fibers and fewer bridging fibers, and hence the reinforcing effect of fibers is reduced. In the limit of $\beta \to 0$, all fibers are bridging fibers, and C_o tends to 0.27 for this two-dimensional case. Bowyer and Bader (1972) used the value of $\frac{1}{3}$ by quoting the result of Cox (1952) for the orientation factor of Young's modulus of a random composite. Cox's value is also shown in Fig. 4.34 by the solid circle.

In the case of a three-dimensional random fiber model, referring to Fig. 4.12, $g(\theta)$ can be expressed as

$$g(\theta)\, d\theta = dS/S$$

where the hemispherical surface area is S. Therefore,

$$g(\theta) = \sin\theta$$

(4.110)

Fig. 4.34. Fiber orientation factor C_o of random fiber array model. —— two-dimensional random array; – – – – three-dimensional random array. Solid and open circles indicate Cox's results. (After Fukuda and Chou 1982.)

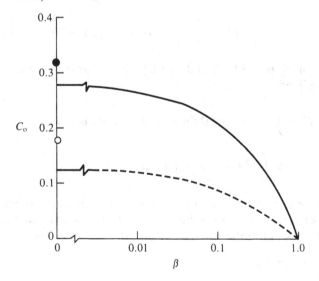

In this case, Eq. (4.107) becomes

$$C_o = \tfrac{1}{8}(1 - \beta^2)(1 + \beta^2)(1 - \beta + \beta \log \beta) \qquad (4.111)$$

This result is shown in Fig. 4.34 by a broken line. In the limit of $\beta \to 0$, C_o becomes $\tfrac{1}{8}$ and this value is again less than Cox's prediction of $\tfrac{1}{6}$ as indicated by the open circle.

4.7   Fracture behavior

Among the various types of short-fiber composites, the fracture behavior of polymer based composites is relatively well understood. The failure of short-fiber composites often initiates at microvoids and microcracks. These defects exist in the reinforcements, the matrix, and the interphase material and are introduced in the fabrication process. The final failure of a short-fiber composite is the result of several micromechanical mechanisms. The macroscopic appearance of the fracture depends on which of these mechanisms dominate the overall fracture process.

According to Friedrich (1985, 1989) and Friedrich and Karger-Kocsis (1989), the major failure mechanisms of short-fiber composites include (a) matrix deformation and fracture, (b) fiber/matrix debonding, (c) fiber pull-out, and (d) fiber fracture. A schematic fracture path through a short-fiber-reinforced polymer is given in Fig. 4.35; the individual failure mechanisms are also demonstrated.

The extent to which a specific failure mechanism occurs depends on the properties of the fiber, matrix, and interphase as well as the geometric form and arrangement of the fibers. As discussed in Sections 4.2.1 and 4.6.1, the efficiency in load transfer between a fiber and its surrounding matrix depends on the length of the fiber relative to its critical length, l_c. If the length of the fiber is shorter than l_c, fiber pull-out and matrix fracture are the dominating mechanisms of energy absorption. On the other hand, when the fiber length is longer than l_c, the fibers will, in some cases, break and in other cases be pulled out; the fiber location and orientation with respect to the crack is an important factor in determining which failure mechanism takes place.

Friedrich (1989) has examined the fracture energy of aligned short-fiber composites and given the following observations:

(1) The matrix material supplies a certain portion of the fracture energy of the composite. For a brittle polymer matrix, this portion is small in comparison to fiber fracture or interfacial failure. Then the fracture energy of the

composite as a result of fiber reinforcement is higher than that of the unfilled matrix. However, in the case of a ductile polymer matrix, the energy absorption in the fracture process is higher than those due to fiber related mechanisms. Thus, the fracture energy decreases as fiber volume fraction increases.

(2) The fiber/matrix interface shear strength, which affects the fiber critical length (Eq. (4.43)), is strongly influenced by the temperature of the environment. Higher temperatures result in higher l_c. Furthermore, the temperature also influences the matrix fracture behavior.

(3) The fracture toughness K_c of a short fiber composite is related to the fracture energy G_c and elastic modulus E by $K_c = \sqrt{(G_c E)}$. Some qualitative observations can be made concerning this relationship. First, the addition of fibers to a brittle polymer matrix enhances K_c due to a simultaneous increase in G_c and E. Second, the addition of fibers to a

Fig. 4.35. Schematic fracture path through a short-fiber-reinforced polymer, and individual mechanisms of failure: (*A*) fiber fracture, (*B*) fiber pull-out, (*C*) fiber/matrix separation, and (*D*) plastic deformation and fracture of the polymer matrix. (After Friedrich 1989.)

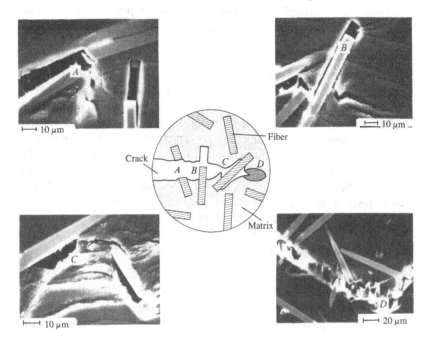

very ductile thermoplastic matrix results in an increase in E but a decrease in G_c. Thus, K_c may decrease or remain unchanged.

It is also noted that the addition of fibers can result in constraining effects on the matrix and a change of the stress state. Consequently, this leads to limited plasticity in the matrix and stress concentrations at fiber ends. The implications of the stress concentration on the fracture of short-fiber composites are discussed below.

Experimental work for identifying fiber end stress concentration was first performed by MacLaughlin (1966), who used a photoelastic method to investigate the effect of fiber end shape and gap size on the shear stress near a single short fiber. MacLaughlin (1968) extended the photoelastic study to a square-ended short fiber flanked by continuous fibers. Photoelastic methods were also used by Chen and Lavengood (1969) to examine the distribution of fiber stress and interfacial shearing stress around a short square-ended fiber.

Theoretical analyses of fiber end stress concentrations have been discussed in Section 4.2. Iremonger and Wood (1967, 1969), Muki and Sternberg (1969, 1970, 1971), Chen and Lewis (1970), Sternberg (1970), Sternberg and Muki (1970), Baker and MacLaughlin (1971) and Takao, Taya and Chou (1981) have also presented analytical solutions with particular emphasis on fiber end separation distance, fiber volume fraction, fiber and matrix modulus ratio, and fiber end geometry. Several general conclusions can be drawn from the analyses: (1) the primary parameters affecting the stress concentrations are gap size, fiber volume fraction and fiber–matrix modulus ratio; (2) square-ended and tapered-end fibers give higher stress concentrations than round-ended fibers; (3) stress concentration increases with decreasing fiber end separation distance; (4) higher stress concentrations exist at the fiber–matrix interface when the end gap is a void as compared to a gap filled with matrix. It is understood that in real composites the fiber ends are usually oblique and uneven and that the concept of fiber end separation distance is difficult to apply to a randomly distributed and misaligned fiber system. A significant finding of the stress analyses surveyed above is that the concentration of stress in the matrix near the discontinuity of a fiber is very severe even under moderate load application. Composite failure initiation, either by fracture of the matrix or by debonding, is likely to occur at these locations.

The experimental work of Curtis, Bader and Bailey (1978) on glass and carbon fiber reinforced polyamide 6.6 has demonstrated the embrittlement effect of short-fiber composites. Theoretical modeling of the fracture of short-fiber composites can be found in the work of Taya and Chou (1981, 1982), Ishikawa, Chou and Taya (1982), Takao, Chou and Taya (1982) and Takao, Taya and Chou (1982). The environmental effect on the fracture of short-fiber composites has been examined by Friedrich, Schulte, Horstenkamp and Chou (1985), Hsu, Yau and Chou (1986), and Yau and Chou (1989).

5   Hybrid composites

5.1   Introduction

The term 'hybrid composites' is used to describe composites containing more than one type of fiber materials. Hybrid composites are attractive structural materials for the following reasons. First, they provide designers with the new freedom of tailoring composites and achieving properties that cannot be realized in binary systems containing one type of fiber dispersed in a matrix. Second, a more cost-effective utilization of expensive fibers such as carbon and boron can be obtained by replacing them partially with less expensive fibers such as glass and aramid. Third, hybrid composites provide the potential of achieving a balanced pursuit of stiffness, strength and ductility, as well as bending and membrane related mechanical properties. Hybrid composites have also demonstrated weight savings, reduced notch sensitivity, improved fracture toughness, longer fatigue life and excellent impact resistance (Chou and Kelly 1980a). Some of the pioneering studies on this topic can be found in the work of Wells and Hancox (1971), Hayashi (1972), Kalnin (1972), Hancox and Wells (1973), Bunsell and Harris (1974), Harris and Bunsell (1975), Walton and Majumdar (1975), Aveston and Sillwood (1976), Bunsell (1976), Harris and Bradley (1976), Zweben (1977), Arrington and Harris (1978), Badar and Manders (1978, 1981a,b), Marom, Fischer, Tuler and Wagner (1978), Rybicki and Kanninen (1978), Summerscales and Short (1978), Aveston and Kelly (1980), Wagner and Marom (1982), Fukuda (1983a–c), and Harlow (1983).

Depending upon the arrangements of fibers and pre-preg layers, hybrids can be categorized into the following types. In the first type the different fiber materials are intimately mixed together and infiltrated with a matrix simultaneously. The hybrid in this case is described as *intermingled* (Aveston and Kelly 1980) or *intraply* (Chamis and Lark 1978) (Fig. 5.1a). The second type of hybrid is made by bonding together separate laminae each containing just one type of fiber in a matrix, and is known as *interlaminated* (Aveston and Kelly 1980) or *interply* (Chamis and Lark 1978) (Fig. 5.1b). The third category of hybrids consists of fabric reinforcements where each fabric contains more than one type of fiber and it can be termed as *interwoven* (Chou and Kelly 1980a) (Fig. 5.1c). Hybrid composites consisting of resin composite plies, metal composite plies and metal foils also have been explained. When a laminated hybrid is composed of plies of different matrices it needs to be fabricated by a consolidation procedure that is compatible with all matrix materials.

Optimization of composite properties can usually be achieved through a suitable combination of fiber types. Reinforcements for hybrids include boron, carbon, glass and aramid fibers. Limited applications of ceramic and metallic filaments have been explored (Renton 1978). Intermediate modulus epoxies, thermoplastics and polyimides are the common polymeric matrices for hybrid composites. Current applications of hybrid composites can be found in aircraft fuselage, wing and tail structures, helicopter rotor blades and automobile parts as well as in an array of sports equipment, ranging from sailboats and racing cars to bicycle frames and hockey sticks.

The fundamental questions pertinent to the study of hybrid composites are (a) how is the load shared among the constituent

Fig. 5.1. Types of hybrid composites: (a) intermingled; (b) interlaminated; and (c) interwoven.

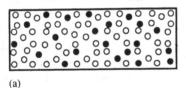

(a)

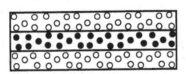

(b)

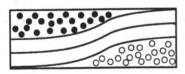

(c)

fibers? (b) are there synergistic effects among the different types of fibers? and (c) will certain combinations of fiber types and microstructure designs produce an overall desirable structural performance? In order to gain a basic understanding of these problems, the various aspects of the mechanical behavior of hybrids are examined. To simplify the consideration of deformation, the following discussions are primarily restricted to unidirectional composites and their laminates. Woven hybrid composites are examined in Chapter 6.

5.2   Stress concentrations

The load redistribution in unidirectional continuous fiber hybrid composite laminae due to fiber breakages is examined in this section. Stress concentration factors are obtained for both intermingled and interlaminated hybrids. The solution techniques are demonstrated for both static and dynamic responses. The terminologies of low modulus (LM) and high modulus (HM) are used to distinguish the two kinds of fibers in the model lamina. For hybrid composites such as glass/carbon and Kevlar/carbon combinations, LM and HM fibers correspond to HE (high elongation) and LE (low elongation) fibers, respectively. The fiber ductility or elongation to break does not enter into the present analysis in an explicit manner. The shear-lag technique demonstrated in Chapters 3 and 4 is again adopted in the following.

5.2.1   *Static case*

Consider a unidirectional lamina composed of HM and LM fibers in alternating positions. Each pair of neighboring HM and LM fibers is designated as the group m. Asterisks (*) are used to denote quantities related to LM fibers.

Fukuda and Chou (1983) have examined the three types of combinations of fiber discontinuity depicted in Fig. 5.2. Let n_1 and n_2 be the number of broken HM and LM fibers, respectively. Thus, in Fig. 5.2, (a) $n_1 = n$, $n_2 = 0$; (b) $n_1 = n$, $n_2 = n - 1$; and (c) $n_1 = n_2 = n$. The counting of broken fibers starts at $m = 0$ and ends at $m = n - 1$.

The axial loads of the mth pair of fibers are denoted by $p_m(x)$ and $p_m^*(x)$; the displacements are $u_m(x)$ and $u_m^*(x)$, and $x = 0$ denotes the plane of fiber fracture. Based upon the assumptions of shear-lag analysis (Hedgepeth 1961; Ji, Hsiao and Chou 1981), the force equilibrium equations of the mth HM and LM fibers are, respectively.

$$Ed\frac{d^2u_m}{dx^2} + \frac{G}{h}(u_m^* + u_{m-1}^* - 2u_m) = 0$$

$$E^*d^*\frac{d^2u_m^*}{dx^2} + \frac{G}{h}(u_{m+1} + u_m - 2u_m^*) = 0$$

(5.1)

Here, E and G denote the fiber extensional modulus and the shear modulus of the matrix, respectively. The lamina is assumed to be of unit thickness; d and h denote, respectively, fiber width and spacing.

Under the assumption of linear elastic deformation, the force–displacement relations become

$$p_m = Ed\frac{du_m}{dx}, \qquad p_m^* = E^*d^*\frac{du_m^*}{dx}$$

(5.2)

Fig. 5.2. Arrays of discontinuous fibers: (a) $n_1 = n$, $n_2 = 0$; (b) $n_1 = n$, $n_2 = n - 1$; (c) $n_1 = n_2 = n$, n_1 and n_2 are, respectively, the number of discontinuous HM and LM fibers. (After Fukuda and Chou 1983.)

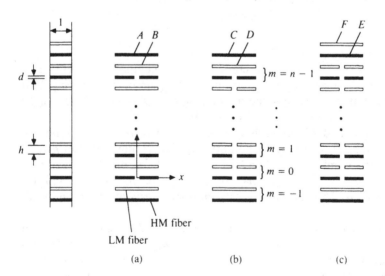

The boundary conditions are

$$p_m(\infty) = p$$

$$p_m^*(\infty) = \frac{E^*d^*}{Ed}\,p$$

$$p_m(0) = 0, \quad p_m^*(0) = 0 \qquad \text{for broken fibers}$$

$$u_m(0) = 0, \quad u_m^*(0) = 0 \qquad \text{for unbroken fibers}$$

(5.3)

To simplify Eqs. (5.1)–(5.3), the following dimensionless parameters are introduced:

$$P_m = \frac{p_m}{p} \qquad\qquad P_m^* = \frac{p_m^*}{p}$$

$$U_m = \frac{u_m}{p}\sqrt{\left(\frac{EdG}{h}\right)} \qquad U_m^* = \frac{u_m^*}{p}\sqrt{\left(\frac{EdG}{h}\right)} \qquad (5.4)$$

$$\xi = \sqrt{\left(\frac{G}{Edh}\right)}x \qquad\qquad R = E^*d^*/Ed$$

Thus, Eqs. (5.1) and (5.2) become

$$\frac{d^2 U_m}{d\xi^2} + U_m^* + U_{m-1}^* - 2U_m = 0$$

$$R\frac{d^2 U_m^*}{d\xi^2} + U_{m+1} + U_m - 2U_m^* = 0$$

(5.5)

$$P_m = \frac{dU_m}{d\xi}$$

$$P_m^* = R\frac{dU_m^*}{d\xi}$$

(5.6)

By adopting the concept of influence functions proposed by Hedgepeth (1961), the dimensionless displacements are expressed as

$$U_m(\xi) = \xi + \sum_{k=-\infty}^{\infty} V_{m-k}(\xi)U_k(0) + \sum_{k=-\infty}^{\infty} W_{m-k}(\xi)U_k^*(0)$$

$$U_m^*(\xi) = \xi + \sum_{k=-\infty}^{\infty} V_{m-k}^*(\xi)U_k(0) + \sum_{k=-\infty}^{\infty} W_{m-k}^*(\xi)U_k^*(0)$$

(5.7)

where V, V^*, W and W^* are the influence functions. Then, from Eqs. (5.5), the following two sets of equations in terms of the influence functions are obtained:

(I)

$$\frac{d^2V_m}{d\xi^2} + V_m^* + V_{m-1}^* - 2V_m = 0$$

$$R\frac{d^2V_m^*}{d\xi^2} + V_{m+1} + V_m - 2V_m^* = 0$$

(5.8)

with the boundary conditions of

$$V_m(0) = 1 \quad (m = 0) \qquad V_m(0) = 0 \quad (m \neq 0)$$

$$V_m^*(0) = 0$$

(5.9)

$$\frac{dV_m(\infty)}{d\xi} = 0 \qquad \frac{dV_m^*(\infty)}{d\xi} = 0$$

(II)

$$\frac{d^2W_m}{d\xi^2} + W_m^* + W_{m-1}^* - 2W_m = 0$$

$$R\frac{d^2W_m^*}{d\xi^2} + W_{m+1} + W_m - 2W_m^* = 0$$

(5.10)

with the boundary condition of

$$W_m(0) = 0$$

$$W_m^*(0) = 1 \quad (m = 0) \qquad W_m^*(0) = 0 \quad (m \neq 0)$$

(5.11)

$$\frac{dW_m(\infty)}{d\xi} = 0 \qquad \frac{dW_m^*(\infty)}{d\xi} = 0$$

Since Eqs. (5.8) and (5.10) are identical in form, only the solution procedure of Eqs. (5.8) is given below. For solving Eqs. (5.8), the following Fourier series expressions are introduced

$$\bar{V} = \sum_{m=-\infty}^{\infty} V_m e^{-im\theta} \qquad \bar{V}^* = \sum_{m=-\infty}^{\infty} V_m^* e^{-im\theta}$$

(5.12)

or, inversely,

$$V_m = \frac{1}{2\pi}\int_{-\pi}^{\pi} \bar{V}e^{im\theta}\,d\theta \qquad V_m^* = \frac{1}{2\pi}\int_{-\pi}^{\pi} \bar{V}^*e^{im\theta}\,d\theta$$

(5.13)

Then, multiplying Eqs. (5.8) by $e^{-im\theta}$ and summing over all m gives

$$\frac{d^2\bar{V}}{d\xi^2} - 2\bar{V} + A\bar{V}^* = 0$$

$$R\frac{d^2\bar{V}^*}{d\xi^2} - 2\bar{V}^* + B\bar{V} = 0$$

(5.14)

where

$$A = 1 + e^{-i\theta} \quad B = 1 + e^{i\theta} \tag{5.15}$$

Also, from Eqs. (5.9),

$$\bar{V}(0, \theta) = 1 \quad \bar{V}^*(0, \theta) = 0$$

$$\frac{d\bar{V}(\infty, \theta)}{d\xi} = 0 \quad \frac{d\bar{V}^*(\infty, \theta)}{d\xi} = 0$$

(5.16)

The solutions of Eqs. (5.14) under the boundary conditions of Eqs. (5.16) are

$$\bar{V} = C_1 e^{-\lambda_1\xi} + C_2 e^{-\lambda_2\xi}$$

$$\bar{V}^* = C_3 e^{-\lambda_1\xi} + C_4 e^{-\lambda_2\xi}$$

(5.17)

where

$$\lambda_1 = \sqrt{[a + \sqrt{(a^2 - b)}]} \qquad \lambda_2 = \sqrt{[a - \sqrt{(a^2 - b)}]}$$

$$a = 1 + \frac{1}{R} \qquad b = \frac{1}{R}(1 - \cos\theta)$$

(5.18)

$$C_1 = \frac{2 - \lambda_2^2}{\lambda_1^2 - \lambda_2^2} \qquad C_2 = -\frac{2 - \lambda_1^2}{\lambda_1^2 - \lambda_2^2}$$

$$C_3 = -\frac{(2 - \lambda_1^2)(2 - \lambda_2^2)}{A(\lambda_1^2 - \lambda_2^2)} \qquad C_4 = -C_3$$

Substituting Eqs. (5.17) into Eqs. (5.13) and considering a, b, λ_1, λ_2, C_1, C_2 and C_3A as even functions with respect to θ, the following results are obtained for the influence functions:

$$V_m = \frac{1}{\pi}\int_0^\pi (C_1 e^{-\lambda_1\xi} + C_2 e^{-\lambda_2\xi})\cos(m\theta)\,d\theta$$

(5.19)

$$V_m^* = \frac{1}{\pi}\int_0^\pi C_3A(e^{-\lambda_1\xi} - e^{-\lambda_2\xi})\frac{\cos(m\theta) + \cos(m+1)\theta}{2(1 + \cos\theta)}\,d\theta$$

Differentiating Eqs. (5.19) and substituting the result for $\xi = 0$ into Eqs. (5.7) and the third condition of Eqs. (5.3), the values of $U_k(0)$ ($0 \le k \le n_1 - 1$) and $U_k^*(0)$ ($0 \le k \le n_2 - 1$) can be obtained. The dimensionless axial loads, P_m and P_m^* are calculated by substituting Eqs. (5.7) into Eqs. (5.6). The stress concentration factor of the mth group of fibers is defined as $P_m(0)/P_m(\infty)$ or $P_m^*(0)/P_m^*(\infty)$.

$$P_m(0)/P_m(\infty) = 1 + \sum_{k=0}^{n_1-1} \frac{\mathrm{d}V_{m-k}(0)}{\mathrm{d}\xi} U_k(0)$$

$$+ \sum_{k=0}^{n_2-1} \frac{\mathrm{d}W_{m-k}(0)}{\mathrm{d}\xi} U_k^*(0)$$

$$P_m^*(0)/P_m^*(\infty) = 1 + \sum_{k=0}^{n_1-1} \frac{\mathrm{d}V_{m-k}^*(0)}{\mathrm{d}\xi} U_k(0) \tag{5.20}$$

$$+ \sum_{k=0}^{n_2-1} \frac{\mathrm{d}W_{m-k}^*(0)}{\mathrm{d}\xi} U_k^*(0)$$

For an HM or an LM fiber adjacent to a discontinuous fiber, the stress concentration factor can be calculated by substituting the corresponding value of m into Eqs. (5.20). For instance, the stress concentration factors for fibers B and A of Fig. 5.2 are, respectively, $P_{n_1-1}^*(0)/P_{n_1-1}^*(\infty)$ and $P_{n_1}(0)/P_{n_1}(\infty)$.

Fukuda and Chou (1981, 1983) have evaluated Eqs. (5.20) and the results are presented for (a) comparisons with the solutions of Hedgepeth (1961) for non-hybrid composites, and (b) hybrid composites. First, for a non-hybrid composite, there is only one type of fiber in the lamina and $R = 1$. Two limiting cases are given below.

(A) $n_1 = 1$ and $n_2 = 0$

Consider, for instance, Fig. 5.2(a). The fiber immediately adjacent to the broken fiber is the one designated as an LM fiber in the $m = 0$ pair. Therefore, the stress concentration factor is

$$P_o^*(0)/P_o^*(\infty) = 1 + \frac{\mathrm{d}V_o^*}{\mathrm{d}\xi} U_o(0) = \frac{4}{3} \tag{5.21}$$

This result coincides with that of Hedgepeth (1961), as expected. The same conclusion can be reached by considering the case of $n_1 = 0$ and $n_2 = 1$ in Fig. 5.2(a).

(B) $n_1 = n_2 = 1$

The stress concentration factor is given by

$$P_1(0)/P_1(\infty) = 1 + \frac{dV_1(0)}{d\xi} U_o(0) + \frac{dW_1(0)}{d\xi} U_o^*(0)$$

$$= \left(\frac{4}{3}\right)\left(\frac{6}{5}\right) \tag{5.22}$$

Next, for the unidirectional hybrid lamina $(R \neq 1)$, Eqs. (5.20) have been solved by numerical integrations using a trapezoidal rule. The results are shown in Figs. 5.3–5.5.

The limiting case of the fracture of one HM fiber $(n_1 = 1,\ n_2 = 0)$ is demonstrated in Fig. 5.3. The fibers adjacent to the broken HM fiber of particular interest are the LM fiber of $m = 0$ and the HM fiber of $m = 1$. The stress concentration factors of these two fibers (K_{LM}, K_{HM}) are plotted in Fig. 5.3 as functions of the stiffness ratio $R = E^*d^*/Ed$. For $R = 1$ (i.e. non-hybrid case), $K_{LM} = 1.33$ is obtained from Eq. (5.21). The limit of $R \to 0$, on the other hand, indicates that the extensional rigidity of the LM fiber is infinitesimal. Then Fig. 5.3 again becomes a model with only one type of fiber. The fiber nearest to the broken fiber in this case is the HM fiber of $m = 1$ and therefore $K_{HM} \to \frac{4}{3}$ at the limit $R \to 0$.

Figure 5.3 also indicates that the stress concentration factor of the

Fig. 5.3. Stress concentration factor vs. stiffness ratio $R = E^*d^*/Ed =$ extensional stiffness of LM fiber/extensional stiffness of HM fiber. (After Fukuda and Chou 1981.)

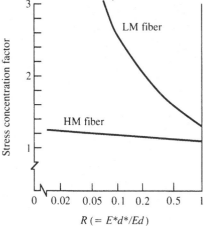

HM fiber in the hybrid lamina is lower than $\frac{4}{3}$ because of the presence of the LM fiber between the discontinuous and continuous HM fibers. For example, at $R = 0.5$, $K_{LM} = 1.67$ and $K_{HM} = 1.11$.

Figure 5.4 depicts the relations between stress concentration factors and the total number of broken fibers, $n_1 + n_2$, for the case of $R = 1$. The letters $A–F$ correspond to fibers $A–F$ in Fig. 5.2. Curves D and E show the stress concentration factors of the fibers immediately adjacent to the broken fibers. Curves C and F give the results for the second nearest fibers to the broken fibers. The cases of A and B in Fig. 5.2 give stress concentration factors insensitive to the number of broken fibers.

Figure 5.5 shows the stress concentration factors for $R = \frac{1}{3}$ which approximately corresponds to carbon/glass hybrid composites. The stress concentration factor of the HM fiber nearest to the discontinuous fibers, i.e. fiber C or E in Figs. 5.2(b) and (c), respectively, is smaller than that of fiber D (Fig. 5.2b) for a fixed number of broken fibers. This means that, as far as the HM fibers are concerned, the stress concentration is reduced in a hybrid composite. Thus, it is possible for the high modulus fibers in a hybrid

Fig. 5.4. Stress concentration factors vs. total number of broken fibers for $R = 1$. (After Fukuda and Chou 1983.)

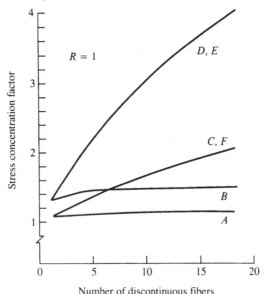

composite to sustain higher loads than in the all-high modulus fiber composite, and a 'hybrid effect' could be realized. The stress concentration factors of the LM fibers are shown by curves D and F in Fig. 5.5. A comparison of curves D of Figs. 5.4 and 5.5 indicates that the stress concentration on the LM fiber increases as R is reduced. This implies that the LM fibers are more susceptible to fracture in a hybrid composite than in a non-hybrid composite.

When the number of fibers in the composite model is high, the solution procedure of the governing equations becomes very complex. Fukunaga, Chou and Fukuda (1984) have evaluated the stress concentration factors using an eigenvector expansion method. Tables 5.1 and 5.2 show the numerical results of their analysis based upon a glass/carbon intermingled hybrid composite ($R = \frac{1}{3}$). The solid and open circles represent HM and LM fibers, respectively.

Fig. 5.5. Stress concentration factors vs. total number of broken fibers for a hybrid composite of $R = \frac{1}{3}$. (After Fukuda and Chou 1983.)

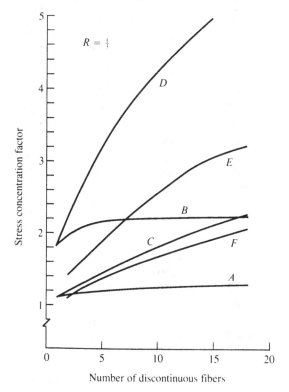

Number of discontinuous fibers

Table 5.1 gives the stress concentration factors for various V_{HM} values, where $K_{ij,k}$, for instance, is the stress concentration factor of the kth fiber due to the breakage of the ith and jth fibers for various fiber relative volume fractions. The fiber arrangements given in Table 5.1 are repeated to generate the composite, but the position

Table 5.1. *Stress concentration factors (SCF) for various* V_{HM} *values. After Fukunaga, Chou and Fukuda (1989)*

	V_{HM}			
	1.0	0.75	0.5	0.25
	●●●●	●●●○	●○●○	○●○○
SCF	1 2 3 4	1 2 3 4	1 2 3 4	1 2 3 4
$K_{1,2}$	1.333	1.356	1.777	1.141
$K_{1,3}$	1.067	1.077	1.121	1.036
$K_{1,4}$	1.029	1.041	1.060	1.017
$K_{2,1}$	1.333	1.347	1.131	1.829
$K_{2,3}$	1.333	1.347	1.131	1.829
$K_{2,4}$	1.067	1.102	1.030	1.208
$K_{3,1}$	1.067	1.077	1.121	1.036
$K_{3,2}$	1.333	1.356	1.777	1.141
$K_{3,4}$	1.333	1.727	1.777	1.311
$K_{4,1}$	1.029	1.007	1.010	1.017
$K_{4,2}$	1.067	1.018	1.030	1.036
$K_{4,3}$	1.333	1.128	1.131	1.318
$K_{12,3}$	1.600	1.654	1.412	2.144
$K_{12,4}$	1.143	1.221	1.135	1.304
$K_{13,4}$	1.419	1.832	1.952	1.340
$K_{13,2}$	1.802	1.772	2.767	1.291
$K_{14,2}$	1.412	1.378	1.817	1.180
$K_{14,3}$	1.412	1.210	1.261	1.359
$K_{23,1}$	1.600	1.654	1.412	2.144
$K_{23,4}$	1.600	2.275	2.038	1.913
$K_{24,1}$	1.419	1.362	1.146	1.886
$K_{24,3}$	1.802	1.494	1.270	2.252
$K_{34,1}$	1.143	1.109	1.172	1.077
$K_{34,2}$	1.600	1.478	2.038	1.258
$K_{123,4}$	1.829	2.776	2.510	2.175
$K_{234,1}$	1.829	1.768	1.522	2.382
$K_{341,2}$	2.022	1.914	3.105	1.418
$K_{412,3}$	2.022	1.822	1.569	2.612

of the fractured fiber (or fibers) is not repeated. In the numerical calculations 80 fibers are used in each composite model.

Table 5.2 presents the effect of fiber bundle size on stress concentration for the relative fiber volume fraction of $V_{HM} = 50\%$. Two (case 2), three (case 3) or four (case 4) fibers of the same type can be placed adjacent to one another besides the alternating arrangement of one HM and one LM fiber (case 1). It is evident that the stress concentration factor is sensitive to the microscopic fiber arrangements. Knowledge of the stress concentration factors in various hybrid fiber arrays is essential to the evaluation of hybrid composite strength.

5.2.2 Dynamic case

The dynamic stress concentration in hybrid composites due to fiber breakage has been examined by Ji, Hsiao and Chou (1981). Figure 5.6 shows an interlaminated hybrid composite for the analytical model; it is composed of a layer of HM fiber and a layer of LM fiber embedded in a common matrix. The fibers are aligned along the x axis, and h_1 and h_2 denote the fiber spacings. A fiber in each array is numbered by an integer n $(-\infty < n < \infty)$. The displacement field of a fiber as a function of location and time is denoted by $u_n(x, t)$ for an HM fiber, and by $u_n^*(x, t)$ for an LM fiber. Similarly, the axial forces in the fibers are denoted by $p_n(x, t)$ and $p_n^*(x, t)$. Ji and colleagues have analyzed the dynamic stress

Table 5.2. *Stress concentration factors for various bundle sizes. After Fukunaga, Chou and Fukuda (1989)*

	1 2 3 4			
Case 1	○ ● ○ ● ¦ ○ ● ○ ● ¦ ○ ● ○ ●			
Case 2	● ● ○ ○ ¦ ● ● ○ ○ ¦ ● ● ○ ○			
Case 3	● ○ ○ ○ ¦ ● ● ● ○ ¦ ○ ○ ● ●			
Case 4	○ ○ ○ ● ¦ ● ● ● ○ ¦ ○ ○ ○ ●			

Fiber location	1	2	3	4
Case 1	1.777	×	1.777	1.121
Case 2	1.376	×	1.762	1.175
Case 3	1.354	×	1.354	1.113
	1.082	1.366	×	1.763
Case 4	1.343	×	1.351	1.112
	1.079	1.365	×	1.764

concentration factor of the fiber $n = 1$ or -1 in the HM fiber array when the fiber $n = 0$ suddenly breaks.

The fundamental equations governing the deformation of the HM and LM fibers are approximated by

$$EA \frac{\partial^2 u_n}{\partial x^2} + \frac{Gh_1}{h_2} (u_{n+1} - 2u_n + u_{n-1}) + \frac{Gh_2}{h_1} (u_n^* - u_n)$$

$$= m \frac{\partial^2 u_n}{\partial t^2}$$

$$E^* A^* \frac{\partial^2 u_n^*}{\partial x^2} + \frac{Gh_1}{h_2} (u_{n+1}^* - 2u_n^* + u_{n-1}^*) + \frac{Gh_2}{h_1} (u_n - u_n^*)$$

$$= m^* \frac{\partial^2 u_n^*}{\partial t^2}$$

(5.23)

for all n ($= -\infty, \ldots, -1, 0, 1, \ldots, \infty$). In Eqs. (5.23), m and m^* are the fiber masses per unit length, E^* and E are the fiber Young's moduli, A^* and A denote fiber cross-sectional areas, and G is the matrix shear modulus.

The boundary conditions are

$$p_o(0, t) = 0$$

$$u_n(0, t) = 0 \quad (n \neq 0) \qquad p_n(\pm\infty, t) = p \quad \text{(all } n\text{)}$$

$$u_n^*(0, t) = 0 \quad \text{(all } n\text{)} \qquad p_n^*(\pm\infty, t) = \frac{E^* A^*}{EA} p \quad \text{(all } n\text{)}$$

(5.24)

Fig. 5.6. A model of interlaminated hybrid composite. (After Ji, Hsiao and Chou 1981.)

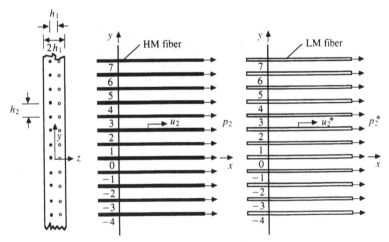

The initial conditions are

$$p_n(x, 0) = p \qquad \frac{\partial u_n}{\partial t}(x, 0) = 0$$

$$p_n^*(x, 0) = \frac{E^* A^*}{EA} p \qquad \frac{\partial u_n^*}{\partial t}(x, 0) = 0 \qquad (5.25)$$

for all n.

The forces and displacements of the fibers are related by

$$p_n = EA \frac{\partial u_n}{\partial t} \qquad (5.26)$$

By introducing the following dimensionless parameters,

$$P_n = \frac{p_n}{p} \qquad P_n^* = \frac{p_n^*}{p}$$

$$U_n = \frac{u_n}{p} \sqrt{\left(\frac{EAGh_1}{h_2}\right)} \qquad U_n^* = \frac{u_n^*}{p} \sqrt{\left(\frac{EAGh_1}{h_2}\right)}$$

$$\xi = \sqrt{\left(\frac{Gh_1}{EAh_2}\right)} x \qquad \tau = \sqrt{\left(\frac{Gh_1}{mh_2}\right)} t \qquad (5.27)$$

$$R = \frac{E^* A^*}{EA} \qquad M = \frac{m^*}{m} \qquad D = \frac{h_2^2}{h_1^2}$$

Eqs. (5.23) can be rewritten as

$$\frac{\partial^2 U_n}{\partial \xi^2} + U_{n+1} - 2U_n + U_{n-1} + D(U_n^* - U_n) = \frac{\partial^2 U_n}{\partial \tau^2}$$

$$R\frac{\partial^2 U_n^*}{\partial \xi^2} + U_{n+1}^* - 2U_n^* + U_{n-1}^* + D(U_n - U_n^*) = \frac{\partial^2 U_n^*}{\partial \tau^2} \qquad (5.28)$$

Also, Eqs. (5.24) become

$$P_0(0, \tau) = 0$$

$$U_n(0, \tau) = 0 \quad (n \neq 0), \qquad P_n(\pm\infty, \tau) = 1 \quad \text{(all } n) \quad (5.29)$$

$$U_n^*(0, \tau) = 0 \quad \text{(all } n), \qquad P_n^*(\pm\infty, \tau) = R \quad \text{(all } n)$$

and Eqs. (5.25) can be written in terms of displacements only as

$$U_n(\xi, 0) = \xi \qquad \frac{\partial U_n}{\partial \tau}(\xi, 0) = 0$$

$$U_n^*(\xi, 0) = \xi \qquad \frac{\partial U_n^*}{\partial \tau}(\xi, 0) = 0 \qquad (5.30)$$

Equations (5.28)–(5.30) can be solved following the approach of Hedgepeth. Three essential steps are involved: the Laplace transform in time, the use of the technique of influence function, and the Fourier series representation of the unknown functions. Ji and colleagues have evaluated the dynamic stress concentration factor of the HM fibers immediately adjacent to the broken fiber. The most severe stress concentration factor $K_1(0, \tau)$ occurs at $\xi = 0$. It is interesting to note that the solution of $K_1(0, \tau)$ is composed of two parts which are related to the HM and LM fibers, respectively. Figures 5.7–5.9 depict the variation of the stress concentration factor $K_1(0, \tau)$ with the dimensionless time τ for $m^*/m = 1$, 2 and 6, respectively. The two components of the solution are denoted by K_1' and K_1''. Also in these figures, $E^*A^*/EA = h_2/h_1 = 1$. Figure 5.7 is for the case of non-hybrid composites and the solution of Hedgepeth for a unidirectional lamina is also shown. The summation of K_1' and

Fig. 5.7. The variation of stress concentration factor K_1 with dimensionless time τ for $m^*/m = E^*A^*/EA = h_2/h_1 = 1$. K_1' and K_1'' are the two components of K_1. (After Ji, Hsiao and Chou 1981.)

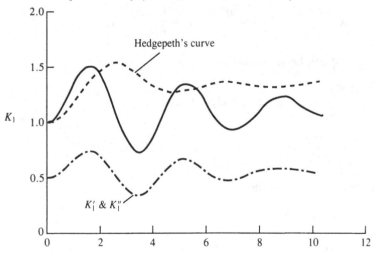

K_1'' gives the total stress concentration factor. As the difference in mass density of the HM and LM fibers increases, the locations of the peak values of K_1' and K_1'' are out of phase. As a result, the maximum value of $K_1' + K_1''$ is reduced (Figs. 5.8 and 5.9). The analysis of Ji and colleagues concludes that the time variations of the stress concentration factors related to the two component fiber materials are always out of phase in a hybrid composite with fibers of different mass densities. The parent HM fiber composite always provides the upper bound for the stress concentration of the hybrid, since this is the case where there is no difference in phase and magnitude of the K_1' and K_1'' values. Furthermore, the magnitudes of the K_1' and K_1'' are determined by the extensional stiffnesses of the component fibers.

5.3 Tensile stress–strain behavior

An idealized stress–strain curve of a hybrid composite containing both high elongation (HE) and low elongation (LE) fibers is depicted in Fig. 5.10 (Aveston and Kelly 1980). For hybrids with good bonding between the component phases, the stress–strain curve is given by *OABC*. The important features of this curve include the elastic behavior indicated by *OA*, the first cracking

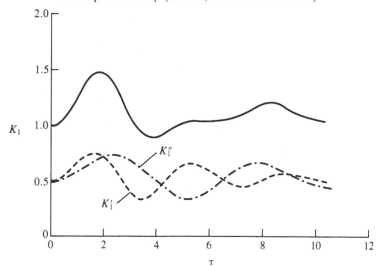

Fig. 5.8. The variation of stress concentration factor K_1 with dimensionless time τ for $m^*/m = 2$, and $E^*A^*/EA = h_2/h_1 = 1$. K_1' and K_1'' are the two components of K_1. (After Ji, Hsiao and Chou 1981.)

Fig. 5.9. The variation of stress concentration factor K_1 with dimensionless time τ for $m^*/m = 6$, and $E^*A^*/EA = h_2/h_1 = 1$. K_1' and K_1'' are the two components of K_1. (After Ji, Hsiao and Chou 1981.)

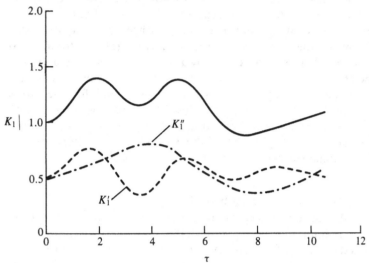

Fig. 5.10. A typical stress–strain curve of hybrid composites. (After Aveston and Kelly 1980.)

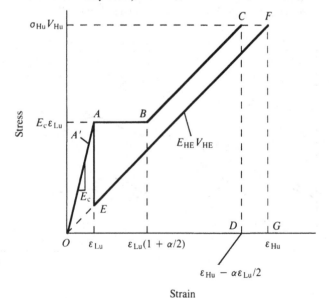

strain ε_{Lu}, the relatively flat portion of the curve *AB,* the subsequent rise of the curve at a smaller slope (*BC*) and the final failure strain of the hybrid given by the point *D.* The subscripts H and L are used to denote parameters related to high and low elongation fibers, respectively. The various features of the stress–strain curve are discussed in the following.

5.3.1 *Elastic behavior*

Theoretical predictions of the elastic moduli of multi-phase short-fiber composites have been performed by Chou, Nomura and Taya (1980) using a self-consistent approach, Nomura and Chou (1984) based upon a bound approach, and Taya and Chou (1984) employing a combination of Eshelby's (1957) equivalent inclusion method and Mori and Tanaka's (1973) back stress analysis.

For unidirectional hybrid composites composed of continuous fibers, Chamis and Sinclair (1979) have examined the elastic properties based upon composite micromechanics approach, linear laminate theory, and finite element analysis. It has been concluded that these methods predict approximately the same elastic properties. The through-the-thickness properties predicted by the micromechanics equations are in good agreement with the finite element results.

For simplicity, the results from the micromechanics approach are introduced below. The analytical model is an intermingled hybrid composite composed of two components, termed *primary composite* and *secondary composite* by Chamis and Sinclair (Fig. 5.11). Since these two components are interchangeable, they can be

Fig. 5.11. An intermingled hybrid composite lamina composed of primary (HM) and secondary (LM) phases.

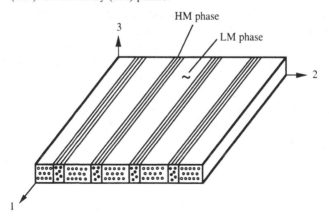

considered the high modulus (HM) and low modulus (LM) components. In the following, the subscripts 1, 2 and 3 denote, respectively, the directions along the fiber, transverse to the fiber and through the thickness in an intermingled hybrid lamina.

The effective longitudinal Young's modulus is approximated by the iso-strain assumption along the 1-direction:

$$E_{11} = E_{11}^{HM} + (E_{11}^{LM} - E_{11}^{HM})V_{LM} \tag{5.31}$$

Here, E_{11}^{HM} and E_{11}^{LM} denote the longitudinal Young's moduli of the HM and LM composites, respectively. V_{LM} is the volume fraction of the LM composite and $V_{LM} + V_{HM} = 1$. Both E_{11}^{HM} and E_{11}^{LM} can be expressed in terms of the properties of the fiber and matrix as given in Eqs. (2.7).

The transverse Young's modulus is obtained by assuming that the HM and LM components are connected in series in the 2-direction

$$E_{22} = E_{22}^{HM} \bigg/ \left[1 + \left(\frac{E_{22}^{HM}}{E_{22}^{LM}} - 1\right)V_{LM}\right] \tag{5.32}$$

Here, E_{22}^{HM} and E_{22}^{LM} are the transverse Young's moduli of the HM and LM composites, respectively. E_{22}^{HM} and E_{22}^{LM} can also be expressed in terms of fiber and matrix elastic properties as well as the fiber volume fraction of each component phase (Eqs. (2.7)).

The effective Young's modulus in the through-the-thickness direction has been modeled by Chamis and Sinclair assuming that the component phases are in parallel in the 3-direction. Thus,

$$E_{33} = E_{33}^{HM} + (E_{33}^{LM} - E_{33}^{HM})V_{LM} \tag{5.33}$$

Here, E_{33}^{HM} and E_{33}^{LM} are the transverse Young's moduli of the unidirectional composites composed of HM and LM fibers, respectively. Since the unidirectional all-HM or all-LM fiber composite is assumed to be transversely isotropic the expression of E_{33} $(=E_{22})$ of Eqs. (2.7) can be used to relate $E_{33}^{HM}(E_{33}^{LM})$ to the constituent material properties. However, it should be noted that for a yarn-by-yarn unidirectional intermingled hybrid lamina, $E_{22} \neq E_{33}$ and thus the composite is not transversely isotropic.

The in-plane shear modulus G_{12} is obtained by an iso-stress approximation

$$G_{12} = G_{12}^{HM} \bigg/ \left[1 + \left(\frac{G_{12}^{HM}}{G_{12}^{LM}} - 1\right)V_{LM}\right] \tag{5.34}$$

Similarly, the interlaminar shear modulus G_{23} is expressed as

$$G_{23} = G_{23}^{HM} \Big/ \left[1 + \left(\frac{G_{23}^{HM}}{G_{23}^{LM}} - 1\right)V_{LM}\right] \qquad (5.35)$$

The interlaminar shear modulus G_{13} is obtained by assuming that the component phases are connected in parallel in the 3-direction. Thus analogous to Eq. (5.31) the following expression can be obtained:

$$G_{13} = G_{13}^{HM} + (G_{13}^{LM} - G_{13}^{HM})V_{LM} \qquad (5.36)$$

in Eqs. (5.34)–(5.36), the shear moduli of the component phases can again be related to the constituent fiber and matrix properties by using Eqs. (2.7).

The Poisson's ratios v_{12} and v_{32} are derived by assuming parallel elements in the 1- and 3-directions.

$$v_{12} = v_{12}^{HM} + (v_{12}^{LM} - v_{12}^{HM})V_{LM} \qquad (5.37)$$

$$v_{32} = v_{32}^{HM} + (v_{32}^{LM} - v_{32}^{HM})V_{LM} \qquad (5.38)$$

For the derivation of v_{13}, the iso-strain and iso-stress states are assumed in the 1- and 3-directions, respectively. The result is

$$v_{13} = v_{13}^{HM} + \frac{V_{LM}(v_{13}^{LM} - v_{13}^{HM})}{(1 - V_{LM})\dfrac{E_{33}^{HM}}{E_{33}^{LM}} + V_{LM}} \qquad (5.39)$$

Experimental measurements of the elastic properties of intermingled hybrid composites can be found in the work of Chamis and Sinclair (1979) and Gruber and Chou (1983). The thermal properties of unidirectional intermingled hybrid composites have also been examined by Chamis and Sinclair (1979). It has been recommended that linear laminate theory be used to predict the thermal expansion coefficients.

5.3.2 *First cracking strain*

The linear portion of the stress–strain curve often extends beyond the failure strain of the pure LE fiber composite (point A' in Fig. 5.10) and the first cracking of the LE fibers occurs at the strain ε_{Lu} (point A in Fig. 5.10). This phenomenon is known as a 'hybrid effect'. The treatment of fiber first cracking strain of Aveston and Kelly (1980) is introduced below.

A unidirectional hybrid composite composed of both high and low elongation components can continue to bear the total load after the first cracking of the low elongation component if the following condition is satisfied:

$$V_{HE} \geq \frac{\sigma_{Lu}}{\sigma_{Hu} + \sigma_{Lu} - \sigma'_H} \qquad (5.40)$$

Here, σ_u denotes the failure stress, V indicates fiber volume fraction, the subscripts L and H denote the LE and HE components, respectively. Also, σ'_H ($= \varepsilon_{Lu} E_H$) is the stress of the HE component at the failure strain of the LE component. For both HE and LE components behaving elastically up to ε_{Lu}, the above condition can also be expressed in terms of the failure strain of the high elongation component

$$\varepsilon_{Hu} \geq \varepsilon_{Lu}(1 + \alpha) \qquad (5.41)$$

where

$$\alpha = \frac{E_{LE} V_{LE}}{E_{HE} V_{HE}} \qquad (5.42)$$

and E denotes the Young's modulus. Obviously the effect of stress redistribution due to fiber breakage is not considered here.

The magnitude of the first cracking strain ε_{Lu} and the extension of the curve between A and B in Fig. 5.10 can be understood using the concept of multiple cracking and constrained failure (Chapter 3). It has been established that the first failure strain is size dependent. The term *size* means essentially the effective diameters of fiber tows and their spacings at a fixed fiber volume content as well as the thickness of lamellae in an interlaminated hybrid.

When an interlaminated hybrid composite is deformed beyond the first cracking strain, parallel cracks will appear in the low elongation phase with the crack planes normal to the loading axis. The spacing of the cracks depends on the bonding (i.e. elastic or frictional) between the component plies, after the initial cracking. If they remain elastically bonded, the crack spacing is determined by the maximum interfacial shear stress at the crack, τ_{max}. On the other hand, if load transfer occurs through frictional bonds between the component phases, the crack spacing is determined by the limiting bond strength τ. In both cases, the formation of a crack through the ply thickness of the low elongation component results in the relaxation of the material in this ply on both sides of the crack. The nearly flat portion of the composite stress–strain curve, i.e. AB in Fig. 5.10, is the consequence of multiple cracking of the low elongation phase and the associated extension of the specimen.

The stress originally carried by the low elongation component on the crack plane has to be shifted to the high elongation phase. The maximum additional stress thrown onto the high elongation phase can be estimated by

$$\Delta\sigma = \frac{\sigma_a}{V_{HE}} - \frac{\sigma_a E_{HE}}{E_c} \tag{5.43}$$

where σ_a is the applied stress and E_c is the Young's modulus of the composite. The first term on the right-hand side of Eq. (5.43) is the stress on the crack plane while the second term gives the stress away from the crack plane in the high elongation component. The additional load carried by the high elongation component induces an extension δl of the specimen ends.

The idea of load transfer between the HE and LE components explains why the strain ε_{Lu}, at which first cracking occurs, depends upon the dispersion of the component phases. When thinner fibers or lamellae are used, the interfacial area per unit volume between the two component phases increases. This also means increased efficiency of load transfer from the HE component bridging the crack back to the LE layers. As a result, the additionally strained length of this component and, hence, the displacement of the specimen ends, δl, are decreased. The product of δl and $E_c\varepsilon_{Lu}$ gives the upper limit of the work available from the loading system to form the crack. Assuming a constant surface work of fracture γ, the decrease in δl will reach such a point that ε_{Lu} must increase above the value for the pure LE composite before the required work of fracture can be extracted from the system. In the case of inter-laminated hybrids with an elastically bonded interface the cracking strain of the low elongation phase of thickness d can be estimated by (Aveston and Kelly 1980)

$$\varepsilon_{Lu} = \sqrt{\left[\frac{2\gamma_{LE}V_{LE}}{E_c\alpha}\sqrt{\left(\frac{E_cG_{LE}}{E_{HE}E_{LE}}\right)\frac{1}{\sqrt{(V_{HE})d}}}\right]} \tag{5.44}$$

where α is defined in Eq. (5.42). Equation (5.44) predicts that ε_{Lu} varies with the inverse square root of ply thickness. This prediction of the hybrid effect is consistent with the experimental observation of carbon/glass sandwich laminates that the carbon ply failure strain is greater when its absolute thickness is smaller.

The theory of multiple cracking in fiber composites has very broad applicability. Multiple cracking occurs in the brittle phase which could be either the matrix or the fiber phase of an aligned fiber composite, the low elongation layers of a non-hybrid laminated composite, or the layers reinforced with LE fibers in an interlaminated hybrid composite. Multiple cracking of non-hybrid composites has been examined in Chapter 3. Aveston and Kelly (1980) have summarized the analytical expressions of the minimum crack spacings, the first cracking strain and the maximum interfacial stress for hybrid and non-hybrid composites with both elastic and sliding friction bonds.

The above discussions have provided the answer to the question posed in Section 5.1 concerning synergistic effects in hybrid composites. The answer to the question of load sharing is also positive. The bond between the fiber and matrix in a hybrid ensures that the LE fiber continues to carry part of the applied load and to contribute to the overall stiffness after first cracking (Bunsell and Harris 1974). The load sharing by the LE fiber is evident from the observation of multiple fractures in well bonded interply hybrids and by the bursts of acoustic emission accompanying the repeated load drops on the stress–strain curve. Finally, the rise of curve *BC* in Fig. 5.10 is attributed to the loading of the high elongation fibers. The failure strain of the hybrid composite (point *D*) is lower than that of the high elongation fiber composite (point *G*). This is because the multiple fracture of the LE fibers and partial debonding between the HE and LE fiber reinforced laminae. Consequently, the HE layers in an interlaminated hybrid composite cannot be stretched uniformly along their length to the ultimate strain level. On the other hand, if the debonding is complete at the first fiber cracking strain, ε_{Lu}, the stress–strain curve follows the path *OAEF*. Note that *EF* and *BC* in Fig. 5.10 have the same slope.

5.3.3   *Differential Poisson's effect*

Another factor that needs to be taken into account in the deformation of hybrid composites is the interlaminar stress induced due to differential Poisson's effect (Aveston and Kelly 1980). It is understood that this effect exists in laminated composites with or without fiber hybridization. To demonstrate the magnitude of the Poisson's strain and its effect on longitudinal splitting of laminated composites, a three-layer non-hybrid cross-ply laminate is considered. The central LE (90°) layer in this case is sandwiched between two HE (0°) layers. Thus the conclusions derived from this example are applicable to interlaminated composites in general. The strain induced by the differential Poisson's effect depends only on the volume fraction of the component phases. However, cracking of the lamina can be minimized by making the LE layers sufficiently thin.

The critical strain for causing longitudinal split due to Poisson's effect can be derived based upon Fig. 3.25. Let the subscripts 1 and 2 denote the longitudinal and transverse fiber directions, respectively, and $x-y-z$ are the reference axes of the cross-ply. Under a simple extension ε_{yy}, the following strains are induced in the 0° and 90° layers, if they are deformed independently:

$$\varepsilon_{xx}(0°) = -v_{12}\varepsilon_{yy}$$

$$\varepsilon_{xx}(90°) = -v_{21}\varepsilon_{yy} \tag{5.45}$$

For the composite laminate subjected to ε_{yy}, the strain induced in the x direction is ε_{xx}. Referring to Fig. 3.25, the transverse strains induced in the 0° and 90° layers due to the Poisson effect are

$$\Delta\varepsilon_{xx}(0°) = \varepsilon_{xx} - \varepsilon_{xx}(0°) = \varepsilon_{xx} + v_{12}\varepsilon_{yy}$$

$$\Delta\varepsilon_{xx}(90°) = \varepsilon_{xx} - \varepsilon_{xx}(90°) = \varepsilon_{xx} + v_{21}\varepsilon_{yy} \tag{5.46}$$

Thus

$$\Delta\varepsilon_{xx}(0°) - \Delta\varepsilon_{xx}(90°) = (v_{12} - v_{21})\varepsilon_{yy} \tag{5.47}$$

The stress σ_{xx} is built up in each layer due to the requirement of compatibility in normal strain in the x direction. These stresses are given approximately by

$$\sigma_{xx}(0°) = \Delta\varepsilon_{xx}(0°)E_{22}$$

$$\sigma_{xx}(90°) = \Delta\varepsilon_{xx}(90°)E_{11} \tag{5.48}$$

The force equilibrium of the laminate in the transverse direction requires

$$\sigma_{xx}(0°)b + \sigma_{xx}(90°)h = \Delta\varepsilon_{xx}(0°)E_{22}b + \Delta\varepsilon_{xx}(90°)E_{11}h = 0 \tag{5.49}$$

From Eqs. (5.47) and (5.49), the transverse strain induced in the 0° layer due to the Poisson's effect is

$$\Delta\varepsilon_{xx}(0°) = \frac{E_{11}h(v_{12} - v_{21})\varepsilon_{yy}}{(h+b)E_c} \tag{5.50}$$

where $E_c = E_{22}(b/h + b) + E_{11}(h/h + b)$ is the effective Young's modulus along the x direction.

If the $0°$ and $90°$ layers in the laminate of Fig. 3.25 are of different materials, then it is necessary to distinguish the elastic constants in Eq. (5.50)

$$\Delta \varepsilon_{xx}(0°) = \frac{E_{11}(90°)h[v_{12}(0°) - v_{21}(90°)]\varepsilon_{yy}}{bE_{22}(0°) + hE_{11}(90°)} \tag{5.51}$$

5.3.4 *Differential thermal expansion*

Additional strain may be induced in a laminated hybrid composite due to differential thermal expansion of the component phases. The carbon/glass hybrid system is a typical case where the axial thermal expansion coefficient of the glass laminae is much larger than that of the carbon layers. Upon cooling down from the stress-free temperature, compressive stress develops in the carbon layers as a result of the differential thermal expansion. This thermally induced compression can partially account for the hybrid effect often observed in carbon/glass hybrids (Bunsell and Harris 1974). The constrained thermal strain in such a hybrid with high and low elongation fibers can be expressed as (Aveston and Kelly 1980)

$$\varepsilon_{LE}^T = \frac{\Delta T E_{HE} V_{HE}}{E_c} (\alpha_{LE} - \alpha_{HE}) \tag{5.52}$$

$$\varepsilon_{HE}^T = \frac{\Delta T E_{LE} V_{LE}}{E_c} (\alpha_{HE} - \alpha_{LE}) \tag{5.53}$$

where ΔT = stress-free temperature – service temperature, and α_{LE} and α_{HE} denote the thermal expansion coefficients. The thermal strain components are independent of fiber or lamina dimension. The relation between the cracking strain and the dimension of the low elongation phase is identical under external load and thermal load. Hence, the treatment of multiple cracking can still be applied in this case.

5.4   Strength theories

Discussions on the strength of hybrid composites begin with an introduction on the rule-of-mixtures type of approach which delineates the contributions of the high elongation and low elongation fibers to the load carrying capacity of the hybrid. In this approach, the fibers are assumed to be of uniform strength and the local stress redistributions due to fiber breakage are not taken into account. This is then followed by the probabilistic strength theories. The synergistic effects between the LE and HE fibers in local stress concentration and fiber strength distributions are modeled to predict the first failure strength and ultimate failure strength of hybrid composites.

5.4.1 *Rule-of-mixtures*

The ultimate tensile strength of a unidirectional hybrid composite can be estimated from the contributions of the component phases at different volume fractions. Consider a binary composite with low elongation fibers. The addition of a small amount of higher elongation fiber decreases the strength of the composite. The ultimate hybrid composite tensile strength, σ_{cu}, is given by (Aveston and Kelly 1980)

$$\sigma_{cu} = \sigma_{Lu}V_{LE} + \varepsilon_{Lu}E_{HE}V_{HE} \tag{5.54}$$

where ε_{Lu} is the failure strain of the LE fiber in the hybrid. The failure of the low elongation fiber leads to the fracture of the hybrid and there is no multiple fracture.

As the content of the high elongation fiber increases, a transition in failure mode occurs when there is sufficient volume of these fibers to carry the load upon the fracture of the low elongation fibers. The fracture mode is multiple fracture of the brittle fibers. The ultimate tensile strength is represented by

$$\sigma_{cu} = \sigma_{Hu}V_{HE} \tag{5.55}$$

The volume fraction of HE fibers should exceed the lower limit, given by Eq. (5.40), to bear the total load at the first cracking strain. Aveston and Kelly (1980) applied Eqs. (5.54) and (5.55) to analyze the experimental data of Kalnin (1972), who studied the failure stress of interlaminated carbon/glass/epoxy hybrids. Figure 5.12 shows the variation of hybrid composite ultimate tensile strength with the relative glass fiber content. The stress is calculated by dividing the load by the cross-sectional area of the fiber and, thus, the contribution of the epoxy matrix is neglected. Aveston and Kelly suggested that for this particular experimental system the fracture of carbon fibers did not produce a large stress concentration leading to the weakening of the glass fibers.

5.4.2 *Probabilistic initial failure strength*

It has been shown in Section 5.3.2 that the phenomena of initial failures of unidirectional hybrid composites and cross-ply laminates share the same basic physical principle. In unidirectional hybrid composites consisting of LE and HE fibers, the failure strain of the LE fibers under tension is often greater than that in the all LE fiber composite. On the other hand, in the non-hybrid $[\pm \theta°/90°]_s$ $(0° < \theta < 90°)$ laminates the failure strength (strain) of the 90° (LE) layer under tension is greater for smaller thickness of the inner 90° layers. It is also known that the failure strength (strain) of the inner 90° layer depends on the material properties of the outer $\pm \theta°$ layers. Thus, in composites which consist of laminae with two different types of material properties such as hybrid composites and $[\pm \theta°/90°]_s$ non-hybrid laminates, the failure strength (strain) of the LE layers is not an intrinsic material property and it depends on the HE material properties and the geometric arrangement of the layers.

Fukunaga, Chou and Fukuda (1984) and Fukunaga *et al.* (1984a&b) have examined the initial failure strength of both hybrid and non-hybrid composites based upon a statistical approach. The hybrid composite under consideration is a sandwiched structure composed of unidirectional glass fiber and carbon fiber laminae (Fig. 5.13a). The non-hybrid composite is a carbon composite with the $[\pm \theta°/90°]_s$ configuration (Fig. 5.13b). Both composites can be depicted by the HE and LE representations of Fig. 5.13(c) where M_{HE} and M_{LE} denote, respectively, the number of HE and LE

Fig. 5.12. Tensile strength of aligned carbon/glass/epoxy hybrids vs. relative fiber content. (After Aveston and Kelly 1980.)

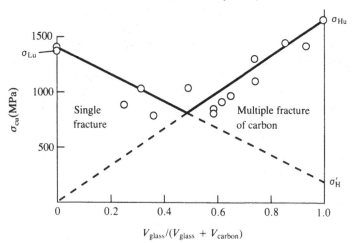

layers. Thus, the total number of layers in the hybrid composite, $M = M_{HE} + M_{LE}$.

Fukunaga and colleagues modeled the composite as a chain of short laminates in series as shown in Fig. 3.17. Each laminate has the length δ equivalent to the ineffective length, and the specimen length $l = N\delta$. In order to obtain the first ply failure strength of the whole laminated composite, the following two-parameter cumulative Weibull distribution functions of failure strain for the LE and HE short layers are assumed

$$F_{LE}(\varepsilon) = 1 - \exp[-(\varepsilon/\varepsilon^*_{LE})^{\beta_{LE}}]$$

$$F_{HE}(\varepsilon) = 1 - \exp[-(\varepsilon/\varepsilon^*_{HE})^{\beta_{HE}}]$$

(5.56)

Here ε^* and β denote the scale and shape parameters, respectively. Then, the cumulative distribution function $H_c(\varepsilon)$ for the first ply failure strain of the composite can be obtained from the weakest link model

$$H_c(\varepsilon) = 1 - [1 - G_c(\varepsilon)]^N$$

(5.57)

where $G_c(\varepsilon)$ is the cumulative distribution function for the first ply failure strain of the short laminate and it is given by

$$G_c(\varepsilon) = 1 - [1 - F_{LE}(\varepsilon)]^{M_{LE}}[1 - F_{HE}(\varepsilon)]^{M_{HE}}$$

(5.58)

Fig. 5.13. (a) Carbon/glass hybrid laminate. (b) $[\pm\theta°/90°]_s$ non-hybrid laminate. (c) Model for analysis. (After Fukunaga *et al.* 1984c.)

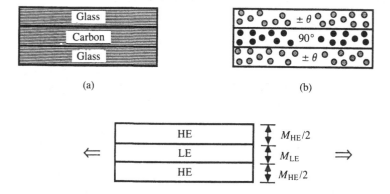

Substitution of Eq. (5.58) into Eq. (5.57) yields

$$H_c(\varepsilon) = 1 - \exp[-NM_{LE}(\varepsilon/\varepsilon_{LE}^*)^{\beta_{LE}} - NM_{HE}(\varepsilon/\varepsilon_{HE}^*)^{\beta_{HE}}]$$

$$(5.59)$$

It should be noted that Eq. (5.59) is concerned with the first ply failure strain of the composite laminate not the LE phase of the composite. This is because, from a probabilistic viewpoint, the failure of the LE layer does not always precede that of the HE layers. Similar to Eq. (5.59) the cumulative distribution function $H_{LE}(\varepsilon)$ for the first ply failure strain of the all LE fiber composite (of the same size as the hybrid composite) is given by

$$H_{LE}(\varepsilon) = 1 - \exp[-NM(\varepsilon/\varepsilon_{LE}^*)^{\beta_{LE}}]$$

$$(5.60)$$

Fukunaga and colleagues have compared the failure strains for the HE/LE/HE laminate and the pure LE laminate for the failure strains at 50% failure probability. From Eqs. (5.59) and (5.60)

$$\frac{M_{LE}}{M}\left(\frac{\varepsilon_c}{\varepsilon_{LE}^*}\right)^{\beta_{LE}} + \frac{M_{HE}}{M}\left(\frac{\varepsilon_c}{\varepsilon_{HE}^*}\right)^{\beta_{HE}} = \frac{\ln 2}{NM}$$

$$\left(\frac{\varepsilon_{LE}}{\varepsilon_{LE}^*}\right)^{\beta_{LE}} = \frac{\ln 2}{NM}$$

$$(5.61)$$

where ε_c and ε_{LE} denote, respectively, the median failure strains of the hybrid and LE composite. In order to obtain an explicit relation between the ratio of median failure strains, $\varepsilon_c/\varepsilon_{LE}$, and the relative volume fraction of the LE material, M_{LE}/M, the case of $\beta_{LE} = \beta_{HE} = \beta$ is considered. Then, Eqs. (5.61) become

$$\frac{\varepsilon_c}{\varepsilon_{LE}} = \left(\frac{M_{LE} + M_{HE}/(\varepsilon^*)^{\beta}}{M}\right)^{-1/\beta}$$

$$(5.62)$$

where $\varepsilon^* = \varepsilon_{HE}^*/\varepsilon_{LE}^*$. The ratio $\varepsilon_c/\varepsilon_{LE}$ in Eq. (5.62) is independent of the size of the composite. By assuming linear stress–strain relations for the HE and LE layers, the first ply failure strength ratio σ_c/σ_{LE} is readily obtained

$$\frac{\sigma_c}{\sigma_{LE}} = \frac{E_c}{E_{LE}}\left(\frac{M_{LE} + M_{HE}/(\varepsilon^*)^{\beta}}{M}\right)^{-1/\beta}$$

$$(5.63)$$

where E denotes the Young's modulus in the loading direction, $E_c = (M_{LE}E_{LE} + M_{HE}E_{HE})/M$.

Numerical calculations of Eqs. (5.62) and (5.63) have been performed for carbon/glass/epoxy unidirectional hybrid composites. It is assumed that $\varepsilon^* = 3.0$ and $E_{\mathrm{HE}}/E_{\mathrm{LE}} = \frac{1}{3}$. Figure 5.14 shows the comparison of Eq. (5.62) with the experimental results of Bader and Manders (1981a&b) for HTS carbon/glass hybrid laminates. The analytical results for $\beta = 10$ seem to show good agreement with experimental results. Figure 5.15 indicates the variation of initial composite failure strength with the relative volume fraction of LE and HE layers. Points A and D in Fig. 5.15 represent the strengths of glass and carbon composites, respectively. Line $BD(\sigma_c/\sigma_{\mathrm{LE}} = E_c/E_{\mathrm{LE}})$ indicates the stress in the hybrid at which failure of the carbon layer takes place. Line $AE(\sigma_c/\sigma_{\mathrm{LE}} = M_{\mathrm{HE}}/(M_{\mathrm{LE}} + M_{\mathrm{HE}}))$ represents the stress in the hybrid assuming that the LE layer carries no load. As the shape parameter β decreases, the first ply failure strain and strength of the hybrid composite increase relative to those of the all carbon fiber composite. When $\beta \rightarrow \infty$, that is for composites without scattering in the failure strains of the LE and HE layers, the first ply failure strain of the hybrid composite is identical to that of the pure LE composite. This relation at $\beta \rightarrow \infty$ in Fig. 5.15 is equivalent to the rule-of-mixtures for the initial failure in hybrids.

Fig. 5.14. Comparisons of the analytical predictions of Fukunaga *et al.* (1984c) and the experimental results of Bader and Manders (1981a&b) for HTS carbon/E-glass hybrid laminates. (After Fukunaga *et al.* 1984c.)

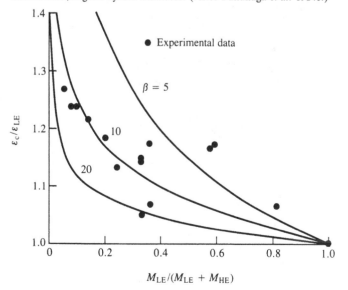

The trend of variation of $\varepsilon_c/\varepsilon_{LE}$ with the relative 90° layer volume fraction for $[\pm\theta°/90°]_s$ laminates is similar to that given in Fig. 5.14. For a given value of shape parameter, $\varepsilon_c/\varepsilon_{LE}$ decreases with the increase in the angle θ.

In summary, the probabilistic strength analysis of Fukunaga, Chou and Fukuda (1984) and Fukunaga *et al.* (1984a) for the initial failure in hybrid and non-hybrid composites has demonstrated a volumetric relation. It has been shown that the initial failure strain or strength is greater in composites composed of low elongation and high elongation materials than in the all low elongation fiber composite. This is the result of a 'size effect'; that is the failure probability is lower in the composite with the smaller size of the low elongation material.

5.4.3   *Probabilistic ultimate failure strength*

In this section three analytic approaches are presented for predicting the ultimate tensile strength of hybrid composites. All these methods consider the variability in fiber strength and the stress redistribution at fiber fracture. First, the analytical model of Zweben (1977) assumes that the LE and HE fibers are arranged in alternating positions in a unidirectional lamina. The fracture of an

Fig. 5.15. Composite initial failure strength σ_c (normalized by σ_{LE}) in glass/carbon/glass laminates. $\beta_{LE} = \beta_{HE} = \beta$, $\varepsilon^*_{HE} = \varepsilon^*_{LE} = 3.0$ and $E_{HE}/E_{LE} = \frac{1}{3}$. (After Fukunaga *et al.* 1984c.)

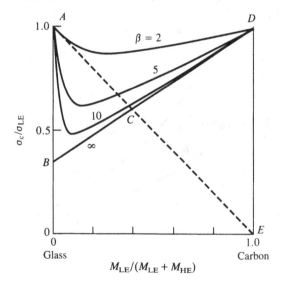

LE fiber induces, in the vicinity of the fiber fracture, a local stress (strain) concentration. The HE fibers will break at the points of stress (strain) concentration. Zweben hypothesizes that the strain level at which the first overstressed HE fiber is expected to break is a lower bound on the ultimate strain of the hybrid composite.

Next, in order to obtain a more realistic view of the multiple fracture of the LE fibers in a hybrid composite, Fukuda and Chou (1982a&b) adopted a Monte-Carlo simulation of the ultimate failure of a unidirectional hybrid composite. The method demonstrates the diffused nature of fracture of the LE fibers. The constraint on the propagation of the LE fiber fractures due to the presence of HE fibers as well as the stress–strain relation of the hybrid composite as influenced by multiple fiber fractures has been demonstrated.

Thirdly, Fukunaga, Chou and Fukuda (1989) have adopted a more rigorous approach by applying the methods of Harlow and Phoenix (1978a&b). The effects on the ultimate strength of hybrid laminates due to the scatter of laminar strengths, relative fiber volume fractions, composite size and laminate stacking sequence have been identified.

The analytical model of Zweben is composed of a single layer of fibers of axial length l with the same arrangement as that shown in Fig. 3.17. High elongation (low modulus) fibers and low elongation (high modulus) fibers are arranged in alternating positions. The total number of fibers in the composite is M, of which $M/2$ are LE, and $M/2$ are HE fibers. The term 'fibers', according Zweben, represents both single fibers and yarns.

It is assumed that the fibers support all of the applied load and the fibers break under load in a random fashion throughout the composite. Attention is focussed on what happens in the vicinity of the breaks in the LE fibers. The strain of the composite is adopted as the independent variable. When an LE fiber breaks at the strain level ε, the two adjacent HE fibers are subjected to a strain concentration of $K_h\varepsilon$, where K_h is the strain concentration factor associated with a single broken LE fiber. Because of the linear elastic deformation assumed in the model, K_h is also the stress concentration factor. Since the model assumes that the fiber axial stress depends only on the axial strain, the stress in these two HE fibers increases from $E_{HE}\varepsilon$ to $K_h E_{HE}\varepsilon$. The axial distance over which the fiber stress is perturbed due to a fiber breakage is known as the ineffective length (Rosen 1964). Zweben denoted the ineffective length associated with a broken LE fiber in the hybrid by δ_h.

Similar to Fig. 3.17, the hybrid composite is composed of a series of links with axial dimension δ_h. The total number of links is $N_h = l/\delta_h$. Failure of the composite results from the propagation of fiber breaks due to local strain concentrations.

The cumulative distribution functions for the failure strains of the LE and HE fibers of length l are assumed in the form of Weibull distributions:

$$F_{LE}(\varepsilon) = 1 - \exp(-pl\varepsilon^q)$$

$$\tag{5.64}$$

$$F_{HE}(\varepsilon) = 1 - \exp(-rl\varepsilon^s)$$

where p, q, r and s are Weibull parameters. The composite strain at which the fracture of the first overstressed HE fibers occurs in the hybrid is given by

$$\varepsilon_h = [Mlpr\delta_h(K_h^s - 1)]^{-1/(q+s)} \tag{5.65}$$

For a composite reinforced with M fibers of the same type, say the LE fibers, Eq. (5.65) becomes

$$\varepsilon = [2Mlp^2\delta_{LE}(K_{LE}^q - 1)]^{-1/2q} \tag{5.66}$$

Here, K_{LE} and δ_{LE} are, respectively, the strain concentration factor and ineffective length of the LE fiber composite.

From Eqs. (5.65) and (5.66), Zweben has obtained the ratio of the lower bounds of failure strain of a hybrid to that of a LE fiber composite of the same length

$$R = \frac{\varepsilon_h}{\varepsilon} = \frac{[Mlpr\delta_h(K_h^s - 1)]^{-1/(q+s)}}{[2Mlp^2\delta_{LE}(K_{LE}^q - 1)]^{-1/2q}} \tag{5.67}$$

Equation (5.67) indicates that the ratio of the failure strains depends on the Weibull strength parameters p, q, r and s, the specimen length l, the ineffective lengths δ_h and δ_{LE}, and the strain concentration factors K_{LE} and K_{HE}. Zweben has derived approximate expressions of these parameters in terms of the fiber elastic properties and the geometric parameters of fiber arrangements in the composites.

For the convenience of assessing the failure characteristics of hybrid composites, Eq. (5.67) can be simplified by assuming that both LE and HE fibers have the same coefficient of variation in tensile failure strain. It can be shown that for this case $q = s$ and Eq. (5.67) becomes

$$R = \left(\frac{r}{p}\right)^{-1/2q}\left[\frac{\delta_h(K_h^q - 1)}{2\delta_{LE}(K_{LE}^q - 1)}\right]^{-1/2q} \tag{5.68}$$

From Eq. (3.54a) the mean strains of the LE and HE fibers for the gauge-length l are given by

$$\bar{\varepsilon}_{LE}(l) = (pl)^{-1/q}\Gamma\left(1 + \frac{1}{q}\right)$$

$$\bar{\varepsilon}_{HE}(l) = (rl)^{-1/s}\Gamma\left(1 + \frac{1}{s}\right)$$

(5.69)

where Γ is a gamma function. Thus, for $q = s$, Eqs. (5.68) and (5.69) yield

$$R = \sqrt{\left(\frac{\bar{\varepsilon}_{HE}}{\bar{\varepsilon}_{LE}}\right)}\left[\frac{\delta_h(K_h^q - 1)}{2\delta_{LE}(K_{LE}^q - 1)}\right]^{-1/2q}$$

(5.70)

Equation (5.70) can be further simplified if the fiber coefficient of variation is small (i.e. 5% or less) and thus the shape parameter is large ($q > 25$). For this case $K_{HE}^q - 1 \approx K_{HE}^q$ and $K^q - 1 \approx K^q$, and Eq. (5.70) is reduced to

$$R = 2^{1/2q}\sqrt{\left(\frac{\bar{\varepsilon}_{HE}}{\bar{\varepsilon}_{LE}}\right)}\left(\frac{\delta_{LE}}{\delta_h}\right)^{1/2q}\sqrt{\left(\frac{K_{LE}}{K_h}\right)}$$

(5.71)

Equation (5.71) indicates that the ratio of the lower bounds of failure strains is sensitive to the mean fiber failure strains and the strain concentration factors and it is less sensitive to the ineffective lengths under the assumption of low fiber coefficient of variations. For the case of an intermingled Kevlar 49/Thornel 300 hybrid composite, Zweben obtained $K_h = 1.462$, $K_{LE} = 1.293$, $\delta_h/\delta_{LE} = 1.573/1.531 = 1.03$, and $\bar{\varepsilon}_{HE}/\bar{\varepsilon}_{LE} = 1.63$. The fiber strain coefficients of variation for both HE and LE fibers are assumed to be about 6% if the Weibull parameter $q = 20$ is adopted. Using these values, Eq. (5.71) gives $R = 1.22$.

In spite of the simplifying assumption used, Zweben's model, in essence, predicts that the introduction of HE fibers into a LE fiber composite enhances the failure strain of the hybrid composite. This effect is attributed to the ability of HE fibers to redistribute the local stress concentration and act like crack arrestors at the micromechanical level.

Fukuda and Chou (1982a) have examined the strength of hybrid composites using the method of Monte Carlo simulation. Figure 5.16 shows an idealized intermingled hybrid composite sheet of unit thickness. The LE and HE fibers assume alternating positions. For the purpose of numerical calculations, a five-fiber region with three LE fibers and two HE fibers is considered. Following the notations of the chain-of-links model given in Fig. 3.17, $M = 5$ and $N = 20$ are adopted for Fig. 5.16. Also the extensional rigidity ratios $R(= E^*d^*/Ed)$ are assumed to be unity (for a non-hybrid composite) and $\frac{1}{3}$ (for simulating a glass/carbon composite).

Using a shear-lag analysis, the stress concentration factors are evaluated. The fiber breakage patterns considered by Fukuda and Chou include all the combinations of one-, two-, three- and four-fiber fracture in a transverse plane. A shear-lag analysis is applied to evaluate the stress concentration factors of all the intact links in the layer where fiber breakage has taken place. Results for two kinds of fiber composites are presented below.

In the first case, there is only one type of fiber; its strength follows a normal distribution with an average normalized strength of unity and a standard deviation of 0.1. A typical sequence of link failure is shown in Fig. 5.17(a), where 0 indicates that the link is intact and the other numbers show the sequence of fracture of links. In this model, the link of $i = 14$, $j = 2$ breaks first; the link of $i = 5$, $j = 4$ breaks second; finally the failure occurs at the transverse plane of $i = 14$. In a total of 100 links, six links are broken. One hundred iterations have been performed by Fukuda and Chou and the number of broken links of each iteration is found to be either five or six. Generally speaking, multiple fiber fractures are not extensive in non-hybrid composites.

Fig. 5.16. Model for the Monte Carlo simulation. (After Fukuda and Chou 1982a.)

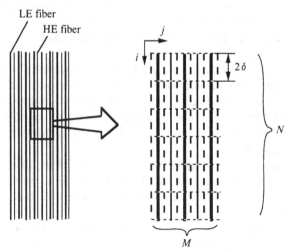

Figure 5.17(b) shows the failure sequence of links in the hybrid composite. It is seen that the LE fibers tend to break at the initial stage of loading and that a total of 16 links are broken for this model to fail at the plane of $i = 10$. A minimum of five and a maximum of 32 link failures are observed during 100 iterations, and the average number of broken links in one iteration is 16.3. The degree of multiple fiber fracture is considerably more extensive in the hybrid composite than the non-hydrid case.

Figure 5.18 shows the examples of stress–strain relations of both non-hybrid and hybrid composites. The stress σ_c and strain ε_c of composites are normalized by the average ultimate stress (σ_{LINK}) and average ultimate strain (ε_{LINK}) of the links of LE fibers. The initial failure strains of the LE fiber composite and the hybrid composite are nearly the same. However, in the hybrid composite, initial multiple failures of the LE fibers are arrested by the HE fibers. As a result, the hybrid composite can withstand more deformation and hence a higher ultimate failure strain. According

Fig. 5.17. Examples of the fiber link failure sequence. (a) Non-hybrid composite. (b) Hybrid composite. (After Fukuda and Chou 1982a.)

0	0	0	0	0		0	0	0	0	0
0	0	0	0	0		0	0	0	0	2
0	0	0	0	0		0	0	0	0	0
0	0	0	0	0		0	0	0	0	0
0	0	0	2	0		0	0	0	0	0
0	0	0	0	0		0	0	0	0	0
0	0	0	0	0		10	0	13	0	14
0	0	0	0	0		0	0	11	0	0
0	0	0	0	0		3	0	4	0	0
0	0	0	0	0		7	16	8	15	1
0	0	0	0	0		0	0	0	0	0
0	0	0	0	0		9	0	0	0	0
0	0	0	0	0		0	0	0	0	0
3	1	4	5	6		0	0	0	0	0
0	0	0	0	0		0	0	0	0	0
0	0	0	0	0		0	0	0	0	0
0	0	0	0	0		0	0	0	0	0
0	0	0	0	0		0	0	0	0	0
0	0	0	0	0		0	0	0	0	0
0	0	0	0	0		12	0	6	0	5
						LE	HE	LE	HE	LE
		(a)						(b)		

to this example, the ultimate strain increases approximately 50% over that of the LE fiber composite.

Although the total number of fibers utilized in this numerical model is extremely small the progressive nature of failure as indicated in the stress–strain curve resembles that of the experimental curves obtained by Bunsell and Harris (1974) of carbon/glass hybrid composites.

The ultimate strength of the hybrid predicted in Fig. 5.18 is lower than those of the LE and HE fiber composites and the ultimate strain of the hybrid is lower than that of the HE fiber composite. This can be understood from the fact that, with the exception of those on the crack plane, the links of the HE fibers are constrained by the surrounding LE links and cannot be deformed to their ultimate strain. As a result, the strength potential of the HE links is not fully realized.

Fukunaga, Chou and Fukuda (1989) have extended the treatment of Harlow and Phoenix (1978a&b) to analyze the ultimate tensile strength of hybrid composites. In their model, the laminate is also treated as a chain of N short segments arranged in series. Each segment has a length δ and the total length of the composite is $l = N\delta$. There are four layers in the laminate. Figure 5.19 shows the stacking sequence of the laminates consisting of LE and HE fiber layers. The strain concentration factors for the various configurations of layer breakage have been evaluated by the eigenvector expansion method (Fukunaga, Chou and Fukuda 1984).

Fig. 5.18. Normalized stress–strain relations of non-hybrid and hybrid composites. (After Fukuda and Chou 1982a.)

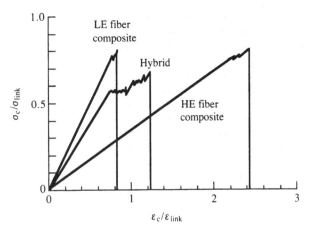

The strength analysis is performed on the basis of the knowledge of the stress redistribution due to layer breakage. It is assumed that each layer follows a two-parameter Weibull probability distribution function and the ultimate failure is defined as the failure of all the layers. The cumulative distribution functions for the ultimate strains of the LE and HE layers are denoted by $F_{LE}(\varepsilon)$ and $F_{HE}(\varepsilon)$, respectively. They are expressed as in Eqs. (5.56). The failure patterns of the four-layered hybrid are shown in Fig. 5.20 where the circles and crosses denote the intact and broken layers, respectively; the possibilities of one, two, three, and four fractured layers are given. According to Fukunaga and colleagues, there are altogether 75 possible failure sequences which can be grouped into eight

Fig. 5.19. Stacking sequences of laminates. (After Fukunaga, Chou and Fukuda 1989.)

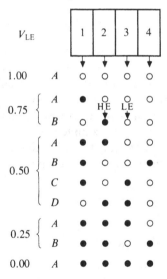

Fig. 5.20. Failure patterns of the four-layered hybrid composite. (After Fukunaga, Chou and Fukuda 1989.)

	b : x o o o	f : x x o o	l : x x x o	
	c : o x o o	g : x o x o	m : x x o x	
a : o o o o	d : o o x o	h : x o o x	n : x o x x	p : x x x x
: 1 2 3 4	e : o o o x	i : o x x o	o : o x x x	
		j : o x o x		
		k : o o x x		

different types. The number of different failure sequences in each type has been identified. A typical failure sequence in each type and the corresponding failure probability are given in Table 5.3 where $F(x)$ stands for $F_{LE}(\varepsilon)$ for the LE layer and $F_{HE}(\varepsilon)$ for the HE layer, and $K_{ij,k}$ denotes the strain concentration factor of the kth layer due to the fracture of the ith and jth layers.

The cumulative distribution function, $G(\varepsilon)$, in failure strain for a laminate segment of length δ is given by the summation of the failure probabilities of failure sequences no. 1 through no. 75. When $G(\varepsilon)$ is given, the cumulative distribution function with respect to the average stress, $G(\sigma)$ can be obtained from the relation $\sigma = E_c \varepsilon$, where E_c is the effective axial Young's modulus of the hybrid laminate. Finally, the cumulative distribution function, $H(\sigma)$, for the hybrid composite of length $l = N\delta$ is given by the weakest link theory as follows:

$$H(\sigma) = 1 - [1 - G(\sigma)]^N \tag{5.72}$$

Two limiting cases are considered. In the case of non-hybrid composites, $F_{LE}(\varepsilon) = F_{HE}(\varepsilon)$, the above results can be reduced to those of Harlow and Phoenix if the local load sharing (equal load sharing) rule is used for the strain redistribution. Another case deserving attention is the situation of high failure probability, for which the first ply failure may trigger the complete failure of the laminate. The cumulative distribution function for the first ply failure strength of a laminate segment is then given by Eq. (5.58)

Table 5.3. *Failure sequence and failure probability of the four-layered hybrid composite*

Type	Nos.	Failure sequence	Failure probability
I	1–24	*abflp*	$F(x)\{F(K_{1,2}x) - F(x)\}\{F(K_{12,3}x) - F(K_{1,3}x)\}$ $\{F(K_{123,4}x) - F(K_{12,4}x)\}$
II	25–36	*ablp*	$F(x)\{F(K_{1,2}x) - F(x)\}\{F(K_{1,3}x) - F(x)\}\{F(K_{123,4}x) - F(K_{1,4}x)\}$
III	37–48	*abfp*	$F(x)\{F(K_{1,2}x) - F(x)\}\{F(K_{12,3}x) - F(K_{1,3}x)\}$ $\{F(K_{12,4}x) - F(K_{1,4}x)\}$
IV	49–60	*aflp*	$F(x)^2\{F(K_{12,3}x) - F(x)\}\{F(K_{123,4}x) - F(K_{12,4}x)\}$
V	61–4	*abp*	$F(x)\{F(K_{1,2}x) - F(x)\}\{F(K_{1,3}x) - F(x)\}\{F(K_{1,4}x) - F(x)\}$
VI	65–70	*afp*	$F(x)^2\{F(K_{12,3}x) - F(x)\}\{F(K_{12,4}x) - F(x)\}$
VII	71–4	*alp*	$F(x)^3\{F(K_{123,4}x) - F(x)\}$
VII	75	*ap*	$F(x)^4$

with ε replaced by σ. Fukunaga and colleagues have shown that $G(\sigma)$ can be expressed on the Weibull probability paper in the following form:

$$\ln\{-\ln(1 - G(\sigma))\} = \beta \ln \frac{\sigma}{\sigma_o} + \beta \ln \frac{M}{M_{LE} + M_{HE}E}$$

$$+ \ln(M_{LE} + M_{HE}E^{\beta}) \qquad (5.73)$$

where $M = M_{LE} + M_{HE}$, $E = E_{HE}/E_{LE}$ and σ_o is the scale parameter.

Numerical results of the hybrid ultimate strength have been obtained for $\beta = \beta_{LE} = \beta_{HE}$, $E_{HE}/E_{LE} = \frac{1}{3}$ and $\varepsilon^*_{HE} = \varepsilon^*_{LE} = 3$. Figure 5.21 shows the median strength for $V_{LE} = 0.5$ and four different laminate configurations. It can be seen that the strength of hybrid laminates with LE layers clustering together (cases A and B in Fig. 5.21) is lower than that for the situation where the LE layers are dispersed more evenly among the HE layers. This obviously results from the higher stress concentration factors in cases A and B than in cases C and D.

Figure 5.22 shows the median strength variation with V_{LE} for the case of $N = 100$. The points A (A') and D (D'), respectively, represent the strengths of an all HE fiber composite and an all LE fiber composite. Line BD ($B'D'$) represents the stress in the

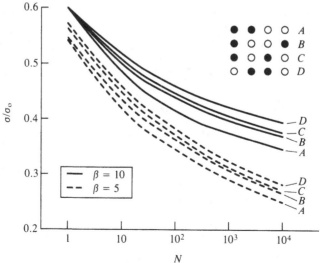

Fig. 5.21. Effects of stacking sequences on the median strength. (After Fukunaga, Chou and Fukuda 1989.)

hybrid at which failure of the LE fiber takes place. Line AC ($A'C'$) represents the stress in the hybrid assuming that the LE fibers carry no load. According to the rule-of-mixtures, line AED ($A'E'D'$) marks the ultimate failure strength of the hybrid. Increases in strength above the rule-of-mixtures are known as a hybrid effect. The present results show that the strength of hybrid laminates is lower than that of the all LE and all HE fiber composites. This finding is consistent with the experimental measurements of hybrid laminate strength (Ji 1982). Some other effects on the strength can also be seen from this figure. These are the scatter of the lamina strengths, the LE fiber relative volume fraction and the laminate stacking sequence. The median strength for $\beta = 10$ is greater than that for $\beta = 5$, whereas the hybrid effect is greater for $\beta = 5$ than that for $\beta = 10$. The strength varies with the laminate stacking sequence as well as the LE fiber relative volume fraction. The symbols in Fig. 5.22 correspond to the four types of fiber arrangements (A, B, C and D) in Fig. 5.21. It can also be seen that the hybrid effect is greatest for the case of $V_{LE} = 0.5$.

The probabilistic analysis of Fukunaga, Chou and Fukuda (1989) has led to the following conclusions: (1) The hybrid composite

Fig. 5.22. The median strength vs. relative fiber volume fraction. (After Fukunaga, Chou and Fukuda 1989.)

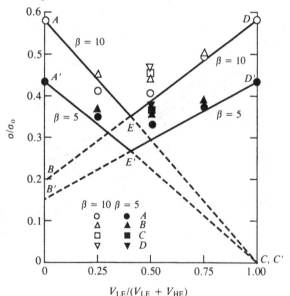

$V_{LE}/(V_{LE} + V_{HE})$

ultimate strength is greater for larger values of shape parameters, whereas the hybrid effect is greater for smaller values of shape parameters. (2) The hybrid effect on the probabilistic ultimate failure strength is most pronounced for $V_{LE} = 0.5$. (3) Both the magnitude and scatter of strengths vary with the length of hybrid composites. (4) The strength is also affected by the lamina stacking sequences because of the difference in stress redistributions resulting from laminar fracture. For a given relative fiber volume fraction, laminates with the LE layers uniformly dispersed among the HE layers are stronger than the laminates with the LE layers clustering together.

5.5   Softening strips

The idea of using softening strips in laminated composites has been developed for the purpose of modifying the local stress state and thus enhancing the load-carrying capability of composite structures. The process of introducing a softening strip involves replacing the low elongation plies with high elongation layers in selected regions, thus forming an interlaminated hybrid composite. Sites of stress concentrations in composites are particularly desirable for the use of softening strips. This concept is demonstrated below for notched laminates.

Sun and Luo (1985) have used three composites for fabrication of the hybrid specimens, i.e. AS4/3501 carbon/epoxy by Hercules, S2/CE9000-9 glass/epoxy by the Ferro Corp., and Scotchply 1002 glass/epoxy by the 3M Co. The base-line carbon/epoxy laminates have the lay-up of $[\pm 45°/0°/0°/0°/\mp 45°]$ and $[\pm 45°/0°]_s$. Each laminate contains a circular hole at the center. To create the softening strips, two plies of S2/CE9000-9 glass/epoxy are used to replace the three 0° carbon/epoxy plies in the $[\pm 45°/0°/0°/0°/\mp 45°]$ laminate, and one ply of Scotchply glass/epoxy is used to replace the two 0° carbon/epoxy layers in the $[\pm 45°/0°]_s$ laminate. Each strip is placed along the axial direction in the center of the laminate. The width of the softening strip is about twice of the hole diameter.

Table 5.4 summarizes the average failure load of all-carbon and hybrid composites. For both laminate systems, there is a significant increase in failure load (22–8%) in the notched hybrid system over the corresponding all-carbon system. However, the unnotched hybrid system shows lower strength than the corresponding all-carbon system as expected. The enhancement in failure load is realized only in the notched specimens. This can be understood from a stress analysis of the hybrid and non-hybrid laminates.

Figure 5.23 presents the results of finite element analysis of the normal stress along the transverse cross-section through the center of the circular hole. The distance in Fig. 5.23 is measured from the edge of the hole. There is a drastic reduction of the stress concentration in the hybridized region due to the presence of the low modulus material. This is believed to be the reason for the higher ultimate strength of the hybrid composite. The stress level outside of the softening strip is elevated.

Table 5.4. *Average maximum load. After Sun and Luo (1985).*

	[±45°/0°/0°/0°/ ∓45°]			[±45°/0°]ₛ		
	all-carbon (kN)	hybrid (kN)	Δ (%)*	all-carbon (kN)	hybrid (kN)	Δ (%)*
unnotched	51.6	41.0	−20.5	30.2	21.4	−29.2
notched	27.3	35.0	28.3	15.6	19.1	22.0
Δ (%)†	−47.0	−14.5		−48.2	−10.9	

* Δ(%) = hybrid maximum load/carbon maximum load −1
† Δ(%) = notched maximum load/unnotched maximum load −1

Fig. 5.23. Finite element analysis of axial normal stress distribution along the transverse section in the notched laminate of [±45°/0°/0°/0°/ ∓45°] through the center of the circular hole. (After Sun and Luo 1985.)

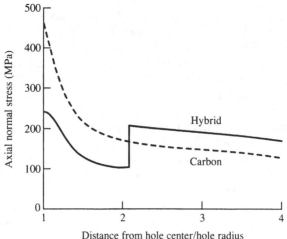

Distance from hole center/hole radius

The combination of low elongation and high elongation materials in interlaminated hybrid composites also offers improved fatigue and crack growth properties. A typical example of this kind is the aramid aluminum laminate (also known as ARALL) originated in the Netherlands. (See, for instance, Marissen 1984; Marissen, Trautmann, Foth and Nowack 1984; Mueller, Prohaska and Davis 1985; Vogelesang and Gunnink 1986; Bucci, Mueller, Schultz and Prohaska 1987; Kenaga, Doyle and Sun 1987; Chen and Sun 1989.) ARALL can be considered as a hybrid of aramid fiber laminae sandwiched in between sheets of high strength aluminum alloy. The hybrid composite is 15% less dense than the monolithic aluminum alloy and shows a 100–1000 fold improvement in fatigue life. The tensile properties of the hybrid are reported to be 15–30% better. Thus, ARALL is desirable for fatigue dominated sheet applications such as in lower wing skins, fuselages and tail skins of aircraft.

Other examples of combinations of high elongation and low elongation sheets of materials can be found in the use of adhesive layers in controlling free edge delamination and impact damage. (See, for instance, Chan, Rogers and Aker 1976; Sun and Norman 1988; Sun and Rechak 1988.)

5.6   Mechanical properties

This section discusses briefly the fracture, impact and fatigue characteristics of hybrid composites. In the case of interlaminated hybrid composites the failure mode is the fracture of the low elongation plies transverse to the loading axis. It has been observed in carbon/glass hybrids that the main transverse fracture is typically of cruciform shape and debonding occurs between the carbon and glass plies outwards from the line where the transverse crack intersects the carbon–glass interface. Ultimately most of the hybrid becomes debonded and the strength and stiffness approach those of the glass plies alone as shown in Fig. 5.10 (Bader and Manders 1978; Pitkethly and Bader 1987).

It is important to note that the initial fracture in the carbon plies does not propagate across the glass plies and load can be progressively diffused back into the carbon plies away from the fracture plane (Section 5.3.2). These processes of multiple cracking in the low elongation phase and the associated debonding have a significant effect on the total work of fracture in hybrid composites. The extension of debonding decreases as the absolute thickness of the carbon layers and volume content of carbon fiber decrease. It also has been suggested that debonding could be inhibited if the thickness of the carbon layers is made sufficiently thin.

The subject of work of fracture has received some attention (McColl and Morley 1977; Kirk, Munro and Beaumont 1978). It has been pointed out that fracture mechanisms at the microscopic level in binary fiber composites can be combined to give different kinds of macroscopic fracture behavior in hybrid composites. Therefore the fracture of the carbon/glass hybrid system, for instance, is expected to show characteristics of the individual composite systems.

The impact resistance of composite materials can be modified through a broad range of methods. Jang *et al.* (1989) have reviewed the major techniques including the control of fiber/matrix interfacial adhesion, matrix modifications, lamination design, through-the-thickness reinforcements, fiber hybridization, and utilization of high-strain fibers. Among these approaches, hybrids offer the benefit of improving the impact resistance of composites based upon high modulus fibers.

The total strain-energy of a composite at its ultimate tensile strength is inversely proportional to the fiber tensile modulus. Thus, high modulus fibers are not desirable for impact resistance from the viewpoint of energy absorption. The same is true when the work of fracture or toughness is considered. Aveston and Kelly (1980) have assessed the effectiveness of hybrid composites in energy absorption. Their analysis is recapitulated in the following. Consider the idealized stress–strain curve of Fig. 5.10, and compare the total energy per unit volume absorbed in a hybrid up to its ultimate tensile strength with the sum of energies of its components. By assuming a mean crack spacing of $1.5x$ (for crack spacing between x and $2x$), the strain at the limit of multiple cracking is $\varepsilon_{Lu}(1 + (\frac{5}{8})\alpha)$ where α is given in Eq. (5.42). The ultimate failure strain of the hybrid is $\varepsilon_{Hu} - (\frac{3}{8})\alpha\varepsilon_{Lu}$, where ε_{Lu} and ε_{Hu} are the ultimate strains of the LE and HE components, respectively.

Based upon these strain values, the area under the stress–strain curve *OABC* of the hybrid composite is

$$U_1 = \tfrac{1}{8}\alpha E_c \varepsilon_{Lu}^2 + \tfrac{1}{2}\varepsilon_{Hu}^2 E_{HE} V_{HE} \tag{5.74}$$

When the HE and LE components are completely debonded, the stress–strain curve follows the path *OAEF*. Then the energy absorbed at ultimate failure is

$$U_2 = \tfrac{1}{2}E_c\varepsilon_{Lu}^2 + \tfrac{1}{2}E_{HE}V_{HE}\varepsilon_{Hu}^2 - \tfrac{1}{2}E_{HE}V_{HE}\varepsilon_{Lu}^2 \tag{5.75}$$

Aveston and Kelly have concluded that for $U_1 > U_2$, α should be greater than three. For the combination of glass and carbon fiber composites, this can be achieved with a carbon fiber volume fraction of over 30%. Furthermore, the condition of multiple cracking (Eq. (5.41)) requires that $\alpha > 7.1$ for $\varepsilon_{Hu} = 2.3\%$ and $\varepsilon_{Lu} = 0.28\%$. This implies $V_{LE} = 48\%$. Thus, toughening through fiber hybridization and multiple cracking can be accomplished in carbon/glass composites with carbon fiber volume fraction between 30% and 48%.

Experimental studies of the impact behavior of hybrid composites can be found in the work of Chamis, Hanson and Serafini (1972), Beaumont, Riewald and Zweben (1974), Adams (1975), Adams and Miller (1975, 1976), Dorey, Sidey and Hutchings (1978); Adams and Zimmerman (1986); and Jang *et al.* (1989). Several test methods have been employed to measure the impact resistance of composite materials. The Charpy impact test and Izod impact test have been performed mainly on unidirectional composites, whereas the drop-weight impact test is usually used for laminated composites. Other tests such as longitudinal impact, transverse impact and pure shear impact tests on unidirectional fiber composites have also been considered.

Both notched and unnotched specimens have been used in Charpy tests. Instrumented Charpy impact tests can be used to determine the maximum load on the specimen, as well as to differentiate between the energy required to initiate damage and the energy absorbed during damage propagation. The relative sizes of the two regions for failure initiation and propagation under the load–time curve provide a qualitative measurement of the ductility of a composite under impact loading. Two materials with the same total Charpy energy may have quite different proportions of the component energies and, thus, distinct mechanisms of failure (Beaumont, Riewald and Zweben 1974). Most composite materials, especially laminated systems, can dissipate a considerable amount of energy in the fracture propagation phase even though the initial fracture may be of a brittle cleavage mode (Adams and Miller 1975). Interlaminated hybrid composites have been found to increase delamination under impact loading. Jang *et al.* (1989) have reported that the impact energies of the interlaminated hybrids generally show a negative hybrid effect, i.e. slightly lower energy dissipation than that predicted by the rule-of-mixtures.

As to the fatigue behavior of hybrids, Phillips (1976) has reported significant improvements in fatigue resistance of glass composites by hybridization with carbon fibers. At the stress level of 300 MPa or about 50% of ultimate, the fatigue life of the three-to-one volume fraction of glass/carbon hybrid is improved by about 100 fold over the all-glass control. This probably arises from the increased stiffness and, hence, the decreased strain for a given stress.

Figure 5.24 shows the stress vs. log life curves for unidirectional carbon/Kevlar hybrids tested in repeated tension (minimum stress/maximum stress = 0.1) reported by Fernando *et al.* (1988). The stresses in Fig. 5.24 are peak stresses. The $\sigma_{max}/\log N$ curves show a uniform variation from the linear form of the curve for plain carbon/epoxy towards the pronounced step-function shape of the curve for the plain Kevlar-49/epoxy composite. Adam et al. (1989) have examined a series of carbon/Kevlar-49/epoxy unidirectional hybrid composites; the fatigue behavior has been established as a function of composition and the ratio of the minimum to maximum stress in cyclic tension and tension/compression. This enables them to represent all data in a single two-parameter fatigue curve.

Other mechanical property data can be found in the literature for the non-linear tensile stress–strain relation (Takahashi and Chou 1987), compressive behavior (Chou, Steward and Bader 1979; Chou and Kelly 1980b; Gruber, Overbeeke and Chou 1982; Kretsis 1987; Yau and Chou 1989), flexural behavior (Fischer and Marom 1987; Marom and Chen 1987) and shear property (Kretsis 1987).

Fig. 5.24. Stress/log life ($\sigma_{max}/\log N$) curves for the family of unidirectional carbon/Kevlar hybrids tested in repeated tension (minimum stress/maximum stress = 0.1). (After Fernando *et al.* 1988.) σ_{max} is the peak stress. The percentages indicate relative fiber volume fractions of carbon and Kevlar.

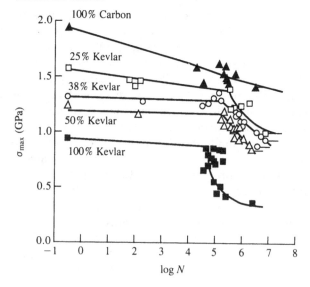

5.7   Property optimization analysis

5.7.1   *Constitutive relations*
A method is presented in this section to determine the concentration of components which can simultaneously optimize certain mechanical, thermal and electrical properties of a hybrid composite. This analysis, developed by McCullough and Peterson (1977) is essential in the pursuit of balanced material properties of a multi-component system.

The basis of this optimization analysis is the assumption that the constitutive relations for several major properties can be cast into simple linear form. Although the hybrids under consideration are restricted to unidirectional composites, the general format of this treatment is applicable to structural elements with more complex fiber arrangements. The constitutive equations for estimating the longitudinal properties of ternary composites are summarized below. Here, the volume fractions of the components are denoted by V, and the weight fraction by w. The subscripts 1, 2 and 3 refer to the components of the ternary system.

Mechanical properties

modulus $E = V_1 E_1 + V_2 E_2 + V_3 E_3$

strength $\sigma = (V_1 E_1 + V_2 E_2 + V_3 E_3)\varepsilon^*$ (5.76)

Poisson's ratio $\nu = V_1 \nu_1 + V_2 \nu_2 + V_3 \nu_3$

Thermal properties

coefficient of $\alpha = [V_1 E_1 \alpha_1 + V_2 E_2 \alpha_2 + V_3 E_3 \alpha_3]/$
expansion $[V_1 E_1 + V_2 E_2 + V_3 E_3]$ (5.77)

thermal conductivity $K = V_1 K_1 + V_2 K_2 + V_3 K_3$

Electrical property

resistivity $\dfrac{1}{\rho} = \dfrac{V_1}{\rho_1} + \dfrac{V_2}{\rho_2} + \dfrac{V_3}{\rho_3}$

Weight density $d = V_1 d_1 + V_2 d_2 + V_3 d_3$ (5.78)

Cost

cost/weight $C = C_1 w_1 + C_2 w_2 + C_3 w_3$ (5.79)

In the strength expression it is assumed that $\varepsilon^* = \min(\varepsilon_1, \varepsilon_2, \varepsilon_3)$ and failure occurs when the composite strain reaches the lowest failure strain of the three components. Clearly, this assumption underestimates the longitudinal strength. In the cases of coefficient of expansion and resistivity, E, α and ρ are treated as property variables. Also, w_i is the weight fraction of the ith component.

It is often useful to examine the performance of a composite in terms of certain 'specific' properties (i.e. the property per unit weight). A specific property $\bar{P}$ can be related to the property P and its weight density d as

$$\bar{P} = \left(\sum_{i=1}^{3} P_i V_i \right) \Big/ \left(\sum_{j=1}^{3} d_j V_j \right)$$

$$= \sum_{i=1}^{3} \left[(d_i V_i) \Big/ \left(\sum_{j=1}^{3} d_j V_j \right) \right] P_i / d_i$$

$$= \sum_{i=1}^{3} \bar{P}_i w_i \tag{5.80}$$

where w_i is defined by the term in the square brackets. The equations for mechanical, thermal and electrical properties, as well as density, have volume fraction as the composition variable while the specific property equations and the cost equation have weight fraction as the composition variable. Volume fraction and weight fraction are related by

$$w_i = d_i V_i / d \tag{5.81}$$

Both weight fraction and volume fraction can be used as concentration variables in the analysis of performance.

A typical property map for ternary systems can be represented by triangular diagrams as shown in Fig. 5.25. Each side of the equilateral triangle is unity in length and it expresses the volume fraction of a component. The volume fraction of a ternary system is represented by a point inside the triangle. For example, the hybrid composite denoted by point R in the property map contains 35%, 25% and 40% of matrix, fiber 1 and fiber 2, respectively. These concentration readings are obtained by drawing lines from point R parallel to the three sides of the triangle. The volume fractions of binary systems and pure components are represented, respectively, by points located on the sides and the apices of the triangle.

A straight line within the triangle of Fig. 5.25 specifies arbitrarily selected property levels available to the ternary system. Thus, line DE represents the various ways of combining the three components to achieve a specified property level, P, of the composite. Point E, for example, indicates that the combination of 55% volume fraction of matrix, 0% volume fraction of fiber 1 and

45% volume fraction of fiber 2 will result in a composite with property *P*. This same property level *P* can be achieved by concentration levels represented by all the other points on line *DE*. Also shown in Fig. 5.25 is the shaded area defined by the concentration line for 20% of matrix material. This particular volume fraction represents a square array of fibers of equal size. By determining the minimum desirable volume fraction of matrix material, the region in the triangle representing a matrix content below the minimal level can be excluded from further consideration in the optimization process.

Figures 5.26(a) and (b) present the various levels of longitudinal specific modulus and longitudinal specific strength of the hybrid system. The lines of constant properties are constructed from the following property data for the carbon/boron/epoxy system: Young's modulus in GPa (345/410/3.4); tensile strength in GPa (2.1/3.1/0.035); density in $10^3 \, \text{kg/m}^3$ (1.66/2.71/1.1); critical strain in % (0.6/0.7/10).

Fig. 5.25. Typical format of a ternary property map. (After McCullough and Peterson 1977.)

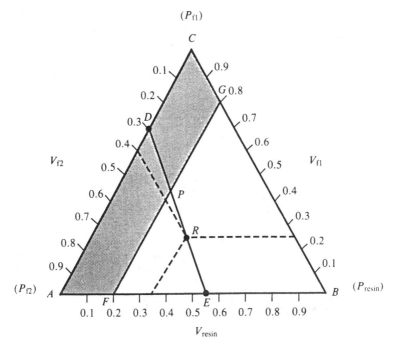

5.7.2   *Graphical illustration of performance optimization*

Figure 5.27 illustrates the property map for two different composite properties P and Q. For clarity only one of the constant property levels of property P is shown. It can be readily demonstrated that a designer can achieve a specified property of the hybrid by simultaneously adjusting the component properties.

Fig. 5.26. Selected property maps for the system carbon/boron/epoxy. (a) specific modulus. (b) Specific strength. (After McCullough and Peterson 1977.)

Specific modulus of carbon/boron composites (unit: 10^6 m)

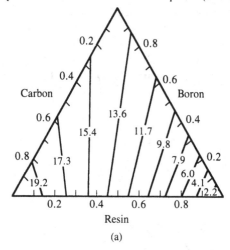

(a)

Specific strength of carbon/boron composites (unit: km)

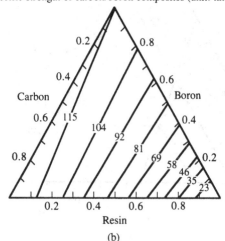

(b)

Any combination of fiber and resin that falls on line P will yield a composite with property P. Similarly, a line denoted by Q_j represents a specified level of property Q achieved by combinations of fiber and resin properties. Any combination of components that falls simultaneously on lines P and Q_j will yield a hybrid composite with properties P and Q_j. Consequently, the intersection of lines P and Q_j uniquely determines the concentrations of the three components to achieve properties P and Q_j.

Suppose that $Q_1 < \cdots < Q_i < \cdots < Q_7$. If a hybrid composite is desired such that for a specified property P (e.g. modulus), the property Q (e.g. strength) is a maximum, then the intersection of P and Q_7 should give the proper combination of concentrations of the components.

It often occurs in the optimization of hybrid composite design that, instead of a specified property, a level of performance is required. For instance, a property level P or greater (or P or less for properties such as density and cost) may be required. Figure 5.28 illustrates superimposed maps for properties P and Q. The requirements are

$$P > P^*$$
$$Q > Q^* \tag{5.82}$$

and

$$V_{\text{matrix}} > V_0$$

Fig. 5.27. Superimposed property maps for properties 'P' and 'Q'. (After McCullough and Peterson 1977.)

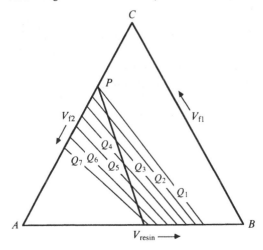

Fig. 5.28. Superimposed property maps illustrating bounding ranges on performance requirements. (After McCullough and Peterson 1977.)

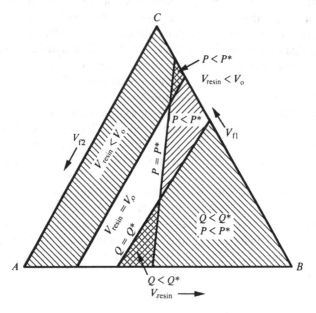

The shaded regions in Fig. 5.28 represent the concentration ranges which fail to meet any one or all of the above requirements. Naturally, the concentration ranges in the unshaded area of the figure will meet or exceed the specified requirements. The optimization of hybrid composite performance can thus be carried out based upon the basic information given in the property maps of the type shown in Fig. 5.26.

This schematic treatment illustrates the basic notions for a performance optimization of multicomponent systems. When the number of components exceeds three, the graphical method is no longer applicable. McCullough and Peterson (1977) have developed algebraic relationships of the properties of multicomponent systems and the optimization procedure has been structured in the form of a classical linear programming problem.

6   Two-dimensional textile structural composites

6.1   Introduction

The term 'textile structural composites' is used to identify a class of advanced composites utilizing fiber preforms produced by textile forming techniques, for structural applications. The recent interest in textile structural composites stems from the need for improvements in intra- and interlaminar strength and damage tolerance, especially in thick-section composites. Textile composites offer the potential of providing adequate structural integrity as well as shapeability for near-net-shape manufacturing (Chou and Ko 1989).

Textile structural composites provide the unique capability that the microstructure of fiber preforms can be designed to meet the needs of the performance of composite structures. Textile structural composites can be fabricated directly to their final shapes or can be assembled or readily machined to specified contours and dimensions. A total system approach is necessary to optimize the composite performance through the consideration of preform availability, cost, ease of processing, needs for secondary work such as machining, joinability of parts, and the overall performance of the composite structure. Chapters 6 and 7 discuss the fundamental characteristics of two- and three-dimensional textile preforms and the analysis of composite behavior based upon these preforms. The following discussions of yarn assembly, as well as textile preforms and characteristics, are based upon the review of Scardino (1989).

The forming of textile preforms requires knowledge of the structure of yarns and fibers. Yarns are linear assemblages of fibers formed into continuous strands having textile characteristics, i.e. substantial strength and flexibility. Figure 6.1 illustrates the idealized models of yarn structures; a yarn may consist of (a) single or (b) multiple continuous fibers, or (c) short (*staple*) fibers, where a substantial amount of twist or entanglement is needed to overcome fiber slippage. Yarns made from staple fibers are referred to as *spun yarns*. Figure 6.1(d) and (e) show that two or more single yarns can be twisted together to form ply or plied yarns. Plied yarns can be further twisted to form multiples (f). Spun yarns can also be combined to form plied yarns. Advanced textile structural composites are mostly based upon continuous filament yarns.

The relative density of fiber packing in the yarn cross-section is quantified by the fiber packing fraction, which is the ratio of fiber specific volume (volume/mass) to yarn specific volume. Fiber packing fractions are determined by a number of factors, including the number of fibers in a yarn, fiber cross-sectional shape, yarn tension, level of yarn twist and yarn manufacturing method. The yarn structures determine the translation of fiber properties into yarn properties. Consider, for example, the axial yarn elastic modulus (E_y). Hearle, Grosberg and Backer (1969) have predicted that $E_y = \cos^2 \theta E_f$, where E_f is the fiber elastic modulus and θ denotes the helical angle of the fibers in a yarn. The translation efficiencies reflect the effect of fiber orientation relative to the yarn axis due to the twist as well as the fiber entanglement in the yarns. The efficiency of fiber packing in a yarn and the fiber-to-yarn strength and modulus translation need to be taken into account in the selection of yarns for textile preforms. Further discussions on the packing of fibers in a yarn are given in Chapter 7.

The selection of fiber preforms as reinforcements for composites requires additional considerations to those at the yarn level. The most basic ones, according to Scardino, are the manipulative requirements in dimensional stability, subtle conformability and

Fig. 6.1. Idealized models of various yarn structures. (After Scardino 1989.)

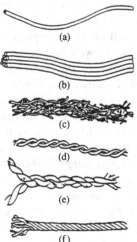

(a)

(b)

(c)

(d)

(e)

(f)

deep draw shapeability. A high degree of dimensional stability is required in pultruded, flat panel or laminated composites. Some conformability is desirable in slightly curved structural parts. Considerable extensibility of the preform is necessary for deep-draw molded composites. These factors not only are pertinent to the selection of composites processing techniques, but also dominate the fiber preform microstructure in the finished product. It should be noted that the orientations of fibers in a preform before and after matrix impregnation can be very different, and can thus have significant implication on composite performance.

6.2   Textile preforms

The major textile forming techniques for composites reinforcement are weaving, knitting, braiding and stitching. There is the lack of a definitive criterion for separating textile preforms into the two-dimensional and three-dimensional types. In Chapters 6 and 7, a rather loose criterion is applied to distinguish these two types, based upon the degree of integration of the yarns as well as the extent of strengthening in the thickness direction of the preform. Consider, for instance, the traditional weaving, knitting and braiding processes; the interaction of yarns (i.e. interlacing, interlooping) in the thickness direction is limited to two or three yarn diameters. As a result, the strengthening effect due to yarn penetration, although higher than that for conventional laminated composites, is fairly small. Therefore, these preforms are considered to be two dimensional. On the other hand, the more recently developed preforms, such as angle-interlock wovens and solid braids, are fully integrated structures, and there is a significant degree of strengthening in the thickness direction. Thus, these preforms can be categorized as three-dimensional. The foregoing definitions are independent of the actual dimensions of the preform.

The uncertainty in the separation of two- and three-dimensional preforms arises when the integration of yarns in the thickness direction is of limited extent and the resulting strengthening is not very significant. An example can be found in multiaxial warp knits. The layers of essentially straight fibers in such a construction are connected by knitting yarns. The degree of strengthening in the thickness direction depends on the type of knitting yarns used.

Figure 6.2 summarizes the major manufacturing techniques for two-dimensional textile preforms. It is feasible to insert laid-in yarns into the basic knitted or braided fabric given in Fig. 6.2, thus significantly modifying the directional stability of the fabric. The great varieties of fabric geometry so induced are not shown in Fig. 6.2 for the reason of simplicity. A brief discussion of wovens, knits and braids for reinforcing composites is given below.

6.2.1   *Wovens*

Woven fabrics, formed on a loom by interlacing two or more sets of yarns, are essentially two-dimensional constructions. When two sets of yarns are interlaced at right angles, the lon-

Fig. 6.2. Manufacturing techniques for two-dimensional textile preforms.

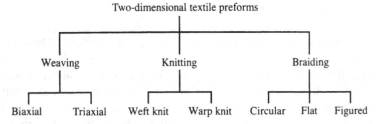

Fig. 6.3. Examples of woven fabric patterns: (a) plain weave ($n_g = 2$); (b) twill weave ($n_g = 3$); (c) four-harness satin ($n_g = 4$); (d) eight-harness satin ($n_g = 8$). (After Ishikawa and Chou 1982a.)

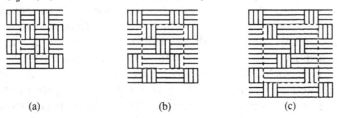

    (a)   (b)   (c)

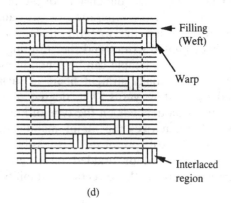

(d)

gitudinal yarns are known as the *warp*, and the widthwise yarns are known as the *filling* or *weft*. The individual yarns in the warp and filling directions are also called an *end* and a *pick*, respectively. Figure 6.3 shows examples of orthogonal woven fabrics. According to Lord and Mohamed (1982) and Schwartz, Rhodes and Mohamed (1982), the manufacture of woven fabrics based upon high speed power looms requires four operations or primary motions: (1) shedding, (2) filling insertion, (3) beat-up, and (4) warp and fabric control. Following Lord and Mohamed, and Schwartz, Rhodes and Mohamed, a brief introduction is given for these four motions.

Shedding involves the movement of the warp yarns to provide a path for the insertion of the weft yarn. One of the techniques of shedding uses heddle wires which are grouped into several frames, known as *harnesses* or *shafts* (Fig. 6.4). Each harness is operated by a separate cam; the purpose of the cam is to lift or lower the harness. As a result of the movement of the harness, the shed is formed. A cam loom is generally limited to designs repeating on six or fewer picks (Schwartz, Rhodes and Mohamed 1982). Besides the cam system, traditional fabrics are often woven on a dobby head loom. A commercially available dobby mechanism uses a maximum of about 24 harnesses, and thus allows the control of interlacing 24 different groups of warp yarns. The Jacquard head provides control

Fig. 6.4. Shedding in fabric weaving. (After Lord and Mohamed 1982.)

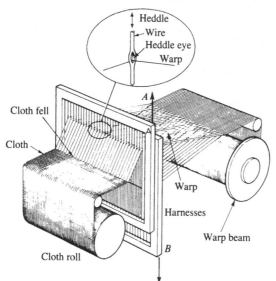

of every individual yarn across the width of the fabric, and it does not have the limitation of the dobby loom. Thus, the yarn interlacing possibilities are greatly enhanced and they are only limited by the number of warp yarns used.

The number of harness frames required for the shedding operation depends on the type of weave. Two harnesses are used for weaving plain fabrics (Fig. 6.3a), and their relative motion is carried out in two weaving cycles. In one cycle, the front harness is in the top position and the back harness is in the bottom position. In the next cycle, the harnesses change positions, and the sequence is repeated. Obviously when the two sheets of warp yarns are at the same level, the shed is closed (Schwartz, Rhodes and Mohamed 1982).

Filling insertion, as the term implies, involves the passing of a filling yarn through the open shed. Figure 6.5 shows schematically the conventional way of filling insertion by a shuttle. In order for the filling insertion motion to take place, the shed has to be sufficiently open and remain open for an adequate period of time. Consequently, the speed of the weaving process is dominated by the speed at which the shuttle travels through the shed. The transit time of the shuttle involves its acceleration and deceleration. Furthermore, it is desirable to remove the filling supply package from the filling carrier so the carrier could be made smaller and the yarn movement in the shedding is reduced. All these considerations provide the impetus for using shuttleless looms.

Fig. 6.5. Filling yarn insertion in fabric weaving. (After Lord and Mohamed 1982.)

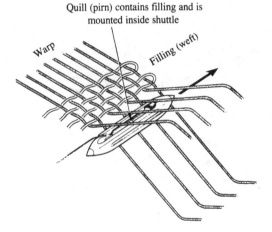

Quill (pirn) contains filling and is mounted inside shuttle

Warp

Filling (weft)

The commonly used shuttless systems include rapiers, gripper projectiles, air jets, and water jets. A rapier is a device made of metal or a composite material with an attachment on the end to carry the filling yarn through the shed. For the case of a single rigid rapier, its length should be at least equal to the loom width. In order to improve the loom speed, double rigid rapiers have been used. These consist of a 'giver', which picks up the filling yarn and carries it to the center of the shed, where the filling yarn is transferred to a 'taker' for transporting the yarn to the other end of the shed. The reduction in carrier traveling time thus doubles the number of picks that can be inserted per unit time. The gripper loom uses a small projectile to transport the filling yarn. It is feasible to use many projectiles which may be initiated from both ends or one end of the loom. In the case of fluid jet looms, the filling yarn is carried by a high pressure air or water jet.

Finally, the purpose of beat-up is for incorporating the filling yarn into the body of the fabric after the filling is inserted. This is accomplished by the use of a wire grate called a *reed*, through which the warp yarns are threaded. The reed is first moved backward to allow the insertion of the filling yarn. When the insertion is finished, the reed moves forward and drives the filling yarn into the fabric. For a continuous operation of the weaving process, it is also necessary to supply the warp yarns continuously, and to remove the fabric from the loom. It should be noted that the repeated actions of shedding and beating induce cyclic tension variations in the yarns (Schwartz, Rhodes and Mohamed 1982). The control of warp and filling yarn tension is essential in the weaving process.

Orthogonal woven fabrics exhibit good dimensional stability in the warp and weft directions. Woven fabrics offer the highest yarn packing density in relation to fabric thickness. The pure and hybrid woven fabrics used in composites are mostly in the forms of plain, basket, twill and satin weaves. Wovens are available in tubular and flat forms.

Woven fabrics provide more balanced properties in the fabric plane than unidirectional laminae; the bidirectional reforcement in a single layer of a fabric gives rise to enhanced impact resistance. The ease of handling and low fabrication cost have made fabrics attractive for structural applications. On the other hand, the limited conformability, poor in-plane shear resistance, and reduced yarn-to-fabric tensile translation efficiency due to yarn crimp are some of the disadvantages of woven fabrics.

Triaxial woven fabrics, made from three sets of yarns which interlace at 60° angles, provide higher isotropy and in-plane shear rigidity than orthogonal wovens. However, no woven fabric construction provides sufficient extensibility for deep-draw molding (Scardino 1989).

6.2.2   *Knits*

A knitted structure is characterized by its interlacing loops. Two basic types of knits can be defined according to the general direction of travel of a looped thread in the fabric (Thomas 1971). In weft knitting, the thread runs widthwise, and the loops are formed by a single weft thread (Fig. 6.6a). The loops in a horizontal

Fig. 6.6. Knitted fabrics: (a) weft knit structure; (b) warp knit structure. (After Thomas 1971.) (c) Knitted fabric with weft and warp laid-ins. (After Wray and Vitols 1982.)

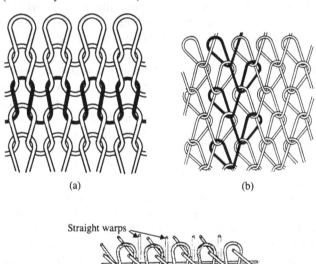

(a)   (b)

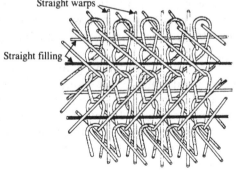

(c)

row are built up one loop at a time. In practice, many weft threads are used simultaneously in weft knitting. In warp knitting, the orientation of a looped thread is warpwise, and all the loops making up a single horizontal row are formed simultaneously (Fig. 6.6b). The principal mechanical elements used in knitting are needles. According to Schwartz, Rhodes and Mohamed (1982), there are three major needle types: the latch needle, the bearded needle and the compound needle. The latch needle has been used most often and it contains a latch which can be closed in the knitting process.

The loops in knitted fabrics are formed essentially on a very similar principle. Following Thomas (1971), the looping process is demonstrated for a single latch needle by the consecutive steps shown in Fig. 6.7. Consider the needle which has at its stem a loop already formed during the course of the knitting process (Fig. 6.7a). A thread is then placed under the hook of the needle. The loop is restrained in its position whereas the needle is allowed to move through it. As the needle moves downward, the existing loop will push the latch and close the hook (Fig. 6.7c). When the top of the hook reaches the level of the existing loop (Fig. 6.7d), this loop is pulled out of the way by the yarn tension. Then as the needle moves upward again, the thread in the hook opens up the latch, and it becomes the next 'existing' loop. More loops are generated as the process repeats. Depending on the type of knitting machine, a variety of needle configurations and looping cycles is available. A detailed discussion of the knitting processes and knit fabrics has been given by Schwartz, Rhodes and Mohamed (1982).

Simple weft and warp knits can provide extensibility in all

Fig. 6.7. The latch needle cycle in fabric knitting. (After Thomas 1971.)

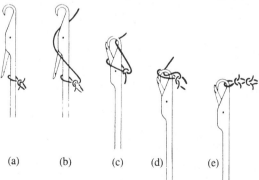

(a)　　(b)　　(c)　　(d)　　(e)

directions and are thus suitable for deep-draw molding techniques. Directional stability can be established by adding laid-in (non-knitting) yarns in the desired directions. According to Scardino, weft inserted warp knits offer flexibility in performance, from complete dimensional stability to engineered directional elongation. Furthermore, weft inserted warp knits with laid-in warp systems offer high yarn-to-fabric translation efficiencies and greater in-plane shear resistance than comparable wovens. An example of a knitted fabric with weft and warp laid-ins is shown in Fig. 6.6(c).

6.2.3   *Braids*

Braided fabrics are constructed from intertwined yarns. In order to understand the characteristics of braided preforms, it is useful to review the basic mechanisms involved in maypole braiders. The paths traced by the carriers of a maypole braider are similar to those of the dancers around the maypole.

Douglass (1964) has explained the operation of some common types of maypole braiders. A simple slide plate machine consists of a deck, a driving mechanism and a superstructure with the take up facility and the braiding guide. The deck has two metal plates. Serpent-like tracks are cut in the upper plate. Between the base plate and the track plate is a train of gears. Each horngear has a circular flanged top which is slotted to engage the bottom driving lugs of the spool carriers. Furthermore, the horngears are so arranged (Fig. 6.8) that the slots in the top flanges of two neighboring horngears will meet at the intersection of the tracks.

Fig. 6.8. Horngears and tracking in a tubular braider. (After Douglass 1964.)

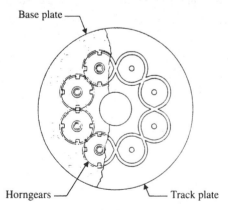

Base plate

Horngears   Track plate

Consider a carrier with its lug engaged in one of the horngear slots and moving in the track; the contour of the track enables the carrier lug to be transferred from one horngear to the next at the intersection. Consequently, this process can be repeated and each carrier can follow a chain of interconnected figure eights in a continuous manner.

According to Douglass (1964), some common types of maypole braiders are the 'Soutache' braider, circular (tubular) braiders and flat braiders. The machine used for Soutache braiding is the simplest of all the braiders, consisting of two horngears which are slotted to take 3, 5, 7, 9, 11, 13 or 17 carriers. Figure 6.9 shows a three-carrier soutache set-up for demonstrating the mechanism of braiding. Circular braiding machines, on the other hand, have an even number of carriers, starting with eight and increasing by steps of four.

Braiders for flat products are characterized by the tracking system, which does not completely encircle the braiding center (Fig. 6.10). The two horngears at the ends of the track have an uneven number of hornslots. Unlike the circular machines, a yarn carrier in this case reverses its path at the terminal gears and as a result flat braids can be accomplished.

The geometric configurations of some two-dimensional braids are given in Fig. 6.11. Figure 6.11(a) shows the braid with a 2/2 intersection repeating pattern, and it is known as a regular, plain or standard braid. Figure 6.11(b) gives a diamond or basket braid

Fig. 6.9. Horngears for a three-carrier 'soutache' braider. (After Douglass 1964.)

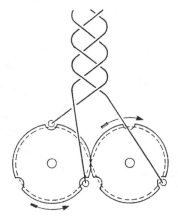

which is characterized by a 1/1 intersection repeating pattern. Figure 6.11(c) shows a regular braid with warp in-laids and it can be made by either a circular or flat braider. There are certain significant similarities and differences between woven and braided fabrics. Both fabrics utilize two sets of yarns; these are the warp and weft yarns in weaving and the yarns moving in clockwise and counter-clockwise directions in circular braiding. As far as interlacing patterns are concerned, they are unlimited in weaving and very limited in braiding (normally 1/1, 2/2 and 3/3). The angle of

Fig. 6.10. Horngears and tracking in a flat braider. (After Douglass 1964.)

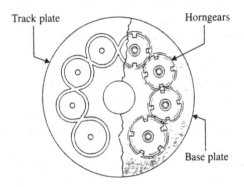

Fig. 6.11. Geometric configurations of (a) flat braid, (b) diamond or basket braid and (c) flat braid with warp in-laids. (After Du, private communication, 1990.)

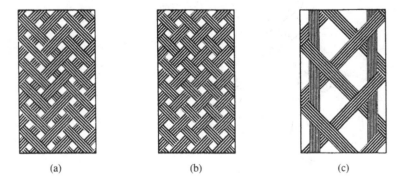

(a)   (b)   (c)

interlacing between the two sets of yarns is 90° in orthogonal woven fabrics and less than 90° in braided fabrics.

The braiding technique is highly versatile and a great variety of geometric patterns can be produced. This is demonstrated by the figured (fancy) braids which have more complex cross-sections (for example, I and T sections) than the traditional braids, or variable cross-section shapes along the axial direction. Figure 6.12 shows a flat-cord-flat fabric, and the tracking system employed for producing such a fabric (Yokoyama *et al.* 1989). The figured braids are categorized as two-dimensional fabrics for the reasons stated earlier. These fabrics could be considered as three-dimensional

Fig. 6.12. (a) A figured braid of flat-cord-flat construction; (b) configuration of the tracking system. (After Yokoyama *et al.* 1989.) Parts *A* and *B* are fabricated by flat braiding, and part *C* is fabricated by a tubular braiding mechanism. The spindles indicated by the open circles move on section *C* of the track and spindles represented by solid circles move through all sections of the track in the sequence (*A*)–(*C*)–(*B*)–(*C*)–(*A*).

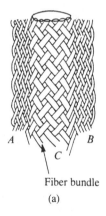

Fiber bundle

(a)

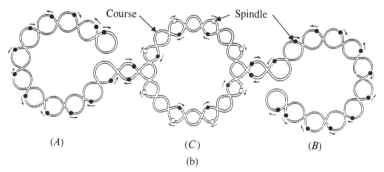

Course   Spindle

(*A*)   (*C*)   (*B*)

(b)

Table 6.1. *Directional behavior of two-dimensional preforms in unjammed configuration*

After Scardino (1989).

Preform construction	Directional stability			Directional conformability			Substantial directional extensibility			Substantial in-plane shear resistance		
	MD	CD	BD	MD	CD	BD	MD	CD	BD	MD	CD	BD
Woven												
Biaxial	×	×										×
Triaxial		×	×	×		×				×	×	×
Knitted												
Weft				×	×	×	×	×	×			
Warp				×	×	×	×	×	×			
Braided												
Circular				×	×					×	×	
(tubular)												
Flat				×	×					×	×	

MD = machine direction; CD = crosswise direction; BD = bias direction.

provided sufficient structural integration and reinforcement are achieved in the thickness direction.

A braided fabric exhibits dimensional stability under tension along the 0° orientation if there are axial laid-in yarns, and along the $\pm\theta°$ directions of the braiding yarns. Without axial laid-in yarns, the dimension of a braid can be changed by applying tension in the 0° and 90° directions. Yarn jamming in a fabric can occur essentially in two ways. First, for a fabric without laid-in yarns, application of a tensile force along the 0° or 90° direction will stretch the fabric until it is jammed and no further movement of the yarns is possible. Second, yarn jamming could occur during fabrication. This condition is characterized by the situation that the yarn covering factor = 1, i.e. there is no void space in between the yarns of a fabric. Thus, the fabric shape cannot be deformed under load. The first concept concerning yarn jamming is useful when the focus is on conformability and large deformation of a fabric. The second concept is important when one is concerned about, for instance, the maximum fiber volume fraction in a composite, and the control of the preforming process. Chapter 7 provides examples of yarn jamming in the fabrication of three-dimensionally braided preforms. The major limitations in machine-made braids at the present time are restricted width, diameter, thickness and shape selection.

There are also non-woven fabrics which are essentially sheet materials composed of randomly oriented fiber segments bonded together. There is a lack of geometrically defined arrangements in non-wovens as compared to wovens, knits and braids. The key methods of fiber bonding in non-wovens are: sticking fibers together as in fiber mats, entangling the fibers to give frictional interaction, stitching through the non-woven web with a textile yarn, and adhesive bonding. Hearle (1989) has given in-depth treatment of the mechanics of non-woven fabrics.

Scardino (1989) has examined the behavior of various fabric structural forms under uniaxial and shear stresses in the machine (0°), crosswise (90°) and bias (±45°) directions. Table 6.1 summarizes the directional characteristics of two-dimensional preforms. Based upon such knowledge it is feasible to design fabric preforms with a specific directional behavior while accommodating certain manipulative requirements in composites manufacturing.

The analysis and modeling of two-dimensional textile structural composites in this chapter focus on woven preforms. The models so developed can be extended to treat braided composites. Composites based upon knitted and non-woven preforms are not considered.

Sections 6.3 to 6.11 are excerpted from the work of Chou and Ishikawa (1989).

6.3   Methodology of analysis

The objective of the analysis in Section 6.3 is to model the thermomechanical behavior of two-dimensional orthogonal woven fabric composites. The fabrics are composed of two sets of mutually orthogonal yarns of either the same material (non-hybrid fabrics) or different materials (hybrid fabrics). Here, the term 'yarns' represents individual filaments, untwisted fiber bundles, twisted fiber bundles or rovings.

An orthogonal woven fabric consists of two sets of interlaced yarns. The length direction of the fabric is known as the *warp*, and the width direction is referred to as the *filling* or *weft*. The various types of fabric can be identified by the pattern of repeat of the interlaced regions, as shown in Fig. 6.3. Two basic geometrical parameters can be defined to characterize a fabric: n_{fg} denotes that a warp yarn is interlaced with every n_{fg}th filling yarn and n_{wg} denotes that a filling yarn is interlaced with every n_{wg}th warp yarn. The present treatment is confined to the case of $n_{wg} = n_{fg} = n_g$ for both hybrid and non-hybrid fabrics. Fabrics with $n_g \geq 4$ are known as satin weaves. As defined by their n_g values, the fabrics in Fig. 6.3 are termed plain weave ($n_g = 2$), twill weave ($n_g = 3$), four-harness satin ($n_g = 4$), and eight-harness satin ($n_g = 8$). The regions in Fig. 6.3 enclosed by the dotted lines define the 'unit cells' or the basic repeating regions for the different weaving patterns. It is also noted that the top sides of the fabrics in Fig. 6.3 are dominated by the filling yarns, whereas the reverse sides are dominated by the warp yarns.

The theoretical basis of the present analysis is the classical laminated plate theory, which is given in Chapter 2. Only the key equations are recapitulated in the following for ease of reference. Under the assumptions of the Kirchhoff hypothesis, the constitutive equations are expressed in the condensed form as

$$\left\{\frac{N}{M}\right\} = \left[\frac{A \mid B}{B \mid D}\right]\left\{\frac{\varepsilon^\circ}{\kappa}\right\} \tag{6.1}$$

Here, N and M are membrane stress resultants and moment resultants, respectively; ε° and κ are the strain and curvature of the laminate geometric mid-plane, respectively. The components of the stiffness matrices A, B and D are evaluated as follows:

$$(A_{ij}, B_{ij}, D_{ij}) = \sum_{k=1}^{n} \int_{h_{k-1}}^{h_k} (1, z, z^2)(\bar{Q}_{ij})_k \, dz \qquad (i, j = 1, 2, 6)$$

(6.2)

where the reduced stiffness constants $\bar{Q}_{ij}$ corresponding to the lamina defined by h_k and h_{k-1} in the thickness direction are used in the calculations. The subscripts 1, 2 and 6 in Eq. (6.2) indicate, in the xyz coordinate system, the x direction, the y direction, and the x–y plane, respectively. More explicitly, Eq. (6.2) can be written as

$$A_{ij} = \sum_{k=1}^{n} (\bar{Q}_{ij})_k (h_k - h_{k-1})$$

$$B_{ij} = \tfrac{1}{2} \sum_{k=1}^{n} (\bar{Q}_{ij})_k (h_k^2 - h_{k-1}^2) \qquad (6.3)$$

$$D_{ij} = \tfrac{1}{3} \sum_{k=1}^{n} (\bar{Q}_{ij})_k (h_k^3 - h_{k-1}^3)$$

The inverted form of Eq. (6.1) is given by

$$\left\{ \begin{matrix} \varepsilon^o \\ \kappa \end{matrix} \right\} = \left[\begin{array}{c|c} A' & B' \\ \hline B' & D' \end{array} \right] \left\{ \begin{matrix} N \\ M \end{matrix} \right\}$$

(6.4)

When the effect of temperature change is taken into account, the constitutive relation of Eq. (6.1) should be written as

$$\left\{ \begin{matrix} N \\ M \end{matrix} \right\} = \left[\begin{array}{c|c} A & B \\ \hline B & D \end{array} \right] \left\{ \begin{matrix} \varepsilon^o \\ \kappa \end{matrix} \right\} - \Delta T \left\{ \begin{matrix} \bar{A} \\ \bar{B} \end{matrix} \right\}$$

(6.5)

where

$$\left\{ \begin{matrix} \bar{A}_x \\ \bar{A}_y \\ \bar{A}_{xy} \end{matrix} \right\} = \sum_{k=1}^{n} \int_{h_{k-1}}^{h_k} \begin{bmatrix} \bar{Q}_{11} & \bar{Q}_{12} & \bar{Q}_{16} \\ \bar{Q}_{12} & \bar{Q}_{22} & \bar{Q}_{26} \\ \bar{Q}_{16} & \bar{Q}_{26} & \bar{Q}_{66} \end{bmatrix}_k \left\{ \begin{matrix} \alpha_{xx} \\ \alpha_{yy} \\ \alpha_{xy} \end{matrix} \right\}_k dz \qquad (6.6)$$

$$\left\{ \begin{matrix} \bar{B}_x \\ \bar{B}_y \\ \bar{B}_{xy} \end{matrix} \right\} = \sum_{k=1}^{n} \int_{h_{k-1}}^{h_k} \begin{bmatrix} \bar{Q}_{11} & \bar{Q}_{12} & \bar{Q}_{16} \\ \bar{Q}_{12} & \bar{Q}_{22} & \bar{Q}_{26} \\ \bar{Q}_{16} & \bar{Q}_{26} & \bar{Q}_{66} \end{bmatrix}_k \left\{ \begin{matrix} \alpha_{xx} \\ \alpha_{yy} \\ \alpha_{xy} \end{matrix} \right\}_k z \, dz \qquad (6.7)$$

ΔT indicates a small uniform temperature change, and α denotes the thermal expansion coefficients. After inversion, Eq. (6.5) becomes

$$\left\{\frac{\varepsilon^{\circ}}{\kappa}\right\} = \left[\frac{A' \mid B'}{B' \mid D'}\right]\left\{\frac{N}{M}\right\} + \Delta T\left\{\frac{\tilde{A}'}{\tilde{B}'}\right\} \tag{6.8}$$

where

$$\left\{\frac{\tilde{A}'}{\tilde{B}'}\right\} = \left[\frac{A' \mid B'}{B' \mid D'}\right]\left\{\frac{\tilde{A}}{\tilde{B}}\right\} \tag{6.9}$$

The constants $\tilde{A}'$ and $\tilde{B}'$ represent, respectively, the in-plane thermal expansion and thermal bending coefficients.

Based upon the iso-stress and iso-strain assumptions, the above constitutive equations can be used to obtain the bounds of the thermoelastic properties. The upper bounds of compliance constants are obtained from the iso-stress assumption; the lower bounds of stiffness constants are then obtained by inverting the compliance constant matrix. Similarly, the upper bounds of stiffness constants are derived from the iso-strain assumption; the lower bounds of compliance constants are then obtained by inverting the stiffness constant matrix. Three techniques for modeling the stiffness and strength properties of fabric composites are introduced in Sections 6.4–6.6 based upon the laminated plate analysis. They are known as the 'mosaic model', 'crimp (fiber undulation) model', and 'bridging model'. The prediction of thermal expansion coefficients of fabric composites is given in Section 6.8 based upon these three models. The analytical techniques so developed are also applied to hybrid fabric composites (Sections 6.9 and 6.10). Finally, the thermoelastic behavior of two-dimensional textile structural composites reinforced with triaxial fabrics is presented in Section 6.11. The following discussions are based on the work of Ishikawa and Chou (1982a–c, 1983a–d), Ishikawa (1981), Chou (1985, 1986, 1989a&b), Yang and Chou (1986, 1987) and Byun and Chou (1989).

6.4   Mosaic model

The basis of idealization of the 'mosaic model' can be seen from Fig. 6.13. Figure 6.13(a) is a cross-sectional view of an eight-harness satin. The consolidation of the fabric with a matrix material is depicted in Fig. 6.13(b), which can be simplified as the mosaic model of Fig. 6.13(c). The key simplification of the mosaic model is the omission of the fiber continuity and crimp (undulation) that exist in an actual fabric.

In general, a fabric composite idealized by the mosaic model can be regarded as an assemblage of pieces of asymmetric cross-ply laminates. Figure 6.14(a) shows the mosaic model of a unit cell for an eight-harness satin composite. The elastic stiffness constants of a cross-ply laminate (Fig. 6.14b) can be derived on the basis of Eqs. (6.3). Assuming that fibers are aligned along the x direction, the stiffness constants, Q_{ij}, of a unidirectional lamina, which has

Fig. 6.13. Idealization of the mosaic model. (a) Cross-sectional view of a woven fabric before resin impregnation; (b) woven fabric composite; (c) idealization of the mosaic model. (After Ishikawa and Chou 1983b.)

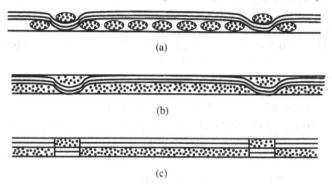

(a)

(b)

(c)

Fig. 6.14. The mosaic model. (a) Repeating region in an eight-harness satin composite; (b) a basic cross-ply laminate; (c) parallel model; (d) series model. (After Ishikawa and Chou 1983b.)

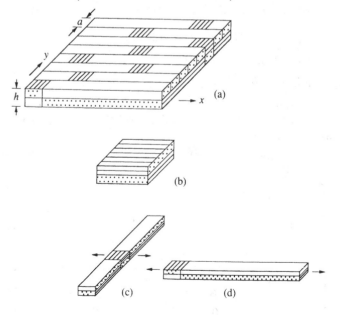

orthotropic symmetry in the x–y plane, are given by

$$Q_{ij} = \begin{bmatrix} E_{11}/D_v & v_{12}E_{22}/D_v & 0 \\ v_{21}E_{11}/D_v & E_{22}/D_v & 0 \\ 0 & 0 & G_{12} \end{bmatrix} \qquad (6.10)$$

where

$$D_v = 1 - v_{12}v_{21} \qquad (6.11)$$

Here, E_{11} and E_{22} are the Young's moduli, G_{12} is the in-plane shear modulus, and v_{12} denotes the Poisson's ratio relating the transverse strain in the x_2 direction and the applied strain in the x_1 direction. The Q_{ij} constants are symmetrical, i.e. $Q_{ij} = Q_{ji}$ (see Chapter 2).

From Eqs. (6.3) and (6.10), the elastic stiffness constants of the cross-ply laminate shown in Fig. 6.14(b) can be derived. The laminate is composed of two unidirectional laminae of thickness $h/2$. The total laminate thickness is h and the x–y coordinate plane is positioned at the geometrical mid-plane of the laminate. Thus, in Eqs. (6.3), $k = 1$ and 2 define, respectively, the laminae with fibers in the y and x directions. The non-vanishing stiffness constants are

$$A_{11} = A_{22} = (E_{11} + E_{22})h/(2D_v)$$
$$A_{12} = v_{12}E_{22}h/D_v$$
$$A_{66} = G_{12}h$$
$$B_{11} = -B_{22} = (E_{11} - E_{22})h^2/(8D_v) \qquad (6.12)$$
$$D_{11} = D_{22} = (E_{11} + E_{22})h^3/(24D_v)$$
$$D_{12} = v_{12}E_{22}h^3/(12D_v)$$
$$D_{66} = G_{12}h^3/12$$

The extension-bending coupling constants B_{11} and B_{22} do not vanish because $E_{11} \neq E_{22}$. Also, it is understood that A_{ij}, B_{ij}, and D_{ij} are symmetrical constants.

Using Eqs. (6.12), Eq. (6.1) can be written in the following explicit form:

$$\begin{Bmatrix} N_x \\ N_y \\ N_{xy} \end{Bmatrix} = \begin{bmatrix} A_{11} & A_{12} & 0 \\ A_{12} & A_{11} & 0 \\ 0 & 0 & A_{66} \end{bmatrix} \begin{Bmatrix} \varepsilon_{xx}^o \\ \varepsilon_{yy}^o \\ \gamma_{xy}^o \end{Bmatrix} + \begin{bmatrix} B_{11} & 0 & 0 \\ 0 & -B_{11} & 0 \\ 0 & 0 & 0 \end{bmatrix} \begin{Bmatrix} \kappa_{xx} \\ \kappa_{yy} \\ \kappa_{xy} \end{Bmatrix}$$

$$(6.13)$$

$$\begin{Bmatrix} M_x \\ M_y \\ M_{xy} \end{Bmatrix} = \begin{bmatrix} B_{11} & 0 & 0 \\ 0 & -B_{11} & 0 \\ 0 & 0 & 0 \end{bmatrix} \begin{Bmatrix} \varepsilon_{xx}^o \\ \varepsilon_{yy}^o \\ \gamma_{xy}^o \end{Bmatrix} + \begin{bmatrix} D_{11} & D_{12} & 0 \\ D_{12} & D_{11} & 0 \\ 0 & 0 & D_{66} \end{bmatrix} \begin{Bmatrix} \kappa_{xx} \\ \kappa_{yy} \\ \kappa_{xy} \end{Bmatrix}$$

Inverting Eqs. (6.13), the following are obtained:

$$
\begin{Bmatrix} \varepsilon_{xx}^o \\ \varepsilon_{yy}^o \\ \gamma_{xy}^o \end{Bmatrix} = \begin{bmatrix} A'_{11} & A'_{12} & 0 \\ A'_{12} & A'_{11} & 0 \\ 0 & 0 & A'_{66} \end{bmatrix} \begin{Bmatrix} N_x \\ N_y \\ N_{xy} \end{Bmatrix}
$$

$$
+ \begin{bmatrix} B'_{11} & B'_{12} & 0 \\ -B'_{12} & -B'_{11} & 0 \\ 0 & 0 & 0 \end{bmatrix} \begin{Bmatrix} M_x \\ M_y \\ M_{xy} \end{Bmatrix}
$$

$$
\begin{Bmatrix} \kappa_{xx} \\ \kappa_{yy} \\ \kappa_{xy} \end{Bmatrix} = \begin{bmatrix} B'_{11} & -B'_{12} & 0 \\ -B'_{12} & -B'_{11} & 0 \\ 0 & 0 & 0 \end{bmatrix} \begin{Bmatrix} N_x \\ N_y \\ N_{xy} \end{Bmatrix}
\tag{6.14}
$$

$$
+ \begin{bmatrix} D'_{11} & D'_{12} & 0 \\ D'_{12} & D'_{11} & 0 \\ 0 & 0 & D'_{66} \end{bmatrix} \begin{Bmatrix} M_x \\ M_y \\ M_{xy} \end{Bmatrix}
$$

In the bound approach, the two-dimensional extent of the fabric composite plate is simplified by considering two one-dimensional models where the pieces of cross-ply laminates are either in parallel or in series as shown in Figs. 6.14(c) and (d). In the parallel model, a uniform state of strain, ε^o, and curvature, κ, in the laminate midplane is assumed as a first approximation. For the one-dimensional repeating region of length $n_g a$, where a denotes the yarn width, an average membrane stress, $\bar{N}_x$, is defined as

$$
\bar{N}_x = \frac{1}{n_g a} \int_0^{n_g a} N_x \, dy
$$

$$
= \frac{1}{n_g a} \left[\int_0^a (A_{11}\varepsilon_{xx}^o + A_{12}\varepsilon_{yy}^o + B_{11}\kappa_{xx}) \, dy \right.
$$

$$
\left. + \int_a^{n_g a} (A_{11}\varepsilon_{xx}^o + A_{12}\varepsilon_{yy}^o + B_{11}\kappa_{xx}) \, dy \right]
$$

$$
= (A_{11}\varepsilon_{xx}^o + A_{12}\varepsilon_{yy}^o) + \frac{1}{n_g a}[aB_{11}^T + (n_g a - a)B_{11}^L]\kappa_{xx}
$$

$$
= A_{11}\varepsilon_{xx}^o + A_{12}\varepsilon_{yy}^o + \left(1 - \frac{2}{n_g}\right)B_{11}^L\kappa_{xx}
\tag{6.15}
$$

The factor $(1 - 2/n_g)$ appears because the terms B_{11} for the interlaced region (B_{11}^T) and non-interlaced region (B_{11}^L) have opposite signs, namely, $B_{11}^T = -B_{11}^L$. It is noted that B_{11}^L is derived for a cross-ply with the same configuration as in Fig. 6.14(b), where the upper surface $(z > 0)$ shows fibers in the x direction. B_{11}^T is for a cross-ply obtained by exchanging the positions of the two laminae in Fig. 6.14(b). Other average stress resultants can be written similar to Eq. (6.15) for uniform mid-plane strain, ε^o, and curvature, κ. The moment resultant, $\bar{M}_x$, for example, is

$$\bar{M}_x = \frac{1}{n_g a} \int_0^{n_g a} M_x \, dy$$

$$= D_{11}\kappa_{xx} + D_{12}\kappa_{yy} + \left(1 - \frac{2}{n_g}\right) B_{11}^L \varepsilon_{xx}^o \qquad (6.16)$$

Let $\bar{A}_{ij}$, $\bar{B}_{ij}$, and $\bar{D}_{ij}$ be the stiffness constant matrices relating the average stress resultant $\bar{N}$ and moment resultant $\bar{M}$ with ε^o and κ. Then

$$\bar{A}_{ij} = A_{ij}$$

$$\bar{B}_{ij} = \left(1 - \frac{2}{n_g}\right) B_{ij}^L \qquad (6.17)$$

$$\bar{D}_{ij} = D_{ij}$$

These components provide upper bounds for the stiffness constants of the fabric composite based upon the one-dimensional model. If these stiffness constants are inverted, lower bounds of the elastic compliance constants can be obtained. All the elastic stiffness constants A, B and D are computed using the basic laminate where the top layer is composed of the filling yarn (Fig. 6.14b).

In the series model, the disturbance of stress and strain near the interface of the interlaced region is neglected. Let the model be subjected to a uniform in-plane force, N_x, in the longitudinal direction. The assumption of constant stress leads to the definition of an average curvature. For instance, the average curvature, $\bar{\kappa}_{xx}$, along the x direction is

$$\bar{\kappa}_{xx} = \frac{1}{n_g a} \int_0^{n_g a} \kappa_{xx} \, dx$$

$$= \frac{1}{n_g a} \left[\int_0^a B_{11}' N_x \, dx + \int_a^{n_g a} B_{11}' N_x \, dx \right]$$

$$= \frac{1}{n_g a} [aB_{11}'^T + a(n_g - 1)B_{11}'^L]N_x$$

$$= \left(1 - \frac{2}{n_g}\right) B_{11}'^L N_x \qquad (6.18)$$

It is also understood that the terms B'_{11} for the interlaced region (B'^T_{11}) and non-interlaced region (B'^L_{11}) are equal and opposite in sign. Other average curvature and mid-plane strain expressions can be written similar to Eq. (6.18) for uniformly applied N and M. Let $\bar{A}'_{ij}$, $\bar{B}'_{ij}$, and $\bar{D}'_{ij}$ be the compliance constant matrices relating the average mid-plane strain, $\bar{\varepsilon}^o$, and curvature, $\bar{\kappa}$, with the stress resultant, N, and moment resultant, M. Thus

$$\bar{A}'_{ij} = A'_{ij}$$

$$\bar{B}'_{ij} = \left(1 - \frac{2}{n_g}\right)B'^L_{ij} \tag{6.19}$$

$$\bar{D}'_{ij} = D'_{ij}$$

Equations (6.19) give the upper bounds for the composite compliance constants and, after inversion, the lower bounds for the stiffness constants.

In summary, both upper and lower bounds for the elastic stiffness and compliance constants can be obtained from the mosaic model. Numerical results demonstrating the relationship between these bounds and $1/n_g$ are shown in Fig. 6.15 for $\bar{A}_{11}$ and $\bar{A}'_{11}$ and in Fig. 6.16 for $\bar{B}'_{11}$. The material properties of a carbon/epoxy composite given in Table 6.2, with fiber volume fraction in the impregnated yarn of 60%, are adopted in the calculations. Bidirectional fiber

Fig. 6.15. Variations of $\bar{A}'_{11}$ and $\bar{A}_{11}$ with $1/n_g$. (After Ishikawa and Chou 1983b.)

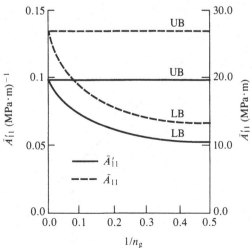

composites are represented by the limiting case of $1/n_g \to 0$ $(n_g \to \infty)$ and the upper and lower bounds of the elastic constants coincide with each other. Plain weaves are represented by the case of $1/n_g = 0.5$. The coupling effects for plain weave composites vanish, as can be seen from Eqs. (6.17) and (6.19), and both the upper and lower bounds of $\bar{B}'_{ij}$ $(\bar{B}_{ij})$ are identical, i.e. zero. However, the bounds of $\bar{A}_{ij}$ $(\bar{A}'_{ij})$ do not coincide for plain weave composites.

6.5   Crimp (fiber undulation) model

The crimp model is developed in order to consider the continuity and undulations of fibers in a fabric composite. Although the formulation of the problem developed in the following is valid for all n_g values, the crimp model is particularly suited for fabrics with low n_g values. The crimp model also provides the basis of analysis for the bridging model (Section 6.6).

Figure 6.17 depicts the geometry of the model where the undulation shape is defined by the parameters $h_1(x)$, $h_2(x)$, and a_u. The parameters $a_o = (a - a_u)/2$ and $a_2 = (a + a_u)/2$ are automatically determined by specifying a_u, which is geometrically arbitrary in the range from 0 to a. Because a pure matrix region appears in

Fig. 6.16. Variations of the average coupling compliance with $1/n_g$. (After Ishikawa and Chou 1983b.)

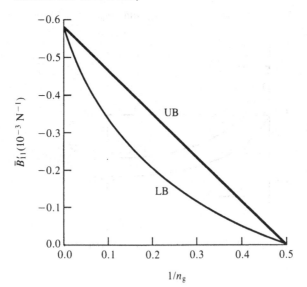

$1/n_g$

Table 6.2. *Material properties of unidirectional laminae*

Material	Fiber volume fraction in impregnated yarns	E_{11}(GPa)	E_{22}(GPa)	G_{12}(GPa)	ν_{12}	ε_2^b	Thickness (mm)	$\alpha_1(10^{-7}/°C)$	$\alpha_2(10^{-5}/°C)$
Carbon/epoxy[1]	60%	113	8.82	4.46	0.3	—	0.4	—	—
	65%	132	9.31	4.61	0.28	—	0.4	−25.0	2.7
Glass/polyester[2]	60%	47.5	15.9	6.23	0.27	0.38%	0.4	—	—
Glass/polyimide[3]	50%	41.2	15.7	5.59	0.3	0.5%	0.244	—	—
Kevlar/epoxy[4]	65%	85.3	5.5	2.54	0.4	—	—	−11.0	3.2

(1) Ishikawa, Koyama and Kobayashi (1977), Ishikawa (1981), Ishikawa and Chou (1983b). (2) Kimpara, Hamamoto and Takehana (1977). (3) Ishikawa and Chou (1982b). (4) Chou and Ishikawa (1989).

the model, the 'overall' fiber volume fraction, V_f, can be different from that in the yarn region.

To simulate the actual configuration, the following form of crimp is assumed for the filling:

$$h_1(x) = \begin{cases} 0 & (0 \le x \le a_o) \\ \left[1 + \sin\left\{\left(x - \frac{a}{2}\right)\frac{\pi}{a_u}\right\}\right]h_t/4 & (a_o \le x \le a_2) \\ h_t/2 & (a_2 \le x \le n_g a/2) \end{cases}$$

(6.20)

The sectional shape of the warp yarn is expressed by

$$h_2(x) = \begin{cases} h_t/2 & (0 \le x \le a_o) \\ \left[1 - \sin\left\{\left(x - \frac{a}{2}\right)\frac{\pi}{a_u}\right\}\right]h_t/4 & (a_o \le x \le a/2) \\ -\left[1 + \sin\left\{\left(x - \frac{a}{2}\right)\frac{\pi}{a_u}\right\}\right]h_t/4 & (a/2 \le x \le a_2) \\ -h_t/2 & (a_2 \le x \le n_g a/2) \end{cases}$$

(6.21)

It is assumed that the laminated plate theory is applicable to each infinitesimal piece of the model along the x axis. Thus, A_{ij}, B_{ij}, and

Fig. 6.17. Fiber crimp model. (After Ishikawa and Chou 1982b.)

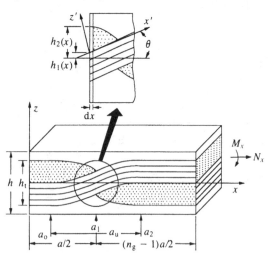

D_{ij} are expressed as functions of x $(0 \le x \le a/2)$ by

$$A_{ij}(x) = \int_{-h/2}^{h_1(x)-h_t/2} Q_{ij}^{M}\, dz + \int_{h_1(x)-h_t/2}^{h_1(x)} Q_{ij}^{F}(\theta)\, dz$$

$$+ \int_{h_1(x)}^{h_2(x)} Q_{ij}^{W}\, dz + \int_{h_2(x)}^{h/2} Q_{ij}^{M}\, dz$$

$$= Q_{ij}^{M}[h_1(x) - h_2(x) + h - h_t/2]$$

$$+ Q_{ij}^{F}(\theta)h_t/2 + Q_{ij}^{W}[h_2(x) - h_1(x)] \qquad (6.22)$$

$$B_{ij}(x) = \tfrac{1}{2}Q_{ij}^{F}(\theta)[h_1(x) - h_t/4]h_t + \tfrac{1}{4}Q_{ij}^{W}[h_2(x) - h_1(x)]h_t$$

$$D_{ij}(x) = \tfrac{1}{3}Q_{ij}^{M}\{[h_1(x) - h_t/2]^3 - h_2^3(x) + h^3/4\}$$

$$+ \tfrac{1}{3}Q_{ij}^{F}(\theta)[h_t^3/8 - 3h_t^2 h_1(x)/4$$

$$+ 3h_t h_1^2(x)/2] + \tfrac{1}{3}Q_{ij}^{W}[h_2^3(x) - h_1^3(x)]$$

where superscripts F, W and M signify the filling yarn, warp yarn and matrix, respectively. Similar expressions can be written for $a/2 \le x \le n_g a/2$.

The local stiffness of the filling yarn, $Q_{ij}^{F}(\theta)$, in the above equations is calculated as a function of the local off-axis angle, $\theta(x)$, which is defined as

$$\theta(x) = \arctan\!\left(\frac{dh_1(x)}{dx}\right) \qquad (6.23)$$

Consider a filling yarn composed of parallel fibers. The fiber direction is denoted as the 1 direction; the 2 and 3 directions are perpendicular to the fiber and they define the transversely isotropic plane. Then, from the Young's moduli (E_{11}, $E_{22} = E_{33}$), shear moduli ($G_{12} = G_{13}$, G_{23}) and Poisson's ratio (v_{12}) of the filling yarn, the elastic constants of the filling yarn with respect to the xyz axes in Fig. 6.17 can be defined (Lekhnitskii 1963). Here, the angle between the 1 and x axes is θ:

$$\frac{1}{E_{xx}^{F}(\theta)} = \frac{\cos^4 \theta}{E_{11}} + \left(\frac{1}{G_{12}} - \frac{2v_{21}}{E_{22}}\right)\cos^2 \theta \sin^2 \theta + \frac{\sin^4 \theta}{E_{22}}$$

$$E_{yy}^{F}(\theta) = E_{22} = E_{33}$$

$$\frac{1}{G_{xy}^{F}(\theta)} = \frac{\cos^2 \theta}{G_{12}} + \frac{\sin^2 \theta}{G_{23}} \qquad (6.24)$$

$$v_{yx}^{F}(\theta) = v_{21}\cos^2 \theta + v_{32}\sin^2 \theta$$

It is also understood from the assumption of transverse isotropy of the filling yarn that $\nu_{12} = \nu_{13}$, $E_{11}/\nu_{12} = E_{22}/\nu_{21}$, $\nu_{23} = \nu_{32}$, and $G_{23} = E_{22}/2(1 + \nu_{23})$.

Thus, the local stiffness constants of the undulated portion of the filling yarn, referring to the xyz coordinate axes, are given as functions of the fiber orientation angle θ

$$Q_{ij}^F(\theta) = \begin{bmatrix} E_{xx}^F(\theta)/D_v & E_{xx}^F(\theta)\nu_{yx}^F(\theta)/D_v & 0 \\ E_{xx}^F(\theta)\nu_{yx}^F(\theta)/D_v & E_{yy}^F(\theta)/D_v & 0 \\ 0 & 0 & G_{xy}^F(\theta) \end{bmatrix}$$

$$(i, j = 1, 2, 6) \quad (6.25)$$

where

$$D_v = 1 - (\nu_{yx}^F(\theta))^2 E_{xx}^F(\theta)/E_{yy}^F(\theta) \qquad (6.26)$$

By substituting Eq. (6.25) into Eqs. (6.22), the local plate stiffness constants can be evaluated. The local compliance constants, $A_{ij}'(x)$, $B_{ij}'(x)$, and $D_{ij}'(x)$ are then obtained by inverting the stiffness constants $A_{ij}(x)$, $B_{ij}(x)$, and $D_{ij}(x)$.

Define the average in-plane compliance of the model under a uniformly applied in-plane stress resultant by

$$\bar{A}_{ij}^{'C} = \frac{2}{n_g a} \int_0^{n_g a/2} A_{ij}'(x)\, dx \qquad (6.27)$$

where the superscript C signifies the crimp model. Since $A_{ij}'(x)$ is a constant within the straight yarn portion of Fig. 6.17, Eq. (6.27) can be rewritten as

$$\bar{A}_{ij}^{'C} = \left(1 - \frac{2a_u}{n_g a}\right) A_{ij}' + \frac{2}{n_g a} \int_{a_0}^{a_2} A_{ij}'(x)\, dx \qquad (6.28)$$

where A_{ij}' in the first term on the right-hand side of Eq. (6.28) denotes the compliance of the straight portion of the yarns, namely a cross-ply laminate, and is independent of x. The other average compliance coefficients $\bar{B}_{ij}^{'C}$ and $\bar{D}_{ij}^{'C}$ are obtained in a similar manner.

$$\bar{B}_{ij}^{'C} = \left(1 - \frac{2}{n_g}\right) B_{ij}' + \frac{2}{n_g a} \int_{a_0}^{a_2} B_{ij}'(x)\, dx \qquad (6.29)$$

$$\bar{D}_{ij}^{'C} = \left(1 - \frac{2a_u}{n_g a}\right) D_{ij}' + \frac{2}{n_g a} \int_{a_0}^{a_2} D_{ij}'(x)\, dx \qquad (6.30)$$

In the case of $n_g = 2$, $\bar{B}_{ij}^{\prime C}$ vanishes because $B_{ij}^{\prime}(x)$ is an odd function with respect to $x = a/2$, the center of undulation, due to the assumed form of $h_1(x)$. Furthermore, Eqs. (6.28)–(6.30) coincide with the upper bounds of the compliance of Eqs. (6.19) as a_u tends to zero. The integrations in Eqs. (6.28)–(6.30) are conducted numerically because of the complexity of the integrands. The final results of the average elastic stiffness, $\bar{A}_{ij}^C$, $\bar{B}_{ij}^C$ and $\bar{D}_{ij}^C$, for the entire strip can be reached by the inversion of $\bar{A}_{ij}^{\prime C}$, $\bar{B}_{ij}^{\prime C}$ and $\bar{D}_{ij}^{\prime C}$. If this procedure is applied in the warp direction, balanced properties such as $\bar{A}_{11}^C = \bar{A}_{22}^C$ can be realized.

Numerical results demonstrating the relationship between the in-plane stiffness, $\bar{A}_{11}$, and $1/n_g$ are given in Fig. 6.18 based upon the unidirectional lamina properties of a carbon/epoxy system (Table 6.2). In Fig. 6.18, UB and LB represent, respectively, the results of the upper and lower bound predictions of the mosaic model; CM denotes the crimp model; circles indicate finite element results. Figure 6.18 demonstrates the reduction in $\bar{A}_{11}$ due to fiber undulation, and the reduction is most severe in plain weave ($1/n_g = 0.5$) as compared to cross-ply laminates ($1/n_g = 0$).

The relationship between the coupling compliance $\bar{B}_{11}^{\prime}$ and $1/n_g$ is

Fig. 6.18. $\bar{A}_{11}^C$ vs. $1/n_g$ for carbon/epoxy composites, $V_f = 60\%$. Finite element results are indicated by ($\bigcirc$) for the mosaic model and by ($\bullet$) for the crimp model. —— mosaic model; – – – – crimp model. (After Ishikawa and Chou 1982b.)

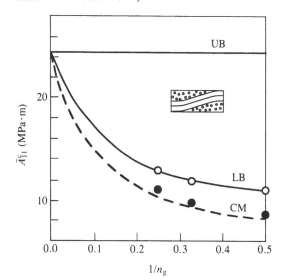

demonstrated in Fig. 6.16. The results from the crimp model coincide exactly with those of the upper bound predictions. This is due to the fact that the second term on the right-hand side of Eq. (6.29) vanishes due to the assumed asymmetrical shape of fiber undulation and, hence, the odd function representation of B'_{ij} with respect to $x = a/2$.

6.6   Bridging model and experimental confirmation

The crimp model which is based upon a single fiber yarn has led to the concept of a bridging model for general satin composites. Such a model is desirable because the interlaced regions in a satin weave are often separated from one another. The hexagonal shape of the repeating unit in a satin weave, as shown in Fig. 6.19, is modified to a square shape (Fig. 6.19b) for simplicity of calculation. A schematic view of the bridging model is shown in Fig. 6.19(c) for a repeating unit which consists of the interlaced region and its surrounding areas. This model is valid only for satin weaves where $n_g \geq 4$. The four regions labeled I, II, IV and V consist of

Fig. 6.19. Concept of the bridging model: (a) shape of the repeating unit of eight-harness satin; (b) modified shape for the repeating unit; (c) idealization for the bridging model. (After Ishikawa and Chou 1982b.)

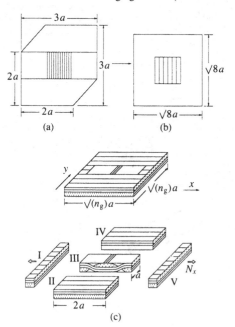

straight filling yarns, and hence can be regarded as pieces of cross-ply laminates of thickness h_t. Region III has an interlaced structure with an undulated filling yarn. Although the undulation and continuity in the warp yarns are ignored in this model, their effect is expected to be small because the applied load is assumed to be in the filling direction.

The in-plane stiffness in region III, where $n_g = 2$, has been derived in Section 6.5 and has been found to be lower than that of a cross-ply laminate. Therefore, regions II and IV carry higher loads than region III; all three of these regions act as bridges for load transfer between regions I and V. It is also assumed that regions II, III and IV have the same average mid-plane strain and curvature. Then, the average stiffness constants for the regions II, III and IV are

$$\bar{A}_{ij} = \frac{1}{\sqrt{n_g}} [(\sqrt{(n_g)} - 1)A_{ij} + \bar{A}_{ij}^C]$$

$$\bar{B}_{ij} = \frac{1}{\sqrt{n_g}} (\sqrt{(n_g)} - 1)B_{ij} \qquad (6.31)$$

$$\bar{D}_{ij} = \frac{1}{\sqrt{n_g}} [(\sqrt{(n_g)} - 1)D_{ij} + \bar{D}_{ij}^C]$$

$\bar{A}_{ij}^C$ and $\bar{D}_{ij}^C$ for the undulated portion III in Fig. 6.19 are obtained from $\bar{A}_{ij}'^C$ and $\bar{D}_{ij}'^C$ of Eqs. (6.28) and (6.30), and $\bar{B}_{ij}'^C = 0$. A_{ij}, B_{ij}, and D_{ij} in Eqs. (6.31) for the cross-ply laminates of regions II and IV in Fig. 6.19 are given in Eqs. (6.12).

It is also postulated that the total in-plane force carried by regions II, III and IV is equal to that by region I or V. Then, the following average compliance constants are derived:

$$\bar{A}_{ij}'^S = \frac{1}{\sqrt{n_g}} [2\bar{A}_{ij}' + (\sqrt{(n_g)} - 2)A_{ij}']$$

$$\bar{B}_{ij}'^S = \frac{1}{\sqrt{n_g}} [2\bar{B}_{ij}' + (\sqrt{(n_g)} - 2)B_{ij}'] \qquad (6.32)$$

$$\bar{D}_{ij}'^S = \frac{1}{\sqrt{n_g}} [2\bar{D}_{ij}' + (\sqrt{(n_g)} - 2)D_{ij}']$$

where $\bar{A}_{ij}'$, $\bar{B}_{ij}'$, and $\bar{D}_{ij}'$ are determined by inverting Eqs. (6.31) and the superscript S denotes properties of the entire satin plane. Finally, $\bar{A}_{ij}^S$, $\bar{B}_{ij}^S$ and $\bar{D}_{ij}^S$ can be obtained by inverting Eqs. (6.32).

The fiber crimp model is effective for plain weave composites whereas the bridging model is valid for satin weave composites. This is because there are no straight yarn regions surrounding an interlaced region in the plain weave. Therefore, no bridging effect is expected in plain weave composites, and the analysis based on the fiber undulation model provides a reasonable prediction of the behavior of plain weave composites.

Numerical results for the relationship between the in-plane elastic stiffness $\bar{A}_{11}^{S}$ and $1/n_g$ are indicated in Fig. 6.20, also using the unidirectional laminar properties of Table 6.2. A prediction by the present theory agrees with experimental results (Ishikawa and Chou 1982b). It should be noted that the overall fiber volume fraction of a fabric composite is slightly less than that of the impregnated yarns due to the resin rich region in the vicinity of the undulation. For instance, for a fiber volume fraction of 65%, the average overall fiber volume fraction in a repeating unit (Fig. 6.19) for $n_g = 8$, $h_t = h$, and $a_u = a$ is around 62%.

Ishikawa, Matsushima, Hayashi and Chou (1985) have conducted experimental verifications of the analytical models for elastic moduli

Fig. 6.20. $\bar{A}_{11}^{S}$ vs. $1/n_g$ for carbon/epoxy composites, $V_f = 65\%$. —— Upper and lower bounds; – – – bridging model solution; (▲, ●) experimental results for a cross-ply laminate and eight-harness satin, respectively. (After Ishikawa and Chou 1982b.)

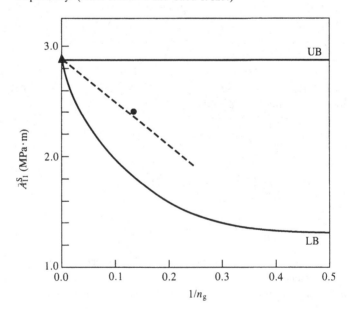

of fabric composites. The experimental materials used include plain weave and eight-harness satin fabric reinforced composites of carbon/epoxy. Ply numbers are 1, 4, 8 and 20 for plain weave fabrics and 2 for eight-harness satin. Yarn orientations are [0°/90°], [15°/−75°], [30°/−60°] and [±45°], as defined by the angles between the loading axis and the yarn direction.

Experimental and theoretical results are compared in Figs. 6.21–6.23. Figure 6.21 presents results of the in-plane stiffness, A_{11}, non-dimensionalized by the corresponding A_{11} of the cross-ply laminate as a function of $1/n_g$. Experimental results of four-ply plain weave and two-ply 8 harness weave composites are given. The symbol ϕ signifies both the averaged value indicated by the solid circle, and the scattering indicated by the horizontal bars. Theoretical predictions of the bridging model (BM) are adopted for $n_g \geq 4$ and the crimp model (CM) for $4 \geq n_g \geq 2$ according to the reason stated earlier. Abbreviations LWC and LWA denote, respectively, the limiting cases where local warping is completely constrained and allowed. Also, UB and LB are, respectively, upper bound and lower bound predictions of the mosaic model.

A good correlation between theory and experiments can be observed for eight-harness satin composites. The experimental data

Fig. 6.21. Relationships between non-dimensionalized in-plane stiffness and $1/n_g$; ϕ experiments. (After Ishikawa *et al.* 1985.)

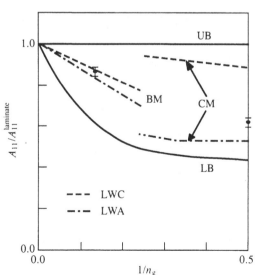

lie in between the LMC and LMA predictions. These results suggest good predictability of the theory based upon the bridging model for satin weave composites. There exists a significant discrepancy between the LWC and LWA curves of plain weave composites even though slight improvement is achieved over the simple bound theory.

Constraint of local warping is another factor governing the in-plane modulus. Neighboring layers in a fabric laminate tend to suppress the warping of one another. Thus, a dependence of elastic moduli on ply number appears for plain weave composites. This effect is demonstrated in Fig. 6.22 where experimental results of specimens of four different ply numbers are indicated. The in-plane stiffness A_{11} of the fabric composite is non-dimensionalized by A_{11} of the cross-ply. Small variations in the theoretical predictions are caused by the scattering of the measured h/a. The in-plane modulus increases from the value for one-ply, which is slightly higher than the LWA prediction, and reaches values slightly lower than the LWC prediction.

The in-plane off-axis elastic moduli results are presented in Fig. 6.23. The off-axis behavior is symmetric with respect to $\phi = 45°$ because it is assumed that the elastic properties in both the filling and warp directions are identical.

In summary, the experimental results of eight-harness satin composites coincide very well with theoretical predictions. There is

Fig. 6.22. Dependence of in-plane stiffness on ply number in plain weave composites: – – – – LWC; — · — LWA; ⬤ experiments. (After Ishikawa *et al.* 1985.)

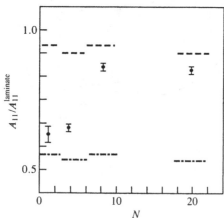

still a discrepancy in the predictions of elastic moduli of plain weave composites based upon two limiting cases: local warping completely prohibited or allowed. All on-axis measured moduli fall in between the two predictions.

6.7 Analysis of the knee behavior and summary of stiffness and strength modeling

Both the crimp model and bridging model described above are now extended to the study of the stress–strain behavior of woven fabric composites after initial fiber failure, known as the *knee phenomenon*. The essential experimental fact for the knee phenomenon is that the breaking strain in the transverse layer, ε_2^b, is much smaller than that of the longitudinal layer in cross-ply laminates. Only the failure of the transverse yarns, which occurs in the warp direction in the present model, is considered. Thus, a failure criterion based upon maximum strain is adopted.

In the following, the crimp model is utilized and attention is confined to the one-dimensional behavior of fabric composites

Fig. 6.23. Off-axis moduli of plain weave and eight-harness satin. $E =$ axial Young's modulus; $G =$ in-plane shear modulus; $- - -$ LWC; $- \cdot -$: LWA. (After Ishikawa *et al.* 1985.)

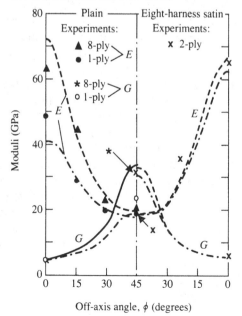

Off-axis angle, ϕ (degrees)

under an applied stress resultant N_x. Then Eq. (6.4) is reduced to

$$\varepsilon_{xx}^{o} = A'_{11}N_x + B'_{11}M_x$$
$$\kappa_{xx} = B'_{11}N_x + D'_{11}M_x \tag{6.33}$$

where M_x is the locally induced moment resultant due to the application of N_x. By assuming first that no bending deflection by the coupling effect is allowed along the x axis,

$$\kappa_{xx} = B'_{11}N_x + D'_{11}M_x = 0 \tag{6.34}$$

This assumption can be realized only if the fabric composite plate is symmetrical with respect to its mid-plane. However, in practical multi-layer fabric composites arranged symmetrically to their mid-planes, this assumption is expected to be approximately true. From Eqs. (6.33) and (6.34)

$$\varepsilon_{xx}^{o} = A''_{11}N_x \tag{6.35}$$

where $A''_{11} = A'_{11} - B'^{2}_{11}/D'_{11}$.

The quantity A''_{11} may be referred to as a *modified in-plane compliance* and it is a function of x. Since N_x is uniform along the x direction, $A''_{11}(x)$ represents a strain distribution before the first transverse matrix cracking. Figure 6.24 depicts two examples of the mid-plane strain distribution relative to that at the point $x = 0$ in Fig. 6.17 and for $a_u = a$. It can easily be seen that the fiber undulation causes local softening and that the maximum strain appears at the center of undulation ($x = a/2$). Also, the strain along the thickness direction at each section is uniform and equal to ε_{xx}^{o} owing to the classical plate theory and the absence of bending. Although the strain distribution calculated from finite element analysis (Ishikawa and Chou 1983b) deviates slightly from the assumed uniform distribution, the present idealization provides a simple method for analyzing the knee phenomenon.

Assume that the region of the highest strain reaches the transverse failure strain ε_2^{b} first, and the damaged area in the warp yarn propagates as the load increases. It is further assumed that classical lamination theory is still valid in this failure process, and that the effective elastic moduli of such a failed region in the warp yarn are much lower than those of a sound area and can be expressed as

$$Q'^{W}_{ij} = \begin{bmatrix} Q^{W}_{11}/100 & Q^{W}_{12}/100 & 0 \\ Q^{W}_{12}/100 & Q^{W}_{22} & 0 \\ 0 & 0 & Q^{W}_{66}/100 \end{bmatrix} \tag{6.36}$$

Here, $Q_{ij}'^W$ denotes the reduced stiffness of the warp yarns after failure, and it is assumed that, with the exception of Q_{22}^W, the Q_{ij} components are reduced by a factor of 1/100 to reflect the weakening effect of transverse cracking. The assumption of the applicability of the classical lamination theory implies that the complex stress and strain fields around the failed region are neglected. Such a successive failure process will continue until the lowest strain in the region reaches ε_2^b. At that time, all the warp regions have failed completely. Beyond this point, the stress–strain curve becomes a straight line again until the final failure of the filling yarns.

Next, consider the case where the restraint on bending is removed. From the classical lamination theory (Chapter 2)

$$\varepsilon_{xx}(z) = \varepsilon_{xx}^o + z\kappa_{xx} \tag{6.37}$$

The strain state under an in-plane stress resultant, N_x, is given by

$$\varepsilon_{xx}(z) = (A_{11}' + zB_{11}')N_x \tag{6.38}$$

Thus, the strain field under the prescribed N_x is determined from A_{11}', B_{11}', and z. Since the strain in a vertical section is distributed linearly according to Eq. (6.37), it is necessary to determine the height, h_3, where the strain reaches the critical value, ε_2^b. If the

Fig. 6.24. Relative strain distribution along the x axis in the fiber crimp model without bending, $a_u = a$; —— carbon/epoxy; – – – glass/polyester. (After Ishikawa and Chou 1982b.)

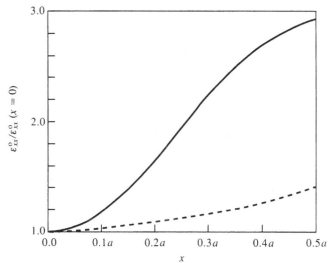

strain at the outer edge of the warp yarns, $\varepsilon_2(h_2)$ according to Eq. (6.37), is larger than ε_2^b, then, for $a_o \leq x \leq a/2$,

$$h_3(x) = h_2 - (h_2 - h_1)\frac{\varepsilon_2(h_2) - \varepsilon_2^b}{\varepsilon_2(h_2) - \varepsilon_2(h_1)} \tag{6.39}$$

Based upon the h_3 value, the plate stiffness in Eqs. (6.22) needs to be modified after the initial failure. For instance, for $a_o \leq x \leq a/2$,

$$A_{ij}(x) = Q_{ij}^M[h_1(x) - h_2(x) + h - h_t/2]$$
$$+ Q_{ij}^F(\theta)h_t/2 + Q_{ij}^W[h_3(x) - h_1(x)] + Q_{ij}'^W[h_2(x) - h_3(x)] \tag{6.40}$$

Modifications similar to Eq. (6.40) are made for B_{ij} and D_{ij} in Eqs. (6.22).

Figure 6.25 presents two numerical examples for a glass/polyester plain weave composite of $a_u = a$ and overall $V_f = 36.8\%$ with and without bending. The finite element analysis and acoustic emission results of Kimpara, Hamamoto and Takehana (1977) are also given. Basic material properties are shown in Table 6.2. The prediction for

Fig. 6.25. Stress–strain curves for plain weave composites of glass/polyester, $V_f = 36.8\%$, and experimental data of acoustic emission; —— analytical results for no bending; – · – · analytical results for unconstrained bending; – – – – finite element simulation; (——) total count in acoustic emission measurement. The vertical arrow indicates the specified value of ε_2^b. (After Ishikawa and Chou 1982b.)

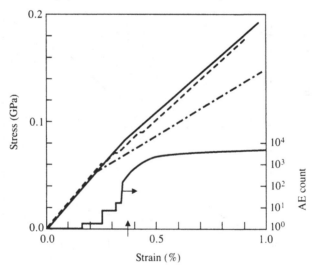

the case without bending compares very favorably with the finite element simulation. It is quite reasonable that the case with bending shows much lower stiffness because it is not subjected to lateral constraints.

In actual plain weave composites, local bending deformation caused by the coupling effect in each interlaced region is constrained by adjacent regions for which the stiffness constants B_{ij} have opposite signs. Therefore, as far as plain weave composites are concerned, one-dimensional analysis for the case without bending should give a reasonable prediction of the knee behavior under in-plane loading.

The bridging model and the process of successive warp yarn failure can be combined to analyze the knee behavior in satin composites. The approaches for plain weave composites are adopted here. First, define the stiffness of the fabric composite under an applied stress resultant N_x without bending to be A_{11}^*. It is noted that $A_{11}^* = 1/A_{11}''$ and the compliance A_{11}'' follows the definition in Eq. (6.35). The average stiffness for regions II, III and IV of Fig. 6.19 is denoted as $\bar{A}_{11}^*$ and calculated by taking the volume average:

$$\bar{A}_{11}^* = (1/\sqrt{n_g})\bar{A}_{11}^{*C} + (1 - 1/\sqrt{n_g})A_{11}^* \qquad (6.41)$$

Then the compliance of the whole satin composite is calculated from an average over its length. Define the compliance $\bar{A}_{11}'' = 1/\bar{A}_{11}^*$. The assumption of uniformity of N_x along the x direction leads to

$$\bar{A}_{11}''^{,S} = (2/\sqrt{n_g})\bar{A}_{11}'' + (1 - 2/\sqrt{n_g})A_{11}''$$

where the superscript S indicates satin composites. Finally, the stiffness of the whole satin composite can be obtained as

$$\bar{A}_{11}^{*S} = 1/\bar{A}_{11}''^{,S} \qquad (6.42)$$

It should be noted that the inversion of the compliance and stiffness constants cannot generally be achieved by merely taking the reciprocal of the respective components (i.e. $C_{ij} \neq 1/S_{ij}$). However, under the assumptions made in the derivations of Eqs. (6.33)–(6.35), Eq. (6.42) is valid for this one-dimensional problem. Expressions similar to Eqs. (6.41) and (6.42) for the case of unconstrained bending can also be obtained but are omitted here. The rest of the procedure for analyzing the knee phenomenon follows that for the plain weave case. The initial failure of the warp yarns occurs at the point of highest strain, for example the center of undulation in the case without bending. Also, since there are regions of uniform strain such as the bridging zones in this model, the entire area of these regions may fail simultaneously, according to the present assumptions.

Figure 6.26 compares numerical and experimental results for stress–strain curves of an eight-harness satin glass fabric/polyimide composite (Table 6.2). Since the test pieces were nearly symmetrical with respect to their mid-planes, the analysis of the case without bending is selected for comparison; the agreement is quite good, particularly for strain values up to the knee point. A theoretical stress–strain curve for a plain weave composite of the same material is also shown in Fig. 6.26. Here, a knee point is defined by a deviation of 0.01% in strain from the linear strain. Then, the knee stress in the eight-harness satin is higher than that of the plain weave, although knee strains are nearly identical. It can be concluded that the elastic stiffness and knee stress in satin composites are higher than those in plain weave composites due to the presence of the bridging zones.

The following is a summary of the stiffness and strength models for two-dimensional orthogonal woven fabric composites:

(1) A fabric composite can be idealized as an assemblage of pieces of asymmetric cross-ply laminates. The upper and

Fig. 6.26. Theoretical and experimental stress–strain curves for glass/polyimide composites, $V_f = 50\%$ in impregnated yarns; —— bridging model solution without bending for eight-harness satin (overall $V_f = 47.7\%$); – – – – fiber undulation model solution without bending for plain weave (overall $V_f = 40.9\%$); —— —— experimental curve; (●) knee points. (After Ishikawa and Chou 1982b.)

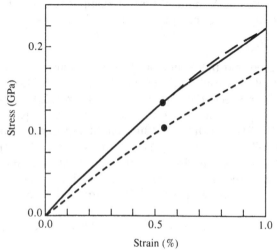

lower bounds of elastic stiffness and compliance of fabric composite plates in such a 'mosaic model' are obtained under the assumption of constant strain and constant stress.

(2) The 'crimp model', which is a one-dimensional approximation and takes into account fiber continuity and undulation, is particularly suited for predicting elastic properties of plain weave composites. The analytical results based upon the crimp model demonstrate that fiber undulation leads to a softening in the in-plane stiffness as compared to the mosaic model. However, fiber undulation has no effect on the coupling constants. Therefore, the solution of the coupling compliance based upon the mosaic model is considered to be reliable.

Both the results of the crimping model and of the mosaic model for the compliance constants $\bar{A}'_{11}$ and $\bar{B}'_{11}$ compare very favorably with the results of a finite element analysis (Ishikawa and Chou 1983b).

(3) In the case of $\bar{D}_{11}$, Ishikawa and Chou (1983b) have adopted a transverse shear deformation theory for a modification of the mosaic model, and examined the response of a fabric composite plate under both cylindrical bending and lateral force. Numerical results of $\bar{D}'_{11}$ based upon the modified transverse shear deformation theory coincide well with the finite element results.

(4) The effect of fiber undulation shapes on $\bar{A}'^c_{11}$ in the crimp model is shown in Fig. 6.27. The geometrical parameters a and h are chosen to be 1.0 and 0.4, respectively. The calculations are performed for the range of a_u/h values from 0 to a/h, where the case $a_u \to 0$ corresponds to the configuration of a mosaic model. The results show that $\bar{A}'^c_{11}$ is susceptible to the shape of undulation, particularly at small n_g values. The highest $\bar{A}'^c_{11}$ value, i.e. the lowest in-plane stiffness, is obtained at around $a_u/h = 1$. On the other hand, the $\bar{A}'^c_{11}$ values at $a_u/h = 0$ and a/h are not far apart. Because in actual fabrics $a_u/h \approx a/h$, the mosaic model ($a_u/h = 0$) seems to be effective in evaluating the in-plane stiffness of a fabric.

(5) The crimp model has been applied to examine the knee phenomenon of plain weave composites. The predicted knee behavior of a glass/polyester composite without bend-

ing shows excellent agreement with the stress–strain curve obtained by using a finite element analysis.

(6) The bound method based upon the mosaic model is useful for a rough estimation of fabric composite stiffness propperties. The crimp model offers better predictability than the mosaic model for the in-plane and bending moduli. However, the crimp model is inadequate for evaluating the elastic properties of satin weave composites with large n_g.

(7) A bridging model has been developed to examine the stiffness and strength of general satin composites. The interlaced regions in a satin fabric are often separated from one another by the non-interlaced regions. Since the regions with straight yarns surrounding an interlaced region have higher in-plane stiffnesses than the latter, they carry higher loads and play the role of load transferring bridges.

(8) The initial elastic stiffness of satin composites can be predicted by the bridging model. The analysis of an eight-harness satin carbon/epoxy composite demonstrates good agreement with experimental data.

Fig. 6.27. Relationship between average in-plane compliance and undulation length. (After Ishikawa and Chou 1983b.)

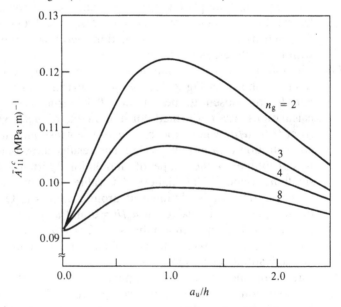

(9) The concept of successive failure of the warp yarns and the bridging idealization have been combined to study the knee behavior in satin composites. The theoretical results for an eight-harness satin glass reinforced polyimide composite compare favorably with the experimental curve. It can be concluded that the bridging regions surrounding an interlaced region are responsible for the higher stiffness and knee stress in strain composites than those in plain weave composites.

6.8 In-plane thermal expansion and thermal bending coefficients

The constitutive equations of a laminated plate taking into account the effects of a small uniform temperature change are given in Eqs. (6.5)–(6.9). In the following, the analytical techniques developed for the mosaic model, crimp model, and bridging model are applied to analyze the thermal problem.

First, for applying the mosaic model, a long strip of the fabric composite (Fig. 6.14a) is again considered. The laminate is free of externally applied load. The average strains and curvatures of a one-dimensional strip of width a along the filling or warp direction due to a uniform temperature change, ΔT, can be expressed in the following forms:

$$\bar{\varepsilon}_{xx}^{o} = \frac{1}{n_g a} \int_0^{n_g a} \Delta T \bar{A}_x'(x) \, dx = \Delta T \bar{A}_1'$$

$$\bar{\varepsilon}_{yy}^{o} = \frac{1}{n_g a} \int_0^{n_g a} \Delta T \bar{A}_y'(y) \, dy = \Delta T \bar{A}_2' \tag{6.43}$$

$$\bar{\kappa}_{xx} = \frac{1}{n_g a} \int_0^{n_g a} \Delta T \bar{B}_x'(x) \, dx = \Delta T \frac{n_g - 2}{n_g} \bar{B}_x'$$

$$\bar{\kappa}_{yy} = \frac{1}{n_g a} \int_0^{n_g a} \Delta T \bar{B}_y'(y) \, dy = \Delta T \frac{n_g - 2}{n_g} \bar{B}_y' \tag{6.44}$$

It should be noted that $\bar{B}_x'$ has opposite signs in the regions $x = 0{-}a$ and $x = a{-}(n_g - 1)a$; the same is true for $\bar{B}_y'$.

Because of the nature of the cross-ply laminates $\bar{A}_{xy}'$ and $\bar{B}_{xy}'$ vanish. From Eqs. (6.43) and (6.44), the average thermal expansion and thermal bending coefficients for the mosaic model are given by

$$\bar{\bar{A}}'_x = \bar{A}'_x, \qquad \bar{\bar{A}}'_y = \bar{A}'_y$$

$$\bar{\bar{B}}'_x = \left(1 - \frac{2}{n_g}\right)\bar{B}'_x, \qquad \bar{\bar{B}}'_y = \left(1 - \frac{2}{n_g}\right)\bar{B}'_y \tag{6.45}$$

Next, the crimp model is applied; the forms of fiber crimp for the filling and warp yarns follow the assumed shapes of Eqs. (6.20) and (6.21), respectively. By assuming no in-plane force and moment and following the derivations of Eqs. (6.43) and (6.44), the fiber crimp model gives

$$\bar{\bar{A}}'^C_x = \left(1 - \frac{2a_u}{n_g a}\right)\bar{A}'_x + \frac{2}{n_g a} \int_{a_0}^{a_2} \bar{A}'_x(x)\,dx$$

$$\bar{\bar{A}}'^C_y = \left(1 - \frac{2a_u}{n_g a}\right)\bar{A}'_y + \frac{2}{n_g a} \int_{a_0}^{a_2} \bar{A}'_y(y)\,dy \tag{6.46}$$

$$\bar{\bar{B}}'^C_x = \left(1 - \frac{2}{n_g}\right)\bar{B}'_x + \frac{2}{n_g a} \int_{a_0}^{a_2} \bar{B}'_x(x)\,dx$$

$$\bar{\bar{B}}'^C_y = \left(1 - \frac{2}{n_g}\right)\bar{B}'_y + \frac{2}{n_g a} \int_{a_0}^{a_2} \bar{B}'_y(y)\,dy \tag{6.47}$$

Here, the superscript C signifies the crimp model. It is understood that $\bar{\bar{A}}'^C_{xy}$ and $\bar{\bar{B}}'^C_{xy}$ vanish for cross-ply constructions. Since $\bar{B}'_x$ and $\bar{B}'_y$ are odd functions of location with respect to the center of undulation (Eq. (6.20)) the integration in Eqs. (6.47) vanishes and

$$\bar{\bar{B}}'^C_x = \left(1 - \frac{2}{n_g}\right)\bar{B}'_x$$

$$\bar{\bar{B}}'^C_y = \left(1 - \frac{2}{n_g}\right)\bar{B}'_y \tag{6.48}$$

The expressions of $\bar{\bar{B}}'$ from Eqs. (6.45) and (6.48) are identical. Thus, fiber crimp has no effect on the thermal bending coefficients. The same conclusion has been obtained for the extension–bending coupling constant in Section 6.5.

For the in-plane thermal expansion coefficient, it is necessary to evaluate the integration in Eqs. (6.46). This is done on the assumption that the classical laminated plate theory is applicable to each infinitesimal piece of width dx of the one-dimensional strip shown in Fig. 6.17. The following steps are taken to obtain $\bar{A}'_x(x)$ and $\bar{A}'_y(y)$. Consider $\bar{A}'_x(x)$ as an example. First, $\bar{A}_x(x)$ and $\bar{B}_x(x)$

are evaluated from Eqs. (6.6) and (6.7) for $0 \le x \le a/2$, and the results are

$$\tilde{A}_x(x) = q_x^M(h_1(x) - h_2(x) + h - h_t/2) + q_x^F(\theta)h_t/2 + q_x^W(h_2(x) - h_1(x)) \tag{6.49}$$

$$\tilde{B}_x(x) = \tfrac{1}{2}q_x^F(\theta)(h_1(x) - h_t/4)h_t + \tfrac{1}{4}q_x^W(h_2(x) - h_1(x))h_t \tag{6.50}$$

where the superscripts F, W and M signify the filling yarn, warp yarn, and matrix region, respectively. Next, from Eq. (6.6), $q_x = \bar{Q}_{11}\alpha_{xx} + \bar{Q}_{12}\alpha_{yy} + \bar{Q}_{16}\alpha_{xy}$; $q_x^F(\theta)$, in particular, is determined from the local stiffness matrix $Q_{ij}^F(\theta)$, following the procedures outlined in Section 6.5. Furthermore, the off-axis thermal expansion coefficients are given by

$$\alpha_{xx}^F(\theta) = \cos^2 \theta \alpha_{11}^F + \sin^2 \theta \, \theta \alpha_{22}^F$$

$$\alpha_{yy}^F(\theta) = \alpha_{22}^F \tag{6.51}$$

$$\alpha_{xy}^F(\theta) = 0$$

where α_{11} and α_{22} denote, respectively, thermal expansion coefficients parallel and transverse to the fiber direction in a unidirectional fiber composite. Thus, $\tilde{A}_x(x)$ and $\tilde{B}_x(x)$ can be determined from Eqs. (6.49) and (6.50). Then, the $\tilde{A}_i'(x)$ and $\tilde{B}_i'(x)$ components are obtained by inverting $\tilde{A}_i(x)$ and $\tilde{B}_i(x)$ as in Eq. (6.9).

Numerical integration of Eqs. (6.46) has been conducted and the results for $\bar{\tilde{A}}_x'^C$ and $\bar{\tilde{B}}_x'^C$ as functions of $1/n_g$ are given in Fig. 6.28. The balanced thermal property such as $\tilde{A}_x' = \tilde{A}_y'$ for a fabric composite can be realized if the above procedure of calculation is conducted for one-dimensional strips along both the filling and warp directions.

Lastly, the bridging model is applied to analyze the thermal properties. It has been noted in Section 6.6 that regions II and IV of Fig. 6.19 are stiffer than the crimped region III and, hence, they carry more load when an external force is applied in the x direction. Regions II, III and IV are termed *bridging regions*. For the thermal property analysis, assuming no mechanical loading, the equilibrium of the bridging regions requires

$$a\left\{\begin{matrix} N^C \\ M^C \end{matrix}\right\} + (\bar{V}(n_g) - 1)a\left\{\begin{matrix} N \\ M \end{matrix}\right\} = 0 \tag{6.52}$$

where the superscript C again denotes the crimped region, and N and M without superscripts are for the cross-ply laminate. Furthermore, under the assumption of uniform strain and curvature in the bridging regions II, III and IV, it is defined that

$$\{\varepsilon^{oC}\} = \{\bar{\varepsilon}^{o}\}$$

$$\{\kappa^{C}\} = \{\bar{\kappa}\}$$

(6.53)

where the bar denotes the average of the bridging regions.

Substituting Eq. (6.5) into Eq. (6.52), and taking into account Eqs. (6.53), the results are expressed in the condensed form

$$\left(\left[\begin{array}{c|c} A^{C} & B^{C} \\ \hline B^{C} & D^{C} \end{array}\right] + (\sqrt{(n_g)} - 1)\left[\begin{array}{c|c} A & B \\ \hline B & D \end{array}\right]\right)\left\{\begin{array}{c} \bar{\varepsilon}^{o} \\ \bar{\kappa} \end{array}\right\} = \Delta T\left(\left\{\begin{array}{c} \bar{\bar{A}}^{C} \\ \bar{\bar{B}}^{C} \end{array}\right\}\right.$$

$$\left. + (\sqrt{(n_g)} - 1)\left\{\begin{array}{c} \bar{\bar{A}} \\ \bar{\bar{B}} \end{array}\right\}\right) \quad (6.54)$$

Fig. 6.28. Variation of the thermal deformation coefficients with $1/n_g$ for carbon/epoxy composites, $V_f = 60\%$ and $a_u/a = 1.0$; ——— $\bar{\bar{A}}'_x$; — — — $\bar{\bar{B}}'_x$; (●) experimental results of $\bar{\bar{B}}'_x$ at 300°K. (After Ishikawa and Chou 1983a.)

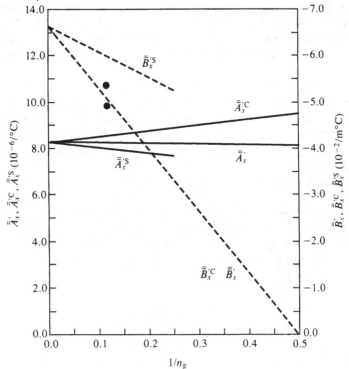

The quantities on the left-hand side of Eq. (6.54) can be related to the average elastic stiffness in the bridging regions as

$$\left[\begin{array}{c|c} A^C & B^C \\ \hline B^C & D^C \end{array}\right] + (\sqrt{(n_g)} - 1)\left[\begin{array}{c|c} A & B \\ \hline B & D \end{array}\right] = \sqrt{n_g}\left[\begin{array}{c|c} \bar{A} & \bar{B} \\ \hline \bar{B} & \bar{D} \end{array}\right] \qquad (6.55)$$

Hence, Eq. (6.54) can be written as

$$\left\{\begin{array}{c} \bar{\varepsilon}^o \\ \bar{\kappa} \end{array}\right\} = \Delta T\left[\begin{array}{c|c} \bar{A}' & \bar{B}' \\ \hline \bar{B}' & \bar{D}' \end{array}\right]\left(\frac{1}{\sqrt{n_g}}\left\{\begin{array}{c} \tilde{A}^C \\ \tilde{B}^C \end{array}\right\}\right.$$

$$\left. + \left(1 - \frac{1}{\sqrt{n_g}}\right)\left\{\begin{array}{c} \tilde{A} \\ \tilde{B} \end{array}\right\}\right) \qquad (6.56)$$

Here, $\bar{A}'_{ij}$, $\bar{B}'_{ij}$, and $\bar{D}'_{ij}$ are obviously obtained by inverting $\bar{A}_{ij}$, $\bar{B}_{ij}$, and $\bar{D}_{ij}$. In comparison to Eq. (6.8) the quantities in the parentheses on the right-hand side of Eq. (6.56) can be regarded as the average values for the bridging regions and hence they are denoted by $\bar{A}_j^*$ and $\bar{B}_j^*$. Thus, we obtain, in index notation

$$\left\{\begin{array}{c} \bar{\bar{A}}'_i \\ \bar{\bar{B}}'_i \end{array}\right\} = \left[\begin{array}{c|c} \bar{A}'_{ij} & \bar{B}'_{ij} \\ \hline \bar{B}'_{ij} & \bar{D}'_{ij} \end{array}\right]\left\{\begin{array}{c} \bar{\bar{A}}_j^* \\ \bar{\bar{B}}_j^* \end{array}\right\} \qquad (6.57)$$

Finally, the whole satin composite of Fig. 6.19 can be regarded as a linkage of regions I, II–III–IV and V in series. The average strain and curvature for the entire model are given in condensed form as

$$\left\{\begin{array}{c} \bar{\varepsilon}^{oS} \\ \bar{\kappa}^S \end{array}\right\} = \frac{1}{\sqrt{n_g}}\left(2\left\{\begin{array}{c} \bar{\varepsilon}^o \\ \bar{\kappa} \end{array}\right\} + (\sqrt{(n_g)} - 2)\left\{\begin{array}{c} \varepsilon^o \\ \kappa \end{array}\right\}\right)$$

$$= \Delta T\left(\frac{2}{\sqrt{n_g}}\left\{\begin{array}{c} \bar{\bar{A}}' \\ \bar{\bar{B}}' \end{array}\right\} + \left(1 - \frac{2}{\sqrt{n_g}}\right)\left\{\begin{array}{c} \tilde{A}' \\ \tilde{B}' \end{array}\right\}\right) \qquad (6.58)$$

where the superscript s signifies the properties of the satin composite, and ε^o and κ denote, respectively, mid-plane strain and curvature for the cross-plies in regions I and V of Fig. 6.19. From Eq. (6.58), the components of the thermal expansion and thermal bending coefficients of the satin composite are expressed as

$$\left\{\begin{array}{c} \bar{\bar{A}}'^S_i \\ \bar{\bar{B}}'^S_i \end{array}\right\} = \frac{2}{\sqrt{n_g}}\left\{\begin{array}{c} \bar{\bar{A}}'_i \\ \bar{\bar{B}}'_i \end{array}\right\} + \left(1 - \frac{2}{\sqrt{n_g}}\right)\left\{\begin{array}{c} \tilde{A}_i \\ \bar{B}'_i \end{array}\right\} \qquad (6.59)$$

Figure 6.28 shows the numerical results of the analysis based upon the elastic properties of Table 6.2. Also, $\alpha_{11} = 0.0$ and $\alpha_{22} = 3.0 \times 10^{-5}/°C$. The general characteristics of the variations of thermal deformation coefficients with $1/n_g$ are very similar to those of the compliance constants $\bar{A}'_{11}$ and $\bar{B}'_{11}$ as discussed earlier. For the thermal bending coefficients, there is considerable discrepancy between the results obtained from the one-dimensional models and the bridging model.

The geometrical shape of the fiber undulation also affects $\bar{\bar{A}}'_x$; this is demonstrated in Fig. 6.29 using the carbon/epoxy properties of Table 6.2. The results indicate that the in-plane thermal expansion coefficient of satin weave composites is less sensitive to a_u/h than that of plain weave composites. Furthermore, the fiber crimp model predicts a larger effect on $\bar{\bar{A}}'_x$ due to a_u/h than the bridging model. In general, the bridging model predictions are also less sensitive to the n_g values than the crimp model predictions. In both models, the maximum in $\bar{A}'_x$ occurs at $a_u/h \approx 1$.

Experimental data on thermal expansion coefficients of fabric

Fig. 6.29. The effect of fiber undulation on $\bar{\bar{A}}'_x$ of carbon/epoxy composites; solid lines: crimp model; broken lines: bridging model. (After Ishikawa and Chou 1983a.)

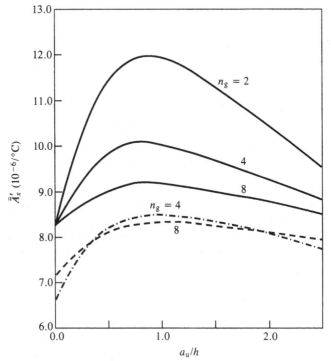

composites are quite limited. Rogers *et al.* (1977, 1981) and Yates *et al.* (1978) have performed measurements of thermal expansion of carbon fiber reinforced plastics. These experiments, however, are based upon thick specimens with 15–25 plies. Due to the constraint of the neighboring layers, an individual ply in the laminate is not free to bend. As a result, modifications to the analysis developed above are necessary for making a meaningful comparison with experiments.

It is assumed that the thermal expansion of a lamina without bending can be realized if there exist bending moments $\{M\}$, under a temperature change, ΔT, and no in-plane force is allowed. Thus,

$$\{N\} = \{\kappa\} = 0 \tag{6.60}$$

Equations (6.8) and (6.60) give

$$[D']\{M\} + \Delta T\{\bar{B}'\} = 0 \tag{6.61}$$

Then,

$$\{M\} = -\Delta T[D']^{-1}\{\bar{B}'\} \tag{6.62}$$

Substituting Eq. (6.62) into Eq. (6.8), and from the expression of ε°, a modified in-plane thermal expansion coefficient for the case without in-plane force and external bending can be defined as

$$\{\bar{A}''\} = \{\bar{A}'\} - [B'][D']^{-1}\{\bar{B}'\} \tag{6.63}$$

Equation (6.63) can be evaluated for the mosaic, crimp and bridging models provided that the appropriate constants are given for a particular model. Also, note the presence of elastic compliance constants in Eq. (6.63). Thus, it is necessary to evaluate, for instance, $\bar{B}_{ij}'^S$ and $\bar{D}_{ij}'^S$ for calculating $\bar{A}''^S$, and $\bar{B}_{ij}'^C$ and $D_{ij}'^C$ for $\bar{A}''^C$. The above modifications are of practical significance because it is desirable to overcome the anti-symmetrical behavior such as that of $\bar{B}_i'$ by suitable stacking in laminate constructions.

Figure 6.30 gives the variation of $\bar{A}_1'$ with $1/n_g$. The theoretical predictions are based upon both the crimp and bridging models using the thermoelastic properties of the unidirectional carbon/epoxy composite of Table 6.2. The experimental results of Rogers *et al.* (1981) for five-harness satin composites are also shown in Fig. 6.30. Two estimated values for a/h were used for the analysis, and a/a_u is assumed to be unity. The bridging model prediction coincides fairly well with experimental results. It is also obvious that the in-plane thermal expansion coefficients are more sensitive to n_g in the case without bending (Fig. 6.30) than in the case of unconstrained bending (Fig. 6.28).

In summary, the following can be stated regarding the thermal property modeling of two-dimensional woven fabric composites:

(1)   The mosaic model provides a simple means for estimating thermal expansion and thermal bending coefficients.

(2)   The one-dimensional crimp model predicts slightly higher in-plane thermal expansion coefficients and the same thermal bending coefficients compared to those obtained from the mosaic model. The limited experimental data on thermal bending coefficients coincide rather well with the predictions of the mosaic and crimp models.

Fig. 6.30. Comparison of theoretical predictions with the experimental results of Rogers *et al.* (1981) for five-harness satin carbon/epoxy composites; —— $a/h = 3.75$; — — — $a/h = 7.5$; $a_u/a = 1.0$; CM and BM indicate fiber crimp and bridging models, respectively; (●) experimental results at 300 K. (After Ishikawa and Chou 1983a.)

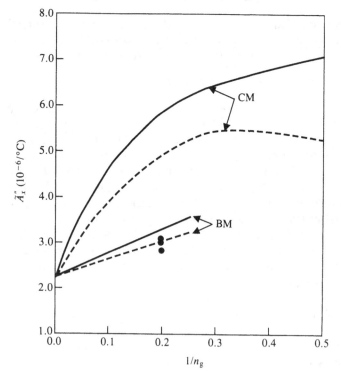

(3) The bridging model is particularly suited for the prediction of thermal expansion constants for satin composites. The experimental results on in-plane thermal expansion coefficients for a five-harness satin composite agree well with the theory.

6.9 Hybrid fabric composites: mosaic model

Hybrid woven fabrics provide a wide variety of material selection for designers with a new degree of freedom in tailoring composites to achieve a better balance of stiffness and strength, increased elongation to failure, better damage tolerance, and significant improvement of cost-effectiveness in fabrication. A basic difference between hybrid and non-hybrid composites is that material variation as well as geometrical variation come into play for the former case.

Figure 6.31 shows an example of a hybrid fabric composite for $n_g = 8$. The front view is dominated by filling yarns and the back side by warp yarns. There are two kinds of fiber materials, denoted by α and β, although there is no restriction regarding the number of fiber materials in a hybrid fabric. For the case of Fig. 6.31, the pattern of arrangement of fiber types in the filling direction repeats

Fig. 6.31. A hybrid woven fabric with $n_g = 8$, $n_{fm} = 2$ and $n_{wm} = 3$. (a) Front view; (b) back view. (After Ishikawa and Chou 1982a.)

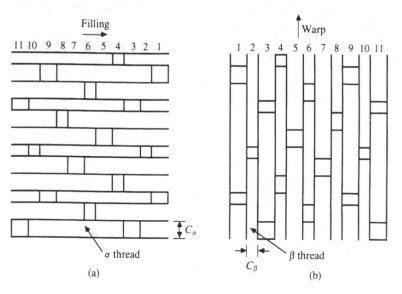

(a)

(b)

for every two warp yarns; thus it is defined that $n_{\text{fm}} = 2$. In the warp direction, the pattern of arrangement of fiber types repeats for every three filling yarns, and $n_{\text{wm}} = 3$. The subscript 'm' indicates a material parameter. The following analysis is limited to fabrics containing only two types of fiber densely woven in both directions, i.e. no gaps are allowed (Ishikawa and Chou 1982a, 1983d).

6.9.1   Definitions and idealizations

It has been adopted that n_{g} ($=n_{\text{fg}} = n_{\text{wg}}$) specifies the fabric geometrical pattern, and n_{fm} and n_{wm} define the fabric material arrangements. The notation n_{m} will be used when consideration of the material parameter is not restricted to any one direction.

In the following, the discussions are first focussed on the pattern of hybrid fabrics in one dimension, along the filling or warp direction. Under the assumption of the mosaic model (Section 6.4), a fabric composite can simply be regarded as an assemblage of pieces of asymmetrical cross-ply laminates.

If n_{g} and n_{m} are numbers not divisible by each other in a given direction, warp or filling, the pattern of the hybrid fabric will repeat in that direction for every $n_{\text{g}} \times n_{\text{m}}$ yarns in the orthogonal fabric. For instance, for $n_{\text{g}} = 5$ and $n_{\text{fm}} = 3$, Fig. 6.32 shows the pattern of fabric repeats in the filling direction for every 15 warp yarns. In general, the pattern of a fabric is repeated in the filling direction after every n_{f} warp yarns, where n_{f} is the least common multiple (LCM) of n_{fg} and n_{fm}, or $n_{\text{f}} = \text{LCM}(n_{\text{fg}}, n_{\text{fm}})$. Similarly, it is defined that $n_{\text{w}} = \text{LCM}(n_{\text{wg}}, n_{\text{wm}})$.

Although the size of a basic repeating unit in the filling direction, for instance, is determined by n_{f}, the detail of fiber arrangement

Fig. 6.32. A fabric where n_{g} and n_{fm} are not divisible by each other in the filling direction; α and β denote two types of fibers. (After Ishikawa and Chou 1982a.)

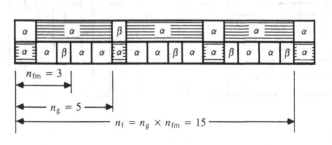

may vary. Figure 6.33 shows two cases of fabric pattern in the filling direction for $n_{fg} = 8$. Here, α and β denote the two types of fiber material of the hybrid. The notation ξ is used to indicate that the filling yarn can be of either α or β type. It is obvious from Fig. 6.33 that the different repeating patterns are generated by continuously shifting the positions of the warp yarns in the filling direction. In general, the number of repeating patterns in the filling direction for a given n_f is equal to the greatest common measure (GCM) of n_{fg} and n_{fm} and is denoted by $n_{fi} = GCM(n_{fg}, n_{fm})$. Naturally, $n_{fi} = 1$ for the case of Fig. 6.32. Again the notation n_i can be used if the discussion is independent of the direction.

Further comments are necessary for identifying the nature of the interlaced regions of a hybrid fabric. In Fig. 6.33 the interlaced region is 'homogeneous' if the yarns are identical or 'heterogeneous' if the yarns are of different types. The notations $HO1^{\alpha}$, $HE3^{\beta}$, etc. are simply for identification purposes pertaining to later discussions. The types of interlacing are termed 'mixed' if both homogeneous and heterogeneous interlacing appear in a repeating pattern. This can occur when n_g and n_m are numbers not divisible (Fig. 6.32) or divisible (for instance $n_g = 8$, $n_{fm} = 4$) by each other.

Fig. 6.33. Homogeneous and heterogeneous interlacings. (a) $n_{fg} = 8$, $n_{fm} = 2$ ($n_{fm}^{\alpha} = n_{fm}^{\beta} = 1$); (b) $n_{fg} = 8$, $n_{fm} = 4$ ($n_{fm}^{\alpha} = 3$, $n_{fm}^{\beta} = 1$). (After Ishikawa and Chou 1982a.)

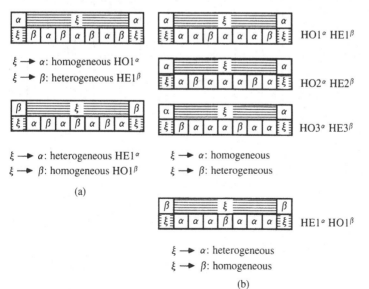

Fig. 6.34. Two-dimensional basic repeating unit of a hybrid fabric for $n_g = 8$ and $n_{fm} = 4$. (a) $n_{wm} = 3$; n_g and n_{wm} are not divisible by each other. (b) $n_{wm} = 4$; n_g and n_{wm} are divisible by each other; (1) mixed interlacing; (2) homogeneous interlacing. (After Ishikawa and Chou, 1982a.)

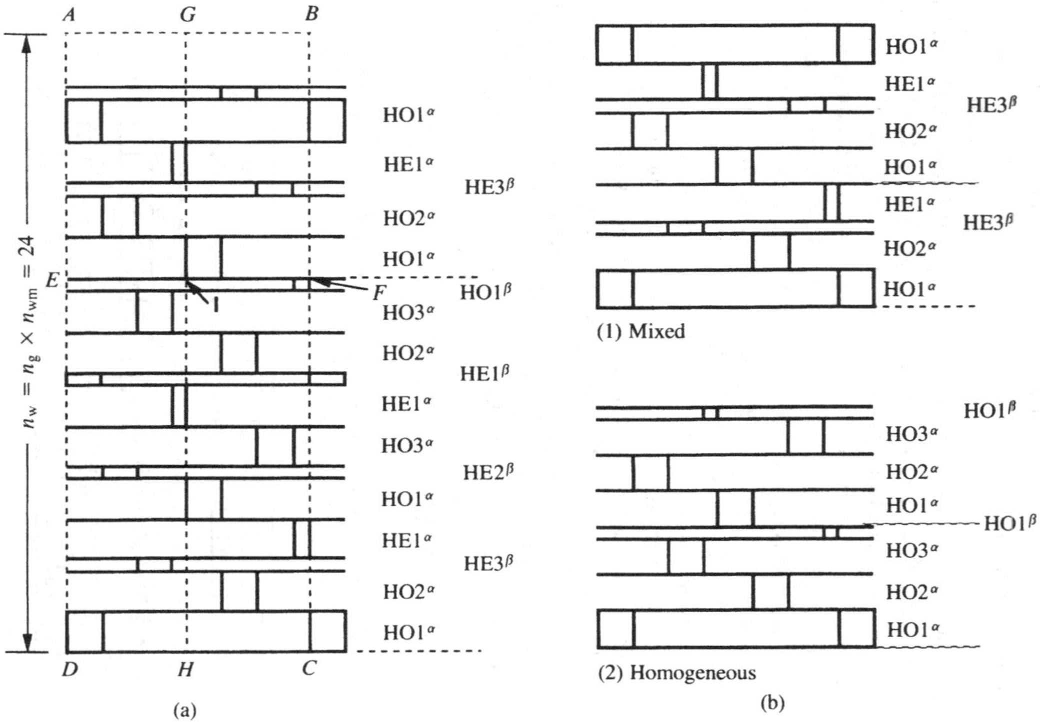

Next, hybrid fabric patterns in two dimensions are identified on the basis of the above definitions established for one-dimensional considerations. Figure 6.34 shows hybrid fabric patterns for $n_g = 8$, $n_{fm} = 4$, and $n_{wm} = 3$ (Fig. 6.34a) and $n_{wm} = 4$ (Fig. 6.34b). It should be noted that all the interlacing patterns of Fig. 6.33(b) appear in Fig. 6.34(a). The area denoted *ABCD* in Fig. 6.34(a) is a possible repeating unit of the fabric. However, there are repetitions in the geometrical and material patterns within this area. The patterns of *AGIE* and *IFCH* are identical. So are the patterns of *GBFI* and *EIHD*. Consequently, the smallest repeating unit of the fabric in two dimensions can be represented by either *AGHD* or *EFCD*. The area *EFCD*, for instance, contains 12 $(n_g n_{wm}/(n_g/n_{fi}))$ filling yarns. It can be further concluded that if n_g and n_m are numbers not divisible by each other in one direction (filling or warp) there exists only one kind of basic repeating unit for defining the two-dimensional fabric. This is true regardless of whether n_g and n_m are numbers divisible or not by each other in the other direction.

On the other hand, there exists more than one type of basic repeating unit in the two-dimensional fabric if n_g and n_m are numbers divisible by each other in both directions. This is illustrated in Fig. 6.34(b). In Fig. 6.34(b1) both homogeneous and heterogeneous interlacing in the filling direction occur and the pattern is considered to be 'mixed' in two dimensions. The pattern is homogeneous in two dimensions for Fig. 6.34(b2). Two other mixed patterns exist: [HO1$^\alpha$, HE2$^\beta$, HO3$^\alpha$, HE1$^\alpha$] and [HE1$^\beta$, HO2$^\alpha$, HO3$^\alpha$, HE1$^\alpha$], and no heterogeneous pattern exists for the geometrical and material parameters given in Fig. 6.34(b).

Let l_f and l_w be the edge lengths of a basic repeating unit in a two-dimensional fabric. If n_g and n_{wm} are numbers not divisible by each other for the repeating unit *AGHD* in Fig. 6.34(a).

$$l_w = (n_{wm}^\alpha C_\alpha + n_{wm}^\beta C_\beta)n_g$$
$$l_f = n_{fm}^\alpha C_\alpha + n_{fm}^\beta C_\beta \tag{6.64}$$

C_α and C_β denote yarn widths as shown in Fig. 6.31. The area of the two-dimensional repeating unit is then given by

$$A_r = n_g(n_{wm}^\alpha C_\alpha + n_{wm}^\beta C_\beta)(n_{fm}^\alpha C_\alpha + n_{fm}^\beta C_\beta) \tag{6.65}$$

This equation is valid for n_g and n_m, which are numbers not divisible by each other in either the filling or the warp direction as well as in both directions.

Alternate expressions for Eqs. (6.64) can be given by considering the repeating unit of *EFCD* in Fig. 6.34(a):

$$l_w = (n^\alpha_{wm} C_\alpha + n^\beta_{wm} C_\beta) n_{fi}$$

$$l_f = (n^\alpha_{fm} C_\alpha + n^\beta_{fm} C_\beta) n_g / n_{fi} \tag{6.66}$$

When n_g and n_m are numbers divisible by each other in both directions

$$A_r = \frac{n_w}{n_{wm}} (n^\alpha_{wm} C_\alpha + n^\beta_{wm} C_\beta)(n^\alpha_{fm} C_\alpha + n^\beta_{fm} C_\beta) \tag{6.67}$$

for $n_w \geq n_f$, and

$$A_r = \frac{n_f}{n_{fm}} (n^\alpha_{wm} C_\alpha + n^\beta_{wm} C_\beta)(n^\alpha_{fm} C_\alpha + n^\beta_{fm} C_\beta) \tag{6.68}$$

for $n_f \geq n_w$.

In summary, the pattern of a regular hybrid satin fabric can be determined by the parameters C_α, C_β, n_g, n_m and, hence, n_i, n_f and n_w. The 'regularity' of fabrics deserves some comment. The concept of regularity is based on the geometrical consideration. For instance, in the regular satin weave of Fig. 6.3(d), the geometrical distribution of the interlaced regions in two dimensions can be determined uniquely by two vectors, i.e. (3, 1) and (1, 3). The vector (3, 1) translates to an interlaced region by three yarns in the filling direction and one yarn in the warp direction. Other combinations of vectors are also possible, for instance (3, 1) and (−2, 2) or (2, −2) and (1, 3). An example of an irregular satin is shown in Fig. 6.3(c), where $n_g = 4$ and a set of two vectors cannot be found to generate the locations of all the interlaced regions. The term 'balanced' hybrid fabric is also used in the analysis. In such a fabric, the total number and arrangements of yarns of each material in the filling and warp directions are identical. Hence, the relations $A_{11} = A_{22}$, $D_{11} = D_{22}$ and $B_{11} = -B_{22}$ hold for a balanced hybrid fabric.

6.9.2   *Bounds of stiffness and compliance constants*

On the basis of the idealizations given in Fig. 6.13, the hybrid fabric composite can be modeled as an assemblage of pieces of cross-ply laminates. It is further assumed that the shear deformation in the thickness direction is neglected.

There exist four different types of material combinations in a cross-ply asymmetrical laminate as depicted in Fig. 6.35, where the upper lamina is assumed to be composed of filling yarns. In the superscripts used in Fig. 6.35 the first Greek letter identifies the upper layer material and the second letter is for the lower layer. The derivations of the components of A_{ij}, B_{ij} and D_{ij} of Eqs. (6.3) for the hybrid fabric composites are straightforward. Also, it is understood that the cross-terms A_{16}, A_{26}, B_{16}, B_{26}, D_{16} and D_{26} for these asymmetrical cross-ply laminates vanish; this is also true when the upper lamina is composed of warp yarns.

6.9.2.1 *Iso-strain*

The distributions of stress resultant (and moment) and strain (and curvature) over the laminate mid-plane vary with location in the hybrid fabric composite. As a first approximation, the assumption of iso-strain in the mid-plane is adopted. Equations (6.3) are then applied to a fundamental region in the laminate. This region, if repeated, should reproduce the geometrical and material arrangements of the entire idealized fabric. Thus, the behavior of the fundamental region should reflect that of the whole laminate. The dimensions of the fundamental region are denoted by l_f and l_w in the filling and warp directions, respectively. It is also defined that $r = C_\beta / C_\alpha$ (see Fig. 6.31).

Fig. 6.35. Material combinations in a cross-ply asymmetrical laminate. The elastic constants for the plies are denoted by: (a) $A_{ij}^{\alpha\alpha}$, $B_{ij}^{\alpha\alpha}$, $D_{ij}^{\alpha\alpha}$, $A_{ij}'^{\alpha\alpha}$, $B_{ij}'^{\alpha\alpha}$ and $D_{ij}'^{\alpha\alpha}$; (b) $A_{ij}^{\alpha\beta}$, $B_{ij}^{\alpha\beta}$, $D_{ij}^{\alpha\beta}$, $A_{ij}'^{\alpha\beta}$, $B_{ij}'^{\alpha\beta}$ and $D_{ij}'^{\alpha\beta}$; (c) $A_{ij}^{\beta\alpha}$, $B_{ij}^{\beta\alpha}$, $D_{ij}^{\beta\alpha}$, $A_{ij}'^{\beta\alpha}$, $B_{ij}'^{\beta\alpha}$ and $D_{ij}'^{\beta\alpha}$; (d) $A_{ij}^{\beta\beta}$, $B_{ij}^{\beta\beta}$, $D_{ij}^{\beta\beta}$, $A_{ij}'^{\beta\beta}$, $B_{ij}'^{\beta\beta}$ and $D_{ij}'^{\beta\beta}$. (After Ishikawa and Chou 1982a.)

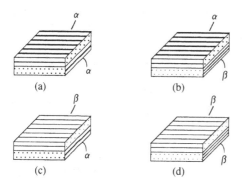

(a)   (b)

(c)   (d)

(A) $\bar{A}_{ij}$ and $\bar{D}_{ij}$

The average stress resultant $\bar{N}_x$, for example, is given as an average over the fundamental region in the $x-y$ plane:

$$\bar{N}_x = \frac{1}{l_f l_w} \int_0^{l_w} \int_0^{l_f} N_x \, dx \, dy$$

$$= \frac{1}{l_f l_w} \int_0^{l_w} \int_0^{l_f} [A_{11}^{\xi\eta} \varepsilon_{xx}^o + A_{12}^{\xi\eta} \varepsilon_{yy}^o + B_{11}^{\xi\eta} \kappa_{xx} + B_{12}^{\xi\eta} \kappa_{yy}] \, dx \, dy$$

$$(6.69)$$

Here, ξ and η stand for the α and β material phases. From Eq. (6.69) the following expressions for the effective stiffness constants of a hybrid fabric composite are given:

$$(\bar{A}_{ij}, \bar{B}_{ij}, \bar{D}_{ij}) = \frac{1}{l_f l_w} \int_0^{l_w} \int_0^{l_f} (A_{ij}^{\xi\eta}, B_{ij}^{\xi\eta}, D_{ij}^{\xi\eta}) \, dx \, dy \qquad (6.70)$$

These averages, in their simple forms, provide upper bounds for the fabric composite stiffness. If these stiffness constants are inverted, lower bounds for the elastic compliance constants can also be obtained.

Both $A_{ij}^{\xi\eta}$ and $D_{ij}^{\xi\eta}$ for the upper ply are identical to those for the lower ply; general expressions of $\bar{A}_{ij}$ and $\bar{D}_{ij}$ can be written regardless of the relative magnitude of n_g and n_m. For instance,

$$\bar{A}_{ij} = \frac{1}{(n_{fm}^\alpha + n_{fm}^\beta r)(n_{wm}^\alpha + n_{wm}^\beta r)}$$
$$\times [n_{fm}^\alpha n_{wm}^\alpha A_{ij}^{\alpha\alpha} + (n_{fm}^\beta n_{wm}^\alpha A_{ij}^{\alpha\beta}$$
$$+ n_{fm}^\alpha n_{wm}^\beta A_{ij}^{\beta\alpha})r + n_{fm}^\beta n_{wm}^\beta A_{ij}^{\beta\beta} r^2] \qquad (6.71)$$

where n_{fm}^α and n_{fm}^β denote the number of α and β yarns, respectively, within the repeating length of n_{fm} yarns in the filling direction. Naturally, $n_{fm}^\alpha + n_{fm}^\beta = n_{fm}$, and $n_{wm}^\alpha + n_{wm}^\beta = n_{wm}$. The expression for $\bar{D}_{ij}$ can be obtained if $A_{ij}^{\xi\eta}(\xi, \eta = \alpha, \beta)$ in Eq. (6.71) are replaced by $D_{ij}^{\xi\eta}$. Finally, it should be noted that $\bar{A}_{ij}$ and $\bar{D}_{ij}$ can be reduced to the special case of non-hybrid fabric composites (Ishikawa and Chou 1983b). The upper bounds of $\bar{A}_{ij}$ and $\bar{D}_{ij}$ thus obtained are identical to those of A_{ij} and D_{ij} of intermingled hybrids in cross–ply laminate form.

(B) $\bar{B}_{ij}$

The $\bar{B}_{ij}$ constants can be obtained with the same approach as for $\bar{A}_{ij}$ and $\bar{D}_{ij}$. However, here it is necessary to distinguish the weaving pattern as indicated by n_g and n_m. Hence, the algebra is more complicated. If n_g and n_m are numbers not divisible by each other in one direction

$$\bar{B}_{ij} = \frac{(n_g - 2)}{n_g} \frac{1}{(n_{fm}^\alpha + n_{fm}^\beta r)(n_{wm}^\alpha + n_{wm}^\beta r)}$$
$$\times [n_{fm}^\alpha n_{wm}^\alpha B_{ij}^{\alpha\alpha} + (n_{fm}^\beta n_{wm}^\alpha B_{ij}^{\alpha\beta}$$
$$+ n_{fm}^\alpha n_{wm}^\beta B_{ij}^{\beta\alpha})r + n_{fm}^\beta n_{wm}^\beta B_{ij}^{\beta\beta} r^2] \qquad (6.72)$$

In the case where n_g and n_m are numbers divisible by each other in both directions, the expression of $\bar{B}_{ij}$ depends upon whether the interlacing is homogeneous, heterogeneous or mixed. For instance, for the case of homogeneous interlacing where $n_f \geq n_w$

$$\bar{B}_{ij} = \frac{1}{n_g(n_{fm}^\alpha + n_{fm}^\beta r)(n_{wm}^\alpha + n_{wm}^\beta r)}$$
$$\times [(n_g n_{fm}^\alpha - 2n_{fi})n_{wm}^\alpha B_{ij}^{\alpha\alpha} + n_g(n_{fm}^\beta n_{wm}^\alpha B_{ij}^{\alpha\beta}$$
$$+ n_{fm}^\alpha n_{wm}^\beta B_{ij}^{\beta\alpha})r + (n_g n_{fm}^\beta - 2n_{fi})n_{wm}^\beta B_{ij}^{\beta\beta} r^2] \qquad (6.73)$$

Similar expressions can be derived for heterogeneous interlacing. In the case of mixed interlacing, the expressions depend upon the details of the material arrangement. However, the differences among the B_{ij}s for homogeneous, heterogeneous and mixed interlacings are, in general, not significant within the usual range of r, around unity. Therefore, Eq. (6.73) can be used as an approximation of $\bar{B}_{ij}$ for such r values when n_g and n_m are numbers divisible by each other in both directions.

6.9.2.2 *Iso-stress*

As another method of estimating the bounds of elastic moduli the assumption of iso-stress is made. Derivations similar to that of Eq. (6.69) can be performed to obtain the average strain expression of the hybrid fabric composite. The average elastic constants are then given by

$$(\bar{A}'_{ij}, \bar{B}'_{ij}, \bar{D}'_{ij}) = \frac{1}{l_f l_w} \int_0^{l_w} \int_0^{l_t} (A_{ij}'^{\xi\eta}, B_{ij}'^{\xi\eta}, D_{ij}'^{\xi\eta}) \, dx \, dy \qquad (6.74)$$

By replacing $A_{ij}^{\xi\eta}$, $B_{ij}^{\xi\eta}$ and $D_{ij}^{\xi\eta}$ in Eqs. (6.71)–(6.73) by $A_{ij}'^{\xi\eta}$, $B_{ij}'^{\xi\eta}$ and $D_{ij}'^{\xi\eta}$, explicit expressions of Eq. (6.74) can be obtained. These are upper bounds for the composite compliance constants; they can be inverted to obtain the lower bounds for the stiffness constants.

6.9.3   *One-dimensional approximation*

The approximate solution presented below is based upon a combination of the series model of Ishikawa (1981) for non-hybrid fabric composites and the mechanics of materials approach for unidirectional composites. The basic assumptions are that the hybrid fabric composite can be divided into repeating regions in the form of one-dimensional strips, and the equilibrium and compatibility conditions are not exactly satisfied. Figure 6.36 shows that the hybrid fabric composite is divided into strips along the filling and warp directions. It is then assumed that the stress resultant (N) is uniform in each strip.

The division of the strips is made according to the elastic moduli under consideration; along the filling (x) direction for A_{11}, B_{11}, D_{11}, A'_{11}, B'_{11} and D'_{11} and along the warp (y) direction for A_{22}, B_{22}, D_{22}, A'_{22}, B'_{22} and D'_{22}. Either the x or the y direction is admissible for determination of all the other non-zero constants.

Evidently the average strain in an α yarn is different from that of a β yarn. The one-dimensional average strain for the α yarn, for instance, can be written by considering the stress and moment resultants in the x direction only:

$$\bar{\varepsilon}_{xx}^{o\alpha} = \frac{1}{l_f} \int_0^{l_f} \varepsilon_{xx}^o \, dx$$

$$= \frac{1}{l_f}\left[N_x \int_0^{l_f} A_{11}'^{\xi\eta} \, dx + M_x \int_0^{l_f} B_{11}'^{\xi\eta} \, dx \right] \tag{6.75}$$

Fig. 6.36. One-dimensional model of hybrid fabric composites. (After Ishikawa and Chou 1982a.)

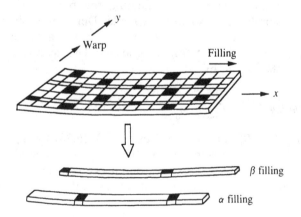

where ξ and η stand for α and β. For the case where n_g and n_{fm} are numbers not divisible by each other in the filling direction, the following expressions for the averaged compliances are obtained for the α-filling yarns:

$$\bar{A}_{ij}^{\prime\alpha} = \frac{1}{(n_{fm}^{\alpha} + n_{fm}^{\beta}r)} \, (n_{fm}^{\alpha} A_{ij}^{\prime\alpha\alpha} + n_{fm}^{\beta} A_{ij}^{\prime\alpha\beta}r)$$

$$\bar{B}_{ij}^{\prime\alpha} = \frac{(n_g - 2)}{n_g} \frac{1}{(n_{fm}^{\alpha} + n_{fm}^{\beta}r)} \, (n_{fm}^{\alpha} B_{ij}^{\prime\alpha\alpha} + n_{fm}^{\beta} B_{ij}^{\prime\alpha\beta}r) \qquad (6.76)$$

$$\bar{D}_{ij}^{\prime\alpha} = \frac{1}{(n_{fm}^{\alpha} + n_{fm}^{\beta}r)} \, (n_{fm}^{\alpha} D_{ij}^{\prime\alpha\alpha} + n_{fm}^{\beta} D_{ij}^{\prime\alpha\beta}r)$$

Naturally, expressions of $\bar{A}_{ij}^{\alpha}$, $\bar{B}_{ij}^{\alpha}$ and $\bar{D}_{ij}^{\alpha}$ for the α-filling yarns can be obtained by inverting $\bar{A}_{ij}^{\prime\alpha}$, $\bar{B}_{ij}^{\prime\alpha}$ and $\bar{D}_{ij}^{\prime\alpha}$ of Eqs. (6.76). A similar procedure can be applied to the β-filling yarns.

Finally, if the average strain and curvature in the α-yarns ($\bar{\varepsilon}^{\alpha}$ and $\bar{\kappa}^{\alpha}$) are not very much different from those in the β-yarn, it is not unreasonable to approximate the entire composite plate with a uniform strain field. Thus, the stiffness constants can be obtained from a volume average. For example,

$$\bar{A}_{11} = \frac{1}{(n_{wm}^{\alpha} + n_{wm}^{\beta}r)} \, (n_{wm}^{\alpha} \bar{A}_{11}^{\alpha} + n_{wm}^{\beta} \bar{A}_{11}^{\beta}r) \qquad (6.77)$$

6.9.4 Numerical results

Consider the numerical example for the case of a carbon/ Kevlar fabric in an epoxy matrix. The basic elastic properties of the constituent unidirectional laminae used in this idealized mosaic model are given in Table 6.2. The fiber volume fraction is chosen to be 65% in order to match that of the experimental systems.

Figure 6.37 shows the relationship between $\bar{A}_{11}/h$ and relative fiber volume fraction for 'balanced fabrics' where $A_{11} = A_{22}$. The carbon and Kevlar yarns are designated as α and β yarns, respectively. Fabric parameters are chosen so as to coincide with those of Zweben and Norman (1976): $n_g = 8$, $n_{fm} = 4$ and $n_{wm} = 4$, while n_{fm}^{α}, n_{fm}^{β}, n_{wm}^{α}, and n_{wm}^{β} vary from 0 to 4. The numbers in parentheses correspond to values of n_{fm}^{α}, n_{fm}^{β}, n_{wm}^{α}, and n_{wm}^{β}. The ratio of the yarn width, r, varies from zero to infinity as the relative fiber volume fraction changes. Since n_g and n_m in this example are numbers divisible by each other, the lower bound predictions are affected by the weaving patterns. Only the lower bound for the case (3, 1; 3, 1) is shown for the full range of relative fiber volume fractions. Also, only homogeneous interlacing types are considered in Fig. 6.37. The upper bounds are identical to one another for these three weaving patterns and are shown by a straight line similar to the predictions of the rule-of-mixtures. The circles and triangles represent the experimental results of Zweben and Norman for carbon/Kevlar hybrid fabrics and laminates composed of unidirectional laminae of the parent components. The fabrics used in the experiments are equivalent to, in the present terminology, the categories of $(3, 1; 3, 1)$ and $(2, 2; 2, 2)$. The relative yarn width is close to $r = 1$ and the interlacing pattern is of the homogeneous type. These results fall in between the bound predictions.

The effect of fabric geometrical patterns of the parent composites on the bound prediction is worth examining. The bound prediction of Kevlar/epoxy is represented in Fig. 6.37 by either $n_{fm}^{\alpha} = n_{wm}^{\alpha} = 0$ or $r = C_{\beta}/C_{\alpha} \to \infty$. Similarly, the carbon/epoxy system corresponds

Fig. 6.37. $\bar{A}_{11}/h$ vs. relative fiber volume fraction. h denotes specimen thickness; UB, upper bound; LB, lower bound. Experimental results of Zweben and Norman (1976); (●) fabric; (▲) laminate. (After Ishikawa and Chou 1982a.)

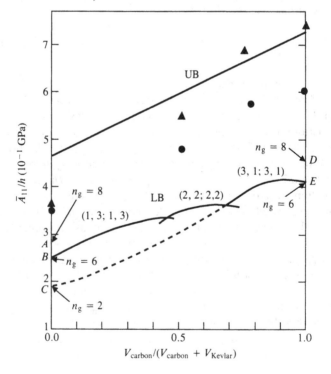

to the case of either $n_{fm}^{\beta} = n_{wm}^{\beta} = 0$ or $r = 0$. The lower bound predictions based on different combinations of n_g and n_m yield different results. Point A in Fig. 6.37 indicates the combination $n_g = 8$ and $n_{fm}^{\alpha} = n_{wm}^{\alpha} = 0$. Point B is for the limiting case of $n_g = 8$, (1, 3; 1, 3) and for $r \to \infty$; this case is equivalent to $n_g = 6$ and $n_{fm}^{\alpha} = n_{wm}^{\alpha} = 0$. Point C is obtained from the case of $n_g = 8$, (3, 1; 3, 1) and $r \to \infty$. The same weaving pattern can be achieved for $n_g = 2$ and $n_{fm}^{\alpha} = 0$. Discussions similar to the above can be made for the case of the carbon/epoxy system. Point D is for $n_g = 8$ and $n_{fm}^{\beta} = n_{wm}^{\beta} = 0$. Point E is for $n_g = 8$, (3, 1; 3, 1) and $r \to 0$; this is equivalent to $n_g = 6$ and $n_{fm}^{\beta} = n_{wm}^{\beta} = 0$. The transition of the geometrical pattern from $n_g = 8$ to either $n_g = 2$ or 6 as r approaches the limiting values can be understood from Figs. 6.38(a) and (b), as well as Eqs. (6.71) and (6.73).

Fig. 6.38. The transition of fabric geometrical pattern as affected by r. (a) $n_g = 8$ (3, 1; 3, 1) and $r = \frac{1}{4}$; this pattern becomes $n_g = 6$ as $r \to 0$. (b) $n_g = 8$, (3, 1; 3, 1) and $r = 8$; this pattern becomes $n_g = 2$ as $r \to \infty$. (c) Homogeneous interlacing for $n_g = 8$, (2, 2; 2, 2) and $r = \frac{1}{4}$; this pattern becomes $n_g = 4$ as $r \to 0$, (d) Heterogeneous interlacing for $n_g = 8$, (2, 2; 2, 2) and $r = \frac{1}{4}$; this pattern becomes a cross-ply laminate as $r \to 0$. (After Ishikawa and Chou 1982a.)

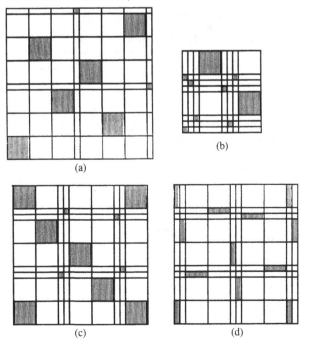

(a)

(b)

(c)

(d)

The relationship between $\bar{B}_{11}/h^2$ and the relative fiber volume fraction is demonstrated in Fig. 6.39 for $n_g = 8$ and $n_{fm} = n_{wm} = 4$. Results for both homogeneous and heterogeneous interlacings are shown. In the case of homogeneous interlacing, i.e. (1, 3; 1, 3), (2, 2; 2, 2) and (3, 1; 3, 1), the basic trends of the lower bounds are similar to the predictions shown in Fig. 6.37. However, the upper bound predictions in this case are also affected by the fabric parameters. In the case of heterogeneous interlacing (2, 2; 2, 2), both the upper and lower bound predictions tend to be very large values when $r = 0$ and $r \to \infty$. As a result, extremely large coupling effects are seen in these limiting cases.

6.10   Hybrid fabric composites: crimp and bridging models

Although the bounds for the elastic properties of hybrid fabric composites can be conveniently estimated by the mosaic model, the upper and lower bounds are rather far apart. An improved analysis based upon the 'crimp model' and 'bridging model' developed for non-hybrid fabric composites is described.

In the following analysis, we specify $n_g = 8$ and the fiber material repeating parameters (n_{fm}^α, n_{fm}^β; n_{wm}^α, n_{wm}^β) are of three types: (3,

Fig. 6.39. $\bar{B}_{11}/h^2$ vs. relative fiber volume fraction; —— upper and lower bound predictions for homogeneous interlacing; —·—·— upper and lower bound predictions for heterogeneous interlacing; ——— one-dimensional approximate solution. (After Ishikawa and Chou 1982a.)

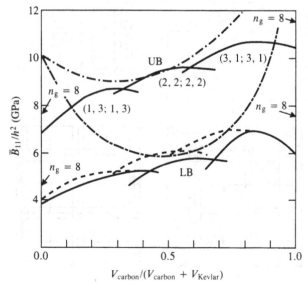

1; 3, 1), (1, 1; 1, 1) and (1, 3; 1, 3). Also, homogeneous interlacing is considered in the analysis. Furthermore, the calculation procedure for the system (1, 3; 1, 3) is the same as that for the system (3, 1; 3, 1) by interchanging the α and β materials. Therefore, only the systems of (3, 1; 3, 1) and (1, 1; 1, 1) need to be considered in the analysis.

6.10.1 Crimp model

The crimp model takes into account fiber continuity; sectional shapes of some typical interlacing regions are shown in Figs. 6.40(a) and (b). The sinusoidal type functions used in Section 6.5 for describing the undulation shapes are adopted here. For the

Fig. 6.40. Typical structures of interlaced regions of hybrid fabric composites; h denotes plate thickness and h_t indicates the total thickness of the yarns. (a) Filling: α material; warp: α or β material. (b) Filling: β material; warp: α or β material. (After Ishikawa and Chou 1983d.)

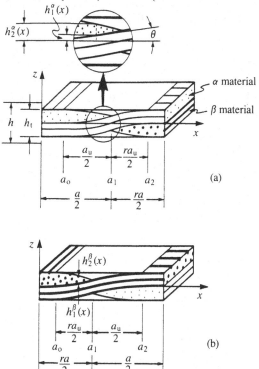

case where the filling yarn is composed of α material (Fig. 6.40a), the height of the filling yarn is given by

$$h_1^\alpha(x) = \begin{cases} 0 & (0 \le x \le a_o) \\ \left\{1 + \sin\left[\left(x - \dfrac{a}{2}\right)\dfrac{\pi}{a_u}\right]\right\}\dfrac{h_t}{4} & (a_o \le x \le a_1) \end{cases} \qquad (6.78)$$

When the filling yarn is composed of β material, the height of the filling yarn is given by

$$h_1^\beta(x) = \begin{cases} 0 & (0 \le x \le ra_o) \\ \left\{1 + \sin\left[\left(x - \dfrac{ra}{2}\right)\dfrac{\pi}{ra_u}\right]\right\}\dfrac{h_t}{4} & (ra_o \le x \le ra_1) \end{cases}$$

$$(6.79)$$

where h_t denotes the total thickness of the yarns.

Corresponding to the cases of Eqs. (6.78) and (6.79), the heights of the warp yarns are given, respectively, by

$$h_2^\alpha(x) = \begin{cases} \dfrac{h_t}{2} & (0 \le x \le a_o) \\ \left\{1 - \sin\left[\left(x - \dfrac{a}{2}\right)\dfrac{\pi}{a_u}\right]\right\}\dfrac{h_t}{4} & (a_o \le x \le a_1) \end{cases} \qquad (6.80)$$

$$h_2^\beta(x) = \begin{cases} \dfrac{h_t}{2} & (0 \le x \le ra_o) \\ \left\{1 - \sin\left[\left(x - \dfrac{ra}{2}\right)\dfrac{\pi}{ra_u}\right]\right\}\dfrac{h_t}{4} & (ra_o \le x \le ra_1) \end{cases}$$

$$(6.81)$$

It should be noted that Eqs. (6.78)–(6.81) are written for the portion of the undulated region where the filling yarn is beneath the warp yarn.

A key assumption made in the fiber crimp model (Ishikawa and Chou 1982b) is that the classical laminated plate theory is applicable to each infinitesimal slice of material of width dx. Then the local plate extensional stiffness coefficients for the portion where the filling yarn is composed of α material, are given by

$$A_{ij}^{\alpha\xi}(x) = Q_{ij}^{M}\left(h - \frac{h_t}{2} + h_1^{\alpha}(x) - h_2^{\alpha}(x)\right)$$

$$+ Q_{ij}^{F\alpha}(x)\frac{h_t}{2} + Q_{ij}^{W\xi}(h_2^{\alpha}(x) - h_1^{\alpha}(x)) \tag{6.82}$$

where the superscripts F, W and M denote the filling yarn region, warp yarn region, and pure matrix material, respectively; ξ stands for α or β material, and h denotes the total laminate thickness, including the pure matrix layers. Furthermore, the first superscript of A_{ij} indicates the filling material and the second one the warp material. This convention is followed for all the stiffness and compliance constants throughout this analysis.

Likewise for the portion of the laminate in Fig. 6.40(b), where the filling yarn is composed of β material,

$$A_{ij}^{\beta\xi}(x) = Q_{ij}^{M}\left(h - \frac{h_t}{2} + h_1^{\beta}(x) - h_2^{\beta}(x)\right)$$

$$+ Q_{ij}^{F\beta}(x)\frac{h_t}{2} + Q_{ij}^{W\xi}(h_2^{\beta}(x) - h_1^{\beta}(x)) \tag{6.83}$$

Similarly, expressions for $B_{ij}^{\alpha\xi}(x)$, $B_{ij}^{\beta\xi}(x)$, $D_{ij}^{\alpha\xi}(x)$ and $D_{ij}^{\beta\xi}(x)$ can also be obtained.

The local thermal deformation coefficients can be obtained by replacing Q_{ij} in Eq. (6.83) by $Q_{ij}\alpha_j$ (Eqs. (6.6) and (6.7)). For instance,

$$\bar{A}_x^{\alpha\xi}(x) = q_x^{M}\left(h - \frac{h_t}{2} + h_1^{\alpha}(x) - h_2^{\alpha}(x)\right)$$

$$+ q_x^{F\alpha}(x)\frac{h_t}{2} + q_x^{W\xi}(h_2^{\alpha}(x) - h_1^{\alpha}(x)) \tag{6.84}$$

where $q_x = Q_{11}\alpha_{xx} + Q_{12}\alpha_{yy} + Q_{16}\alpha_{xy}$. Explicit expressions of off-axis properties in the filling yarn region, $Q_{ij}^{F\xi}(x)$, are given in Section 6.5. The local compliance constants $A_{ij}'^{\xi\eta}(x)$, $B_{ij}'^{\xi\eta}(x)$ and $D_{ij}'^{\xi\eta}(x)$ are obtained by inverting $A_{ij}^{\xi\eta}(x)$, $B_{ij}^{\xi\eta}(x)$ and $D_{ij}^{\xi\eta}$, where ξ and η indicate α or β material. Then, the thermal coefficients $\bar{A}_i'^{\xi\eta}$ and $\bar{B}_i'^{\xi\eta}$ can be obtained from Eq. (6.9).

Finally, consider again the one-dimensional idealized model of a hybrid laminate. The average extensional compliance for the portion containing α yarns is defined as

$$\bar{A}_{ij}^{\prime C \alpha \xi} = \frac{2}{a} \int_0^{a/2} A_{ij}^{\prime \alpha \xi}(x) \, dx = \left(1 - \frac{a_u}{2}\right) A_{ij}^{\prime \alpha \xi} + \frac{2}{a} \int_{a_o}^{a_1} A_{ij}^{\prime \alpha \xi}(x) \, dx$$

$$(6.85)$$

For the case of β filling yarns,

$$\bar{A}_{ij}^{\prime C \beta \xi} = \left(1 - \frac{a_u}{a}\right) A_{ij}^{\prime \beta \xi} + \frac{2}{ra} \int_{ra_o}^{ra_1} A_{ij}^{\prime \beta \xi}(x) \, dx \qquad (6.86)$$

The superscript 'C' in Eq. (6.85) signifies the fiber crimp model. The other averaged compliance constants $\bar{B}_{ij}^{\prime C \alpha \xi}$, $\bar{B}_{ij}^{\prime C \beta \xi}$, $\bar{D}_{ij}^{\prime C \alpha \xi}$ and $\bar{D}_{ij}^{\prime C \beta \xi}$ can be obtained in a similar manner. Expressions for the averaged in-plane thermal expansion coefficients can be obtained from Eqs. (6.85) and (6.86) by replacing $A_{ij}^{\prime}$ by the appropriate $\bar{A}_i^{\prime}$. Also, the thermal bending coefficients can be easily obtained. However, it should be noted that $\bar{B}_{ij}^{\prime C \xi \eta}$ and $\bar{\bar{B}}_i^{\prime C \xi \eta}$ do not vanish when the integrations in Eqs. (6.85) and (6.86) are carried out over the entire length of $a(1 + r)/2$ (Fig. 6.40), unlike the cases of non-hybrid fabrics. This fact is caused by the difference in yarn width and properties of the constituent fibers of the fabric. Finally, the averaged stiffness constants $\bar{A}_{ij}^{C \xi \eta}$, $\bar{B}_{ij}^{C \xi \eta}$ and $\bar{D}_{ij}^{C \xi \eta}$ can be obtained by inverting these averaged compliance constants. Then the averaged thermal constants $\bar{\bar{A}}_i^{C \xi \eta}$ and $\bar{\bar{B}}_i^{C \xi \eta}$ are obtained from the inverted form of Eq. (6.9). It should be noted that the thermoelastic constants derived here are based upon the definitions of $h_1(x)$ and $h_2(x)$ given in Eqs. (6.78)–(6.81), i.e. the filling yarn is situated beneath the warp yarn. Thus, the coupling stiffness constants for the right-hand portions of Figs. 6.40(a) and (b) for instance, are denoted by $-\bar{B}_{ij}^{C \alpha \beta}$ and $-\bar{B}_{ij}^{C \beta \alpha}$, respectively.

6.10.2 *Bridging model*

The case of fabrics with $n_g = 8$, $(n_{fm}^{\alpha}, n_{fm}^{\beta}; n_{wm}^{\alpha}, n_{wm}^{\beta}) = (3, 1; 3, 1)$ and homogeneous interlacing pattern is considered first. A possible shape of the minimum repeating unit is indicated in Fig. 6.41 as the area $ABCD$. The three-dimensional view of this repeating unit showing the interlaced configurations of the α and β yarns is given in Fig. 6.42, which consists of five regions R_1, R_2, R_3, R_4 and R_5, arranged in series along the loading direction. However, other choices for the division in regions are possible. It is assumed in the following analysis that the resultant force in the loading direction of every region is identical.

To exemplify the analysis of the bridging model, region R_2 is considered. Region R_2 consists of four sub-regions labeled R_2^1, R_2^2, R_2^3 and R_2^4 (see Fig. 6.43). By assuming an iso-strain condition for the sub-region, the average compliance constant of each region can be found. Then, the averaged stiffness constants of each sub-region are obtained by inverting the corresponding compliance constants. On the basis of the assumption of iso-strain, the average stiffness of the entire region R_2 can be determined. The elastic constants of the other regions can also be determined following this procedure.

For the fabric composite of Fig. 6.42, it is assumed that each

Fig. 6.41. A hybrid fabric with homogeneous interlacing, for $n_g = 8$, $n_{fm} = n_{wm} = 4$; α and β indicate two types of yarn material; $ABCD$ and $EFGD$ denote two choices of repeating units. (After Ishikawa and Chou 1983d.)

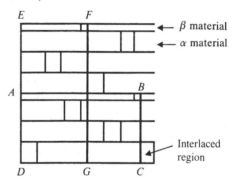

Fig. 6.42. A bridging model for $n_g = 8$ and the (3, 1; 3,1) case (region $ABCD$ of Fig. 6.41). (After Ishikawa and Chou 1983d.)

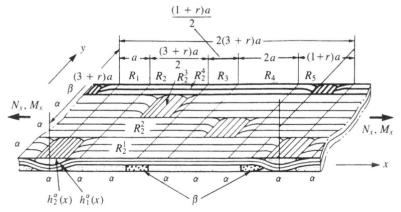

region, R_1, R_2, R_3, R_4, or R_5 carries the same load N_x. Thus, the compliance constants of the entire composite can be regarded as the volume average of the compliances of the individual regions. Then, the inversion of the compliance constants gives the stiffness coefficients of the entire composite unit cell, $\bar{A}_{ij}$, $\bar{B}_{ij}$ and $\bar{D}_{ij}$. The basic idea of the analysis briefly outlined above is identical to that of the 'bridging model' of Section 6.6 in which only non-hybrid composites are considered. The details of the derivations of elastic and thermal deformation constants can be found in Ishikawa and Chou (1983d).

Both regions $ABCD$ and $EFGD$ of Fig. 6.41 can be treated as repeating regions for the entire fabric composite. A three-dimensional view of region $EFGD$ can also be found in Ishikawa and Chou (1983d). As to the case of the $(1, 3; 1, 3)$ material combination, the thermoelastic constants can be obtained from the above procedure by simply interchanging the α and β materials. Ishikawa and Chou (1983d) have also examined the case of a fabric of $n_g = 8$ with homogeneous interlacing and material repeating parameters $(2, 2; 2, 2)$. The cases of $(3, 1; 3, 1)$, $(1, 3; 1, 3)$ and $(2, 2; 2, 2)$ give all possible fiber material combinations for homogenous interlacing in hybrid fabrics with the given fabric parameters.

6.10.3 *Numerical results and summary of thermoelastic properties*

Numerical work has been performed to examine the thermoelastic properties of a carbon/Kevlar/epoxy hybrid fabric composite. The basic material properties of unidirectional laminae of carbon/epoxy and Kevlar/epoxy are given in Table 6.2. The fiber volume fraction of all the unidirectional laminae is assumed to be

Fig. 6.43. Detailed view of region R_2 in Fig. 6.42. (After Ishikawa and Chou 1983d.)

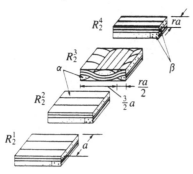

65%, which is slightly higher than the total fiber volume fraction of the fabric composite due to the presence of pure matrix layers. Figure 6.44 shows the predictions of the extensional stiffness of the bridging model as well as those from the bound approach (Fig. 6.37) of Ishikawa and Chou (1982a) for a carbon/Kevlar/epoxy system of $n_g = 8$. Three different material repeating parameters are presented and the theoretical curves are obtained by changing r, the yarn width ratio of α and β materials. Because values of r far from unity are impractical, the curves in Fig. 6.44 are truncated. The predictions based upon the bridging concept fall in between the upper and lower bounds and compare very favorably with the experimental data of Zweben and Norman (1976).

The following is a summary of the analysis of thermoelastic properties of hybrid woven fabric composites:

(1) The structural characteristics of woven hybrid fabrics have been identified by the material parameter n_m (n_{fm} and n_{wm}) as well as the geometrical parameter n_g (n_{fg} and n_{wg}). If the

Fig. 6.44. $\bar{A}_{11}/h$ vs. relative fiber volume fraction of carbon/Kevlar/epoxy composites with $n_g = 8$; — — — bound theory; —— bridging model; (● and ▲ experimental data for fabric and cross-ply laminate composites, respectively; ($h = h_t$, $h/a = 0.4$). (After Ishikawa and Chou 1983d.)

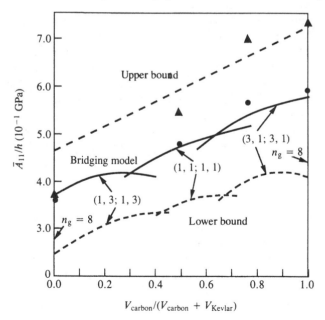

n_g and n_m of a fabric are numbers not divisible by each other in one or both directions (filling and warp) there is a unique interlacing pattern. There is more than one type of interlacing pattern if n_g and n_m are numbers divisible by each other in both directions.

(2) In the analysis of the mosaic model, the fabric composite is regarded as an assemblage of asymmetrical cross-ply laminates. Upper and lower bounds for the elastic stiffness and compliance of hybrid composites have been obtained assuming iso-strain and iso-stress, respectively. The influence of fabric parameters on elastic properties can be assessed using this model.

(3) The magnitude of the coupling terms of B_{ij} and B'_{ij} depends on whether n_g and n_m are numbers divisible by each other. In the case where n_g and n_m are numbers divisible by each other in both the warp and the filling directions, the upper and lower bounds of B_{ij} and the lower bounds of A_{ij} and D_{ij} are influenced by the interlacing types.

(4) The transition of n_g from one value to another occurs as the ratio of yarn width of the component fibers approaches zero or infinity. In such extreme cases, the magnitude of the coupling terms becomes very large, especially for heterogeneous interlacing. The distinct interlacing types for given n_g and n_m, however, render nearly identical solutions for the bounds when the yarn width ratio is around unity.

(5) The one-dimensional fiber undulation or crimp concept has been modified to treat the interlacing of two different types of fibers, and it has been incorporated into a general 'bridging model' for predicting thermoelastic properties of hybrid fabric composites.

(6) The predicted values of the axial elastic stiffness constant are insensitive to the choice of a repeating unit of the fabric material.

6.11   Triaxial woven fabric composites

6.11.1   *Geometrical characteristics*

Biaxial woven fabrics exhibit relatively low elastic moduli or low resistance to extension when deformed along the bias direction (45° to warp and filling) as compared with deformation in the warp and filling directions. A triaxial woven fabric (Doweave fabric), is composed of three sets of yarns (two sets of warp yarns and one set of filling yarn), which intersect and interlace with one another at 60° angles as shown in Fig. 6.45.

For the purpose of identification, the warp yarns are called 'one o'clock' and 'eleven o'clock' warps. The filling yarn is horizontal and is interwoven with the warp yarns in different sequences depending on the fabric style. The geometry of the fabric can vary from a very open but stable construction, such as basic weave, and stuffed basic weave (which has additional yarns in the filling direction), to a tightly packed construction, such as the bi-plane weave. Figure 6.46(a) shows a schematic diagram of the stuffed basic weave. The bi-plane weave is quite similar to the basket weave of biaxial woven fabrics. As shown in Fig. 6.46(b), the filling yarns in a bi-plane weave are woven both over and under two sets of warp yarns to form a closed construction.

With the load bearing yarns arranged in three instead of two directions, the triaxial woven fabrics yield more isotropic responses to both tensile and shear deformations, offering an alternative to the inherent structural weakness of conventional biaxial fabrics. The ability of the triaxial woven fabrics to maintain structural integrity

Fig. 6.45. Triaxial woven fabric. (After Yang and Chou 1989.)

Fig. 6.46. The geometries of (a) stuffed basic triaxial woven fabric and (b) bi-plane weave triaxial woven fabric. (After Yang and Chou 1989.)

(a)   (b)

even with a very open construction is quite unique among textile structures.
The filling yarn of a triaxial woven fabric may be composed of bundles of different size and material from the warp yarns. Therefore, hybrid fiber constructions are available for triaxial woven fabrics as for biaxial woven fabrics. Also, by proper selection of material combinations, yarn sizes and fabric weaving patterns, a wide range of geometrical and mechanical properties can be engineered in triaxial woven fabrics.

Although considerable effort has been made to investigate the mechanical behavior of triaxial woven fabrics (see Dow 1969; Dow and Tranfield 1970; Skelton 1971; Scardino and Ko 1981; Schwartz, Fornes and Mohamed 1982; Schwartz 1984), the properties of composites reinforced with triaxial woven fabrics have not been adequately evaluated. Dow (1982) developed an analytical method; the geometrical model used for the calculation of the fiber volume fraction and elastic properties of the triaxial woven fabric composite resembles the crimp model of Fig. 6.17. The undulated yarns are divided into segments and each of these segments is treated as an off-axis short-fiber composite lamina. The elastic properties of the triaxial fabric composite unit cell are calculated by averaging the contribution from each of the short-fiber composites.

In the following, a more refined analytical model is developed to predict the thermoelastic properties of triaxial fabric composites. An outline of the methodology of analysis is given first. It is then extended to biaxial, non-orthogonal woven fabric composites. Numerical results of the thermoelastic properties are presented as a function of the fabric construction parameters. The contents of Sections 6.11.2 and 6.11.3 are excerpted from Yang and Chou (1989).

6.11.2 *Analysis of thermoelastic behavior*

For the purpose of analyzing the thermoelastic constitutive relations of triaxial woven fabric composites, a unit cell of basic triaxial weave is identified, as shown in Fig. 6.47, which contains three impregnated yarn bundles oriented in space and interstitial matrix regions. Repeating the unit cell in the fabric plane obviously reproduces the complete triaxial woven structure. This methodology can easily be extended to treat other types of weaving patterns.

The concept of the 'crimp model' (Ishikawa and Chou 1982b) is extended to the following analysis. In this model, each impregnated yarn bundle is further idealized as an undulated unidirectional lamina as shown in Fig. 6.48. The geometrical configuration of each undulated lamina can be simulated as follows. First, consider the lamina of filling yarns. The upper boundary for the undulated configuration is given by (Fig. 6.48a)

$$H(x_1) = \left[1 + \sin\frac{\pi x_1}{l_1}\right]\frac{H_t}{2} \qquad (0 \le x \le 2l_1) \qquad (6.87)$$

Here, x_1 coincides with the x axis and H_t is the thickness of the undulated lamina.

Next, the form of fiber undulation in the one o'clock warp lamina as shown in Fig. 6.48(b) is

$$H(x_2) = \left[1 - \sin\left(x_2 - \frac{l_2}{2}\right)\frac{\pi}{l_2}\right]\frac{H_t}{2} \qquad (0 \le x_2 \le 2l_2) \qquad (6.88)$$

Here, x_2 is in the direction of 60° from the x axis. Similarly, in the eleven o'clock warp lamina (Fig. 6.48c), the form of fiber undulation is

$$H(x_3) = \left[1 + \sin\left(x_3 - \frac{l_3}{2}\right)\frac{\pi}{l_3}\right]\frac{H_t}{2} \qquad (0 \le x_3 \le 2l_3) \qquad (6.89)$$

where x_3 is in the direction of −60° from the x axis.

The crimp in the undulated laminae reduces the composite stiffness as compared with that of straight reinforcements. The local off-axis angle of each undulated lamina along the x_1, x_2 and x_3

Fig. 6.47. Unit cell structure of the basic triaxial woven fabric composite. (After Yang and Chou 1989.)

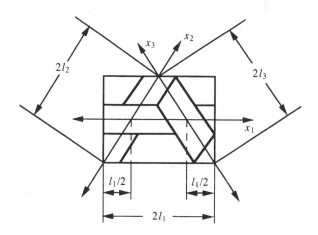

directions can be obtained by

$$\theta = \tan^{-1} \frac{dH(x_i)}{dx_i} \qquad (i = 1, 2 \text{ or } 3) \qquad (6.90)$$

The effective thermoelastic properties of each undulated lamina can be derived through the following procedures. First, the undulated lamina can be regarded as an assemblage of many small pieces

Fig. 6.48. Geometrical configurations of undulated filling and warp laminae. (After Yang and Chou 1989.)

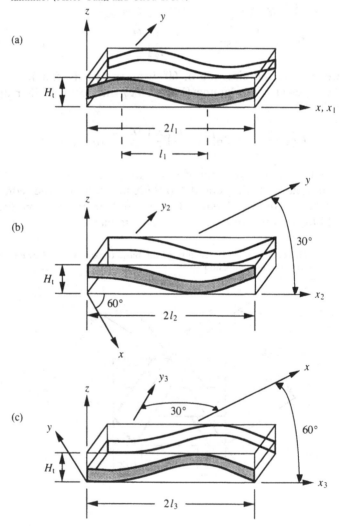

of unidirectional lamina. Each of these segments is uniquely characterized by an off-axis angle as defined in Eq. (6.90). The reduced effective thermoelastic properties in the x direction for the filling lamina are the same as those given in Eqs. (6.24) and (6.51).

By assuming that each of these short composite laminar segments is subjected to the same stress, the strain in each segment is

$$\varepsilon_{xx}(\theta) = \frac{\sigma_{xx}}{E_{xx}(\theta)}$$

$$\varepsilon_{yy}(\theta) = -\nu_{xy}(\theta)\frac{\sigma_{xx}}{E_{xx}(\theta)} \tag{6.91}$$

The normal strains averaged over the length $2l_1$ along the x direction are

$$\bar{\varepsilon}_{xx} = \frac{1}{2l_1}\int_0^{2l_1} \varepsilon_{xx}(\theta)\,\mathrm{d}x$$

$$\bar{\varepsilon}_{yy} = \frac{1}{2l_1}\int_0^{2l_1} \varepsilon_{yy}(\theta)\,\mathrm{d}x \tag{6.92}$$

The average longitudinal Young's modulus, transverse Young's modulus and Poisson's ratio can be obtained as

$$E_{xx} = \frac{\sigma_{xx}}{\bar{\varepsilon}_{xx}} \qquad E_{yy} = E_{22} \qquad \nu_{xy} = -\frac{\bar{\varepsilon}_{yy}}{\bar{\varepsilon}_{xx}} \tag{6.93}$$

The average in-plane shear modulus can be obtained by assuming that each of these segments is subjected to the same shear strain. Thus,

$$G_{xy} = \frac{1}{2l_1}\int_0^{2l_1} G_{xy}(\theta)\,\mathrm{d}x \tag{6.94}$$

Thus, the averaged stiffness constants of the undulated filling lamina can be obtained by using Eq. (6.10).

The average thermal expansion coefficients along the x and y directions are defined as

$$\alpha_{xx} = \frac{1}{2l_1}\int_0^{2l_1} (\alpha_{11}\cos^2\theta + \alpha_{22}\sin^2\theta)\,\mathrm{d}\theta$$

$$\alpha_{yy} = \frac{1}{2l_1}\int_0^{2l_1} \alpha_{yy}(\theta)\,\mathrm{d}x = \alpha_{22} \tag{6.95}$$

$$\alpha_{xy} = \frac{1}{2l_1}\int_0^{2l_1} \alpha_{xy}(\theta)\,\mathrm{d}x = 0$$

The procedures outlined above can be applied to obtain the effective thermoelastic properties of both one o'clock and eleven o'clock warp laminae along the x_2 and x_3 directions, respectively. However, the x_2 and x_3 directions are, respectively, at $60°$ and $-60°$ off-axis orientations with respect to the x axis. The effective properties of these two warp laminae in the $x-y$ plane can be obtained by the following coordinate transformation (Jones 1975):

$$\bar{Q}_{11} = Q_{11}\cos^4\phi + 2(Q_{12} + 2Q_{66})\sin^2\phi\cos^2\phi$$
$$+ Q_{22}\sin^4\phi$$

$$\bar{Q}_{12} = (Q_{11} + Q_{22} - 4Q_{66})\sin^2\phi\cos^2\phi$$
$$+ Q_{12}(\sin^4\phi + \cos^4\phi)$$

$$\bar{Q}_{22} = Q_{11}\sin^4\phi + 2(Q_{12} + 2Q_{66})\sin^2\phi\cos^2\phi + Q_{22}\cos^4\phi$$

$$\bar{Q}_{16} = (Q_{11} - Q_{12} - 2Q_{66})\sin\phi\cos^3\phi$$
$$+ (Q_{12} - Q_{22} + 2Q_{66})\sin^3\phi\cos\phi$$

$$\bar{Q}_{26} = (Q_{11} - Q_{12} - 2Q_{66})\sin^3\phi\cos\phi$$
$$+ (Q_{12} - Q_{22} + 2Q_{66})\sin\phi\cos^3\phi \tag{6.96}$$

$$\bar{Q}_{66} = (Q_{11} - Q_{22} - 2Q_{12} - 2Q_{66})\sin^2\phi\cos^2\phi$$
$$+ Q_{66}(\sin^4\phi + \cos^4\phi)$$

$$\bar{\alpha}_{xx} = \alpha_{xx}\cos^2\phi + \alpha_{yy}\sin^2\phi$$

$$\bar{\alpha}_{yy} = \alpha_{xx}\sin^2\phi + \alpha_{yy}\cos^2\phi$$

$$\bar{\alpha}_{xy} = (\alpha_{xx} - \alpha_{yy})\sin\phi\cos\phi$$

$$\bar{q}_x = \bar{Q}_{11}\bar{\alpha}_{xx} + \bar{Q}_{12}\bar{\alpha}_{yy} + \bar{Q}_{16}\bar{\alpha}_{xy}$$

$$\bar{q}_y = \bar{Q}_{12}\bar{\alpha}_{xx} + \bar{Q}_{22}\bar{\alpha}_{yy} + \bar{Q}_{26}\bar{\alpha}_{xy}$$

$$\bar{q}_{xy} = \bar{Q}_{16}\bar{\alpha}_{xx} + \bar{Q}_{26}\bar{\alpha}_{yy} + \bar{Q}_{66}\bar{\alpha}_{xy}$$

Here, ϕ represents $+60°$ and $-60°$, respectively, for one o'clock and eleven o'clock warp yarns.

Upon knowing the effective thermoelastic properties of each undulated lamina in the $x-y$ plane, the composite properties can be derived under the assumption that each of these undulated composite laminae is subjected to the same strain along the x direction. Thus, the effective in-plane thermoelastic properties of the triaxial fabric composite unit cell are given as (Rosen, Chatterjee and

Kibler 1977)

$$Q_{ij}^* = \sum_{n=1}^{3} V^{(n)} \bar{Q}_{ij}^{(n)}$$

$$q_x^* = \sum_{n=1}^{3} V^{(n)} \bar{q}_x^{(n)}$$

$$q_y^* = \sum_{n=1}^{3} V^{(n)} \bar{q}_y^{(n)}$$
(6.97)

$$q_{xy}^* = \sum_{n=1}^{3} V^{(n)} \bar{q}_{xy}^{(n)}$$

where V is volume fraction and (n) denotes the yarns in the x_1, x_2 and x_3 directions. The thermal expansion coefficients of the triaxial woven fabric composite are found from

$$\alpha_{xx}^* = S_{11}^* q_x^* + S_{12}^* q_y^* + S_{16}^* q_{xy}^*$$

$$\alpha_{yy}^* = S_{12}^* q_x^* + S_{22}^* q_y^* + S_{26}^* q_{xy}^*$$
(6.98)

$$\alpha_{xy}^* = S_{16}^* q_x^* + S_{26}^* q_y^* + S_{66}^* q_{xy}^*$$

Here, S_{ij}^* is the inversion of Q_{ij}^* of Eqs. (6.97).

By assuming that the yarns have a circular cross-section with diameter d, and $l_1 = l_2 = l_3 = l$ for the unit cell of Fig. 6.47, the highest fiber volume fraction that can be obtained for a basic triaxial weave is about 43%. For the yarn spacing/diameter ratios (l/d) of 2, 3, 4, 5 and 6, the fiber volume fraction (V_f) values are, respectively, 42.5, 24, 17.5, 14.2 and 11. Higher volume fractions can be obtained by changing the weave pattern to stuffed basic weave or bi-plane weave. As the l/d ratio increases, the fiber volume fraction decreases and the crimp can be minimized. Thus, the unit cell structure approaches a $[0°/\pm60°]$ laminate composite with straight reinforcements.

Figures 6.49(a)–(c) demonstrate the variation of longitudinal Young's modulus, in-plane shear modulus and longitudinal thermal expansion coefficient of triaxial woven carbon fabric reinforced epoxy composites with yarn spacing/diameter ratios. The results of $[0°/\pm60°]$ laminate composites as functions of fiber volume fraction can also be found in these figures. These results all indicate that as l/d increases, the difference in thermoelastic constants between woven structures and straight laminae is reduced as expected.

Even though the stiffness reduction of triaxial woven fabric composites as compared with [0°/±60°] laminates is quite severe when l/d is small, it is feasible to place additional laid-in yarns (non-crimp yarns) in the filling direction to enhance the axial properties as shown in Fig. 6.46(a). Furthermore, fiber hybridiza-

Fig. 6.49. Comparisons of the predicted thermoelastic properties of triaxial woven fabric composite (carbon/epoxy) and [0°/±60°] angle-ply laminate composite (carbon/epoxy) as functions of yarn spacing ratio (l/d). (a) Longitudinal Young's modulus. (b) In-plane shear modulus. (c) Longitudinal coefficient of thermal expansion (CTE). (After Yang and Chou, 1989.)

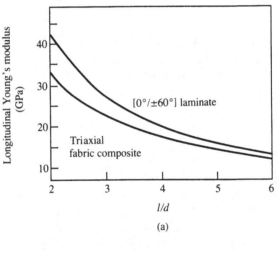

(a)

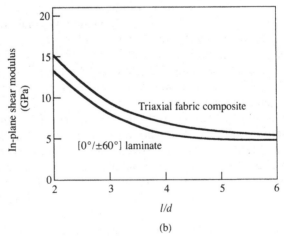

(b)

tion allows considerable design flexibility in meeting the requirements of high performance composites. Thus, by the proper selection of material combinations and fabric structural geometry, a wide range of mechanical properties can be engineered.

6.11.3 *Biaxial non-orthogonal woven fabric composites*

Biaxial non-orthogonal woven fabric composites can be produced by flat braiding, or they could occur in the fabrication of bi-axial orthogonal woven fabric composites. The flow of matrix material and the curvature of the mold surface could distort orthogonal yarns into non-orthogonal positions. The geometry of a non-orthogonal woven fabric is depicted in Fig. 6.50. It can be treated simply as a triaxial woven fabric without the filling yarn. Consequently, the methodology developed for the triaxial woven fabric composites can be readily applied. The composite unit cell is composed of two undulated laminae interlaced together at any angle, the magnitude of which depends upon the braiding pattern or the distortion of the fabric.

Figure 6.51 illustrates the variation of Young's modulus with the braiding angle or the angle of a biaxial non-orthogonal woven fabric composite. Yang and Chou (1989) have also shown that as 2θ decreases below the right angle, the thermal expansion coefficient increases along the y direction and decreases along the x direction; the in-plane shear modulus decreases with the decrease in the bias angle.

Fig. 6.49 (cont.).

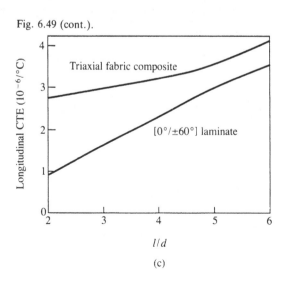

(c)

6.12   Nonlinear stress–strain behavior

The nonlinear stress–strain behavior of fabric composites due to transverse crackings initiated in warp yarns has been discussed in Section 6.7. Although transverse cracking can account for the nonlinearity at small strains, both the filling yarns and the matrix rich regions contribute to the overall nonlinear behavior of fabric composites.

Ishikawa and Chou (1983c) first adopted a one-dimensional (crimp) model to examine the material nonlinearities in the filling yarn and the pure matrix region. This approach is then extended to

Fig. 6.50. A non-orthogonal woven fabric composite and its unit cell for analysis. (After Yang and Chou 1989.)

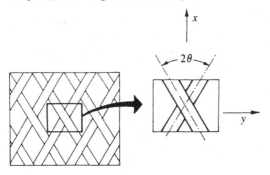

Fig. 6.51. The predicted longitudinal Young's modulus of non-orthogonal carbon fabric/epoxy composites as a function of bias angle ($V_f = 60\%$). (After Yang and Chou 1989.)

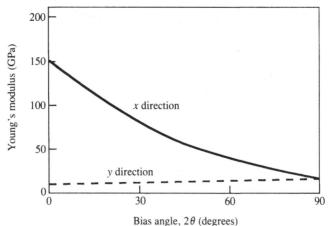

the bridging model for satin weave composites, and combined with the analysis of transverse matrix cracking to provide a more comprehensive description of the nonlinear elastic stress–strain behavior of fabric composites. The essence of the treatment of Ishikawa and Chou can be understood by considering the filling yarn depicted in Fig. 6.17. Segments of this yarn are subjected to off-axis loading in the $x–z$ plane due to fiber undulation. Thus, nonlinear shear deformation is induced in the filling yarn due to the axial load. Following Hahn and Tsai (1973), the nonlinear shear strain–stress relation is assumed to be

$$\varepsilon_{zx} = S_{55}\sigma_{zx} + S_{5555}(\sigma_{zx})^3 \tag{6.99}$$

Here, S_{55} and S_{5555} represent, respectively, the linear and nonlinear compliance constants. As to the nonlinear shear behavior of the matrix material under tensile loading, the constitutive relation is assumed to follow the same form as Eq. (6.99)

$$\varepsilon_{xx}^{M} = S_{11}^{M}\sigma_{xx}^{M} + S_{1111}^{M}(\sigma_{xx})^3 \tag{6.100}$$

Ishikawa and Chou have performed a numerical analysis of the stress–strain relation for glass/polyimide. The basic properties of a unidirectional lamina are given in Table 6.2 and $S_{5555} = 37.0$ $(1/\text{GPa}^3)$. The major ambiguity of the analysis lies in the value of the nonlinear shear compliance, S_{5555}. Because of the lack of experimental data, an estimated value based upon the stress–strain curve for a glass/epoxy composite (Jones 1975) is used. The elastic properties of polyimide are $E = 4.31$ GPa and $\nu = 0.36$. The nonlinear extensional compliance $S_{1111}^{M} = 9.88$ $(1/\text{GPa}^3)$ is also assumed to be the same as that of epoxy. Other assumptions are that $a = a_{u} = 0.4$ mm, $h = h_{t} = 0.244$ mm, and the bending-free state of deformation is valid.

Figure 6.52 indicates the numerical results of this nonlinear analysis (solid line) and the result from the consideration of transverse cracking only (dashed line) for the glass/polyimide composite. Both eight-harness satin ($n_{g} = 8$) and plain weave ($n_{g} = 2$) composites are indicated. The experimental stress–strain data of an eight-harness satin as indicated by the dots are included. The nonlinear analysis compares better with the experiment in the range of large strain than the results given by Ishikawa and Chou (1982b) for matrix cracking only. It is also observed that the contribution from shear nonlinearity increases at higher stress levels and for lower n_{g} values. Furthermore, the effect of nonlinearity on the composite behavior from the filling yarn far exceeds that from the pure matrix region.

6.13   Mechanical properties

The microstructure of two-dimensional woven fabric composites is responsible for some unique mechanical properties which are not found in their equivalent cross-ply laminates. The tension–tension fatigue behavior of woven fabric composites has been examined by Schulte, Reese and Chou (1987). In the following, the structure–property relationships are demonstrated in terms of the friction and wear behavior, and the notched strength of woven fabric composites.

6.13.1   *Friction and wear behavior*

When two surfaces interact, contact is made at their asperities. With the application of a normal load and relative motion, plastic deformation at the asperity contact zones occurs. As a result, adhesive junctions are formed which, under the influence of motion, tend to get fractured. Fracture occurs not at the original point of contact, but at some point within the softer material.

Fig. 6.52. Non-linear stress–strain relations of glass/polyimide fabric composites with $a = a_u = 0.4$ mm and $h = h_t = 0.244$ mm. (After Ishikawa and Chou 1983c.)

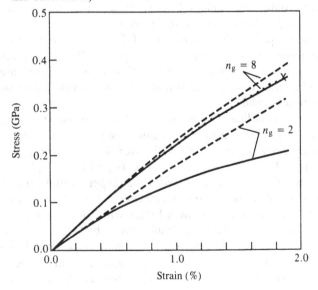

Hence material is transferred from one surface to the other. Subsequently, these transferred particles come loose due to the repeated contact.

In sliding wear, material loss is dominated primarily by adhesive mechanisms and secondarily by surface fatigue and abrasion; the abrasive component increases with increasing surface roughness. As compared with the abrasive wear conditions, the sliding wear process is much milder and is, consequently, extremely sensitive to the microstructure of the surface being worn. This is especially true for composite systems (Mody, Chou and Friedrich 1988).

In sliding wear, the sliding velocity (v) effects are manifested in frictional heating generated at the sliding interface. At some critical velocity, steady-state wear will no longer prevail, and the coefficient of friction and/or the wear rate will increase sharply. Reinforcing fibers usually increase the critical velocity of polymeric matrices. The influences of contact pressure (p) on sliding wear and of temperature on limiting pv values are also of major concern. Other factors include humid environments, counterface properties (i.e. surface roughness, density and height of the asperities), and the state of sliding interface (i.e. lubricants, films). The issue of fiber reinforcement raises additional important parameters, such as the type of fiber preforms, volume fraction and fiber orientation.

Woven forms of fiber reinforcement have demonstrated superior wear characteristics for self-lubricating bearings. Mody, Chou and Friedrich (1988) have investigated the sliding friction and wear of a neat thermoplastic matrix (PEEK), and examined the changes achieved by the incorporation of unidirectional continuous and two-dimensional woven carbon fibers. In their experiments, a pin-on-disc type wear testing machine is used; the specimen temperature, the torque generated at the sliding interface, the sliding velocity (in terms of revolutions per minute), the sliding distance (in terms of the number of revolutions made), and the sliding time are monitored. The sliding counterface is a polished steel surface.

The dimensionless wear rate (w), in the units of μm/m (depth worn per unit distance slid), is computed by using the measured mass loss (Δm) and density (ρ), along with the apparent contact area (A) and the sliding distance (L) in the following form:

$$w = \Delta m/(AL\rho) \tag{6.101}$$

The wear resistance of a material is the reciprocal of the wear rate (w^{-1}). The experiments show that, initially, wear progresses in a non-linear fashion. Later, as a definite sliding interface is established, the steady-state condition prevails, and the mass loss increases linearly with increases in sliding time.

Because of the anisotropic nature of fiber composites, it is important to identify the sliding directions relative to the fiber orientations. Three principal directions for the unidirectional continuous fiber composite have been identified, as shown in Fig. 6.53(a). Fibers in the plane of sliding and parallel to the direction of sliding are termed *parallel* (P). In-plane fibers oriented transverse to the direction of sliding are termed *anti-parallel* (AP), and fibers that stand normal to the plane of sliding are designated as *normal* (N). Following Mody, Chou and Friedrich (1988), six sliding directions are defined for a five-harness satin composite (Fig. 6.53b). The warp direction of the fabric, which has 80% of the fibers oriented in the direction of sliding, is referred to as the *parallel direction* (P). On the other hand, the filling direction of the fabric, which has 20% parallel to the sliding direction, is referred to as the *anti-parallel direction* (AP). Having thus defined the P and AP directions for the woven fabric system, consider a face perpendicular to the warp direction. This face will have a combination of fibers that stand normal to it, and parallel or transverse to it, depending on the direction of sliding on this face. Similarly, for the face orthogonal to the filling orientation of the fabric, the same reasoning prevails. From Fig. 6.53(b) it can also be concluded that the pair $N_{P(N,P)}$ and $N_{P(N,AP)}$ is the same as the pair $N_{AP(N,P)}$ and $N_{AP(N,AP)}$, if the warp and filling fiber yarns are the same fiber type. (The notations within the parentheses represent fibers of those orientations which are being slid.)

Sliding wear experiments have been conducted by Mody and

Fig. 6.53. Sliding directions with respect to the fiber orientation for (a) a unidirectional continuous fiber composite, and (b) a two-dimensional woven fabric composite. (After Mody, Chou and Friedrich, 1988.)

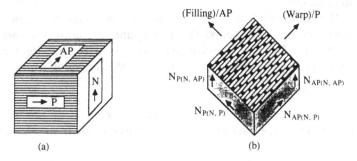

colleagues for unreinforced PEEK matrix, unidirectional continuous fiber composites, and two-dimensional woven fabric composites at three temperatures (50, 150 and 240°C) and three pv values (0.3, 0.6 and 0.9 MPa m/s). Here p denotes contact pressure and v is sliding velocity. The variations of wear rate for these three material systems at 50°C and $pv = 0.3$ MPa m/s are summarized in Fig. 6.54. The wear rate of unreinforced PEEK is relatively high. In the case of unidirectional carbon/PEEK composites, the wear rates are highly anisotropic with the AP direction showing nearly twice the wear rate of the P and N orientations. For two-dimensional woven fabric composites, owing to the equivalence of the sliding directions $N_{P(N,P)}$ to $N_{AP(N,P)}$, and of $N_{P(N,AP)}$ to $N_{AP(N,AP)}$, four unique sliding directions are identified. These include the P-oriented surface, the AP-oriented surface, the surface containing a combination of N- and P-oriented fibers (N, P), and the fourth, which has a combination of N- and AP-oriented fibers (N, AP). Wear rates of these four surfaces at 50°C turn out to be quite uniform, and thus only their average value is indicated in Fig. 6.54. Models for the wear mechanisms of composites as functions of fiber orientation have been presented by Mody, Chou and Friedrich (1988, 1989).

6.13.2 *Notched strength*

Curtis and Bishop (1984) and Bishop (1989) have assessed the strength behavior of woven fabric composites. It has been

Fig. 6.54. Comparisons of the sliding wear rates of unreinforced PEEK, as well as unidirectional and two-dimensional fabric carbon/PEEK composites, at 50°C and $pv = 0.3$ MPa m/s. (After Mody, Chou and Friedrich 1988.)

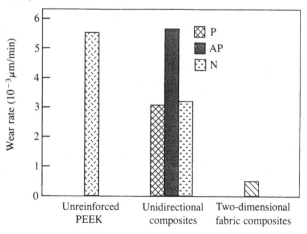

concluded that the woven fabrics are effective in limiting the growth of damage in laminated composites. It is suggested that woven fabrics be utilized in the 45° layers of a [0°/±45°] laminate with the unidirectional non-woven layers providing the needed stiffness and strength in the loading direction. Bishop has devised a scheme for laying up balanced fabric laminates without warping and unnecessary residual stresses; the line of crimped fibers in the fabric is an important parameter in the design of the lay-ups. The mechanical performance of plain and notched laminates under tensile, compressive and fatigue loadings has been reported by Bishop (1989).

To further demonstrate the damage tolerance of woven fabric composites, the example of molded-in holes is discussed below. The process of molding holes into the fabric at the laminate fabrication stage, instead of drilling the holes in the finished laminate, takes advantage of the microstructure of the woven preform. As a result, in the vicinity of the hole the fiber volume fraction is increased at regions where the stress concentrations are high, and the continuity of fiber is maintained.

Chang, Yau and Chou (1987) and Yau and Chou (1988) have examined the notched strength in tension and compression for Kevlar/epoxy and carbon/Kevlar/epoxy hybrid laminates. Specimens with molded-in holes exhibit tensile failure strengths which are up to nearly 40% higher than those of drilled specimens. Figure

Fig. 6.55. Molded-in holes in a carbon–Kevlar/epoxy [0°]$_{4s}$ laminate. (After Chang, Yau and Chou 1987.)

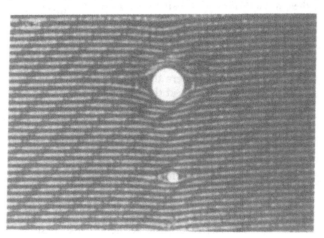

6.55 shows the fiber geometry around molded-in holes in a carbon–Kevlar/epoxy $[0°]_{4s}$ hybrid laminate. The compression behavior of woven carbon fiber reinforced epoxy composites with molded-in holes can be found in the work of Ghasemi Nejhad and Chou (1990a&b).

7 Three-dimensional textile structural composites

7.1 Introduction

Three-dimensional textile preforms are fully integrated continuous-fiber assemblies with multi-axial in-plane and out-of-plane fiber orientations (Chou, McCullough and Pipes 1986; Ko 1989a). Composites reinforced with three-dimensional preforms exhibit several distinct advantages which are not realized in traditional laminates. First, because of the out-of-plane orientation of some fibers, three-dimensional preforms provide enhanced stiffness and strength in the thickness direction. Second, the fully integrated nature of fiber arrangement in three-dimensional preforms eliminates the inter-laminar surfaces characteristic of laminated composites. The superior damage tolerance of three-dimensional textile composites based upon polymer, metal and ceramic matrices has been demonstrated in impact and fracture resistance. Third, the technology of textile preforming provides the unique opportunity of near-net-shape design and manufacturing of composite components and, hence, minimizes the need for cutting and joining the parts. The potential of reducing manufacturing costs for special applications is high. The overall challenges and opportunities in three-dimensional textile structural composites are very fascinating.

Three-dimensional textile preforms can be categorized according to their manufacturing techniques. These include braiding, weaving, knitting and stitching, as shown in Fig. 7.1.

There are three basic braiding techniques for forming three-dimensional preforms, namely 2-step, 4-step and solid braidings. In the case of 2-step braiding, the axial yarns are stationary and the braider yarns move among the axials. Thus, the axial yarns are responsible for the high stiffness and strength in the longitudinal direction and relatively low Poisson contraction. A high degree of flexibility in manufacturing can be achieved in 2-step braiding by varying the material and geometric parameters of the axial and braider yarns.

Flexibility in the manufacturing of 4-step braids is somewhat less than that of 2-step braids. All yarn carriers change their positions in the braiding process and do not maintain a straight configuration. As a result, the preforms exhibit relatively high Poisson contractions. In order to enhance the longitudinal stiffness and strength, straight laid-in yarns are often employed.

It can be demonstrated that 4-step and 2-step braidings are merely variations of a general braiding scheme. By inserting some axial yarns and placing braiding yarns at proper locations on the braiding machine, a 4-step braiding process can be converted to a 2-step braiding process.

Besides the more recently developed 2-step and 4-step braidings, which involve the sequential and discrete movement of yarn carriers, the maypole type braiding technique is also capable of producing three-dimensional solid braids. Both square and circular shapes are feasible. The technology of solid braiding has been well developed, and commercial machines are available with the maximum number of carriers currently limited to 24. The application of solid braids to composite materials is limited to simple shapes.

In woven preforms, there are two major categories. The angle-interlock multi-layer weaving technique requires interlacing the yarns in three dimensions. The warp yarn in this three-dimensional construction penetrates several weft layers in the thickness direction, and therefore the preform structure is highly integrated. In orthogonal wovens, the yarns assume three mutually perpendicular orientations in either a Cartesian coordinate system or a cylindrical coordinate system. The yarns in the Cartesian weave are not wavy, and as a result matrix rich regions often appear in the composites.

The process of stitching is mainly based upon an existing technology for converting two-dimensional preforms to three-dimensional ones. Because of the simplicity of the stitching operation, it is feasible to join composite parts continuously in a cost-effective manner. Both lock stitch and chain stitch have been

Fig. 7.1. Three-dimensional textile preforms.

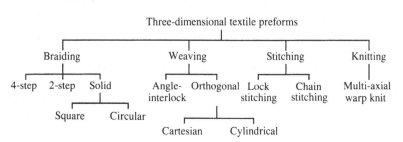

utilized. Major concerns of the stitching operation include depth of penetration of the stitching yarns and, hence, the thickness of two-dimensional preforms that can be stitch-bonded, as well as the degree of sacrifice of the in-plane properties due to the damage to in-plane yarns.

The technique of knitting is particularly desirable for producing preforms with complex shapes because the variability of the geometric forms is almost unlimited. The large extensibility and conformability of the preforms enable them to be designed and manufactured for reinforcing composites subject to complex loading conditions. The versatility of knitted preforms offers a new dimension in textile structural composites technology.

In this chapter the discussion of knitting is focussed on the conversion of two-dimensional structures (for example, unidirectional laminae) to three-dimensional ones through knit-loop-bonding. In this process, the two-dimensional layers or structures are formed at the same time when they are bonded. The technology of multi-directional multi-layer warp knit, for instance, is attractive because it enables the bonding of the unidirectional lamina by knitting yarns whereas the yarns in an individual lamina remain straight. In other words, unlike the stitch-bond of woven fabrics, the yarns in the two-dimensional structure are not wavy and hence do not sacrifice their stiffness and strength in the principal material directions. The manufacturing process is highly integrated, and the properties in the through-the-thickness direction depend upon the density and material of the knitting yarn. The potential of knitting in producing cost-effective thick laminates is attractive.

7.2   Processing of textile preforms

This section outlines the processing techniques of braiding, weaving, stitching and knitting for making three-dimensional textile preforms, with particular emphasis on braiding and weaving. According to Du, Popper and Chou (1991), braiding can form shapes either by overbraiding mandrels in conventional circular machines or by using new braiding patterns to form solid shapes directly. Weaving can be done by using either conventional looms with multi-layer constructions or entirely new equipment. Knitting can be used to interconnect fiber arrays that have been arranged by other techniques. Stitching has been used to interconnect layers of two-dimensional fabrics for achieving desired thickness and inter-laminar strength.

7.2.1 *Braiding*

Three-dimensional braids have been produced on traditional horn-gear machines. At the present time, horn-gear based braiding machines use a small number of yarn carriers (≤ 24) and cannot form complex shapes. Their applicability is therefore limited. A number of new machines have been developed to create complex shapes. These newer braiding processes include 2-step (Popper and McConnell 1988), AYPEX (Weller 1985), interlock twiner (Cole 1988), and row and column (Florentine 1982), which is also referred as *Omniweave, Magnaweave*, or *4-step* in the literature.

A schematic view of a set-up for the three-dimensional braiding process is shown in Fig. 7.2. Axial yarns, if present in a particular braid, are fed directly into the structure from packages located below the track plate. Braiding yarns are fed from bobbins mounted on carriers that move on the track plate. The pattern of motion of the braiders and the presence/absence of axial yarns determine the type of braids, as well as the microstructure. The processes of 2-step and 4-step braiding are introduced below.

7.2.1.1 *2-step braiding*

The preform structure of a 2-step braid includes a large number of parallel (axial) yarns aligned for efficient reinforcement and a smaller number of braiding yarns (braiders) that interconnect

Fig. 7.2. A set-up for three-dimensional braiding. (After Du, Popper and Chou 1991.)

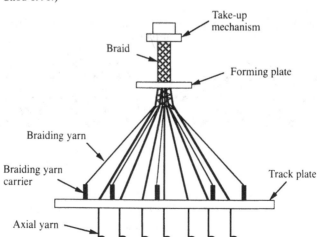

the axial yarns and form the fabric shape. The axial array can be arranged in essentially any shape, including I-beams, box beams, circular tubes, etc., whereas the braiders are arranged around the perimeter of the axial array as shown in Fig. 7.3. In the braiding process, the braiders move through the axial array in two sequential steps. In the first, the braiders all move in one diagonal line but in alternating directions (Fig. 7.3a). In the second, they move along the other diagonal line (Fig. 7.3b).

Although the machine action consists of only two steps, each braider moves through a larger portion of the structure. This can be seen by tracing the path of a single braider subjected to the repeated 2-step machine action. The paths followed by all braiders will completely intercinch the axial yarns and lock them in the desired shape.

Compared with other three-dimensional braiding processes, 2-step braiding has several distinct advantages. A relatively simple sequence of braider motions can form a wide range of shapes. During each step of the process, all the braiders are simultaneously outside of the axial array, and thus it is possible to add various inserts to the structure or even rearrange the axial array geometry to change the preform cross-section. Furthermore, this structure can be made with a high level of fiber packing and a large number of axially oriented fibers as needed in many applications (Du, Popper and Chou 1989, 1991).

The 2-step process has motivated a number of researchers. Li and El Shiekh (1988) modeled the microgeometry using idealized

Fig. 7.3. 2-Step braiding pattern showing the relative motion of yarns. (After Du, Popper and Chou 1991.)

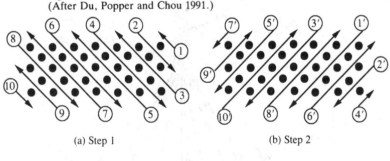

(a) Step 1   (b) Step 2

● - Axial yarn

(n) - Braider yarn

circular yarns. Ko, Soebroto and Lei (1988) and Whitney (1988) have evaluated the mechanical properties of consolidated 2-step composites.

7.2.1.2 *4-step braiding*

The 4-step braiding process, so named by Li, Kang and El Shiekh (1988), requires four distinct Cartesian motions of the yarns in the fabric cross-sectional plane in each machine cycle. Following G. W. Du (private communication, 1990), the 4-step braiding process is depicted in Fig. 7.4 for a 1×1 set-up in which the yarn carriers are arranged in a rectangular plane with eight columns ($m = 8$) and four layers ($n = 4$). Here, the yarn carriers are indicated by the circles, and can move along the y and z direction tracks. The process is termed 1×1 if the distance traveled by a carrier in each machine step is equal to the inter-yarn spacing in the y or z direction. Other braiding patterns (i.e. 1×2, etc.) are feasible, which require machine set-ups different from that of Fig. 7.4. It is noted that the carriers occupy alternating positions on the perimeter of the set-up. The total number of carriers in the $m \times n$ rectangular slab for the 1×1 braiding pattern is (Li, Kang and El Shiekh 1988):

$$N = mn + m + n = (m + 1)(n + 1) - 1 \qquad (7.1)$$

Thus, for the 8×4 array, there are 44 carriers.

Consider the starting carrier positions as shown in Fig. 7.4(a). In step-1 of the machine cycle, all the rows of carriers move in the y direction; adjacent rows move in opposite directions as indicated by the arrows. In step-2 of the machine cycle (Fig. 7.4b), all the columns of carriers move vertically; adjacent columns move in opposite directions as indicated by the arrows. Note that in step-1 and step-2 movements, the carriers on the perimeter of the set-up remain stationary. The displacement of an individual carrier can be identified (for example, carriers marked A and B in Fig. 7.4a). Step 3 (Fig. 7.4c) is similar to step-1 except that the directions of movement of the same row are opposite to each other. The same comparison can be made between step-2 and step-4 (Fig. 7.4d). These four steps comprise a machine cycle, because at the end of the cycle (Fig. 7.4e) the carrier arrangement is the same as that at the beginning of the machine cycle, although the individual carriers have changed their locations.

It is interesting to note that the 44 carriers in the slab of Fig. 7.5(a) can be divided into four groups. These are denoted as groups

Fig. 7.4. Yarn carrier configurations and movements in a 4-step braiding set-up. (After G. W. Du, private communication, 1990.)

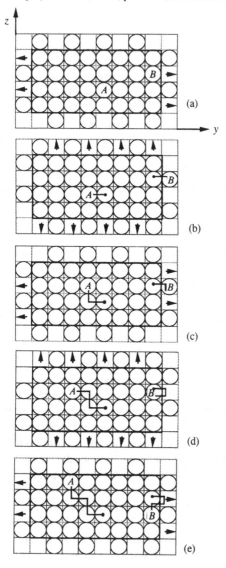

Fig. 7.5. The four yarn carrier groups in an 8 × 4 slab. Each group defines a unique yarn path. (a) Yarn carrier location; (b) carrier path for group 1; (c) carrier path for group 2; (d) carrier path for group 3; and (e) carrier path for group 4. (After G. W. Du, private communication, 1990.)

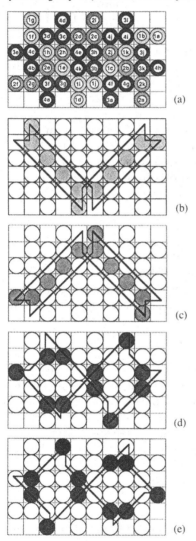

(a)

(b)

(c)

(d)

(e)

1, 2, 3 and 4. Within each group the carriers are labeled in alphabetical order from *a* to *k*. The characteristic of each group is that all the carriers within the group share the same path of motion. For example, carrier 1*a* in group 1, moves along the path of $1a \rightarrow 1b \rightarrow 1c \rightarrow 1d \rightarrow 1e \rightarrow 1f \rightarrow 1g \rightarrow 1h \rightarrow 1i \rightarrow 1j \rightarrow 1k \rightarrow 1a$ (Fig. 7.5b). All the other carriers in this group follow the same path. The paths of groups 2, 3 and 4 are indicated in Figs. 7.5(c), (d) and (e), respectively. The path for group 1 carriers in Fig. 7.5(b) is not marked directly on the carriers to avoid overlapping and confusion; the same is true for Fig. 7.5(c). The movement of a carrier, for instance, from position 1*a* to 1*b*, or 1*b* to 1*c*, etc. requires one machine cycle which comprises the steps as shown in Fig. 7.4. The complete cycle of movement of a carrier, i.e. $1a \rightarrow 1b \rightarrow \cdots \rightarrow 1a$ (returning to the original position) is termed a *repeat*.

Li, Kang and El Shiekh (1988) have shown that the number of yarn groups in an $m \times n$ slab is given by

$$G = mn/\text{LCM}(m, n) \qquad (7.2)$$

where $\text{LCM}(m, n)$ denotes the least common multiple of m and n. Furthermore, each group has the same number of carriers, which is N/G. The number of machine cycles required for all the carriers to return to their original positions is thus also equal to N/G. It should be noted that the above discussions are valid only for the 1×1 braiding pattern.

7.2.1.3. *Solid braiding*

The term *solid braiding* is used here to describe the category of three-dimensional preforms produced by the continuous intertwining of yarns in the maypole fashion. Figure 7.6(a) shows the horn-gear set-up for square braiding. The longitudinal and cross-sectional views of some square braids are given in Fig. 7.6(b). Solid braids with circular cross-sections are also available. However, it is not feasible to produce three-dimensional preforms with complex shapes using solid braiding.

7.2.2 *Weaving*

Advances in textile manufacturing technology are rapidly expanding the number, type and complexity of preforms which offer reinforcements in the through-the-thickness direction. The traditional weaving technique for producing two-dimensional fabrics has been modified to achieve a much higher degree of integration in fiber geometry in the thickness direction. Angle-interlock weaving and orthogonal weaving are the two distinct techniques by which the fibers are incorporated at an angle and parallel to the thickness direction, respectively.

7.2.2.1 *Angle-interlock multi-layer weaving*

Angle-interlock multi-layer woven fabrics for thick section composite applications can be produced on either a dobby loom or a Jacquard loom. The cam-system is limited to fabricating double- or triple-layer cloth. Yarns or fibers in angle-interlock multi-layer wovens are interlaced in a manner similar to two-dimensional woven structures, except that warp fibers may penetrate more than

Fig. 7.6. (a) Horngear set-up for square braiding. (After Ko 1989a.) (b) Examples of square braids. (After Steeger 1989.)

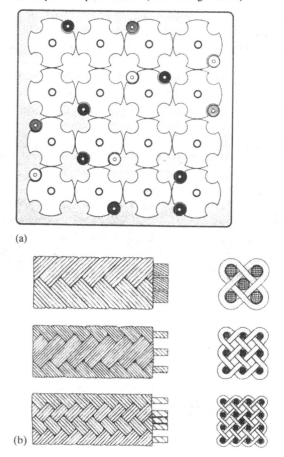

(a)

(b)

one layer of weft yarns. The warp direction again coincides with the machine direction, just as in two-dimensional wovens, whereas the filling yarn insertion takes place in the transverse direction. Many other preform configurations are possible, such as those with laid-in non-crimp yarns (to reduce Poisson's effect), or a combination of different fiber materials within the same preform (Whitney and Chou 1988, 1989).

Many variations in the basic geometry of angle-interlock preforms are feasible, depending on the number of layers interlaced, the pattern of repeat, and the presence of laid-in yarns. Whitney (1988) has discussed the fiber architectures in which all warp yarns interlace the same number of weft yarns. In order to demonstrate the geometric variability of angle-interlock fabrics, a highly idealized example is given in the following. Discussions are based on the fabric structure of the 1×1 pattern, i.e. the warp yarn orientation can be represented by one inter-yarn spacing in the horizontal direction and one inter-yarn spacing in the vertical direction, as shown in Fig. 7.7.

Following Byun, Leach, Stroud and Chou (1990a), the key geometric parameters for identifying the preform microstructure include the number of weft yarns in the thickness direction (n_f), as

Fig. 7.7. Three-dimensional angle-interlock woven preforms as identified by $[n_f, n_{ft}]$: (a) [5, 2], (b) [5, 4], and (c) [6, 6]. (After Byun *et al.* 1990.)

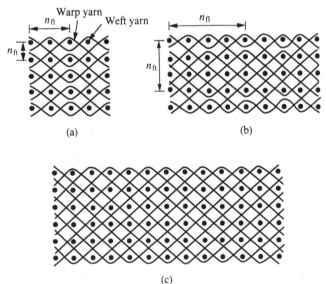

(a)

(b)

(c)

well as the number of weft yarns interlocked by a warp yarn in the thickness direction (n_{ft}) and in the length direction (n_{fl}). Parametric relations are obtained based on the preform structures which have the following restrictions: (1) The fabric structure is symmetric in the thickness direction with respect to the mid-plane. (2) The weft yarns have the same degree of interlocking by warp yarns. (3) The number of weft yarns in the thickness direction is the same along the warp direction. Employing the notation of $[n_f, n_{ft}]$, the woven preforms of Figs. 7.7(a), (b) and (c) can be identified as [5, 2], [5, 4] and [6, 6].

The following relationship needs to be satisfied to ensure the interlocking of weft yarns by warp yarns for the 1×1 pattern:

$$n_{ft} = n_{fl} \tag{7.3}$$

With the above condition, a maximum number of warp yarns can be achieved in the preform. For a $[n_f, n_{ft}]$ weave, the total number of warp yarns (n_w) is $2n_f$. However, not every warp yarn interlocks with n_{ft} weft yarns. This can be seen from Figs. 7.7(a) and (b) where the warp yarns at the top and bottom faces only interlace with the weft yarns in the surface layers. The degree of reinforcement in the thickness direction is related to the number of warp yarns (n_{wi}) that interlock with the n_{ft} weft yarns. When $n_f = n_{ft}$, all the warp yarns interlock with all n_{ft} weft yarns, i.e. $n_{wi} = 2n_{ft}$. When $n_f \neq n_{ft}$, n_{wi} is given as follows:

$$n_{wi} = 2kn_{ft} \qquad \text{for } k < n_f/n_{ft} < k + 1 \tag{7.4}$$

$$n_{wi} = (2k - 1)n_{ft} \qquad \text{for } n_f/n_{ft} = k \quad (k \geq 2) \tag{7.5}$$

where k is an integer. For the fabrics of Figs. 7.7(a), (b) and (c), the n_{wi} values are, respectively, 8, 8 and 12.

Thus, the number of warp yarns (n_{wn}) which do not interlace n_{ft} weft yarns is

$$n_{wn} = 2n_f - n_{wi} \tag{7.6}$$

In Figs. 7.7(a) and (b), $n_{wn} = 2$ and in these cases each warp yarn at the surface only interlace with one layer of weft yarns. However, the n_{wn} warp yarns can also interlace with the multi-layer of weft yarns. Consider a fabric preform with the [5, 3] weave pattern and $n_{wn} = 4$. Figures 7.8(a) and (b) show the two possible configurations of the warp yarns near the free surfaces.

When the condition of Eq. (7.3) is not satisfied, the resulting fabric is not highly integrated and there are non-interlaced yarns. In the following, the case of three-dimensional weaves with non-interlaced weft yarns is discussed. An additional geometric parameter, n_{ws}, is identified; it denotes the number of weft rows shifted by adjacent warp planes. Using the notation of $[n_f, n_{ft}, n_{ws}]$, the fabrics of Figs. 7.9(a), (b) and (c) can be identified as $[6, 2, 1]$, $[5, 3, 0]$ and $[5, 3, 1]$, respectively.

Finally, the total number of warp yarns can be obtained as

$$\begin{aligned} n_w &= n_f + 1, & n_f \geq 3 && \text{for } n_{ft} = 2 \\ n_w &= n_f + n_{ws} - 1, & n_f = 2n_{ft} + 1 && \text{for } n_{ft} \geq 3 \end{aligned} \tag{7.7}$$

Thus, the number of warp yarns interlaced through the n_{ft} weft yarns is

$$n_{wi} = n_w - 2n_{ws} \tag{7.8}$$

It should be noted that the parametric relations for a three-dimensional weave in which every weft yarn is interlocked with warp yarns can also be obtained in the case that Eq. (7.3) is not satisfied.

Fig. 7.8. Two variations of the $[5, 3]$ weave with $n_{wn} = 4$. the two warp yarns near the surface interlace with (a) one weft yarn layer or (b) two weft yarn layers. (After Byun *et al.* 1990.)

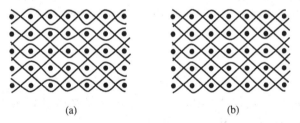

(a)   (b)

Fig. 7.9. Three-dimensional angle-interlock woven preforms as identified by $[n_f, n_{ft}, n_{ws}]$: (a) $[6, 1, 1]$, (b) $[5, 3, 0]$, and (c) $[5, 3, 1]$. (After Byun *et al.* 1990.)

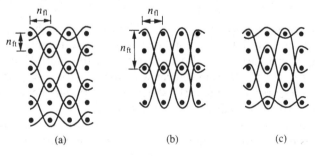

(a)   (b)   (c)

7.2.2.2 *Orthogonal weaving*

Figure 7.10 shows an orthogonal woven fabric where the yarns are placed in three mutually orthogonal directions. Because of the nature of fiber placement, matrix rich regions are created in composites reinforced with orthogonal woven preforms. Since the thickness direction yarns are incorporated into the preform in the weaving process, they do not cause damage to the in-plane fibers. This is different from the case of stitching bonding of two-dimensional fabrics. Orthogonal woven fabrics can be fabricated by maintaining one stationary axis either by predeposition of the yarn system or a space rod which is subsequently retracted and replaced by axial yarns. The two sets of yarns in the plane perpendicular to the axial yarns are then inserted in an alternating manner (Ko 1989a). Both Cartesian and cylindrical woven fabrics are available.

7.2.3 *Stitching*

The process of stitching for making three-dimensional preforms is relatively simple. The basic needs include a sewing machine, needle and stitching thread. The processing variables are stitch density (stitch/unit length), the size of the stitch thread, and types of stitch. Both lock stitch and chain stitch are available (Fig. 7.11). A lock stitch becomes unbalanced if the tension in either the bobbin thread or the needle thread is higher than that in the other thread. The necessary clearance between the feed and dog as well as the length of the needle stroke in the case of lock stitching, for instance, are

Fig. 7.10. An orthogonal woven fabric. (After Chou, McCullough and Pipes 1986.)

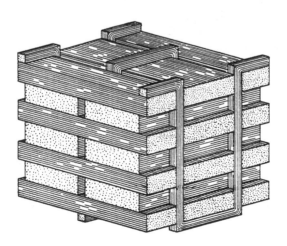

Fig. 7.11. (a) Lock stitch and (b) chain stitch seams. (After Ogo 1987.)

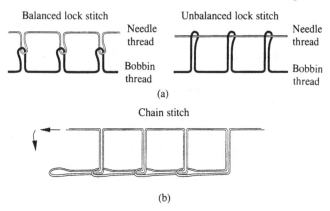

(a)

(b)

Fig. 7.12. Schematic of the lock stitch process. (After Ogo 1987.)

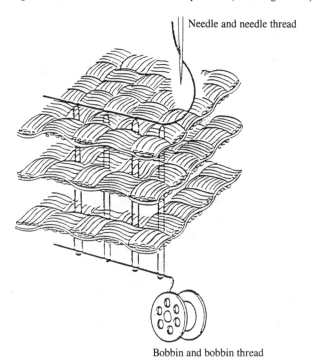

determined by the thickness of the two-dimensional preform to be stitch-bonded. Figure 7.12 shows the schematic of a lock-stitching process for bonding fabric layers. The needle thread needs to be abrasion resistant and can be bent to small curvature in the needle hole.

7.2.4 *Knitting*

Three-dimensional knitted fabrics can be produced by either a weft knitting or warp knitting process. For additional strengthening in the 0° and 90° directions, laid-in yarns can be placed inside the knitting loops. Figure 7.13 shows a weft knit fabric with laid-in weft and warp yarns.

The most promising knitted preform which provides a high degree of structural integration in the thickness direction is perhaps the multi-axial warp knit. It consists of warp (0°), weft (90°) and bias ($\pm\theta$) yarns held together by a chain of tricot stitch through the thickness of the assembly (Fig. 7.14). Different kinds of multi-axial warp knits have been developed. The main attraction of the knitted construction is that it possesses the advantage of unidirectional laminates while also providing enhanced stiffness and strength in the thickness direction (Ko, Pastore, Yang and Chou 1986).

7.3 Processing windows for 2-step braids

The purpose of the following discussions is to demonstrate that knowledge of the microgeometry and structure of textile preforms provides the basis for understanding flexibility in processing. The work of Du, Popper and Chou (1991) in 2-step braiding is recapitulated as an example of such an approach. The

Fig. 7.13. Weft knit with laid-in weft and warp yarns. (After Ko 1989a.)

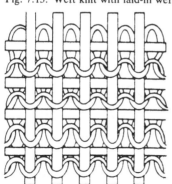

key inputs of the analysis are (1) the size, type and shape of braiders and axial yarns, (2) the braid pattern (size of axial yarn array), and (3) the advance rate during braiding. The key outputs are braid dimensions, fiber orientation, inter-yarn void content, fiber volume fraction, and geometric limits imposed by yarns jamming against each other. The modeling work is for preforms of rectangular cross-section. However, the methodology regarding yarn cross-sections, unit cells, and yarn jamming can be used to analyze more complex shapes, as well as other types of three-dimensional fabrics.

The major assumptions are: (1) Multi-filament yarns are used for both braiders and axial yarns. These yarns are composed of a large number of fibers, and their cross-sections can be readily deformed to prismatic shapes. (2) Fiber cross-section is round. (3) Fibers are parallel along the yarn length, i.e. zero twist. (4) Yarn tension is high enough to ensure a straight yarn path, except for the braider yarns, which are bent around the braid surface. (5) Filaments are inextensible.

7.3.1   *Packing of fibers and yarn cross-sections*

The fiber volume fraction of a three-dimensional preform depends on the level that fibers pack against one another in a yarn and the level to which yarns pack against one another in the structure. Two methods for estimating inter-fiber packing are described in this section. Section 7.3.3 discusses yarn packing in preforms.

The geometry of inter-fiber packing in yarns has been studied primarily for textile applications (see Hearle, Grosberg and Backer

Fig. 7.14. Mult-axial warp knit fabric. (After Chou, McCullough and Pipes 1986.)

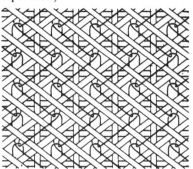

1969). Two basic idealized packing forms can be identified: open packing (Fig. 7.15a) and close packing (Fig. 7.15b), in which the fibers are arranged in concentric and hexagonal patterns, respectively.

In open-packed yarns the packing fraction, defined as the fiber-to-yarn area ratio, has been computed as a function of the number of fibers. If the outer ring is completely filled and the fibers are circular, the yarn packing fraction is

$$\kappa_o = \frac{3N_r(N_r - 1) + 1}{(2N_r - 1)^2} \tag{7.9}$$

where N_r is the number of rings, and its relationship to the number of fibers, N_f, is given by

$$N_r = \tfrac{1}{2} + \sqrt{(\tfrac{1}{4} + \tfrac{1}{3}(N_f - 1))} \tag{7.10}$$

For a large number of fibers, κ_o approaches 0.75.

In close-packed yarns, for any number of circular fibers if the outer layer is completely filled, the yarn packing fraction equals the

Fig. 7.15. Fiber packing in yarns. (a) Open packing in a circular yarn. (b) Close packing in a hexagonal yarn. (c) Open packing in a diamond-shaped yarn. (d) Close packing in a diamond-shaped yarn. (After Du, Popper and Chou 1991.)

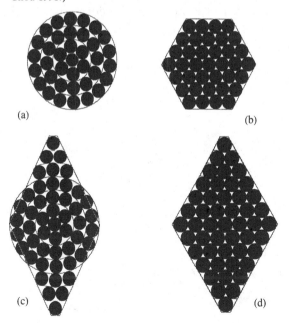

(a)

(b)

(c)

(d)

area ratio of a circle to the hexagon in which the circle is inscribed:

$$\kappa_c = \frac{\pi}{2\sqrt{3}} = 0.91 \qquad (7.11)$$

The yarn packing fractions predicted by the two models assume circular and hexagonal yarn cross-sections. However, as shown in Figs. 7.15(c) and (d), they apply equally well to other shapes if the number of fibers is sufficiently large. Factors that affect both the packing of fibers in a yarn and the packing of yarns in a preform include yarn tension, inter-yarn contact, yarn twist, fiber cross-section, fiber straightness, manufacturing method, and preform geometry.

In addition to the level of yarn packing fraction, the yarn cross-sectional shape plays a significant role in determining how many fibers can be packed into a fabric. In the textile literature, the yarns are often assumed to have a circular cross-section (see Peirce 1937; Brunnschweiler 1954). However, it has been shown that the cross-section of even highly twisted yarns deviates significantly from a circular shape and the yarn cross-section varies considerably for different types of preforms. Many attempts have been made to develop more realistic geometric models for yarns in fabrics by assuming elliptical and race-track cross-sections (Hearle, Grosberg and Backer 1969).

Fig. 7.16. Cross-sections of axial yarns in a rectangular braided preform before consolidation. (After Du, Popper and Chou 1991.)

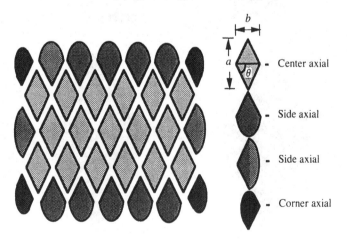

In the following, a model for the yarn cross-section in 2-step braids is developed. It is observed from composite specimens that the yarns in the preform have cross-sections as shown in Fig. 7.16. After matrix addition and consolidation in a mold, the fabric is observed to be flattened, as shown in Fig. 7.17. The axial yarns have different cross-sections depending on their locations in the preform. Central yarns, which form the bulk of the structure, are diamond-shaped. Axial yarns on the side and corners of the preform are pentagonal. Braiding yarns, which occupy the space in between the axial yarns, are rectangular.

The aspect ratio of axial yarns, f_a, is related to the inclination angle of the braiders (Figs. 7.16 and 7.18) and is given by

$$f_a = \frac{a}{b} = \tan \theta \qquad (7.12)$$

where f_a is influenced by braider yarn tension or external lateral compression applied at the forming point during the process. It can also be changed by compacting the entire braided preform during matrix consolidation. These aspect ratios affect the shape of the final braid as well as the braider yarn orientation angle (α) and fiber volume fraction (V_f). With unit axial aspect ratio ($\theta = \pi/4$), the cross-section of the center axial yarns becomes square. In this special case, the fiber volume fraction is at a maximum.

The axial yarn dimensions can be calculated from the cross-section of the consolidated braid in Fig. 7.17. These relations are

Fig. 7.17. Cross-sections of axial yarns in a rectangular braided composite. (After Du, Popper and Chou 1991.)

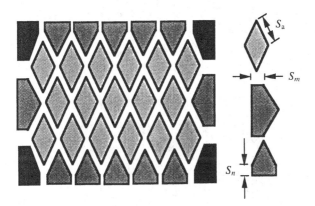

given in terms of yarn area and inclination angle. The yarn area is in turn evaluated from its linear density and the fiber packing fraction and fiber density:

$$S_a = \sqrt{\left(\frac{\lambda_a}{\rho_a \kappa_a \sin(2\theta)}\right)}$$

$$S_m = \frac{0.5\lambda_a}{2S_a\rho_a\kappa_a \sin\theta} \qquad (7.13)$$

$$S_n = \frac{0.5\lambda_a}{2S_a\rho_a\kappa_a \cos\theta}$$

The parameters λ_a and ρ_a are the linear density and the fiber density of axial yarns, respectively. The packing fraction, κ_a, is assumed to be constant for all axial yarns.

Fig. 7.18. Path of one braiding yarn in the fabric: (a) braid pattern ; (b) top view; and (c) front view. (After Du, Popper and Chou 1991.)

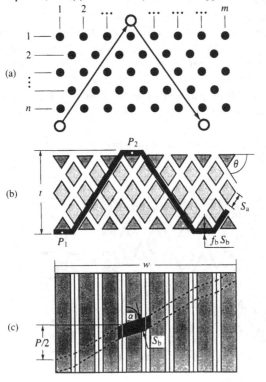

For, only one braider yarn is shown in Fig. 7.18 for two steps of the braiding process. These yarns are assumed to be rectangular with aspect ratio f_b. The aspect ratio is usually much less than unity because of compression by the axial yarns. The dimension of a braider yarn can also be calculated from its packing fraction (κ_b), yarn linear density (λ_b), and fiber density (ρ_b):

$$S_b = \sqrt{\left(\frac{\lambda_b}{\rho_b \kappa_b f_b}\right)} \tag{7.14}$$

7.3.2 *Unit cell of the preform*

In order to understand the microscopic arrangements of yarns, it is necessary to identify the 'unit cell' of the fabric preform. By definition, a unit cell constitutes the smallest repeating entity in the structure. The complexity of three-dimensional preform structures often makes the identification of unit cells a difficult task.

The unit cell of the 2-step braid is composed of four sub-cells, labelled A, B, C and D in Fig. 7.19. The repeat of these four sub-cells will generate the whole braided structure. Because of geometric similarity, any one of these four can be utilized to derive the basic structural characteristics. The length of the unit cell in Fig. 7.19 is the length of braid formed in one machine step. This length is actually half of the fabric pitch length (P), as shown in Figs. 7.18–7.20. Five layers are shown in Fig. 7.19. The number of columns has been assumed to be very large so that the rather complicated edge configuration of the preform can be avoided. Figure 7.20 shows the difference between finite and infinite columns and their effects on braider paths. Figure 7.20(a) shows the yarn path on the surface of a specimen seven columns wide. Note that the trace of the braider yarns lies on an inclined line. In an infinitely wide specimen (Fig. 7.20b) the trace of the braider yarns is perpendicular to the braiding direction.

The width and thickness of the braided preform can be computed from Figs 7.17 and 7.18 in terms of m (number of axial columns) and n (number of axial layers) as well as yarn geometric and material parameters:

$$w = (m - 1)\left(2S_a \cos\theta + \frac{f_b S_b}{\sin\theta}\right) + 2(S_m + f_b S_b) \tag{7.15}$$

$$t = (n - 1)\left(S_a \sin\theta + \frac{f_b S_b}{2\cos\theta}\right) + 2(S_n + f_b S_b) \tag{7.16}$$

From Eqs. (7.15) and (7.16), the aspect ratio of the braided preform can be obtained as

$$f = \frac{t}{w} \tag{7.17}$$

The braider yarn orientation can be determined by computing the *projected* length (i.e. segment $P_1 P_2$ in Fig. 7.18b) of one braider over one half of the pitch length. Note that the angle between a braider and the axial yarns (α) appears to vary on the front view in Fig. 7.18(c). This apparent variation occurs because a segment of a braider yarn in the interior of the preform has a different projected angle compared to a segment on the preform surface. The projected

Fig. 7.19. Unit cell model of a 2-step braided preform showing four sub-cells. Each sub-cell includes a braider yarn and a number of axial yarns. (After Du, Popper and Chou 1991.)

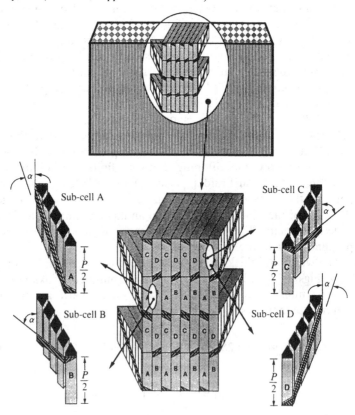

length L_p of the segment P_1P_2 in the axial direction is

$$L_p = t \csc \theta + 2S_n(1 - \csc \theta) + 2S_a \cos \theta \qquad (7.18)$$

The braider yarn angle is then given by

$$\alpha = \tan^{-1}\left(\frac{2L_p}{P}\right) \qquad (7.19)$$

The total length of one braider yarn in a unit cell is

$$L_b = \frac{L_p}{\sin \alpha} = \frac{P}{2 \cos \alpha} \qquad (7.20)$$

Then the volume of the braider yarn (v_b) and the volume of axial yarns (v_a) in a unit cell and the total volume of a unit cell (v_t) can be determined:

$$v_b = L_b f_b S_b^2 \qquad (7.21)$$

$$v_a = \frac{P}{2} S_a[(n - 1)S_a \sin(2\theta) + 4S_n \cos \theta] \qquad (7.22)$$

$$v_t = v_a + \frac{P}{2} f_b S_b[(n - 1)S_a + L_p] \qquad (7.23)$$

The fiber volume fraction V_f (total fiber volume/unit cell volume), the braider fiber volume fraction V_b (braider fiber volume/unit cell volume), and the volume fraction of the void V_v (volume of

Fig. 7.20. Effect of fabric width on braid geometry: (a) finite-width preform; (b) infinite-width preform. (After Du, Popper and Chou 1991.)

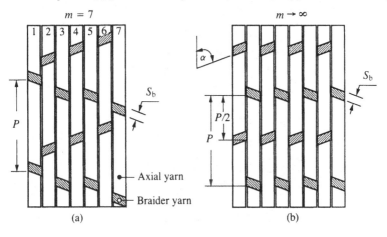

inter-yarn voids/unit cell volume), can then be obtained:

$$V_f = \frac{v_a \kappa_a + v_b \kappa_b}{v_t} \qquad (7.24)$$

$$V_b = \frac{v_b \kappa_b}{v_t} \qquad (7.25)$$

$$V_v = 1 - \frac{v_a + v_b}{v_t} \qquad (7.26)$$

7.3.3   *Criterion for yarn jamming*

The allowable microstructural states of a fabric preform are limited by the condition at which the yarns jam against one another. Knowledge of yarn jamming is essential in identifying the processing windows of fabric preforms. Although jamming is discussed frequently in the textile literature, it is often neglected in the analysis of composites.

In 2-step braids, the braider angle becomes very small as the braider yarns become parallel to the axial yarns. However, as the pitch length is reduced, the braider angle increases, and a limiting state is reached in which the yarns jam against one another. If all other parameters remain constant, the pitch length cannot be reduced further. The state of jamming is illustrated in Fig. 7.21 where the yarn-to-yarn contact is shown. This rather complex

Fig. 7.21. Surface geometry of braid at jamming. (After Du, Popper and Chou 1991.)

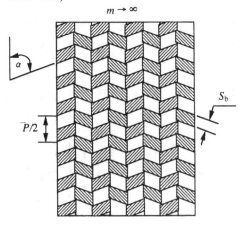

limiting state, however, can be described simply by

$$P_j = \frac{2S_b}{\sin \alpha} \tag{7.27}$$

where P_j denotes the pitch length at jamming. Equation (7.27) is applicable to specimens of finite width.

Due to the edge effect of finite-width structures, the orientation angle, α, of all braiders are not equal. In a braiding step, the braiders on the side surface of the preform will have shorter length than those in the interior. However, since all yarns advance at the same pitch length, the 'edge' braiders will lie at a somewhat lower angle than those passing through the center of the structure.

Du, Popper and Chou (1991) have conducted experiments to measure the geometric and material parameters of 2-step braids. A comparison of measured and predicted values of two samples are given in Table 7.1. Braid I is rectangular in cross-section, consisting of 12-column by five-layer axial yarns (Kevlar-49 with a linear

Table 7.1 *Material and geometric parameters of 2-step braided preforms. After Du, Popper and Chou (1991)*

Parameters	Braid I		Braid II	
	Measured	Computed	Measured	Computed
λ_a (g/m)	9.57		3.33	
λ_b (g/m)	2.39		0.25	
m	12		7	
n	5		5	
P (mm)	17.8		4.98	
f_a	0.78		1.32	
f_b	0.12		0.05	
κ_a	0.70		0.78	
κ_b	0.70		0.78	
t (mm)	9.0	8.9	7.0	7.3
w (mm)	63.0	62.9	15.0	14.6
α (°)	65.0	65.9	75.0	76.7
V_f (%)	56.0	56.8	73.0	73.6

Braid I: bare fiber preform.
Braid II: infiltrated with matrix.

density of 9.57×10^{-3} kg/m) and 15 braider yarns (Kevlar-49 with a linear density of 2.39×10^{-3} kg/m). Braid II is also rectangular in cross-section, with seven-column by five-layer axial arrays and ten braiders. In Braid II, all axial yarns are made of Kevlar-29 with a linear density of 3.33×10^{-3} kg/m: Kevlar-49 is used for the braider yarns which are much finer than the axial yarns with a linear density of 2.53×10^{-4} kg/m. Braid II was impregnated with an epoxy by resin transfer molding.

Based upon the relations between process variables and fabric geometry, it has been shown that the range of allowable fabric structures is dictated by effects such as yarn jamming and fiber packing. Figure 7.22 demonstrates the processing window for 2-step braids when the braider yarn orientation angle and pitch length, total fiber volume fraction, and yarn linear densities are considered. The processing window is bounded by two limiting states: yarn jamming and zero braider angle. Preform constructions corresponding to the curved 'jamming' line are at their tightest possible state, and constructions corresponding to the $\alpha = 0$ curve have infinite pitch length. As the dimensionless pitch length P/S_a increases, the

Fig. 7.22. Fiber volume fraction vs. braider-to-axial linear density ratio. The allowable process window is shown ($\kappa_a = 0.8$, $\kappa_b = 0.8$, $\theta = 38°$, $f_b = 0.2$, $n = 5$). (After Du, Popper and Chou 1991.)

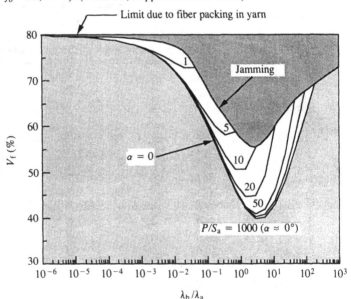

fiber volume fraction decreases. Increasing the ratio of braider-to-axial yarn linear density causes the maximum allowable fiber volume fraction to go through a minimum. At fixed levels of pitch length, an increase in λ_b/λ_a first reduces the fiber volume fraction because the inclusion of larger braider yarns creates more void space. However, at large λ_b/λ_a ratios, a higher fiber volume fraction is realized. At a ratio of about 200, the fabric reaches a limiting state in which the braider yarn angle approaches zero due to the infinite pitch length. The fiber packing in the yarns, taken as 0.8, limits the maximum fiber volume fraction in the fabric.

A 'microstructure map' of 2-step braids, which gives the relationship among fiber volume fraction, pitch length, braider yarn orientation angle and braider yarn volume fraction, is shown in Fig. 7.23. The minimum allowable fiber volume fraction increases with a reduction in braider yarn pitch length. For a fixed pitch length and above the minimum allowable fiber volume fraction, both V_f and α increase with an increase in braider fiber volume fraction. This map demonstrates that a wide range of orientation angle and fiber volume fraction can be achieved by varying the pitch length and the amount of braider yarns relative to the axials. Maps of microstructures provide guidance in designing preforms for a specific application.

Fig. 7.23. Property volume fraction (V_f) vs. fiber orientation angle (α) for various pitch length and volume fraction of braider yarns ($\kappa_a = 0.8$, $\kappa_b = 0.8$, $\theta = 38°$, $f_b = 0.2$, $n = 5$). (After Du, Popper and Chou 1991.)

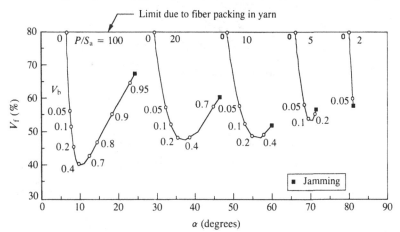

7.4   Yarn packing in 4-step braids

Knowledge of yarn packing in three-dimensional structures is essential for determining the unit cell configuration of a fiber preform as well as the condition for yarn jamming. In a 4-step braiding process, the braiding yarn carriers move in a two-dimensional grid with two sets of perpendicular tracks (Fig. 7.4). For the sake of simplicity, the following discussions are restricted to 4-step braids without laid-in yarns.

7.4.1   *Unit cell of the preform*

When the specimen cross-sectional area is large, the dominant unit cell configuration can be represented by a parallelepiped (Fig. 7.24) with the size of $P_a \times P_b \times P_c$. The braiding axis is assumed to coincide with the x axis. Obviously, for a 1×1 braid, $P_b = P_c$. A unit cell contains four yarns situated along the diagonals. It is understood that in Fig. 7.24 the yarns are idealized as geometric lines and, thus, they intersect at the center of the unit cell.

The details of the yarn arrangement in a 1×1 braid can be visualized by taking the $123'4'$ cross-section of the unit cell. This is

Fig. 7.24. Unit cell of a 4-step braided preform. (After Yang, Ma and Chou 1986).

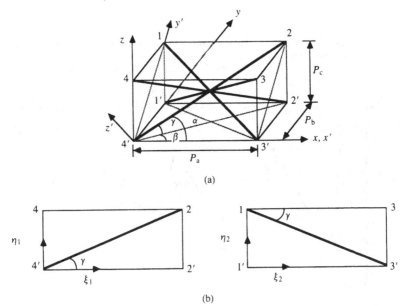

(a)

(b)

shown in Fig. 7.25. Referring to Fig. 7.24, the yarns along the diagonals 13′ and 4′2 are contained in the cross-section, and are at an angle α to the braiding axis. Yarns of type 4′2 are shown in Fig. 7.25 by the inclined sections. Yarns of types 1′3 and 2′4 show elliptical sections. Yarns of type 13′ are also parallel to the cross-section and they are blocked by the other three types of yarns in the cross-sectional view of Fig. 7.25.

It should be noted that Fig. 7.25 is valid for $\alpha < 58°$, which is the critical angle for yarn jamming (Section 7.4.2). The pitch length, P, which is the preform take-up length for one machine cycle (four steps), is defined in Fig. 7.25, along with the braider yarn orientation angle α.

7.4.2 *Criterion for yarn jamming*

The condition for yarn jamming in a 4-step braided preform can be understood from the yarn geometric arrangements. The following assumptions are made in the derivation of the yarn jamming criterion: (1) the braiding yarns are circular in cross-section, with diameter d, (2) the yarns are in a stable configuration, namely, each yarn in Fig. 7.24 is in contact with the other three, and (3) the braid is of the 1×1 type.

Fig. 7.25. Fiber configuration in the cross-section 4′3′21 of Fig. 7.24. (After G. W. Du, private communication, 1990.)

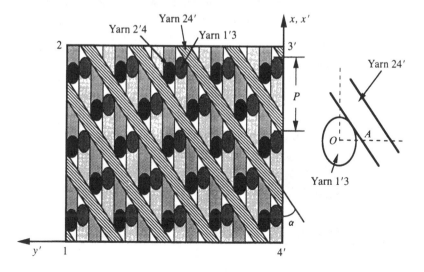

Figure 7.25 shows the yarn configuration of the cross-sectional plane 123'4' before yarn jamming. The pitch length and fiber diameter are denoted by P and d, respectively. From the relation of tangency between an ellipsoid (yarn 1'3) and a line (yarn 24'), the distance OA can be obtained:

$$OA = \frac{d}{2} \sqrt{(1 + \sec^2 \alpha)} \tag{7.28}$$

Another geometric relation for this yarn configuration is

$$P \tan \alpha = 4d = d(1 + \sec \alpha + \sqrt{(1 + \sec^2 \alpha)}) \tag{7.29}$$

which yields the braider yarn angle and aspect pitch length (P/d):

$$\alpha = 41.4°; \qquad P/d = 4.54 \tag{7.30}$$

As compaction of preform continues, the yarn configuration finally reaches a limiting state where both the yarns 2'4 and 1'3 touch the yarn 24'. Figure 7.26 shows the yarn configuration at jamming. The geometric relation for this case is:

$$P \tan \alpha = 4d = d(\sec \alpha + \sqrt{(1 + \sec^2 \alpha)}) \tag{7.31}$$

Thus, the criteria for yarn jamming are

$$\alpha = 57.8°; \qquad P/d = 2.52 \tag{7.32}$$

and the conditions $\alpha > 57.8°$ and $P/d < 2.5$ are physically not feasible. It is interesting to note that jamming in 4-step braids occurs at a unique yarn orientation angle, which is independent of the yarn material and processing parameters.

Fig. 7.26. Yarn configuration at jamming (After G. W. Du, private communication, 1990.)

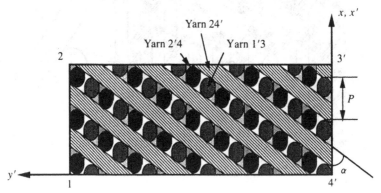

The consequence of such a characteristic of 4-step braiding is the absence of a processing window for providing the flexibility of manufacturing. However, such a window can be created by using two or more types of braiding yarns and inserting laid-in yarns, thus expanding the ranges of fiber geometric and material parameters.

7.5 Analysis of thermoelastic behavior of composites

The analysis of thermoelastic behavior of three-dimensional fabric composites can be made based upon the knowledge of the microstructure of the preforms. For the preforms reviewed in Section 7.2, their unit cell structures are sufficiently well established.

For braided composites, the unit cells of both 2-step and 4-step braids are well known, whereas the unit cells for solid braids depend on the specific preform designs. In the case of weaving, the unit cell of an angle-interlock woven may occupy the entire preform thickness. This is true for the preforms shown in Fig. 7.7 where there are no repeating units in the thickness direction. The unit cell structures of orthogonal wovens and stitch-bonded preforms are similar. Because the thickness direction yarns in both cases are normal to the free surfaces, they can be considered as limiting cases of the angle-interlock configuration.

The knitting yarns in a multi-axial wrap knit are severely curved. Because the knitting yarns usually have fine dimension and low stiffness, their contributions to the composite thermoelastic properties are perhaps negligible. When high stiffness knitting yarns are used, their contributions to the thickness direction properties need to be taken into account. Because of the low volume fraction of knitting yarns, relative to that of the in-plane fibers, it is not unreasonable to neglect the in-plane behavior of the knitting yarns in the composite.

Unlike the case of unidirectional laminates, the thermoelastic behavior of three-dimensional composites is complicated by the fiber configuration in the thickness direction. In the following, three different analytical approaches are outlined. Among them, the energy approach considers the elastic strain energies due to the interaction of yarns at an interlock. The fiber inclination model is based upon the lamination analogy, whereas the macro-cell approach utilizes stiffness tensor transformation and an averaging technique.

Besides these three modeling techniques, a micro-cell approach has been adopted by Whitney (1988) and Whitney and Chou (1989) for analyzing angle-interlock woven composites. This technique is also based upon the lamination analogy. In view of the large geometric variability of angle-interlock wovens, their elastic behavior perhaps can be more efficiently analyzed by the macro-cell approach.

7.5.1   *Elastic strain-energy approach*

An elastic strain-energy approach has been adopted by Ma, Yang and Chou (1986) to derive the elastic stiffness of three-dimensional textile structural composites. Although their analysis is for a 4-step braided composite, the methodology has general applicability.

In the general case, the unit cell structure of Fig. 7.24 can be considered as composed of three sets of mutually orthogonal yarns as well as yarns assuming the diagonal positions. The unit cell is centered on an 'interlock' of these yarns. The analytical model then then focuses on the interaction of the yarns at the center of the unit cell.

The following assumptions are made in the analysis: (1) The baseline and diagonal yarns are regarded as 'composite rods' after being impregnated with matrix materials. The stiffness and strength of the composite are mainly derived from the three-dimensional composite rod structure. (2) The composite rods are homogeneous and linearly elastic, and have uniform circular cross-sections that do not flatten under external loading. (3) The composite rods possess tensile, compressive, and bending rigidities. (4) A jamming force exists at the region of contact between two interlocking composite rods. The rods can be treated as either compressible or incompressible under the action of jamming forces.

Because of the complexity of the yarn configurations at their interlocking positions, the model does not simulate each individual 'lock' separately. The interactions among the yarns are dealt with in approximate fashion by projecting the yarn positions onto a set of mutually orthogonal planes. Within each two-dimensional projection, the interactions between two yarns are taken into account.

Consider, for instance, the interaction of two baseline composite rods (Fig. 7.27). Three types of elastic strain energies in the composite rods are taken into account. These include the strain energies due to bending, extension and compression over the region of fiber contact. Based upon the knowledge of the elastic strain energy of the baseline and diagonal composite rods, the elastic properties of the composite can be obtained through the application of an energy principle.

7.5.2 Fiber inclination model

The fiber inclination model developed by Yang, Ma and Chou (1986) can be understood also by considering the yarn arrangements in a 4-step braided preform. Consider again the unit cell structure based upon the yarns oriented along the four body diagonals in a 4-step braided fabric (Fig. 7.24). The three-dimensional composite can thus be regarded as an assemblage of unit cells as shown in Fig. 7.28(a), where the emphasis is placed on the yarn orientation rather than the interaction among yarns. Here only one set of diagonal yarns in the composite is shown for clarity. The zig-zagging yarn segments are not confined to one layer only. Each yarn in the composite extends through the whole length of the material and changes its orientation at the interlocks. Furthermore, straight laid-in yarns along the edges of the unit cell can be added in the present formulation.

The methodology for the analysis of the fiber inclination model is based upon a modification of the classical laminated plate theory. The following geometrical characteristics are assumed by Yang, Ma and Chou (1986): (1) All the yarn segments parallel to a diagonal direction in the layer *ABCD* (Fig. 7.28a), for instance, are treated as forming an inclined lamina (Fig. 7.28b) after matrix impregnation. (2) Fibers within a lamina are considered to be straight and

Fig. 7.27. Yarn interaction at the point of interlock.

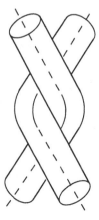

unidirectional. Fiber interlocking and bending due to the change of orientation from one diagonal direction to another at the corners of the unit cell are not taken into account. (3) A unit cell in Fig. 7.24(a) can be further considered as an assemblage of four inclined unidirectional laminae. The intersections among the four inclined laminae are ignored. Each unidirectional lamina is characterized by a unique fiber orientation and all the laminae have the same thickness. Furthermore, the fiber volume fraction of each lamina is assumed to be the same as that of the composite.

The laminate approximation of the unit cell structure is shown schematically in Fig. 7.29. The geometrical configuration and stacking sequence of the inclined laminae composed of yarns in four diagonal directions in the unit cell are given below. First, the $\xi_1 - \zeta_1 - \eta_1$ and $\xi_2 - \zeta_2 - \eta_2$ coordinate systems are assigned to

Fig. 7.28. (a) The idealized zig-zagging yarn arrangement in the braided preform. (b) Schematic view of the inclined laminae representing the diagonal yarns of the 'fiber inclination model'. (After Yang, Ma and Chou 1986.)

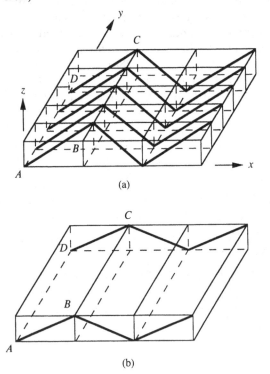

(a)

(b)

lamina 4'2'24 and lamina 1'3'31, respectively, as shown in Fig.
7.24(b). Referring to Figs. 7.24 and 7.29, the equations describing
the height of each lower surface of laminae 1 and 3 and the height
of each upper surface of laminae 2 and 4 measured from the base
plane ($z = 0$) of the unit cell are:

$$\text{lamina 1 (yarn 4'2): } H_1(\xi_1) = \frac{P_c\xi_1}{L} \qquad (0 \le \xi_1 \le L)$$

(7.33)

$$\text{lamina 2 (yarn 1'3): } H_2(\xi_2) = \frac{P_c\xi_2}{L} \qquad (0 \le \xi_2 \le L)$$

(7.34)

$$\text{lamina 3 (yarn 42'): } H_3(\xi_1) = P_c\left(1 - \frac{\xi_1}{L}\right) \qquad (0 \le \xi_1 \le L)$$

(7.35)

$$\text{lamina 4 (yarn 13'): } H_4(\xi_2) = P_c\left(1 - \frac{\xi_2}{L}\right) \qquad (0 \le \xi_2 \le L)$$

(7.36)

where $L = \sqrt{(P_a^2 + P_b^2)}$.

The yarn orientation angles α, β and γ in Fig. 7.24 are defined as

$$\alpha = \tan^{-1}\sqrt{\left(\frac{P_b^2 + P_c^2}{P_a}\right)}$$

$$\beta = \tan^{-1}\frac{P_b}{P_a} \qquad\qquad (7.37)$$

$$\gamma = \tan^{-1}\frac{P_c}{L}$$

With the above geometrical relations and assumptions, the
three-dimensional braided composite of Fig. 7.24 can be modeled
based upon the classical lamination theory. The approach is
essentially an extension of the fiber crimp model of Section 6.5.
Thus, the constitutive equations of a laminated plate follow Eq.
(6.1). The stiffness constants, Q_{ij}, are given by Eq. (6.10).

Since the undirectional yarns in each of the four laminae of Fig.
7.29 are at an angle, γ, with respect to the ξ-direction (see Fig.
7.24b), the effective elastic properties of lamina 1 in the $\xi-\zeta$ plane,

for example, are given by Eq. (6.74) as

$$E_{\xi\xi}(\gamma) = \left(\frac{\cos^4\gamma}{E_{11}} + \left(\frac{1}{G_{12}} - \frac{2\nu_{21}}{E_{22}}\right)\cos^2\gamma\sin^2\gamma + \frac{\sin^4\gamma}{E_{22}}\right)^{-1}$$

$$E_{\zeta\zeta}(\gamma) = E_{22} = E_{33}$$

$$G_{\xi\zeta}(\gamma) = \left(\frac{\cos^2\gamma}{G_{12}} + \frac{\sin^2\gamma}{G_{23}}\right)^{-1} \tag{7.38}$$

$$\nu_{\zeta\xi}(\gamma) = \nu_{21}\cos^2\gamma + \nu_{32}\sin^2\gamma$$

The transverse isotropy in the plane perpendicular to the yarn direction has been taken into account. Then the stiffness matrix, $Q_{ij}(\gamma)$, similar to Eq. (6.25) can be written in terms of $E_{\xi\xi}$, $E_{\zeta\zeta}$, $G_{\xi\xi}$, $\nu_{\zeta\xi}$, and $D_\gamma = 1 - \nu_{\zeta\xi}^2(\gamma)E_{\xi\xi}(\gamma)/E_{\zeta\zeta}$.

For lamina 1, the yarn segments in an inclined lamina also form an off-axis angle, β, with respect to the braiding direction (x axis). Thus, the effective laminar elastic properties in the x direction are

Fig. 7.29. Four unidirectional laminae representing the inclined yarns. (After Yang, Ma and Chou 1986.)

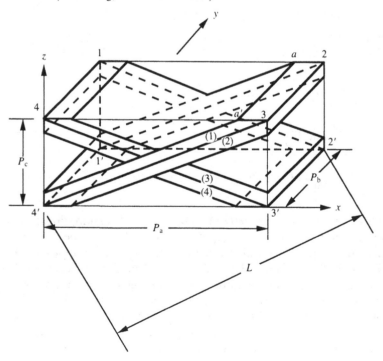

further reduced, and the stiffness constants of the laminae become

$$
\bar{Q}_{ij}(\beta, \gamma) = \begin{bmatrix} \bar{Q}_{11} & \bar{Q}_{12} & \bar{Q}_{16} \\ \bar{Q}_{12} & \bar{Q}_{22} & \bar{Q}_{26} \\ \bar{Q}_{16} & \bar{Q}_{26} & \bar{Q}_{66} \end{bmatrix} \tag{7.39}
$$

where

$$
\bar{Q}_{11} = \left[\frac{E_{\xi\xi}(\gamma)}{D_\gamma} \right] \cos^4 \beta + 2 \left[\frac{E_{\xi\xi}(\gamma) v_{\zeta\xi}(\gamma)}{D_\gamma} + 2G_{\xi\zeta}(\gamma) \right]
$$
$$
\times \cos^2 \beta \sin^2 \beta + \left[\frac{E_{\zeta\zeta}}{D_\gamma} \right] \sin^4 \beta
$$

$$
\bar{Q}_{12} = \left[\frac{E_{\xi\xi}(\gamma)}{D_\gamma} + \frac{E_{\zeta\zeta}}{D_\gamma} - 4G_{\xi\zeta}(\gamma) \right] \cos^2 \beta \sin^2 \beta
$$
$$
+ \left[\frac{E_{\xi\xi}(\gamma) v_{\zeta\xi}(\gamma)}{D_\gamma} \right] [\cos^4 \beta + \sin^4 \beta]
$$

$$
\bar{Q}_{16} = \left[\frac{E_{\xi\xi}(\gamma)}{D_\gamma} + \frac{E_{\xi\xi}(\gamma) v_{\zeta\xi}(\gamma)}{D_\gamma} - 2G_{\xi\zeta}(\gamma) \right] \cos^3 \beta \sin \beta
$$
$$
+ \left[\frac{E_{\xi\xi}(\gamma) v_{\zeta\xi}(\gamma)}{D_\gamma} - \frac{E_{\zeta\zeta}}{D_\gamma} + 2G_{\xi\zeta}(\gamma) \right] \cos \beta \sin^3 \beta
$$

$$
\bar{Q}_{22} = \left[\frac{E_{\xi\xi}(\gamma)}{D_\gamma} \right] \sin^4 \beta + 2 \left[\frac{E_{\xi\xi}(\gamma) v_{\zeta\xi}(\gamma)}{D_\gamma} + 2G_{\xi\zeta}(\gamma) \right] \tag{7.40}
$$
$$
\times \cos^2 \beta \sin^2 \beta + \left[\frac{E_{\zeta\zeta}}{D_\gamma} \right] \cos^4 \beta
$$

$$
\bar{Q}_{26} = \left[\frac{E_{\xi\xi}(\gamma)}{D_\gamma} - \frac{E_{\xi\xi}(\gamma) v_{\zeta\xi}(\gamma)}{D_\gamma} - 2G_{\xi\zeta}(\gamma) \right] \cos \beta \sin^3 \beta
$$
$$
+ \left[\frac{E_{\xi\xi}(\gamma) v_{\zeta\xi}(\gamma)}{D_\gamma} - \frac{E_{\zeta\zeta}}{D_\gamma} + 2G_{\xi\zeta}(\gamma) \right] \cos^3 \beta \sin \beta
$$

$$
\bar{Q}_{66} = \left[\frac{E_{\xi\xi}(\gamma)}{D_\gamma} + \frac{E_{\zeta\zeta}}{D_\gamma} - 2 \frac{E_{\xi\xi}(\gamma) v_{\zeta\xi}(\gamma)}{D_\gamma} - 2G_{\xi\zeta}(\gamma) \right]
$$
$$
\times \cos^2 \beta \sin^2 \beta + G_{\xi\zeta}(\gamma)[\cos^4 \beta + \sin^4 \beta]
$$

Knowing the effective laminae properties with respect to the x–y coordinate system, the local plate stiffness matrices $A_{ij}(x)$, $B_{ij}(x)$ and $D_{ij}(x)$ can be calculated from the lamination theory:

$$
[(A_{ij}(x), B_{ij}(x), D_{ij}(x)] = \sum_{m=1}^{n} \int_{h_{m-1}}^{h_m} \bar{Q}_{ij}(\beta, \gamma)[1, z, z^2]\, \mathrm{d}z
$$
$$
\tag{7.41}
$$

The integration is performed through the thickness of the unit cell of Fig. 7.29. By neglecting the contribution of the pure matrix region, the extensional stiffness matrix $A_{ij}(x)$, for instance, can be evaluated as follows:

$$A_{ij}(x) = \int_{H_1(\xi_1)}^{H_1(\xi_1)+h'} \bar{Q}_{ij}^{(1)}(\beta, \gamma) \, dz + \int_{H_2(\xi_2)-h'}^{H_2(\xi_2)} \bar{Q}_{ij}^{(2)}(\beta, \gamma) \, dz$$
$$+ \int_{H_3(\xi_1)}^{H_3(\xi_1)+h'} \bar{Q}_{ij}^{(3)}(\beta, \gamma) \, dz + \int_{H_4(\xi_2)-h'}^{H_4(\xi_2)} \bar{Q}_{ij}^{(4)}(\beta, \gamma) \, dz$$

$$(7.42)$$

where the superscripts (1), (2), (3) and (4) correspond to the laminae in Fig. 7.29. Also, $h' = h/\cos\gamma$, where h is the thickness of a lamina. It should be noted that the signs of the angles β and γ of the laminae 2, 3 and 4 depend on the fiber orientations. In order to avoid over-estimation of the composite properties, the portions of the laminae which lie outside of the unit cell (such as the region above $aa'32$ in Fig. 7.29) have been excluded from the integration in Eq. (7.41). The lamina thickness is so determined that the total cross-sectional area of laminae (1), (2), (3) and (4) in the x–z plane is equal to that of the unit cell.

The inversion of the local stiffness matrices $A_{ij}(x)$, $B_{ij}(x)$ and $D_{ij}(x)$ of Eq. (7.41) yields the local laminate compliance matrices $A_{ij}'(x)$, $B_{ij}'(x)$ and $D_{ij}'(x)$. The average in-plane compliances of the unit cell under a uniformly applied in-plane stress resultant are

$$\bar{A}_{ij}' = \frac{1}{P_a} \int_0^{P_a} A_{ij}'(x) \, dx$$
$$\bar{B}_{ij}' = \frac{1}{P_a} \int_0^{P_a} B_{ij}'(x) \, dx \qquad\qquad (7.43)$$
$$\bar{D}_{ij}' = \frac{1}{P_a} \int_0^{P_a} D_{ij}'(x) \, dx$$

Then, the averaged stiffness matrices $\bar{A}_{ij}$, $\bar{B}_{ij}$, and $\bar{D}_{ij}$ for the unit cell can be obtained by the inversion of $\bar{A}_{ij}'$, $\bar{B}_{ij}'$, and $\bar{D}_{ij}'$. Finally, effective laminate engineering constants E_{xx}, E_{yy}, v_{xy} and G_{xy} can be expressed as functions of the stiffness constants $\bar{A}_{ij}$ and the unit cell thickness.

Figures 7.30 and 7.31 show the comparisons of theoretical calculations for the axial Young's modulus and Poisson's ratio with experimental data for three-dimensional braided carbon/epoxy composites (Yang, Ma and Chou 1986). The basic material properties are $E_f = 234.4$ GPa and $v_f = 0.22$ for Celion 12K carbon fiber, and $E_m = 3.4$ GPa and $v_m = 0.34$ for epoxy matrix. The average yarn angle in the braided preform is denoted by α.

As the braiding angle becomes smaller, the performance of the inclined laminae approaches that of the unidirectional laminae. The interchange of the stacking sequence of the four inclined laminae in the unit cell does not affect the effective in-plane properties. The

Fig. 7.30. Predicted axial elastic moduli of three-dimensional braided composites as functions of fiber volume fraction, V_f, and fiber orientation angle, α. ●, ▲, ■ and ×: experimental data. (After Yang, Ma and Chou, 1986.)

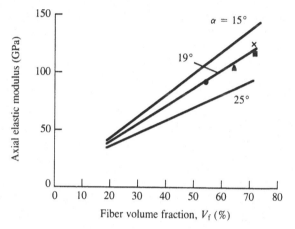

Fig. 7.31. Predicted Poisson's ratios of three-dimensional braided composites as functions of fiber volume fraction, V_f, and fiber orientation angle, α. ●, ■, and ×: experimental data. (After Yang, Ma and Chou, 1986.)

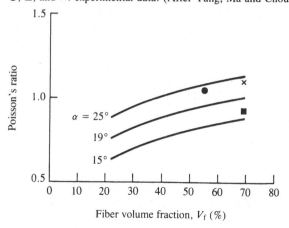

Poisson's ratio of three-dimensional composites based upon 4-step braiding is considerably higher than that of a unidirectional composite with the same fiber volume fraction. The Poisson's contraction can be minimized by introducing laid-in yarns in the axial direction.

7.5.3   *Macro-cell approach*

The approach of the macro-cell model is different from that of the unit cell. Instead of considering the smallest repeating unit in a preform, the macro-cell is established for the entire cross-section of the specimen. It takes into account the arrangements of the yarns around the edges of the specimen. However, the most distinct advantage of this approach perhaps is its capability of dealing with specimens of 'thick' cross-sections, since the elastic properties are derived from tensor transformations. In order to apply such a model, it is necessary to have detailed knowledge of the fiber geometric configurations.

In the following, the macro-cell model (Byun, Du and Chou 1991) is applied to the analysis of elastic properties of 2-step braided fabric composites. The treatment is excerpted from Byun, Whitney, Du and Chou (1991).

7.5.3.1 *Geometric relations*

Consider the 2-step braided composite depicted in Fig. 7.17. The variation in the braider yarn orientation along its length is taken into account by introducing the average yarn orientation angle. The average is identified by considering one braider yarn which travels through the length of the macro-cell. For an m-column by n-layer braided preform, a braider yarn travels $m + (n + 1)/2$ pitch lengths before it repeats its spatial position. Thus, for the preform of Fig. 7.17 the repeating length is ten pitch lengths. Figure 7.32 shows the schematic view of the braider yarn location and

Fig. 7.32. Schematic view of a braider yarn extending through ten pitch lengths. The numbers indicate the braider yarn carrier locations in Fig. 7.3. (After Byun *et al.* 1991.)

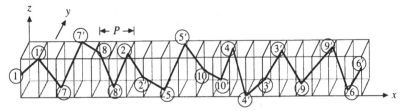

orientation; the numbers indicate the positions of braider yarn carriers in Fig. 7.3.

In order to identify the reinforcing direction of braider yarns, the yarn segments generated due to the carrier movement from position 1 to 7' in Fig. 7.32 are projected onto the $y-z$ (Fig. 7.33) and $z-x$ (Fig. 7.34) planes. Thus, all the yarn segments are identified according to their directions with respect to the $x-y-z$ coordinate. L_{bi}, L_{by} and L_{bz} denote the total projected length of braider yarns which are inclined to the x, y and z axes, parallel to the $x-y$ plane, and parallel to the $z-x$ plane, respectively. Then, from Fig. 7.17 and the parameters defined in Eqs. (7.12)–(7.14),

$$L_{bi} = 4(m-1)(n-1)\left(S_a + \frac{S_b f_b}{\sin 2\,\theta}\right) \tag{7.44}$$

$$L_{by} = 2[(n-1)S_m + 2(m-1)S_a \cos \theta] \tag{7.45}$$

$$L_{bz} = 2[2mS_n + (n-1)S_a \sin \theta] \tag{7.46}$$

where m and n denote the column and layer numbers, respectively.

Fig. 7.33 Projections of the yarn segment 11′77′ of Fig. 7.32 onto the $y-z$ plane. (After Byun *et al.* 1991.)

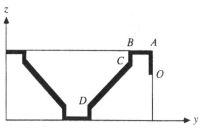

Fig. 7.34. Projections of the yarn segment 11′77′ of Fig. 7.32 onto the $x-z$ plane. (After Byun *et al.* 1991.)

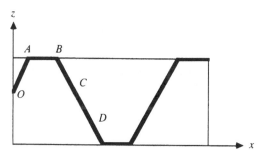

The average angle between the braider yarn and the braid axis is given by

$$\bar{\alpha} = \tan^{-1}\left(\frac{2L_t}{(2m+n+1)P}\right) \tag{7.47}$$

where $L_t (= L_{bi} + L_{by} + L_{bz})$ is the projected length of the braider yarn onto the y–z plane. Based upon the average braider yarn orientation angle, the total length of braider yarn (L_b) is approximated as

$$L_b = \frac{L_t}{\sin \bar{\alpha}} \tag{7.48}$$

The lengths of braider yarns inclined to the xyz axes and parallel to the x–y and z–x planes can be obtained in a similar manner. Thus the volumes of braider yarns of these three orientations are given by

$$V_{bi} = S_b^2 f_b(L_{bi}/\sin \bar{\alpha})$$
$$V_{by} = S_b^2 f_b(L_{by}/\sin \bar{\alpha}) \tag{7.49}$$
$$V_{bz} = S_b^2 f_b(L_{bz}/\sin \bar{\alpha})$$

The axial yarns have three different cross-sections as shown in Fig. 7.17. The total volume of the axial yarns is

$$V_a = h[(m-1)(n-1)S_a^2 \sin(2\theta) + 4(m-1)S_a S_n \cos\theta$$
$$+ 2(n-1)S_a S_m \sin\theta + 4S_m S_n] \tag{7.50}$$

Using Eqs. (7.15) and (7.16), the total macro-cell volume is

$$V_t = wt \tag{7.51}$$

The fiber volume fractions of braider yarns of different orientations can therefore be obtained from Eqs. (7.49)–(7.51); they are used for evaluating the volume average of the stiffness constants.

7.5.3.2 Elastic constants

For the purpose of predicting the composite elastic properties, the yarns are treated as unidirectional composite rods. The direction cosines between the reference coordinate system, xyz, and the 123 coordinate system associated with the unidirectional composite can be established by setting the 2 axis perpendicular to the z axis (Fig. 7.35):

$$
\begin{array}{lll}
l_{1x} = \cos\beta \cos\gamma & l_{2x} = -\sin\beta & l_{3x} = -\cos\beta \sin\gamma \\
l_{1y} = \sin\beta \cos\gamma & l_{2y} = \cos\beta & l_{3y} = -\sin\beta \sin\gamma \\
l_{1z} = \sin\gamma & l_{2z} = 0 & l_{3z} = \cos\gamma
\end{array} \tag{7.52}
$$

Considering the average angle of braider yarn $\bar{\alpha}$ instead of α in Fig. 7.35, the angles β and γ can be expressed in terms of $\bar{\alpha}$ and the aspect ratio of axial yarns, f_a, as

$$\beta = \tan^{-1}[\cos\theta\tan\bar{\alpha}] = \tan^{-1}\left[\frac{\tan\bar{\alpha}}{\sqrt{(1+f_a^2)}}\right] \tag{7.53}$$

$$\gamma = \tan^{-1}\left[\frac{\sin\theta}{\sqrt{(\cot^2\bar{\alpha}+\cos^2\theta)}}\right] = \tan^{-1}\left[\frac{f_a\sin\bar{\alpha}}{\sqrt{(1+f_a^2\cos^2\bar{\alpha})}}\right] \tag{7.54}$$

Using these direction cosines, the compliance matrix (S) of the unidirectional composite (OO$'$ in Fig. 7.35) referring to the 123 coordinate system can be transformed to that referring to the xyz coordinate system:

$$S'_{ijmn} = l_{pi}l_{qj}l_{rm}l_{sn}S_{pqrs} \qquad (i, j, m, n, p, q, r, s = 1, 2, 3) \tag{7.55}$$

From symmetry conditions and using contracted notation, Eq. (7.55) is reduced to a simple form:

$$S'_{ij} = q_{mi}q_{nj}S_{mn} \qquad (i, j, m, n = 1\text{--}6) \tag{7.56}$$

where q_{ij} denotes the element belonging to the ith row and jth column of the transformation matrix (see Lekhnitskii 1963). For a unidirectional composite with transverse isotropy, the compliance matrix has five independent constants.

In order to determine the effective stiffness matrix of the composite, the compliance matrix is inverted and then averaged

Fig. 7.35. Orientation of the braider yarn (OO$'$). (After Byun *et al.* 1991.)

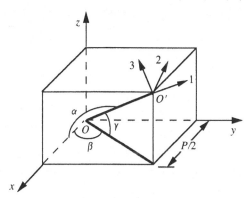

over the macro-cell volume. The average should include all four yarn orientations in Fig. 7.33, namely the axial yarn ($\beta = 0$, $\gamma = 0$), braider yarn (BA) parallel to the χ–y plane ($\beta = \bar{\alpha}$, $\gamma = 0$), braider yarn (BC) parallel to the z–x plane ($\beta = 0$, $\gamma = \bar{\alpha}$) and inclined braider yarn (CD) in the interior of the macro-cell ($\beta = f(\bar{\alpha}, f_a)$, $\gamma = g(\bar{\alpha}, f_a)$). Thus, the effective stiffness of the composite C_{ij}^c is

$$C_{ij}^c = \sum_{n=1}^{4} (C_{ij})_n \frac{V_n}{V_t} \qquad (7.57)$$

where $(C_{ij})_n$ and V_n/V_t are, respectively, the stiffness matrix and volume fraction of the unidirectional composite for an individual reinforcing direction. Finally, the stiffness matrix of the composite is inverted to obtain the compliance matrix S_{ij}^c. The engineering elastic constants are then obtained from the compliance matrix. For example, $E_{xx} = 1/S_{11}^c$, $E_{yy} = 1/S_{22}^c$ and $v_{xy} = -S_{12}^c/S_{22}^c$, etc.

Experimental measurements of the elastic properties of 2-step braided composites have been reported by Byun *et al.* (1991). The composites are the same as for Braid II given in Table 7.1. Experimental observations of specimen cross-section confirm the yarn shapes assumed in the analysis. Based upon the input data of Table 7.1, the macro-cell model predicts the following composite geometric parameters: thickness (t) = 7.1 mm, width (w) = 15.3 mm, average braider angle ($\bar{\alpha}$) = 71.1° and fiber volume fraction (V_f) = 73.2%. Table 7.2 shows the comparison of elastic properties based upon the macro-cell predictions and experiments.

Table 7.2. *Comparisons of composite elastic properties from the macro-cell model predictions and experiments. After Byun et al. (1991).*

Elastic constants	Macro-cell model	Experiment
E_{xx} (GPa)	48.4	52.4 (5*)
E_{yy} (GPa)	7.83	
E_{zz} (GPa)	7.95	
G_{xy} (Gpa)	2.58	1.45 (4*)
G_{yz} (Gpa)	2.68	
G_{xz} (GPa)	2.59	
v_{xy}	0.33	0.53 (3*)
v_{yz}	0.35	
v_{xz}	0.36	

* Number of tests.

The analytical predictions deviate significantly from experimental results for the in-plane shear modulus and Poisson's ratio. Some reasons of uncertainty in the measurements of fabric composite elastic properties are discussed in Section 7.7.

7.6   Structure–performance maps of composites

Considerable effort has been devoted by researchers to evaluate the effectiveness of various reinforcement concepts. However, the analyses and experiments performed on advanced composites are usually reported for individual systems; it is thus difficult to acquire a more comprehensive view. Chou and Yang (1986) and Chou (1989), motivated by the concept of deformation mechanism maps of Ashby, Gandi and Taplin (1979), Gandi and Ashby (1979) and Frost and Ashby (1982) as well as the work of Dow (1984) have integrated the results of studies in the modeling of thermoelastic behavior of unidirectional laminated composites, as well as two-dimensional (2-D) and three-dimensional (3-D) textile structural composites. Through the construction of structure–performance maps, the relative effectiveness and uniqueness of various reinforcement concepts can be assessed. These maps provide guidance in material selection for structural design, and in identifying the needs of future work.

In order to assess the capability of various reinforcement configurations with different fibers and matrix combinations, Chou and Yang conducted parametric studies of the structure–performance relationship. The geometric parameters considered include fiber orientation in unidirectional laminated constructions, weaving parameters in two-dimensional fabrics, and braiding parameters in three-dimensional constructions. The material parameters are fiber and matrix thermoelastic properties. Four types of reinforcement forms are presented below: laminated angle-plies based upon unidirectional layers with the off-axis angle (θ) ranging from 0° to 90°; [0°/90°] cross-plies; two-dimensional woven fabrics with n_g ranging from 2 (plain weave) to 8 (eight-harness satin); and 2-step and 4-step braided composites. For 2-step braids, yarn linear density, pitch length and aspect ratio are allowed to change. The braiding angle between a fiber segment and braiding axis in the case of 4-step braids varies from 15° to 35°. The analytical tools employed in the construction of these maps include the lamination theory for cross-ply and angle-ply laminates, the crimp and bridging models (Chapter 6) for two-dimensional fabrics, the fiber inclination model for 4-step braided composites, and the macro-cell model for 2-step braided composites.

Several maps are presented here to illustrate the correlation between reinforcement configurations and the thermoelastic behavior of composites. The fiber volume fractions of the composites are assumed to be 73% for 2-step braided composites and 60% for all other composites. Figures 7.36–7.39 present the thermoelastic behavior of carbon, Kevlar and glass reinforced epoxy composites. Figure 7.40 shows the variation of thermal expansion coefficients for PEEK matrix composites. Figure 7.41 gives the elastic properties of C, SiC and Al_2O_3 fiber reinforced Mg matrix composites. Figure 7.42 demonstrates the elastic properties of glass matrix composites reinforced with C, SiC and Al_2O_3 fibers. The three-dimensional

Fig. 7.36. E_{xx} vs. E_{yy} for carbon/epoxy (■ unidirectional angle-ply; △ two-dimensional woven; + three-dimensional braided), Kevlar/epoxy (▲ unidirectional angle-ply; □ two-dimensional woven; × three-dimensional braided), and glass/epoxy (● unidirectional angle-ply; ○ two-dimensional woven; ▼ three-dimensional braided) composites. (p) = plain weave; (s) = eight-harness satin. (After Chou and Yang 1986.)

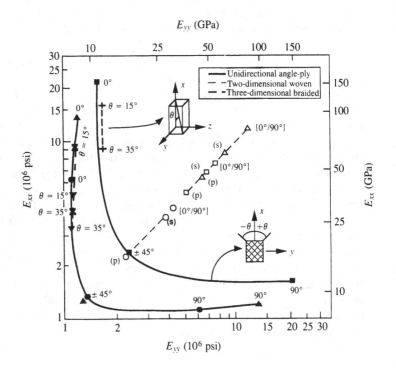

preforms discussed in Figs. 7.36–7.42 are based upon 4-step braiding.

Figures 7.43 and 7.44 show the variations of elastic properties of 2-step braided composites of Kevlar/epoxy with fabric geometric and processing parameters. The linear density ratio of axial and braider yarns, the pitch length of braider yarns, and the aspect ratios of axial and braider yarns are considered. The structure–performance maps are constructed by starting with a set of values of these parameters, and then varying each parameter independently while keeping the other parameters at their original values. The ranges of these values are denoted on the curves in Figs. 7.43 and

Fig. 7.37. E_{xx} vs G_{xy} for carbon/epoxy (■ unidirectional angle-ply; △ two-dimensional woven; + three-dimensional braided), Kevlar/epoxy (▲ unidirectional angle-ply; □ two-dimensional woven; × three-dimensional braided), and glass/epoxy (● unidirectional angle-ply; ○ two-dimensional woven; ▼ three-dimensional braided) composites. (p) = plain weave; (s) = eight-harness satin. (After Chou and Yang 1986.)

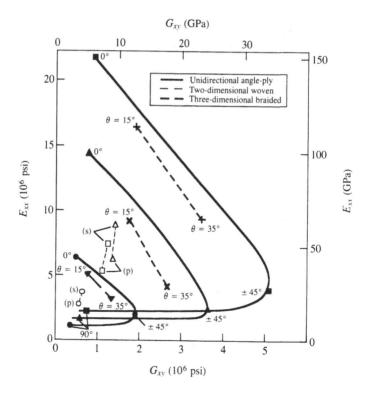

7.44 where the calculations are performed at equal intervals as indicated by the symbols on each curve (Byun *et al.* 1991).

It is noted from Figs. 7.43 and 7.44 that the Young's moduli and shear moduli are insensitive to the axial yarn aspect ratios between 1 and 3. The maximum volume fraction of axial yarns is achieved when the yarn aspect ratio is around unity, which also gives the maximum E_{xx} and G_{yz}. Increases in the linear density ratio of axial yarn to braider yarn result in an increase in axial yarn volume fractions while the braider yarn volume fraction becomes smaller.

Since the stiffness increases of the 2-step braided composite in the longitudinal and transverse directions are primarily due to the contribution of the axial yarns and braider yarns, respectively, the increase of the axial yarn volume fraction improves E_{xx}. In the meantime, E_{yy} and G_{yz} become smaller due to the reduction in braider yarn volume fraction. Furthermore, the increase in the aspect ratio of the braider yarns results in an increase in their thickness, which in turn gives a larger volume of matrix pockets in the composite. Since the total fiber volume fraction is reduced due to

Fig. 7.38. E_{xx} vs. v_{xy} for carbon/epoxy (■ unidirectional angle-ply, △ two-dimensional woven, + three-dimensional braided) composites. (After Chou and Yang 1986.)

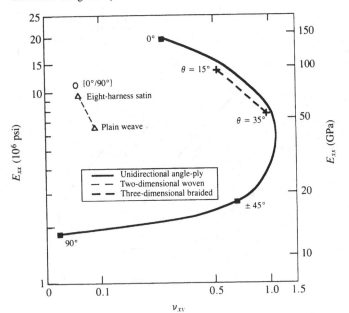

the increase in volume of the composite, all the components of Young's moduli and shear moduli in Figs. 7.43 and 7.44 are reduced as the braider yarn aspect ratio becomes bigger. Finally, longer braider yarn pitch length gives a small volume fraction and orientation angle of braider yarns. Consequently, E_{yy} and G_{yz} are reduced as the braider yarn pitch length increases.

Chou and Yang (1986) have compared the unique thermal and elastic characteristics among various reinforcement configurations. In general, the in-plane thermoelastic properties of unidirectional lamina depend strongly on the fiber orientation. The unidirectional reinforcement provides the highest elastic stiffness along the fiber direction. The Young's moduli of off-axis unidirectional laminae are lower than that of the unidirectional lamina.

The [0°/90°] cross-ply yields identical thermoelastic properties in 0° and 90° orientations. Their in-plane shear rigidity is poor. The longitudinal Young's modulus of an angle-ply laminate is lower than

Fig. 7.39. α_{xx} vs. α_{yy} for carbon/epoxy (■ unidirectional angle-ply; △ two-dimensional woven; + three-dimensional braided), Kevlar/epoxy (▲ unidirectional angle-ply; □ two-dimensional woven; × three-dimensional braided), and glass/epoxy (● unidirectional angle-ply; ○ two-dimensional woven; ▼ three-dimensional braided) composites. (p) = plain weave; (s) eight-harness satin. (After Chou and Yang 1986.)

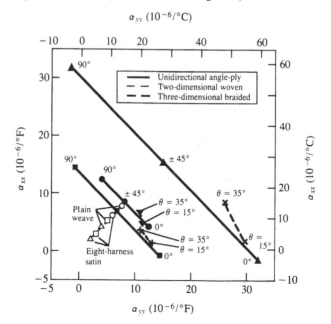

that of a unidirectional lamina. But better transverse elastic property and in-plane shear resistance can be achieved through the stacking of the unidirectional laminae with different fiber orientations. For ±45° angle-ply, the in-plane stiffness drops to a minimum, while the shear modulus reaches its maximum.

The two-dimensional biaxial woven fabric composites can provide balanced in-plane thermoelastic properties within a single ply. They behave similar to [0°/90°] cross-plies, although the fiber waviness tends to reduce the in-plane efficiency of the reinforcements. As the fabric construction changes from plain weave to eight-harness satin, the frequency of crimp due to fiber cross-over is reduced, and the fabric structure approaches that of [0°/90°] cross-plies.

The thermoelastic properties of braided composites also show a strong dependence on fiber orientation. Three-dimensionally braided composites have demonstrated good in-plane properties, which are comparable to those of unidirectional angle-plies with the

Fig. 7.40. α_{xx} vs. α_{yy} for carbon/PEEK (■ unidirectional angle-ply; △ two-dimensional woven; + three-dimensional braided), Kevlar/PEEK (▲ unidirectional angle-ply; □ two-dimensional woven; × three-dimensional braided), and glass/PEEK (● unidirectional angle-ply; ○ two-dimensional woven; ▼ three-dimensional braided) composites. (After Chou and Yang 1986.)

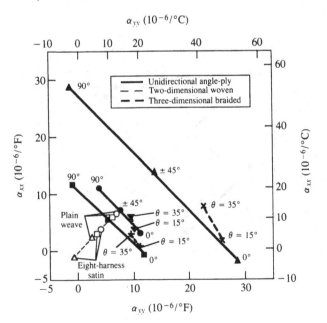

same range of fiber orientation. The longitudinal Young's moduli and in-plane shear rigidities of three-dimensional braided composites with braiding angles ranging from 15° to 35° are better than those of two-dimensional woven fabric composites. But the transverse Young's moduli are lower and the major Poisson's ratios higher than those of two-dimensional woven fabric composites. However, three-dimensional braided composites are unique in providing both stiffness and shear rigidity along the thickness direction. Also, because of the integrated nature of the fiber arrangement, there are no interlaminar surfaces in three-dimensional composites.

Furthermore, a comparison of their elastic behavior indicates that the performance of 2-step braided composites is much more versatile than that of 4-step braided composites. The presence

Fig. 7.41. E_{xx} vs. E_{yy} carbon/magnesium (● unidirectional angle-ply; △ two-dimensional woven; ▼ three-dimensional braided), SiC/magnesium (▲ unidirectional angle-ply; ○ two-dimensional woven; × three-dimensional braided), and Al_2O_3/magnesium (■ unidirectional angle-ply; □ two-dimensional woven; + three-dimensional braided) composites. (After Chou and Yang 1986.)

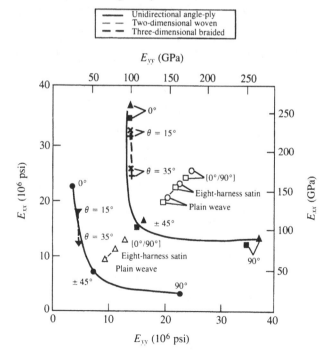

of both axial and braider yarns in 2-step braids allows much flexibility in the design of the preform microstructure. However, such flexibility can be achieved in 4-step braided composites if laid-in axial yarns are used.

The structure–performance maps can form the basis for material selection and component design; these findings can be easily extended to generate a wider range of information. Take the woven fabric composite as an example; although the properties shown in the maps are primarily along the filling and warp directions, the off-axis properties can be readily obtained through proper tensor transformation. Upon knowing these properties, it would be feas-

Fig. 7.42. E_{xx} vs. G_{xy} for carbon/borosilicate glass (● unidirectional angle-ply; △ two-dimensional woven; ▼ three-dimensional braided), SiC/borosilicate glass (■ unidirectional angle-ply; ○ two-dimensional woven; + three-dimensional braided), and Al$_2$O$_3$/borosilicate glass (▲ unidirectional angle-ply; □ two-dimensional woven; × three-dimensional braided) composites and three-dimensional carbon/carbon composites (◆). (After Chou and Yang 1986.)

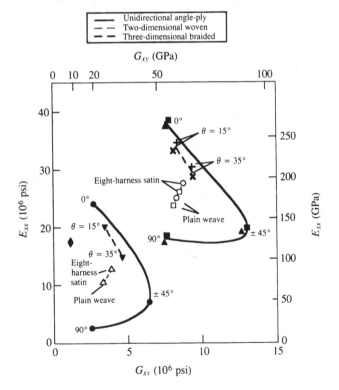

Fig. 7.43. The variations of E_{xx} and E_{yy} with material and processing parameters (intervals of the parameters, axial yarn aspect ratio: 1, pitch length:2, linear density ratio: 0.05, braider yarn aspect ratio: 0.02). (After Byun *et al.* 1991.)

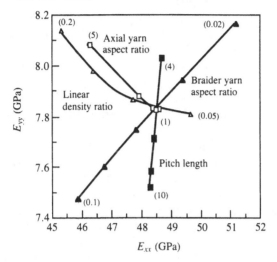

Fig. 7.44. The variations of G_{xy} and G_{yz} with material and processing parameters (intervals of the parameters, axial yarn aspect ratio: 1, pitch length: 2, linear density ratio: 0.05, braider yarn aspect ratio: 0.02). (After Byun *et al.* 1991.)

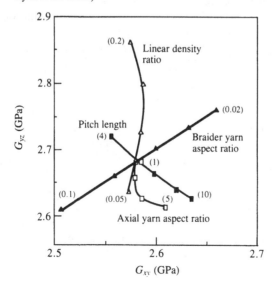

ible to tailor composite structures with various combinations of reinforcement forms, or with different material combinations such as hybrid unidirectional laminate, hybrid woven fabric structures, or hybrid laid-in three-dimensional structures.

It should be noted that the analytical modeling techniques employed in the construction of the structure–performance maps assume 'defect-free' composites, i.e. perfect fiber/matrix interfacial bonding, perfect fiber alignment, void-free matrix materials, etc. It is understood that in actual composites defects are frequently introduced in the fabrication and handling process. Limited studies in this regard have been made, including the effect of fabrication induced fiber distortion on the thermoelastic properties of two-dimensional fabric composites (Yang and Chou 1989), the effects of fiber/matrix interfacial debonding on the effective elastic properties (Takahashi and Chou 1986), void content of as-fabricated polymeric matrix composites (Yoshida, Ogasa and Hayashi 1986), cracking of polymeric matrices in fabric composites (Ishikawa and Chou 1982), and the effect of fiber bundle size and distribution on the behavior of three-dimensional braided Al_2O_3/Al–Li composites (Majidi, Yang and Chou 1986). It is expected that with the advancement in mathematical modeling and experimental techniques, performance maps for strength and failure of various two- and three-dimensional composites can also be constructed.

7.7   Mechanical properties of composites

This section summarizes the strength, fracture, and damage tolerance behavior of three-dimensional fabric composites. The material systems cited here include both polymer and metal based composites.

7.7.1   *Tensile and compressive behavior*

Majidi, Yang, Pipes and Chou (1985) examined the tensile and compressive behavior of 4-step braided composites of alumina fiber in an aluminum–lithium matrix. The continuous, polycrystalline α-alumina yarn (Fiber FP manufactured by the Du Pont Co.) contains 210 filaments of approximately 20 μm diameter. The properties of Fiber FP are: tensile strength = 1380 MPa, tensile modulus = 345–79 GPa, elongation to failure = 0.4%, density = 3.90 g/cm^3, and melting point = 2045°C (Dhingra, Champion and Krueger 1975). The aluminum matrix is alloyed with 2–3 wt% lithium for an enhanced chemical bond between the fiber and the matrix.

Figure 7.45 depicts the tensile stress–strain curves of FP/Al–Li composites of unidirectional laminates and three-dimensional braided composites at $V_f = 17\%$ and 36%; the tensile behavior of the pure Al–Li matrix is also given. A bilinear behavior is observed for all composites; the 'knees' on the stress–strain curves occur at about 0.02%. Yielding of the matrix appears to be responsible for the bilinearity. Since the bilinear behavior has also been observed in the unidirectional composites, it is believed to be a material property rather than an effect caused by the braided structure. The *in situ* strength of the matrix may well be higher than that measured for the bulk material. Such a phenomenon has been discussed by Kelly and Macmillan (1966). When the fiber spacing is very small ($\leq 10\,\mu$m), as is the case in the material studied by Majidi and colleagues, the yield stress of the matrix is controlled by the Orowan stress and it is higher than that of the bulk matrix. The yield stress and work hardening increase with decreasing spacing between the fibers. While the yield stress goes up, the strain at which the matrix starts yielding in the composite drops for very small fiber spacing (Kies 1962).

The ultimate tensile strengths of three-dimensional braided composites with a braiding angle of about 20° are 189 MPa and 383 MPa for $V_f = 17\%$ and 36%, respectively. These values are slightly lower than those predicted for $\pm 20\%$ angle-ply laminates of

Fig. 7.45. Axial tensile stress–strain responses of (a) unidirectional FP/Al–Li composite ($V_f = 0.50$), (b) three-dimensional FP/Al–Li composite ($V_f = 0.36$), and (c) the unreinforced matrix. (After Majidi and Chou 1987.)

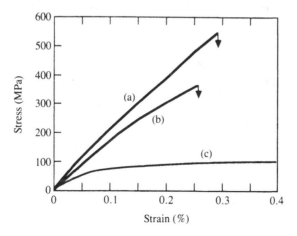

the same fiber volume fraction. Microscopic examination of the fracture surfaces reveals brittle fracture of fibers and considerable deformation in the matrix between the fibers.

The measured initial Young's moduli are 97 GPa and 171 GPa for $V_f = 17\%$ and 36%, respectively, which agree well with theoretical predictions based upon the fiber inclination model (Yang, Ma and Chou 1986). The secondary Young's moduli can be approximated by assuming that the contribution of the matrix to the composite modulus is negligible after yielding. The measured Poisson's ratios are 0.30 ($V_f = 17\%$) and 0.27 ($V_f = 36\%$). It should be noted that a certain degree of damage to brittle fibers often occurs in the braiding process. Thus, the *in situ* fiber stiffness and strength properties need to be estimated from, for instance, those measured on unidirectional composites.

Figure 7.46 shows the compressive stress–strain behavior of the same material as in Fig. 7.45(b). The curve demonstrates an initial linear region up to a strain of about 0.15% followed by nonlinear behavior. Other compressive properties include a failure strain of about 1.8% and a major Poisson's ratio of 0.3. Kinking appears to be the primary mode of failure in compression.

In transverse tension of three-dimensional braided FP/Al–Li composites the stress–strain curve is highly nonlinear. The onset of nonlinearity is at a strain of about 0.5%. The ultimate strength is

Fig. 7.46. Compressive stress–strain responses of FP/Al–Li composite. (After Majidi *et al.* 1985.)

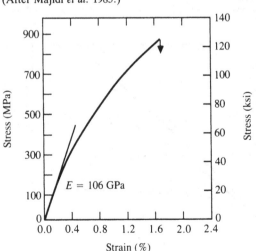

considerably lower than that for the axial specimen and, thus, significantly smaller than that of the unreinforced matrix material. Transverse cracks initiate within the matrix-rich regions between fiber bundles, most likely at microscopic voids in the matrix.

7.7.2 Shear behavior

Majidi *et al.* (1985) have examined the shear behavior of FP/Al–Li composites with a 4-step braided preform. Both intralaminar (in-plane) and interlaminar shear measurements are made. It has been shown that the short beam shear test (*ASTM Standards*, D2344-84 1987) causes premature failure by a flexural mode on the tensile surface of the specimen even for a fairly small specimen span-to-depth ratio. It is due to the high ratio of interlaminar shear strength to tensile strength in three-dimensional braided composites. The two-rail shear test (*ASTM Standards*, D4255-83 1987) also proved inadequate. The tests suitable for the measurement of the shear strength of three-dimensional composites are the Iosipescu shear test originally proposed by Iosipescu (1967) for isotropic materials and applied to composite laminates by Walrath and Adams (1983a&b), and the double-notch shear test.

The in-plane shear strength parallel to the braiding direction measured with the Iosipescu shear test for $V_f = 17\%$ is 139.6 MPa, which is comparable to the theoretical shear strength of 151 MPa for [±20°] angle-ply laminates of the same material and fiber volume fraction. Majidi, Rémond and Chou (1987) reported the shear properties of three-dimensional braided FP/Al–Li composite tubes, with the tube axis parallel to the braiding axis and braiding angles of approximately ±20°. The in-plane shear strengths measured from torsion tests are 141.8 MPa for $V_f = 17\%$ and 102.1 MPa for $V_f = 36\%$. The shear moduli are 36.6 GPa and 39.0 GPa for $V_f = 17\%$ and 36%, respectively.

The interlaminar shear strength measured by the double-notched shear test is 144.5 MPa. The calculated interlaminar shear strength for 0° unidirectional laminates is 100.8 MPa. This improved interlaminar shear property in the thickness direction of three-dimensional braided composites gives much improved fracture and impact resistance over the conventional laminates. Some dfficulties in the testing of three-dimensional textile structural composites exist. Machining of specimens should be avoided as much as possible because it destroys the integrated nature of the fiber preform and thus results in lower strength. It is also difficult to obtain meaningful readings of the shear strain and, hence, shear modulus from the Iosipescu specimens. This is due to the relatively small size of the region of pure shear, the highly non-homogeneous three-dimensional braided structure based upon large bundles and the small size of the strain gages used. All these factors indicate the difficulties in the testing of textile structural composites in general.

7.7.3   Fracture behavior

7.7.3.1 In-plane fracture

The fracture and toughness characteristics of unidirectional and three-dimensional braided FP/Al–Li composites have been examined by Majidi, Yang and Chou (1986, 1988). Metal matrix composites, particularly those incorporating ceramic fibers, offer very high strength and stiffness, but often significantly lower fracture toughness than unreinforced metallic matrices. The reduced toughness is due to the restriction of plastic deformation in the presence of the stiff fibers and a strong fiber/matrix bond which eliminates or restricts fiber debonding and pullout.

Majidi and colleagues have measured fracture toughness using compact tension tests on the basis of the linear elastic fracture mechanics approach and the notched three-point bend test (Tattersal and Tappin 1966), which involves measurement of the work of fracture (fracture surface energy averaged over the whole fracture process).

Figures 7.47(a) and (b) show the load versus crack opening displacement (COD) curves for repeated loading and unloading of a three-dimensional braided composite and a unidirectional laminate under compact tension tests. The stress intensity factors are calculated from P_Q and the maximum load P_{max} indicated in Fig. 7.47(a). P_Q is determined by drawing a straight line with a slope 5% less than the slope of the linear part of the load–COD curve and finding the corresponding load at the intersection of this line with the curve. For Figs. 7.47(a) and (b), the crack propagation is perpendicular to the braiding axis of the three-dimensional braided composite and the fiber direction of the unidirectional laminate, respectively. In both cases, the first part of the load–COD curve is highly non-linear and, therefore, P_Q is considerably lower than P_{max}. For subsequent loading cycles, however, the sharpened crack removes the non-linearity in the curve and P_Q approaches P_{max}. Then, the stress intensity factors K_Q and K_{max} can be calculated from P_Q and P_{max}, respectively (*Annual Book of ASTM Standards,* E399 1978). Since the K_{max} values are reasonably constant within the range of crack length to specimen width ratio, Majidi, Yang and Chou (1986) adopted the average K_{max} as the critical stress intensity factor, K_c. Rémond (1987) has characterized the fracture behavior of three-dimensional braided metal matrix composites with $V_f = 36\%$. For the notch perpendicular to the principal reinforcement axis, the unidirectional laminate ($K_{max} = 30.7\,\mathrm{MPa}\sqrt{m}$) appeared tougher than the three-dimensional braided composite ($K_{max} = 27.3\,\mathrm{MPa}\sqrt{m}$). For the notch parallel to the principal reinforcement axis, the three-dimensional braided composite ($K_{max} = 21.5\,\mathrm{MPa}\sqrt{m}$) is tougher than the unidirectional laminate ($K_{max} = 19.0\,\mathrm{MPa}\sqrt{m}$). The [±20°] angle-ply laminate appears less tough than the two other composites, the longitudinal toughness being $K_{max} = 24.6\,\mathrm{MPa}\sqrt{m}$ and the transverse toughness $K_{max} = 16.4\,\mathrm{MPa}\sqrt{m}$.

Electron microscopy investigations of the fracture surfaces indicate virtually no pull-out of the individual fibers. However, occasionally the whole fiber bundle has been pulled out over a small length of 1–2 mm. This is accompanied by some debonding between the fiber bundle and the surrounding matrix. The mechanism of crack propagation perpendicular to the braiding axis is believed to be the fracture of fibers and eventual fracture of the fiber bundle ahead of the crack tip. The above behavior differs greatly from that of the unidirectional FP/Al–Li composite, which shows a rapid

Fig. 7.47. Load–crack opening displacement (COD) curves obtained from compact tension tests on (a) three-dimensional braided FP/Al–Li composite ($V_f = 17\%$), and (b) unidirectional FP/Al–Li composite ($V_f = 34\%$). (After Majidi, Yang and Chou 1986.)

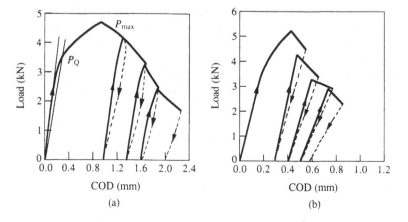

crack propagation from the outset and the fracture surface has a flat, brittle appearance with no macroscopic dimples.

In the case of notched three-point bend tests, unidirectional composites fracture in a much more brittle and less controlled manner than braided composites, and the load–deflection curve shows the sharp drop of load. The work of fracture, γ_f, which is measured from the total energy absorbed for the complete fracture of the specimen, or the area under the load–deflection curve, are $7.92 \pm 1.27 \, \text{kJ/m}^2$ for the braided composites and $4.56 \pm 0.44 \, \text{kJ/m}^2$ for unidirectional laminate for $V_f = 17\%$.

The difference in the strength and fracture behavior between textile structural composites and traditional laminated composites has been further demonstrated for the FP/Al–Li composites by Majidi, Yang and Chou (1986). In the case of unidirectional composites, the contributions from fiber debonding and pull-out are negligible since the critical load transfer length, l_c, is only $0.34 \, \text{mm}$. This is calculated from the equation $l_c = \sigma_f d/(2\tau)$ where σ_f is the fiber ultimate strength (1380 MPa), d is the fiber diameter (20 μm), and τ is assumed to be equal to the shear yield strength of the matrix (~40 MPa). The strong fiber/matrix interface also reduces the length on either side of the broken fiber over which the matrix deforms plastically (see Cooper and Kelly 1967). For unidirectional composites with $V_f = 34\%$ and matrix tensile strength of 160 MPa, this length is only $0.04 \, \text{mm}$. Therefore, plastic deformation is severely restricted in the unidirectional system, and the lack of fiber pull-out and the limited plastic deformation in the matrix are responsible for the planar fracture and the low γ_f.

In three-dimensional braided composites, fibers are not uniformly distributed in the matrix as in the unidirectional laminae, and each fiber bundle can be regarded as an individual reinforcement. The volume fraction of fibers within the bundle is approximately 50%, and the volume fraction of bundles in the composite is approximately 40%. Using the diameter of 2 mm and tensile strength of 586 MPa for the bundles, it is found that $l_c = 12.6 \, \text{mm}$ and the length over which the matrix deforms plastically is 4 mm. Although factors such as fiber inclination and interactions among bundles are not considered above, these values illustrate the beneficial effect of fiber clustering on the extent of matrix plastic deformations and on the pull-out and debonding mechanisms. The non-homogeneous microstructure, therefore, appears to be at least partially responsible for the higher work of fracture of the three-dimensional braided composites as compared with the unidirectional laminate which shows a catastrophic planar fracture.

Majidi, Yang and Chou (1986) have also examined the effect of thermal treatment on the fiber/matrix interface strength and, hence, the fracture toughness of three-dimensional braided composites. The fiber/matrix interface deteriorates after isothermal heating at 500°C. There is a decrease in the fracture load and an increase in the amount of bundle pull-out which results in larger work of fracture ($\gamma_f = 20.30 \pm 14.35 \, \text{kJ/m}^2$ after 72 hours of thermal treatment). This reflects the weak nature of the interfacial reaction zone which grows intergranularly towards the center of the fiber.

Guénon, Chou and Gillespie (1989) reported the in-plane fracture toughness, K_{Ic}, of carbon/epoxy composites with a three-dimensional orthogonal interlock fabric preform. The K_{Ic} values for three-dimensional fabric composites are 28.56 MPa$\sqrt{m}$ and 29.45 MPa$\sqrt{m}$ in two principal material directions, which are higher than that of laminates (21.22 MPa$\sqrt{m}$). The through-the-thickness yarns are beneficial to the in-plane toughness by arresting and deviating the crack. The interaction between a crack and inhomogeneities, simulating fiber arrays, has been examined by Fowser and Chou (1989, 1990a&b).

7.7.3.2 *Interlaminar fracture*

Traditional laminated composites exhibit low interlaminar fracture toughness and are susceptible to delamination when subjected to interlaminar stress concentrations. Improvements in damage tolerance to date have focussed on utilizing tougher matrices (Hunston 1984) or interleafing concepts (Masters 1987). Through-the-thickness reinforcement provides an alternative approach to substantially increasing the resistance to delamination (see Whitney, Browning and Hoogsteden 1982; Guess and Reedy 1985; Mignery, Tan and Sun 1985; Dexter and Funk 1986; Fowser 1986; Guénon, Chou and Gillespie 1987; Ogo 1987).

The orthogonal interlock fabric architecture (Fig. 7.10) retains the in-plane performance while enhancing out-of-plane properties, by including a small amount of through-the-thickness reinforcement. The 'z direction' fibers are also known to be detrimental to the in-plane tensile and compressive properties. The interlocking process avoids the cutting of fibers, as it occurs in the stitching process. However, it creates matrix pockets that reduce the volume fraction of the in-plane fibers relative to the analogous two-dimensional laminates. The z direction fibers also tend to be deformed in the processing of the fabric composites.

Guénon, Chou and Gillespie (1989) have studied the effect of fiber geometry on the interlaminar and in-plane fracture behavior of orthogonal interlocked fabric composites. The experimental work is based upon a T300/3501-6 carbon/epoxy system. Referring to Fig. 7.10, the fabric preform can be described as a [0°/90°] laminate in which some through-the-thickness yarns are interlaced. The in-plane yarns contain 6000 filaments per yarn and the through-the-thickness yarns have 1000 filaments per yarn. The spacing between two z direction yarns in both plate directions is 2.8 mm and the plate contains about 13 z direction yarns/cm². The total number of plies is 27, with 14 and 13 plies in two mutually orthogonal directions. The overall volume fraction is 50%, while the volume fraction of the z direction is 1%.

(A)   Mode I interlaminar fracture

Guénon, Chou and Gillespie (1989) have adopted two test methods for Mode I interlaminar fracture, the double cantilever beam (DCB) test and the 'tabbed DCB', which uses long aluminum tabs bonded along both sides of the specimen to prevent the deviation of crack propagation from a self-similar manner. Both types of specimens are pin-loaded in tension in displacement-controlled mode.

The load–deflection curves for the three-dimensional composite specimens show a nonlinear unloading sequence and an appreciable permanent deformation after unloading. The crack tip did not completely close after unloading. These features can be explained by the crack closure process of the three-dimensional fabric composite. Most of the z direction yarns do not break in the plane of the crack. Instead, they fracture near the outer surface of the specimen where they are curved by the weaving process, and then debonded and pulled out. Figure 7.48 shows the fracture surface with the z direction yarn protruding out of the plane of fracture. During unloading, the pulled-out yarns do not resume their initial locations and therefore progressively undergo compressive stresses that lead to a nonlinear unloading behavior and a permanent deflection of the specimen after a zero load is reached.

Two data reduction methods have been adopted by Guénon, Chou and Gillespie (1989) for the three-dimensional fabric composites. These are the area method, based upon energy considerations, and the compliance method, based upon the linear elastic beam theory. The interlaminar critical strain energy release rate, G_{Ic}, values from the area method are 0.307 kJ/m² (two-dimensional regular DCB), 0.286 kJ/m² (two-dimensional tabbed DCB) and 3.85 kJ/m² (three-dimensional tabbed DCB). The compliance method gives G_{Ic} values of $0.235 \, \text{kJ/m}^2$ (two-dimensional regular DCB), $0.179 \, \text{kJ/m}^2$ (two-dimensional tabbed DCB) and $2.66 \, \text{kJ/m}^2$ (three-dimensional tabbed DCB). The compliance method only takes into account the energy of crack initiation. In the case of two-dimensional unidirectional laminates, the crack propagation energy is generally equal to the initiation energy and therefore both methods give similar results. In three-dimensional fabric composites, the fracture, debonding and pull-out of z direction yarns as well as the bridging of the crack by the z direction yarns dissipate energy. Therefore, the area method gives a higher and more accurate result of interlaminar fracture toughness.

The mode I delamination problem of three-dimensional orthogonal interlock fabric composites has been further examined by Byun, Gillespie and Chou (1990b) using a finite element analysis. The material systems and specimen geometries including two-dimensional regular DCB, two-dimensional tabbed DCB and three-dimensional tabbed DCB (see Guénon, Chou and Gillespie 1989) are simulated in this numerical work. Specifically, the mode I fracture behavior of carbon/epoxy composites is examined for

Fig. 7.48. Fracture surface of the orthogonal interlock fabric composite showing a pulled-out yarn. (After Guénon, Chou and Gillespie 1989.)

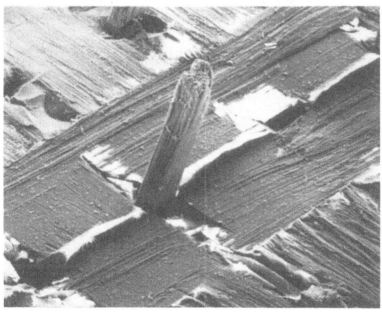

various initial crack lengths. The strain energy release rates, G_I, are evaluated based upon the crack closure method (Rybicki and Kanninen 1977) to ascertain the influence of through-the-thickness fibers on crack driving force. Byun and colleagues also have considered the effect of progressive debonding of the z axis yarns on the strain-energy release rate.

In the finite element model, the length of an element side is the crack increment utilized in the study ($\Delta a = 0.5562$ mm). The initial crack length is 25.4 mm and the locations of the through-the-thickness fibers are $3\Delta a$, $8\Delta a$, $13\Delta a$ and $18\Delta a$. A vertical unit displacement of 1 mm is applied to simulate the displacement controlled loading conditions used in the experimental work of Guénon and colleagues. Three types of through-the-thickness fiber debonding are modeled: perfect bonding, moderate bonding, where the z axis fiber is debonded over 25% of the specimen thickness, and complete debonding, where the load is carried by the fibers only. The z axis yarns are assumed to be initially perfectly bonded; partial or complete debonding does not occur until the crack front passes the reinforcement. Also, fiber fracture is not considered by Byun and colleagues.

Figure 7.49 demonstrates the effect of through-the-thickness yarns and bonding conditions on the strain-energy release rate. The strain-energy release rate for the two-dimensional laminate monotonically decreases with increasing crack length as one would expect

Fig. 7.49. Numerical strain-energy release rates of two-dimensional laminated and three-dimensional orthogonal woven composites as functions of crack length: — two-dimensional laminate; ○ fully debonded; + moderately bonded; ● perfectly bonded. (After Byun, Gillespie and Chou 1990.)

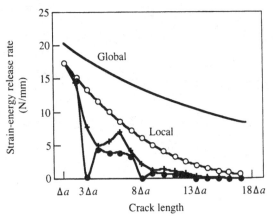

under fixed grip conditions. The introduction of through-the-
thickness yarns reduces the local strain-energy release rate significantly in the case of perfect bonding condition as the crack
approaches the first array of z axis yarns at $3\Delta a$. The crack driving
force for interlaminar crack growth decreases because the load is
transferred to the z axis yarns. The crack opening displacement is
also reduced by the presence of the vertical fibers. As debonding
occurs, the decrease of strain-energy release rate is less significant
because the crack tip opening displacement increases.

In the numerical analysis, Byun and colleagues assume that the
crack continues to propagate without fiber fracture. Consider the
perfect bonding case of Fig. 7.49 when the crack propagates beyond
the first array of yarns to $4\Delta a$. The strain-energy release rate is
observed to increase slightly but is still significantly less than the
two-dimensional specimen. This is due to the increase in deformation of the z axis yarns as load transfer occurs, which results in an
increase in the crack tip opening displacement. Due to the presence
of the z axis yarns at $8\Delta a$, the strain-energy release rate begins to
diminish as the crack tip approaches the next site of through-the-
thickness fibers where a second reduction in crack driving force is
observed. The process continues until the strain-energy release rate
is identically zero.

Figure 7.50 shows the tensile stress in the through-the-thickness
yarn as a function of crack length. As the crack approaches the first
yarn at $3\Delta a$, the load in the fiber increases rapidly. Additional crack

Fig. 7.50. Tensile stresses in the z axis fiber arrays as functions of crack
length for perfect fiber–matrix bonding: — first array; ○ second array; ●
third array (After Byun, Gillespie and Chou 1990.)

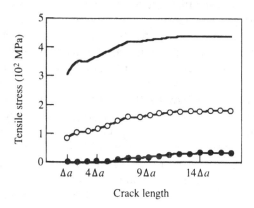

Crack length

growth results in an asymptotic value of load in the first yarn as the second yarn begins to carry load. The process continues until the stress in the next array of yarns is zero. At this point, the applied load is being carried exclusively by multiple z axis yarns bridging the crack surface and the crack driving force is identically zero. The information presented in Figs. 7.49 and 7.50 demonstrates the load bearing and transferring mechanisms in the interlaminar fracture process of a three-dimensional fabric composite. This information enabled Byun and colleagues to predict the macroscopic critical load for mode I interlaminar fracture. More importantly, understanding of the load redistribution at the microscopic level is beneficial to the design of fabric preform structure for enhanced damage tolerance.

(B)   Mode II interlaminar fracture

The mode II interlaminar fracture toughness of three-dimensional orthogonal fabric composite has been assessed by Liu and Chou (1989). The mode II fracture toughness tests are performed on both three-dimensional composites and two-dimensional laminates of the same carbon/epoxy system using end-notch-flexural (ENF) specimens.

Byun, Gillespie and Chou (1989), following the work of Liu and Chou (1989), have conducted a finite element analysis for evaluating the mode II strain-energy release rate. Similar to the mode I interlaminar fracture, the crack driving force for mode II interlaminar crack growth decreases as the crack approaches the z axis yarns because the load is being transferred to these yarns.

7.7.4   *Impact*

Majidi and Chou (1986) reported the impact behavior of both three-dimensional braided and unidirectional FP/Al–Li composites. The average fiber volume fractions are 17% and 34% for three-dimensional braided and unidirectional composites, respectively. Instrumented drop-weight impact tests have been carried out on un-notched impact panels, which do not require machining and, hence, do not sustain damage to the integrated fiber structure. Figure 7.51 compares the load–deflection traces obtained from through-the-thickness penetration impact tests. The three-dimensional braided composites absorb significantly higher energy and show larger deflection than the unidirectional composite during both damage initiation and propagation stages. By definition, the initiation energy is the area under the load–deflection curve up to

Fig. 7.51. Load–deflection traces obtained from through-the-thickness penetration impact tests. (a) Al–Li alloy; (b) three-dimensional braided FP/Al–Li composite; (c) unidirectional FP/Al–Li composite. (After Majidi and Chou 1986.)

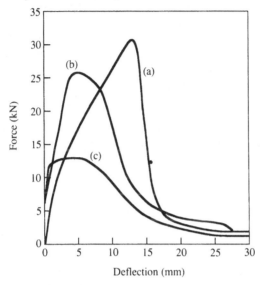

Fig. 7.52. Cross-sectional views of the FP/Al–Li composite specimens impacted at 54 J of incident energy. (a) Three-dimensional braided, $V_f = 0.17$; (b) three-dimensional braided, $V_f = 0.36$; (c) [±20°] angle-ply; (d) unidirectional laminate. (After Majidi and Chou 1987.)

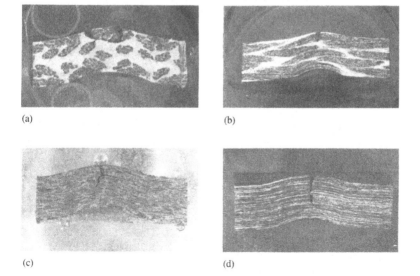

the maximum load. The total impact energy absorbed by the three-dimensional braided composites is close to that of the matrix material. The unidirectional composite fractures like a brittle material with cracks propagating through the entire specimen, while in the braided composite damage is restricted to a small region, and the specimen shows a ductile type of behavior. A comparison of the cross-sectional views of the FP/Al–Li composite specimen impacted at 54 J of incident energy is shown in Fig. 7.52.

Additional information on the mechanical behavior of three-dimensional fabric composites can be found in the work of Kregers and Teters (1982), Ko and Pastore (1985), Crane and Camponeschi (1986), Ko (1986), Yau, Ko and Chou (1986), Simonds, Stinchcomb and Jones (1988), Ko (1989b), and Whitcomb (1989).

8   Flexible composites

8.1   Introduction

The term 'flexible composites' is used hereinafter to identify composites based upon elastomeric polymers of which the usable range of deformation is much larger than those of the conventional thermosetting or thermoplastic polymer-based composites (Chou and Takahashi 1987). The ability of flexible composites to sustain large deformation and fatigue loading and still provide high load-carrying capacity has been mainly analyzed in pneumatic tire and conveyor belt constructions. However, the unique capability of flexible composites is yet to be explored and investigated. This chapter examines the fundamental characteristics of flexible composites.

Besides tires and conveyor belts, flexible composites can be found in a wide range of applications. Coated (with PVC, Teflon, rubber, etc.) fabrics have been used for air- or cable-supported building structures, tents, parachutes, decelerators in high speed airplanes, bullet-proof vests, tarpaulin inflated structures such as boats and escape slides, safety nets, and other inexpensive products. Hoses, flexible diaphragms, racket strings, surgical replacements, geotextiles, and reinforced membrane structures in general are examples of flexible composites.

Following Chou (1989, 1990), the nonlinear elastic behavior of three categories of materials is examined: pneumatic tires, coated fabrics, and flexible composites containing wavy fibers. These materials provide the model systems of analysis with elastic behaviors ranging from small to large deformations.

The performance characteristics of pneumatic tires are primarily controlled by the anisotropic properties of the cord/rubber composite. The low modulus, high elongation rubber contains the air and provides abrasion resistance and road grip. The high modulus, low elongation cords carry most of the loads applied to the tire in service. According to Walter (1978), the first quantitative study of cord/rubber elastic properties in the tire industry was published in Germany by Martin (1939), who analyzed bias ply aircraft tires using thin shell theory to approximate toroidal tire behavior.

Martin's analysis of the orthotropic composite elastic constants assumes that the fibers are inextensible and the matrix stiffness is negligibly small; this approach has been referred to here as the *classical netting analysis*. Studies of the cord/rubber properties became active worldwide in the 1960s as represented by the work of Clark (1963a&b, 1964) in the USA, Gough (1968) in Great Britain, Akasaka (1959–64) in Japan, and Biderman *et al.* (1963) in the Soviet Union.

The existing analysis on tire mechanics is primarily based upon the well developed anisotropic theory of rigid laminated composites for small linear elastic deformation. Thus, the problems of viscoelasticity, strength behavior, fatigue and large non-linear behavior are often ignored.

In the case of coated fabrics, limited attention has been given to the material stress–strain response to arbitrary loading paths and histories. Experimental studies of the biaxial stress–strain behavior can be found in the works of Skelton (1971), Alley and Fairslon (1972) and Reindhardt (1976). Attempts have also been made by Akasaka and Yoshida (1972) and by Stubbs and Thomas (1984) to analytically model the elastic and inelastic properties of coated fabrics under biaxial loading. Some of these results are briefly recapitulated in this chapter.

Section 8.4 focusses on the understanding of the large nonlinear deformation of flexible composites. To this end, model material systems for analytical purposes need to be identified. The large nonlinear deformation could originate from two sources, i.e. matrix and fiber. In order to fully realize the ability of the elastomeric matrix to sustain large deformation, the fibers must be able to deform accordingly with the matrix. This can be achieved by (a) using short fibers, (b) arranging continuous fibers in such an orientation that they are allowed to rotate as the load increases, and (c) using reinforcements in woven, knitted, braided, or other wavy forms.

Possibility (c) is particularly interesting in that it utilizes the waviness of the fibers. The gradual straightening of the wavy fibers under external loading results in enhanced stiffness with an increase in deformation. The linear and nonlinear elastic behavior of two- and three-dimensional textile structural composites has been examined by Ishikawa and Chou (1983), Chou (1985), and Chou and Yang (1986) based on small deformation theory. The nonlinear finite deformation analyses of flexible composites are presented in Chapter 9.

8.2 Cord/rubber composites

Cord/rubber composites for pneumatic tires are examined in this section from the viewpoint of the mechanics of anisotropic materials. Cord/rubber composites are complex elastomeric composites composed of (a) the rubber matrix of usually quite low modulus and high extensibility, (b) the reinforcing cord of much higher modulus and lower extensibility than the matrix, and (c) the adhesive film which bonds the cord to the matrix. The combination is subjected to (a) fluctuating loads, mostly tensile but on occasion compressive, (b) temperatures as high as 125°C, and (c) moisture. Obviously substantial stress develops at the cord-rubber interface. Some of the discussions presented herein on the materials and mechanics aspects of pneumatic tires are based upon the review articles of Walter (1978) and Clark (1980).

The construction of tires involves calendering sheets of rubber around an array of parallel textile cords to form a flat, essentially two-dimensional anisotropic sheet. The cords usually have substantial twist and often are made up of two or three oppositely twisted yarns. These composite sheets are then assembled into various tire configurations. Figure 8.1(a) shows the typical bias or angle-ply design which utilizes two or more, usually an even number, of plies laid at alternate diagonal angles to one another. Figure 8.1(b) depicts a typical radial tire construction involving radially oriented cords while the tread area is reinforced by a belt structure of relatively small angle with respect to the tire center line. The radial tire construction provides stiff longitudinal reinforcement for the tread area (and, hence, is less subject to slip) and flexibility for the vertical deflection. In the terminology of laminated composites, bias

Fig. 8.1. (a) Bias tire. (b) Radial ply tire. (After Clark 1980.)

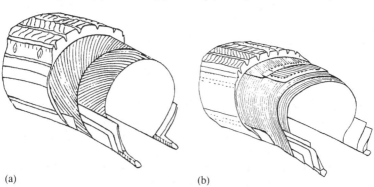

(a)   (b)

and radial tires can be categorized as laminates with $[+\theta/-\theta]$ and $[+\theta/-\theta/90°]$ orientations with respect to the tire center line.

8.2.1   *Rubber and cord properties*

For relatively small strain, rubber may be treated as a homogeneous and isotropic material. The Young's modulus, determined from the initial slope of the stress–strain curve, may be as low as 0.69 MPa (100 psi) for non-reinforced (unfilled) elastomers to as high as 689 MPa (100 000 psi) for highly vulcanized (high sulfur) compounds such as ebonite. The Young's modulus of rubber is affected by the conditions of physical testing (i.e. strain rate, temperature, cyclic load history) and chemical vulcanization parameters (i.e. the compounding ingredients, state of cure) (see Clark 1980).

The assumption of negligible volume change of rubber leads to the following values of Poisson's ratio (v), bulk modulus (K), Young's modulus (E) and shear modulus (G):

$$v \approx \tfrac{1}{2}$$

$$K \to \infty \tag{8.1}$$

$$E \approx 3G$$

Rubbers used in calendered plies of tires have E values of 5.51 MPa (800 psi) for textile body ply, 20.67 MPa (3000 psi) for textile tread ply and 13.78 MPa (2000 psi) for steel tread ply. The v value for these materials is 0.49.

The Young's moduli for tire cords vary with cord constructions. The following values are for belt ply: 109.55 GPa (15.9×10^6 psi) for steel, 24.8 GPa (3.6×10^6 psi) for Kevlar, and 11.02 GPa (1.6×10^6 psi) for rayon. The values for body ply are: 3.96 GPa (575×10^3 psi) for polyester, and 3.45 GPa (500×10^3 psi) for nylon. Experiments have shown that textile cords can carry some load in compression, although compressive loads are believed to be the source of many textile failures and should be avoided whenever possible.

Twisting of the cord is needed in order to provide adequate cord fatigue life under service conditions. However, twisting of fiber into tire cord can result in as much as a one-third decrease in tensile Young's modulus for belt ply cords, and a one-half decrease in Young's modulus for body ply cord. It has been predicted that the axial Young's modulus of a single twisted fiber yarn is approximately equal to $1/(1 + 4\pi^2 R^2 T^2)$ of that of the untwisted yarn. Here, R and T denote yarn radius and twist (number of turns per unit length), respectively (Hearle, Grosberg and Backer 1969). The twisted and multi-plied cords should be considered as transversely isotropic, although they are commonly approximated as isotropic. Textile cords normally show substantial nonlinearity in their stress–strain behavior. However, since the rubber behavior is relatively elastic in the small strain range, and the cords in a laminate are often aligned at an angle to the load direction, the composite acts more like a linearly elastic solid than the cord itself. Figure 8.2 shows the stress–strain curve of a tubular specimen using rayon yarn in a rubber matrix. The fibers in this specimen are in angle-ply arrangement. According to Clark (1980), most pneumatic tires do not operate with strain much in excess of 10%.

8.2.2 Unidirectional composites

The linear elastic behavior of a unidirectional cord/rubber composite can be easily deduced from the basic equations given in Section 2.2. By assuming that

$$E_{\rm f} \gg E_{\rm m}$$
$$v_{\rm m} = 0.5$$

$$(8.2)$$

Fig. 8.2. Load–strain curve of a cylindrical tube with rayon yarns in a rubber matrix. (After Clark 1980.)

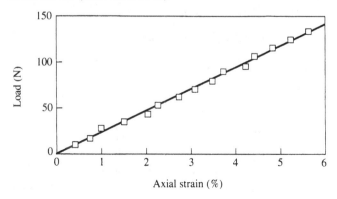

The following relations can be obtained:

$$E_{11} = E_f V_f \gg E_{22}$$

$$\nu_{21} = 0$$

$$E_{22} = c_1 \frac{E_m}{V_m} = \frac{(1 + 1.3 V_f) E_m}{(1 + 0.5 V_f) V_m} \qquad (8.3)$$

$$G_{12} = c_2 \frac{G_m}{V_m} = (1 + V_f) \frac{G_m}{V_m}$$

where c_1 and c_2 denote two coefficients.

Akasaka (1989) considered the same assumptions as Eqs. (8.2) and obtained the simpler form slightly different from Eqs. (8.3), with the coefficients $c_1 = \frac{4}{3}$ and $c_2 = 1$. Then,

$$E_{22} \approx \frac{4}{3} \frac{E_m}{V_m}$$

$$G_{12} \approx \frac{G_m}{V_m} \approx \frac{E_{22}}{4} \qquad (8.4)$$

Akasaka (1989) has noted that the relation of $G_{12} \approx E_{22}/4$ is independent of cord volume fraction and has good predictability as compared to existing formulas and experimental results (Walter and Patel 1979; Clark 1980). Also, $c_1 = c_2 = 1$ has been used by Jones (1975).

Based upon Eqs. (8.4), the variation of lamina transformed reduced stiffness with cord off-axis angle θ follows from Eqs. (2.16) and can be approximated as (Akasaka and Hirano 1972):

$$\bar{Q}_{11} \approx E_{22} + E_{11} \cos^4 \theta$$

$$\bar{Q}_{22} \approx E_{22} + E_{11} \sin^4 \theta$$

$$\bar{Q}_{66} \approx E_{22}/4 + E_{11} \sin^2 \theta \cos^2 \theta$$

$$\bar{Q}_{12} \approx E_{22}/2 + E_{11} \sin^2 \theta \cos^2 \theta \qquad (8.5)$$

$$\bar{Q}_{16} \approx E_{11} \sin \theta \cos^3 \theta$$

$$\bar{Q}_{26} \approx E_{11} \sin^3 \theta \cos \theta$$

When a unidirectional cord/rubber sheet is subjected to simple tension, an interesting deformation behavior occurs, which is not observed in most of the rigid composites. This can be elucidated by using Eq. (2.17) for the relation between γ_{xy} and the applied σ_{xx} as well as the approximations of Eqs. (8.4)

$$\gamma_{xy} = \bar{S}_{16}\sigma_{xx} \approx \frac{-2\sin\theta\cos^3\theta}{E_{22}}(2 - \tan^2\theta)\sigma_{xx} \tag{8.6}$$

Thus, the stretching-shear coupling vanishes at $\theta \approx 54.7°$, and $\gamma_{xy} < 0$ for $\theta < 54.7°$ and $\gamma_{xy} = 0$ for $\theta > 54.7°$.

8.2.3 *Laminated composites*

The constitutive equations for the laminated cord/rubber composites are of the same general form as Eqs. (2.25)–(2.30). However, they can be simplified by using the approximated expressions of Eqs. (8.5) for $\bar{Q}_{ij}$. Also, the engineering elastic constants, referring to the x–y coordinate system, for the angle-ply laminated composite can be deduced. Using the results of Eqs. (8.4), the following expressions of engineering elastic constants of a $\pm\theta$ laminate in terms of the properties of fiber and matrix as well as the fiber volume fraction are obtained under the assumptions of $E_f \gg E_m$ and $v_m = 0.5$ (See Akasaka 1989, and Clark 1963a&b):

$$E_{xx} = E_f V_f \cos^4\theta + \frac{4G_m}{1 - V_f}$$
$$- \frac{[E_f V_f \sin^2\theta\cos^2\theta + 2G_m/(1 - V_f)]^2}{E_f V_f \sin^4\theta + 4G_m/(1 - V_f)}$$

$$E_{yy} = E_f V_f \sin^4\theta + \frac{G_m}{1 - V_f}$$
$$- \frac{[E_f V_f \sin^2\theta\cos^2\theta + 2G_m/(1 - V_f)]^2}{E_f V_f \cos^4\theta + 4G_m/(1 - V_f)}$$

$$G_{xy} = E_f V_f \sin^2\theta\cos^2\theta + \frac{G_m}{1 - V_f} \tag{8.7}$$

$$v_{xy} = \frac{E_f V_f \sin^2\theta\cos^2\theta + 2G_m/(1 - V_f)}{E_f V_f \sin^4\theta + 4G_m/(1 - V_f)}$$

$$v_{yx} = \frac{E_f V_f \sin^2\theta\cos^2\theta + 2G_m/(1 - V_f)}{E_f V_f \cos^4\theta + 4G_m/(1 - V_f)}$$

The approach for obtaining Eqs. (8.7) based upon the assumptions of $\pm\theta$ cord angles and specially orthotropic symmetry is known as the *modified netting analysis*.

The classical netting analysis which assumes inextensible cords $(E_f \rightarrow \infty)$ simplifies Eqs. (8.7)

$$E_{xx} = 4G_m(1 - V_f)(\cot^4 \theta - \cot^2 \theta + 1)$$
$$E_{yy} = E_{xx}(\pi/2 - \theta)$$
$$G_{xy} = E_f V_f \sin^2 \theta \cos^2 \theta + G_m/(1 - V_f) \qquad (8.8)$$
$$v_{xy} = \cot^2 \theta$$
$$v_{yx} = \tan^2 \theta$$

Figures 8.3–8.5 show the results of analytical predictions based upon Eqs. (8.5) for E_{xx}, G_{xy}, and v_{xy}, respectively, as functions of the off-axis angle θ. These results coincide very well with the experimental data, as reported by Clark (1963a&b) and based upon $E_{11} = 1440\,\text{MPa}$ and $E_{22} = 6.9\,\text{MPa}$. It is evident that Poisson's ratios well in excess of one-half exist in cord/rubber composites.

Because of the incompressibility of the rubber matrix and the relatively small volume change associated with the cord materials, due to its high stiffness, it can be assumed that the cord/rubber composite is incompressible. Thus, for small strain, $\varepsilon_{xx} + \varepsilon_{yy} + \varepsilon_{zz} = 0$, and

$$v_{xz} = -\varepsilon_{zz}/\varepsilon_{xx} = 1 + \varepsilon_{yy}/\varepsilon_{xx} = 1 - v_{xy} \qquad (8.9)$$

Figure 8.6 indicates the analytical results of Eq. (8.9) and the experimental data of v_{xz} as a function of θ (Clark 1980) for

Fig. 8.3. Young's modulus, E_{xx}, vs. cord angle, θ, for a two-ply laminate. — Eqs. (8.7); (×) experimental data. (After Clark 1963a.)

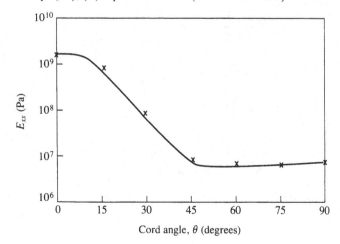

Cord angle, θ (degrees)

$E_{11} = 294$ MPa and $E_{22} = 6.6$ MPa. One of the solid lines is based upon the v_{xy} expression of Eqs. (8.7), and the simplifying expression of Eqs. (8.4), namely

$$v_{xy} = \frac{E_{11} \sin^2 \theta \cos^2 \theta + E_{22}/2}{E_{11} \sin^4 \theta + E_{22}} \tag{8.10}$$

The other solid line is based upon the v_{xy} expression of Eqs. (8.8).

Fig. 8.4. Shear modulus, G_{xy}, vs. cord angle, θ, for a two-ply laminate. — Eqs. (8.7); ($\times$) experimental data. (After Clark 1963a.)

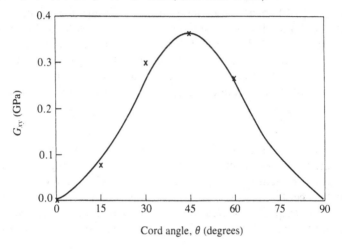

Fig. 8.5. Poisson's ratio, v_{xy}, vs. cord angle, θ, for a two-ply laminate. — Eqs. (8.7); ($\times$) experimental data. (After Clark 1963a.)

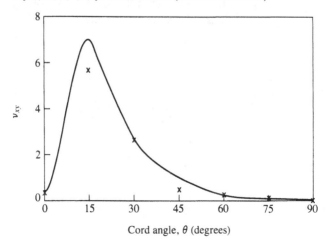

It is interesting to note that for a range of θ values, v_{xz} is negative; the laminate becomes thicker under axial load.

The interlaminar stresses σ_{zz}, τ_{zx} and τ_{zy} are not considered in the classical lamination theory. These stresses and their corresponding strains do exist in appreciable magnitude which promote a reduction in the apparent stiffness of cord/rubber laminates. As a result, the composite becomes more flexible and exhibits lower natural frequencies of vibration and static buckling loads (Walter 1978).

Walter (1978) reviewed the work of Kelsey, who considered a two-ply [$\pm \theta$] cord/rubber laminate, simulating the behavior of the belt in a radial tire. Assuming the belt of finite width in the y direction is loaded in the x (circumferential) direction, γ_{yz} vanishes due to symmetry and ε_{zz} is assumed to be negligibly small. γ_{xz} is maximum at the free edge of the belt and can be approximated, for the case of inextensible cords ($E_f \rightarrow \infty$), by the simple expression

$$\gamma_{xz} = \varepsilon_{xx}(2\cot^2\theta - 1) \tag{8.11}$$

Equation (8.11) indicates that γ_{xz} vanishes when the two plies are oriented at $\theta = \pm\cot^{-1}\sqrt{(1/2)} = \pm 54.7°$. The magnitude of γ_{xz} decays exponentially away from the free edge and vanishes along the belt center-line ($y = 0$). It is interesting to note that $\theta = 54.7°$ is also the angle for which the normal stress and shear strain are uncoupled and each off-axis ply behaves as specially orthotropic.

Fig. 8.6. Poisson's ratio, v_{xz}, vs. cord angle, θ, for a two-ply laminate. — Eqs. (8.8) and (8.10); ($\times$) experimental data. (After Akasaka 1989.)

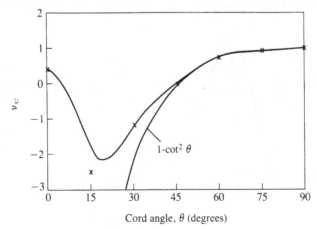

Cord angle, θ (degrees)

Interlaminar shear strains have been observed by inserting straight pins normal to the ply surface in a cord/rubber belt system and observing its rotation under extensional load (Bohm 1966) or by scribing a straight line on the edge of the specimen and monitoring the rotation of the line under load. Figure 8.7 shows the interlaminar shear strain measured by X-ray technique for a two-ply polyester–rubber as a function of cord angle θ (Lou and Walter 1978). The solid line is based upon the predictions of Eq. (8.11). The importance of interlaminar shear decreases as the number of plies increases.

Walter (1978) has presented values of the 18 elastic constants of A_{ij}, B_{ij} and D_{ij} for bias, belted-bias and radial constructions; the material combinations of nylon and rayon body plies with steel, PVA and rayon belt plies are included. For the case of a specially orthotropic laminate $(A_{16} = A_{26} = D_{16} = D_{26} = B_{ij} = 0)$ with respect to the x–y axes, the out-of-plane flexural rigidities are

$$(EI)_x = A_{11}h^2/12 = E_{11}h^3/12(1 - v_{xy}v_{yx})$$
$$(EI)_y = A_{22}h^2/12 = E_{22}h^3/12(1 - v_{yx}v_{xy})$$

(8.12)

where I is the area moment of inertia, and h denotes ply thickness.

8.2.4 *Cord loads in tires*

According to Clark (1980) the key to good tire design is long fatigue life. The loads on typical textile cords in pneumatic

Fig. 8.7. Interlaminar shear strain, γ_{xz}, vs. cord angle, θ, for a two-ply polyester rubber. — Eq. (8.11); ($\times$) experimental data. (After Lou and Walter 1978.)

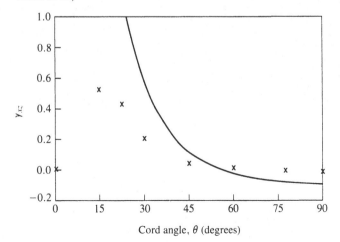

Cord angle, θ (degrees)

tires are extremely complex and the sources of loads can be identified as follows: (a) inflation load, (b) vertical load, (c) steering forces, (d) road irregularities, (e) camber, (f) speed, and (g) torque.

The tensile cord load due to inflation pressure can be predicted with some certainty by considering the axisymmetric nature of inflation and approximating the tire geometry as a thin toroidal shell. However, this task is complicated by the fact that the tire does not maintain a constant geometry during inflation. Furthermore, the membrane forces obtained from the thin shell analysis may not adequately represent the force distributions in the bead and tread regions. Figure 8.8 shows schematically the cross-section of a pneumatic tire and the designation of the locations (Clark 1980).

The measurement of cord loads is important to the analysis and design of tires. Various techniques have been employed; these include the use of grid or elongation marks for outer plies, X-ray photography relying on metal markers for inner plies, and resistance foil strain gages imbedded in the tire for direct cord load measurement in a tire under operating conditions. The force transducers using resistance foil strain gages are much smaller than the clip gages, the rubber–wire gages, or the liquid–metal gages. Details of these measurement techniques can be found in Clark and Dodge (1969), Patterson (1969), and Walter and Hall (1969).

The measurements of tire cord loads have indicated that the loads

Fig. 8.8. Location description in a cord/rubber pneumatic tire. (After Clark 1980.)

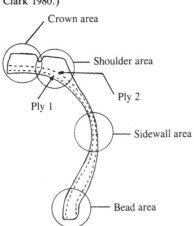

induced by normal direct inflation account for about 10–15% of cord strength. Another simple type of cord load is induced due to the load carried by the tire. The cord load at a given location can fluctuate fairly widely as the tire rolls. Also, the typical cord load cycle varies with the locations on the tire, i.e. crown, sidewall or shoulder region. Steering induces additional loads. Relatively small amounts of steer could induce very large increases in the cord loads. Figure 8.9 shows the basic characteristics of cord load fluctuation in a rolling tire (Clark 1980). It should be noted that compressive cord loads are possible. The characteristics of other cord loads due to road irregularities, speed and torque are even more difficult to quantify in a systematic manner.

The measurement of tire cord loads provides the basis of analysis of the response of cord/rubber composites to the specified boundary conditions. The netting theory, which only takes into account the deformation of the cord and neglects completely the contribution of the matrix rubber, was adopted in the earlier research on bias constructions. The uncertainty of the orientation of the cord in the net structure at different stress levels of inflation has limited the applicability of this theory.

The theory of laminated composites has undoubtedly provided an efficient means of analysis of cord/rubber composites. It is understood that the theory has its limitations due to the following reasons:

(1) Textile cord strains of several per cent could develop at some locations in the tire; even larger and nonlinear strains could develop in the rubber.

Fig. 8.9. Basic characteristics of cord load fluctuation in a rolling tire. (After Clark 1980.)

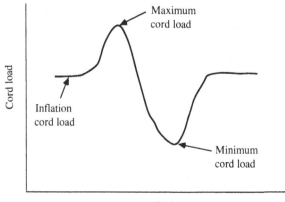

Position

(2) Interlaminar deformations are not taken into account in the theory, assuming plane stress condition.

(3) Cord/rubber composites usually exhibit bimodulus behavior (Bert and Kumar 1981; Bert and Reddy 1982).

(4) The viscoelastic behavior is assumed to be small and is often neglected in the analysis.

(5) Perfect cord/rubber interfacial bonds are assumed.

(6) The membrane forces in the bead and tread regions may be very complex.

(7) Fatigue and hygrothermal loadings may also complicate the problem.

However, in spite of its limitations, the lamination theory has been applied with some success for investigating a number of tire mechanics problems including stress analysis, obstacle enveloping, treadwear, and vibration. It is thus an efficient tool based upon linearly elastic, homogeneous and anisotropic material properties for the representation of nonlinear viscoelastic, heterogeneous calendered plies of cord/rubber tire composites (Walter 1978). The large nonlinear deformation of flexible composites is treated in Chapter 9.

8.3   Coated fabrics

Coated fabrics used in load bearing environments, for instance, those for air- or cable-supported building structures, tents, and inflated structures such as escape slides, must exhibit specific mechanical properties. Some of the general requirements include retaining flexibility over a wide temperature range, sufficient tensile and tear strength, low air permeability, and sufficient dimensional stability (Skelton 1971).

It has been recognized that coated fabrics generally exhibit nonlinear stress–strain behavior due to straightening of the crimped yarns under uniaxial or biaxial tension. As noted by Akasaka (1989), the microscopic deformation behavior of the woven yarns embedded in the matrix and subjected to membrane loading is very complex. Thus, modeling of the strength behavior of these materials requires reasonably precise knowledge of the deformation of the yarns as a function of load configuration and magnitude.

The linear elastic properties of laminates composed of coated fabrics can be readily derived based upon the lamination theory of Section 2.3. Akasaka and Yoshida (1972) presented explicit expressions for elastic moduli of laminates of coated fabrics; the analytical predictions were compared with experimental data of laminates of canvas.

Skelton (1971), among others, reported the biaxial stress–strain behavior of coated orthogonal fabrics. It is concluded that the stress–strain response at various stages of manufacture of coated fabrics is dependent mainly on the crimp in the two sets of yarns.

Fig. 8.10. (a) A section of the fabric along warp yarns in off-loom (top), heat set (middle) and coated (bottom) states. (b) A section of the fabric along filling yarns in off-loom (top), heat set (middle) and coated (bottom) states. (c) Surface feature of the fabric in heat set state. (After Skelton 1971 © ASTM. Reprinted with permission.)

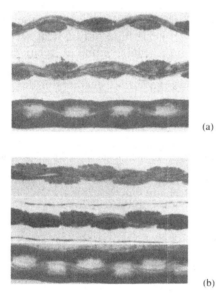

(a)

(b)

(c)

The balance of crimp is determined by the restraints imposed on the fabric during the heat setting process, which precedes the coating operation. If the fabric is set under tension in the warp direction, the warp yarns tend to become straight and the yarns in the filling direction become highly crimped. Thus, when such a fabric is subjected to biaxial loading, it is almost inextensible in the warp direction. Consequently, Skelton concluded that if a balanced fabric is required with similar biaxial tensile behavior in the warp and filling directions, the fabric must be heat set with both warp and filling directions under restraint.

It is interesting to recapitulate the experimental observations of Skelton (1971) for the biaxial testing of coated fabrics. Figures 8.10(a) and (b) show, respectively, the section views of a plain weave fabric based upon high tenacity polyester. Since the fabric is set under tension along the warp direction during heat setting, the warp crimp is minimum and the filling crimp is relatively high. Three stages, i.e. off-loom, heat set and coated state, are demonstrated. Figure 8.10(c) shows the surface features of the fabric in the heat set state.

Figure 8.11 shows the biaxial load–elongation curves for this

Fig. 8.11. Bi-axial load–elongation curves for a fabric; load ratio (warp/fill) = 1:2. WL = warp direction, loom state; FL = filling direction, loom state; WH = warp direction, heat set; FH = filling direction, heat set; WC = warp direction, coated; FC = filling direction, coated. (After Skelton 1971 © ASTM. Reprinted with permission.)

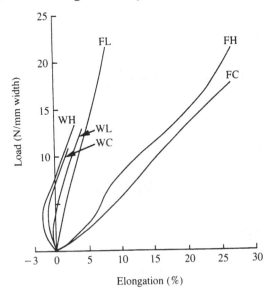

fabric with load ratio (warp/filling) = 1:2. The biaxial behavior can be understood by bearing in mind that in the heat set state the crimp is unbalanced; the warp yarns are essentially straight and the filling yarns are highly crimped. Thus, according to Skelton, the extension of the highly crimped direction of the filling yarns brings about an increase in crimp and reduction in width in the warp direction, in spite of the applied load in that direction. Consequently, the load–elongation curve shows negative elongation in the warp direction at low load level.

The elastic and inelastic responses of coated fabrics have been studied by Stubbs and Thomas (1984) and Stubbs (1988) using a space truss model. The model is capable of accounting for arbitrary loading sequences.

8.4 Nonlinear elastic behavior – incremental analysis

The flexible composites discussed in this section are also composed of continuous fibers in an elastomeric matrix. Because of the low shear modulus of the matrix and the highly anisotropic nature ($E_{11} \gg E_{22}$) of the composites, their effective elastic properties are very sensitive to the fiber orientation. The geometric nonlinearity of the flexible composite is mainly caused by the reorientation of fibers. The material nonlinearity is also pronounced in elastomeric composites under large deformation.

In order to fully realize the ability of the elastomeric matrix composite to sustain large deformation, Takahashi and Chou (1986), Takahashi, Kuo and Chou (1986), Chou and Takahashi (1987), and Takahashi, Yano, Kuo and Chou (1987) have predicted the nonlinear constitutive relation of flexible composites with sinusoidally shaped fibers based upon a step-wise incremental analysis and the classical lamination theory. In this section, the work of Chou and Takahashi is recapitulated. Both fiber geometric nonlinearity and matrix material nonlinearity have been taken into account. Because of the superposition of the infinitesimal solutions from lamination theory, the limitation of this approach is obvious. However, being a well established analytical technique in the composites field, the lamination theory does provide a convenient tool for discerning the basic characteristics of flexible composites.

Comparisons are made between the analytical predictions and experimental data for tire cord/rubber, and glass and Kevlar/silicone-elastomer flexible composite laminae. Since composites with fibers in wavy form have been used as a model system, the geometric aspects of curved fibers are examined first.

8.4.1   *Geometry of wavy fibers*

To demonstrate the effect of fiber extensibility from geometric design, a flexible composite composed of continuous fibers with sinusoidal waviness in a ductile matrix is used as a model system. Perfect bonding between the fibers and matrix is assumed. The geometric relations among the wavelength (λ), amplitude (a), and fiber length (s) of a sinusoidally shaped fiber are identified first. Then, two types of fiber arrangements are considered: the iso-phase model, and the random-phase model. The fibers are assumed to maintain the sinusoidal shape of which the geometric parameters λ, a and s vary with the increase of applied load.

The spatial position of a typical fiber in the xyz coordinates is given by:

$$y = a \sin \frac{2\pi x}{\lambda} \tag{8.13}$$

where the parameters a and λ of the curved fiber are shown in Fig.

Fig. 8.12. Geometrical relationships between a/λ, s/λ and θ_{max}, where θ_{max} is the maximum angle between the fiber and x axis. (After Chou and Takahashi 1987.)

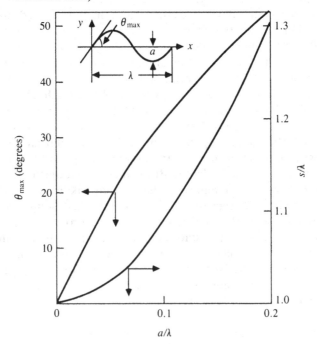

8.12. The angle θ between the tangent to the fiber and x axis is a function of x:

$$\tan \theta = \frac{dy}{dx} = \frac{2\pi a}{\lambda} \cos \frac{2\pi x}{\lambda} \tag{8.14}$$

The length of fiber, ds, between x and $x + dx$ is

$$ds = \sqrt{(dx^2 + dy^2)} = dx \sqrt{\left[1 + c \cdot \cos^2\left(\frac{2\pi x}{\lambda}\right)\right]} \tag{8.15}$$

where

$$c = (2\pi a/\lambda)^2$$

Obviously, the maximum value of $\tan \theta$ occurs at

$$|\theta_{\max}| = \tan^{-1}\left(\frac{2\pi a}{\lambda}\right) \tag{8.16}$$

The fiber length, s, between $x = 0$ and λ is given by

$$s = \int ds = \frac{\lambda}{2\pi} \int_0^{2\pi} \sqrt{(1 + c \cdot \cos^2 \beta)} \, d\beta \tag{8.17}$$

By the use of an elliptic integral of the second kind,

$$s = \lambda \sqrt{(1 + c)}\left(1 - \frac{1}{2^2} k^2 - \frac{1^2 \cdot 3}{2^2 \cdot 4^2} k^4 - \frac{1^2 \cdot 3^2 \cdot 5}{2^4 \cdot 4^2 \cdot 6^2} k^6 - \cdots\right) \tag{8.18}$$

where

$$k^2 = \frac{c}{1 + c} \tag{8.19}$$

Equation (8.18) can be rewritten as

$$\frac{s}{\lambda} = \frac{1}{\sqrt{(1 - k^2)}}\left(1 - 2\left(\frac{k^2}{8}\right) - 3\left(\frac{k^2}{8}\right)^2 - 10\left(\frac{k^2}{8}\right)^3\right.$$
$$\left. - \frac{175}{4}\left(\frac{k^2}{8}\right)^4 - \frac{441}{2}\left(\frac{k^2}{8}\right)^5 - \cdots\right) \tag{8.20}$$

By the use of Taylor expansion, we have

$$\frac{s}{\lambda} = 1 + 2\left(\frac{k^2}{8}\right) + 13\left(\frac{k^2}{8}\right)^2 + 90\left(\frac{k^2}{8}\right)^3$$
$$+ 644\left(\frac{k^2}{8}\right)^4 + 4708.5\left(\frac{k^2}{8}\right)^5 + \cdots \tag{8.21}$$

In the following analysis, terms up to $(k^2/8)^5$ in Eq. (8.21) are taken into account, and the range of a/λ is limited to below $\frac{1}{5}$. The relationship between a/λ and s/λ is shown in Fig. 8.12 where $\theta_{\max}$ is the maximum angle between the fiber and the x axis. For example, for $\theta_{\max} = 20°$, $a/\lambda = 0.058$ and $s/\lambda = 1.032$. The curved fiber composite with $a/\lambda = 0.10$ can be extended up to 9.23% of its original length only by the straightening of the fiber, if the matrix stiffness is negligible.

Two kinds of arrangements of the curved fibers in the composite have been considered: the iso-phase model and random-phase model. The iso-phase model is defined in Fig. 8.13, where all the fibers are in the same phase in the x direction. The distance between the fibers in the y direction is assumed to be constant. In the random phase model (Fig. 8.14), the axial locations of sinusoidal shaped fibers do not assume any regular pattern.

8.4.2   *Axial tensile behavior*

The nonlinear tensile stress–strain behavior of flexible composites containing wavy fibers has been investigated according to the iso-phase and random-phase models. The lamination theory described in Section 2.3 is the basis of this analysis. The applied load is parallel to the axes of the sinusoidally shaped fibers.

8.4.2.1 *Iso-phase model*

The linear elastic stress–strain relations are derived first. Consider Fig. 8.13; each volume element between x and $x + dx$ is

Fig. 8.13. Iso-phase model. (After Chou and Takahashi 1987.)

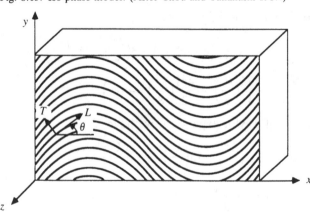

approximated by a unidirectional straight fiber composite, in which fibers are inclined at an angle θ to the x axis, as defined by Eq. (8.14). The transformation of coordinates between the composite reference axes (xyz) and the fiber local axes (LTz) is given by:

$$\begin{pmatrix} L \\ T \\ z \end{pmatrix} = \begin{pmatrix} \cos\theta & \sin\theta & 0 \\ -\sin\theta & \cos\theta & 0 \\ 0 & 0 & 1 \end{pmatrix} \begin{pmatrix} x \\ y \\ z \end{pmatrix} \tag{8.22}$$

The positive direction of θ is defined in Fig. 8.13. Under the uniaxial tension, σ_{xx}, Eq. (2.17) gives

$$\varepsilon_{xx} = \bar{S}_{11}\sigma_{xx}$$

$$\varepsilon_{yy} = \bar{S}_{12}\sigma_{xx} \tag{8.23}$$

$$\gamma_{xy} = \bar{S}_{16}\sigma_{xx}$$

It is interesting to note the stretching–shear coupling represented by $\bar{S}_{16}$. Figure 8.15 shows schematically the γ_{xy} induced by an applied stress σ_{xx}.

The average tensile strain of the iso-phase composite, ε_{xx}^*, is

$$\varepsilon_{xx}^* = \frac{1}{\lambda} \int_0^\lambda \varepsilon_{xx}\, \mathrm{d}x \tag{8.24}$$

Fig. 8.14. Random-phase model. (After Chou and Takahashi 1987.)

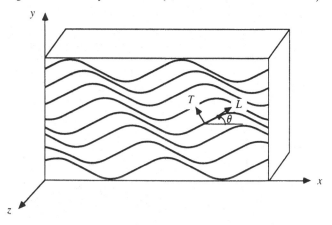

From Eqs. (8.14) and (8.23),

$$
\varepsilon_{xx}^* = \left[\frac{1 + \dfrac{c}{2}}{(1+c)^{3/2}} S_{11} - \left(\frac{1 + \dfrac{3}{2}c}{(1+c)^{3/2}} - 1 \right) S_{22} \right.
$$

$$
\left. + \frac{\dfrac{c}{2}}{(1+c)^{3/2}} (2S_{12} + S_{66}) \right] \sigma_{xx} \tag{8.25}
$$

The effective Young's modulus of the iso-phase model in the x direction is given by

$$
E_{xx}^* = \frac{(1+c)^{3/2}}{\left(1 + \dfrac{c}{2}\right)S_{11} - \left(1 + \dfrac{3}{2}c - (1+c)^{3/2}\right)S_{22} + \dfrac{c}{2}(2S_{12} + S_{66})} \tag{8.26}
$$

In a small volume element between x and $x + dx$, the tensile strain of the fiber along its axial direction is expressed by

$$
\varepsilon_L = \varepsilon_{xx} \cos^2 \theta + \varepsilon_{yy} \sin^2 \theta + \gamma_{xy} \sin \theta \cos \theta \tag{8.27}
$$

Substituting Eqs. (8.23) into Eq. (8.27) and integrating over s, the average fiber axial strain is

$$
\varepsilon_L^* = \frac{1}{s} \int_0^s \varepsilon_L \, ds = [(S_{11} - S_{12})F(k) + S_{12}]\sigma_{xx} \tag{8.28}
$$

where

$$
F(k) = 1 - \frac{1}{2}k^2 - \frac{3}{16}k^4 - \frac{3}{32}k^6 - \frac{111}{2048}k^8 - \frac{141}{4096}k^{10} \tag{8.29}
$$

Fig. 8.15. Schematic illustration of the deformed shape of the iso-phase model under uniaxial tension σ_{xx}. (After Chou and Takahashi 1987.)

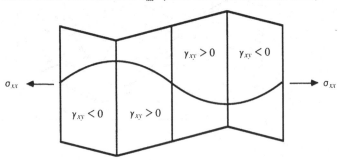

Here, the relations among s, x and λ, Eqs. (8.14)–(8.20), and the elliptic integral are used in the derivations.

8.4.2.2 *Random-phase model*

In the case of the iso-phase model, the stretching–shear coupling constants $\bar{S}_{16}$ and $\bar{S}_{26}$ do not vanish. This coupling effect could be eliminated through the random positioning of wavy fibers along the x-direction:

$$y = a \sin(2\pi(x - d)/\lambda) \tag{8.30}$$

where d is the translation of the fiber in the x direction. A random distribution of d $(0 \le d \le \lambda)$ is assumed in this model. That is, in each infinitesimal section, dx, fibers with any arbitrary orientation angles exist with the same probability. Therefore, it is assumed that ε_{xx} is uniform throughout the sample under uniaxial tension. The stress in a fiber segment depends on its orientation, θ:

$$-\frac{2\pi a}{\lambda} \le \tan \theta \le \frac{2\pi a}{\lambda}$$

By these assumptions, the classical lamination theory can again be applied.

The stress–strain relations of a unidirectional lamina consisting of straight fibers are given by Eq. (2.13) with the reduced stiffness Q_{ij} given by Eqs. (2.14). The transformed stress–strain relations of an off-axis lamina, referring to the x–y coordinate system, are given by Eqs. (2.15) and (2.16). The small element of the random-phase composite situated between the sections at x and $x + dx$ is treated as a laminate with different orientations. The fibers with the orientation angle θ which lies in the range defined by

$$0 \le \theta \le \tan^{-1}\left(\frac{2\pi a}{\lambda}\right) \tag{8.31}$$

have the probability dx/λ.

Therefore, the stress–strain relation of the laminate can be rewritten as

$$\begin{pmatrix} \sigma_x \\ \sigma_y \\ \tau_{xy} \end{pmatrix} = \begin{pmatrix} C_{11}^* & C_{12}^* & C_{16}^* \\ C_{12}^* & C_{22}^* & C_{26}^* \\ C_{16}^* & C_{26}^* & C_{66}^* \end{pmatrix} \begin{pmatrix} \varepsilon_x \\ \varepsilon_y \\ \gamma_{xy} \end{pmatrix} \tag{8.32}$$

where

$$C_{mn}^* = \frac{1}{\lambda} \int_0^\lambda Q_{mn}(\theta) \, dx \tag{8.33}$$

The average stiffness constants of Eq. (8.33) are

$$C_{11}^* = \frac{1}{(1+c)^{3/2}} \left[Q_{11}\left(1+\frac{c}{2}\right) + (Q_{12}+2Q_{66})c \right.$$

$$\left. + Q_{22}\left((1+c)^{3/2} - \left(1+\frac{3}{2}c\right)\right) \right]$$

$$C_{22}^* = \frac{1}{(1+c)^{3/2}} \left[Q_{11}\left((1+c)^{3/2} - \left(1+\frac{3}{2}c\right)\right) \right.$$

$$\left. + (Q_{12}+2Q_{66})c + Q_{22}\left(1+\frac{c}{2}\right) \right]$$

$$C_{12}^* = \frac{1}{(1+c)^{3/2}} \left[(Q_{11}+Q_{22}-4Q_{66})\frac{c}{2} + Q_{12}((1+c)^{3/2}-c) \right]$$

$$C_{66}^* = \frac{1}{(1+c)^{3/2}} \left[(Q_{11}+Q_{22}-2Q_{12}-2Q_{66})\frac{c}{2} \right.$$

$$\left. + Q_{66}((1+c)^{3/2}-c) \right]$$

$$C_{16}^* = C_{26}^* = 0$$

(8.34)

Inversion of Eq. (8.32) leads to

$$\begin{pmatrix} \varepsilon_{xx} \\ \varepsilon_{yy} \\ \gamma_{xy} \end{pmatrix} = \begin{pmatrix} S_{11}^* & S_{12}^* & 0 \\ S_{12}^* & S_{22}^* & 0 \\ 0 & 0 & S_{66}^* \end{pmatrix} \begin{pmatrix} \sigma_{xx} \\ \sigma_{yy} \\ \tau_{xy} \end{pmatrix}$$

(8.35)

where

$$S_{11}^* = (C_{22}^* C_{66}^* - C_{26}^{*2})/D$$
$$S_{22}^* = (C_{11}^* C_{66}^* - C_{16}^{*2})/D$$
$$S_{12}^* = (C_{16}^* C_{26}^* - C_{12}^* C_{16}^*)/D$$
$$S_{66}^* = (C_{11}^* C_{22}^* - C_{12}^{*2})/D$$
$$D = C_{11}^* C_{22}^* C_{66}^* - C_{12}^{*2} C_{66}^*$$

(8.36)

Following Eqs. (2.9), the Young's modulus and Poisson's ratio in the x direction of the random-phase model are given by:

$$E_{xx}^* = 1/S_{11}^*$$
$$v_{xy}^* = -S_{12}^*/S_{11}^*$$

(8.37)

If the random-phase model is subjected to uniaxial tension, σ_{xx}, the strain components are

$$\varepsilon_{xx} = \sigma_{xx}/E_{xx}^*$$
$$\varepsilon_{yy} = -(v_{xy}^*/E_{xx}^*)\sigma_{xx} \qquad (8.38)$$
$$\gamma_{xy} = 0$$

The strain of the fiber in its axial direction is calculated by substituting Eqs. (8.38) into Eq. (8.27) and averaging over s (Eq. (8.28))

$$\varepsilon_L^* = (\varepsilon_{xx} - \varepsilon_{yy})F(k) + \varepsilon_{yy} \qquad (8.39)$$

where $F(k)$ is given by Eq. (8.29).

8.4.2.3 Nonlinear tensile stress–strain behavior

The nonlinear axial (x direction) tensile stress–strain behavior of the flexible composite is examined using the stepwise incremental analysis of Petit and Waddoups (1969). Consider an incremental tensile strain Δe_{xx}, applied on either the iso-phase or random-phase model. Here, $\Delta e_{xx} = \Delta l/l$; Δl and l are the incremental length and the current length, respectively. Using the initial Young's modulus E_{xx}^*, the first stress increment, $\Delta \sigma_{xx}$, is calculated by the linear elastic relation:

$$\Delta \sigma_{xx} = E_{xx}^* \Delta e_{xx} \qquad (8.40)$$

where the expressions of E_{xx}^* are given by Eqs. (8.26) and (8.37) for the iso-phase and random-phase models, respectively. The nth stress increment is added to the previous stress state after $n-1$ increments to determine the current total stress:

$$(\sigma_{xx})_n = (\sigma_{xx})_{n-1} + (\Delta \sigma_{xx})_n \qquad (8.41)$$

For the iso-phase model, the average tensile strain increment of the fiber along its axial direction, Δe_L^*, is obtained by substituting Eq. (8.40) into Eq. (8.28):

$$\Delta e_L^* = [(S_{11} - S_{12})F(k) + S_{12}] \Delta \sigma_{xx} \qquad (8.42)$$

For the random-phase model, the transverse strain increment, Δe_{yy}, is determined from Δe_{xx} and v_{xy}^*:

$$\Delta e_{yy} = -v_{xy}^* \Delta e_{xx} \qquad (8.43)$$

Then, the tensile strain increment of the fiber is calculated by substituting Δe_{xx} and Δe_{yy} into Eq. (8.39):

$$\Delta e_L^* = (\Delta e_{xx} - \Delta e_{yy})F(k) + \Delta e_{yy} \qquad (8.44)$$

The total strain, referring to the current specimen length, after n increments is

$$e_{xx} = \sum_{i=1}^{n} (\Delta e_{xx})_i = \sum_{i=1}^{n} \left(\frac{\Delta l}{l} \right)_i \qquad (8.45)$$

Replacing Δl by the infinitesimal increment, dl, it follows:

$$e_{xx} = \int_{l_o}^{l} \frac{dl}{l} = \ln \frac{l}{l_o} = \ln(1 + \varepsilon_{xx}) \qquad (8.46)$$

Here, ε_{xx} is the tensile strain referred to the initial specimen length l_o:

$$\varepsilon_{xx} = \frac{l - l_o}{l_o} \qquad (8.47)$$

In the range of large strain, the use of ε_{xx} is more convenient than the summation of Δe_{xx}. From Eq. (8.46)

$$\varepsilon_{xx} = \exp(e_{xx}) - 1 \qquad (8.48)$$

Then, the total strain, after the nth increment, in the axial direction (ε_{xx}), transverse direction (ε_{yy}) and the fiber (ε_L^*) are given by:

$$(\varepsilon_{xx})_n = \exp\left[\sum_{i=1}^{n} (\Delta e_{xx})_i \right] - 1 \qquad (8.49)$$

$$(\varepsilon_{yy})_n = \exp\left[\sum_{i=1}^{n} (\Delta e_{yy})_i \right] - 1 \qquad (8.50)$$

$$(\varepsilon_L^*)_n = \exp\left[\sum_{i=1}^{n} (\Delta e_L^*)_i \right] - 1 \qquad (8.51)$$

Finally, the change of fiber shape under loading needs to be taken into account. Due to the tensile loading in the x direction, the wavelength of the curved fiber is changed to

$$\lambda = \lambda_o(1 + \varepsilon_{xx}) \qquad (8.52)$$

where λ and λ_o are, respectively, the current and initial values of the wavelength, and the total strain ε_{xx} is given by Eq. (8.49). The current value of the fiber length is

$$s = s_o(1 + \varepsilon_L^*) \qquad (8.53)$$

where s_o is the initial fiber length and ε_L^* is the total fiber strain given by Eq. (8.51).

In order to determine the shape of the fiber, it is assumed that the fiber maintains a sinusoidal waviness during deformation while varying its amplitude (a) and wavelength (λ). The current value of the amplitude, a, can be determined by Fig. 8.12 from the given current values of λ and s. The values of $k^2 = c/(1 + c)$, $c = (2\pi a/\lambda)^2$, E_{xx}^* and v_{xy}^* after the nth step are determined from the current values of λ and a, and these values are used in the $(n + 1)$th step of the incremental analysis. Eqs. (8.41) and (8.49) give the uniaxial tensile stress–strain relation of the flexible composite.

The elastic constants of fibers (Chamis 1984) and matrices (*Modern Plastics Encyclopedia* 1983) used in the numerical calculations of Chou and Takahashi (1987) are shown in Table 8.1. Linear elastic stress–strain relations are assumed for glass and Kevlar fibers. Rubber elasticity (James and Guth 1943; Treloar 1973) is assumed for PBT and the other elastomeric polymers:

$$\sigma = \frac{E_{\mathrm{m}}^{\mathrm{o}}}{3}\left(\alpha - \frac{1}{\alpha^2}\right) \tag{8.54}$$

where $E_{\mathrm{m}}^{\mathrm{o}}$ is the initial Young's modulus of the matrix, and α is the extension ratio:

$$\alpha = 1 + \varepsilon_{xx} \tag{8.55}$$

The secant Young's modulus of the matrix, E_{m}, is determined from the current tensile strain, ε_{xx}, (Eqs. (8.49) and (8.55)):

$$E_{\mathrm{m}} = \frac{\mathrm{d}\sigma}{\mathrm{d}\varepsilon_{xx}} = \frac{E_{\mathrm{m}}^{\mathrm{o}}}{3}\left(1 + \frac{2}{\alpha^3}\right) \tag{8.56}$$

Numerical examples of the stress–strain relations predicted by the incremental analysis are shown in Figs. 8.16 and 8.17. The results indicate that Kevlar is less effective than glass fiber in contributing

Table 8.1. *Elastic constants and elongations* (*Chou and Takahashi* 1987)

	E_{L} (GPa)	E_{T} (GPa)	G_{LT} (GPa)	v_{LT}	v_{TT}	ε_{b} (%)
Glass fiber	72.52		29.7	0.22		4
Kevlar	151.6	4.13	2.89	0.35	0.35	3.5
PBT matrix		2.156	0.77		0.4	50–300

Isotropic relation $G = E/2(1 + v)$ is assumed.

Fig. 8.16. Comparisons of the effects of glass and Kevlar fibers on the tensile stress (σ_{xx})–strain (ε_{xx}) curves for an iso-phase composite at various E_m^o. Rubber elasticity is assumed for the matrix. Crosses (×) show average fiber axial tensile strain; ε_L^* reaches 4% and 3.5% for glass and Kevlar fibers, respectively. $v_m = 0.4$, $V_f = 50\%$, $a/\lambda = 0.1$. (After Chou and Takahashi 1987.)

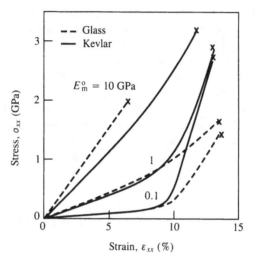

Fig. 8.17. Tensile stress (σ_{xx})–strain (ε_{xx}) curves of Kevlar/PBT polymer composites predicted by using the iso-phase (solid line) and random-phase (dotted line) models. $E_m^o = 1$ GPa, $v_m = 0.4$ and $V_f = 50\%$. Crosses (×) show average fiber axial tensile strain; ε_L^* reaches 3.5%. (After Chou and Takahashi 1987.)

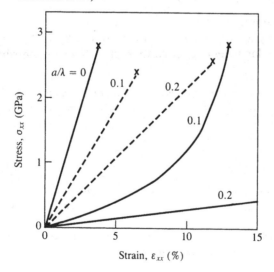

to the stiffness of curved fiber composites, because the transverse Young's modulus of Kevlar is lower than that of glass. After the wavy fibers are stretched, however, Kevlar becomes increasingly more effective with regard to stiffness and strength (Fig. 8.16). For a given wavy fiber composite, the random-phase model predicts higher Young's modulus and lower elongation than those of the iso-phase model (Fig. 8.17).

8.4.3   *Transverse tensile behavior*

The transverse tensile behavior of wavy fiber composites has been analyzed for both iso-phase and random-phase models by Kuo, Takahashi and Chou (1988). The lamination theory is again the basis of the incremental analysis.

8.4.3.1 *Iso-phase model*

Consider the small volume element situated between y and $y + dy$ in the iso-phase model as shown in Fig. 8.13. It is assumed that the transverse stress σ_{yy} is uniformly distributed along one wavelength λ. Then an element of the size $dy\,dx$ can be treated as an off-axis unidirectional lamina. From Eq. (2.17) and plane stress condition, the strain components in this element are

$$\varepsilon_{xx} = \bar{S}_{12}\sigma_{yy} \qquad \varepsilon_{yy} = \bar{S}_{22}\sigma_{yy} \qquad \gamma_{xy} = \bar{S}_{26}\sigma_{yy} \qquad (8.57)$$

Then the transverse strain averaged over the wavelength of the iso-phase model is

$$\varepsilon_{yy}^* = \frac{1}{\lambda} \int_0^\lambda \varepsilon_{yy}\, dx \qquad (8.58)$$

The effective Young's modulus in the y direction is

$$E_{yy}^* = \frac{\sigma_{yy}}{\varepsilon_{yy}^*}$$

$$= \frac{(1+c)^{3/2}}{\left((1+c)^{3/2} - 1 - \dfrac{3c}{2}\right)S_{11} + \left(1 + \dfrac{c}{2}\right)S_{22} + \dfrac{c}{2}(2S_{12} + S_{66})} \qquad (8.59)$$

Following the approach of Section 8.4.2.1, the average tensile strain along the fibers due to transverse tension is obtained by substituting Eqs. (8.57) into Eq. (8.27) and averaging over the length s:

$$\varepsilon_L^* = [(S_{12} - S_{11})F(k) + S_{11}]\sigma_{yy} \qquad (8.60)$$

$F(k)$ is given by Eq. (8.29).

Kuo, Takahashi and Chou (1988) also analyzed the transverse tensile behavior based upon the constant strain assumption. This assumption is validated by the observation during transverse tension experiments that the elongation of the specimen is uniform throughout its width away from the specimen ends. Although the constitutive relations are not of the same form for constant stress and constant strain analyses, the numerical calculations in Kuo, Takahashi and Chou (1988) yield the same result. This is the direct consequence of the approaches, namely the stress (or strain) is considered in the average sense along the x direction.

8.4.3.2 *Random-phase model*

The transverse Young's modulus and minor Poisson's ratios are given by

$$E_{yy}^* = 1/S_{22}^*$$
$$v_{yx}^* = -S_{12}^*/S_{22}^*$$

(8.61)

Fig. 8.18. Comparisons between theoretical predictions and experimental data of transverse tension of an iso-phase model. Specimen initial $a/\lambda = 0.05$–0.07 and $V_f = 1.337\%$ for Thornel-300/silicone elastomer composites. (After Kuo, Takahashi and Chou 1988.)

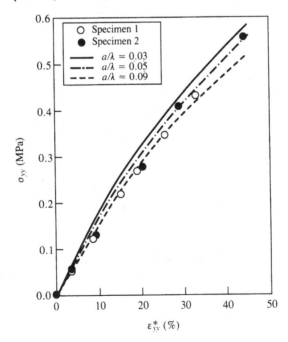

Under the transverse stress, σ_{yy}, the strain components are

$$\varepsilon_{xx} = -(v_{yx}^*/E_{yy}^*)\sigma_{yy}$$
$$\varepsilon_{yy} = \sigma_{yy}/E_{yy}^* \qquad\qquad (8.62)$$
$$\gamma_{xy} = 0$$

Again, the average tensile strain along the fiber is obtained from Eq. (8.39).

Figure 8.18 depicts the comparison between theoretical curves and experimental data of an iso-phase model under transverse tension. The experimental material system reported by Kuo, Takahashi and Chou (1988) is based upon Sylgard 184 silicone elastomer reinforced with Thornel-300 carbon fiber. A loose fiber bundle contains 1000 filaments, with a filament diameter of $7\,\mu$m. The specimen fabrication technique follows that given by Luo and Chou (1988). The initial a/λ values of the specimens are in the range of 0.05–0.07. The fiber volume fraction is very low, about 1.34%.

9 Nonlinear elastic finite deformation of flexible composites

9.1 Introduction

Flexible composites, which are described in Chapter 8, behave very differently from conventional rigid polymer composites in the following ways:

(1) Flexible composites are highly anisotropic (i.e. longitudinal elastic modulus/transverse elastic modulus $\gg 1$). Figure 9.1 compares the normalized effective Young's modulus (E_{xx}/E_{22}) vs. fiber orientation for two types of unidirectional composites. The upper curve obtained from Kevlar-49/silicone elastomer shows that the stiffness of the elastomeric composite lamina is very sensitive to the fiber orientation. At a 5° off-axis fiber orientation, for example, a 1° change in fiber angle causes the effective stiffness to change by 53%. The lower curve obtained from Kevlar-49/epoxy shows less than 7% change at the same off-axis angle.

(2) Flexible composites show low shear modulus and hence large shear distortion, which allows the fibers to change their orientations under loading.

(3) Flexible composites have a much larger elastic deformation range than that of conventional rigid polymer composites. Thus, the geometric changes of the configuration (i.e. area, direction, etc.) need to be taken into consideration.

(4) The nonlinear elastic behavior with stretching–shear coupling, due to material and geometrical effects, is pronounced in flexible composites under finite deformation.

Therefore, the conventional linear elastic theory, based on the infinitesimal strain assumption for rigid matrix composites, may no longer be applicable to elastomeric composites under finite deformation.

The theories of non-linear and finite elasticity made a major advancement during the Second World War, in response to the development of the rubber industry. M. Mooney, in 1940, advanced his well-known strain–energy function. Rivlin and colleagues (for example, Rivlin 1948a&b; Rivlin and Saunders 1951; Ericksen and

Rivlin 1954), in a series of publications starting in 1948, successfully predicted the large deformation of rubber-like incompressible isotropic material. These works have greatly enhanced and stimulated the development of nonlinear finite elasticity. The fundamental aspects of finite elasticity can be found in advanced text books (for example, Truesdell 1966; Fung 1977; Malvern 1969; Spencer 1972; Lai, Rubin and Krempl 1978).

To predict the large deformation of fiber reinforced rubber material, Adkins and Rivlin (1955) treated the nonlinear, anisotropic, and finite deformation problem by using the 'ideal fiber reinforced material theory'; the assumptions of volume incompressibility and fiber inextensibility are basic to the analysis. Further developments of this theory can be found in the work of Rivlin (1964), Pipkin and Rogers (1971) and Spencer (1972). Difficulties often arise in applying this theory to composites with complicated fiber geometries and in cases where the extension of the fibers cannot be neglected.

Fig. 9.1. Variation of effective Young's modulus with fiber orientation. (After Luo, 1988.)

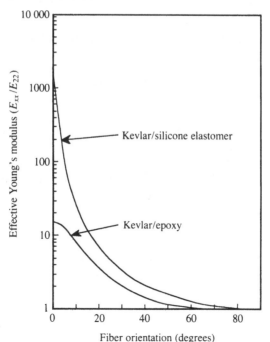

The ability of flexible composites to sustain large deformation and fatigue loading, and still provide high load carrying capacity, has been mainly analyzed in textile cord/rubber composites and coated fabrics. However, most of the existing analyses on the mechanics of pneumatic tires are primarily based on the composite lamination theory for small linear deformation. Chou (1989) has provided a review of the mechanics of flexible composites.

In recent years, the constitutive relation of biological materials has been a subject of considerable research interest. A variety of biological materials are incompressible, viscoelastic, and anisotropic; they often demonstrate nonlinear behavior with a large deformation range (Fung 1981). For instance, Aspden (1986) considered the influences of fiber reorientation in biological materials during finite deformation by using a fiber orientation distribution function and assuming that the fiber carries only axial tension. However, the finite deformation and the rigid body rotation of fibers, as well as the shear property of the matrix material, which greatly influence the fiber reorientation during deformation, are not adequately considered in the analysis. Humphrey and Yin (1987) presented a constitutive model based upon a pseudostrain–energy function, and compared the theoretical analysis with both uniaxial and biaxial experimental results. The parameters used in the energy function are dependent on the experimental data; the fiber spatial arrangements, which are responsible for the geometric nonlinearity, are ignored in their analysis.

Various response functions have been proposed to represent the experimentally determined nonlinear stress–strain curves in principal material directions. Petit and Waddoups (1969) employed the increment method. Hahn (1973) and Hahn and Tsai (1973) used the complementary energy density to derive the stress–strain relation, which is nonlinear in shear but linear in tensile properties. Jones and Morgan (1977) used an orthotropic material model in which the nonlinear mechanical properties are functions of the elastic energy density. The nonlinear elastic behavior of textile structural composites has been examined by Ishikawa and Chou (1983). However, these analyses are restricted to a small strain range.

In an effort to provide a rigorous treatment of the finite deformation problem, two analytical approaches, considering both geometric and material nonlinearities, have been employed in this chapter to predict the constitutive relations of flexible composites (R. S. Rivlin, private communication, 1986; Luo and Chou, 1988a, 1990a&b).

(1) In the first method (Section 9.3), a closed form representation of the constitutive equations has been derived based on the Lagrangian description. The strain–energy density is assumed to be a function of the Lagrangian strain components referring to the initial principal material coordinate $\bar{X}$ (Fig. 9.2a).

(2) In the second approach (Section 9.4), a nonlinear constitutive relation has also been developed based upon the Eulerian description where the deformed configuration of the composite is used as the reference state. A stress–energy function, referring to the moving principal material

Fig. 9.2. A rectangular element of composite lamina before and after loading (a) in the Lagrangian system, (b) in the Eulerian system. (After Luo and Chou 1988a.)

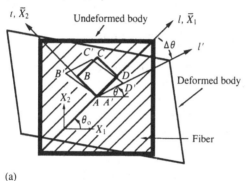

(a)

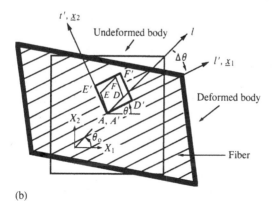

(b)

coordinate x (Fig. 9.2b), provides the basis for deriving the constitutive relations; and an iterative calculation method is employed.

The constitutive relations obtained from Sections 9.3 and 9.4 have been applied to study the nonlinear elastic behavior of flexible composites with wavy fibers in Section 9.5. Section 9.2, which is based upon Luo (1988), provides the basis for the theoretical treatment of this chapter.

9.2   Background

9.2.1   *Tensor notation*
Some brief descriptions of the notations and operations of tensors are shown in this section. These are taken from various sources, including Fung (1965, 1977), Rivlin (1970) and Lai, Rubin and Krempl (1978). It is not intended to provide a comprehensive coverage of tensor analysis. Only the subjects that are relevant to the present work are described. For simplicity, only Cartesian tensors are used, and thus the distinction between contravariance and covariance disappears and all indices of the tensor components can be written as subscripts. Furthermore, tensors are printed in bold-faced letters.

Einstein summation convention
The following three equations have the same meaning:

$$y_i = a_{ij}x_j$$

$$= \sum_{j=1}^{n} a_{ij}x_j$$

$$= a_{i1}x_1 + a_{i2}x_2 + a_{i3}x_3 + \cdots + a_{in}x_n \qquad (9.1)$$

The first line of Eq. (9.1) follows the rule of Einstein summation. Here, j is known as the dummy index, which repeats once, denoting a summation with respect to that index over its range, and i is a free index, which appears once in every term of the equation, assuming the numbers of 1, 2 or 3. The following are two other examples:

(1)   For the two vectors $\mathbf{a} = a_i\mathbf{e}_i$, and $\mathbf{b} = b_i\mathbf{e}_i$ $(i = 1,2,3)$, the scalar product is defined by

$$c = \mathbf{a} \cdot \mathbf{b} = \sum_{i=1}^{3} \sum_{j=1}^{3} a_i b_j (\mathbf{e}_i \cdot \mathbf{e}_j) \equiv a_i b_j (\mathbf{e}_i \cdot \mathbf{e}_j) \qquad (9.2)$$

(2) For the matrices $\mathbf{a} = [a_{ij}]$ and $\mathbf{b} = [b_{ij}]$ ($i = 1,2,3$, $j = 1,2,3$), the product of these two matrices is

$$[c_{ij}] = \mathbf{ab} = \sum_{k=1}^{3} [a_{ik}b_{kj}] \equiv [a_{ik}b_{kj}] \tag{9.3}$$

Kronecker delta

The Kronecker delta δ is defined as

$$\delta_{ij} = \begin{cases} 1 & \text{for } i = j \\ 0 & \text{for } i \neq j \end{cases} \tag{9.4}$$

or

$$\delta = \begin{bmatrix} \delta_{11} & \delta_{12} & \delta_{13} \\ \delta_{21} & \delta_{22} & \delta_{23} \\ \delta_{31} & \delta_{32} & \delta_{33} \end{bmatrix} = \begin{bmatrix} 1 & 0 & 0 \\ 0 & 1 & 0 \\ 0 & 0 & 1 \end{bmatrix} \tag{9.5}$$

The following relations are useful:

(1) $\delta_{ij} = 3$ $\tag{9.6}$
(2) $\delta_{im} T_{mj} = T_{ij}$ $\tag{9.7}$
(3) For the mutually perpendicular unit vectors $\mathbf{e}_1$, $\mathbf{e}_2$ and $\mathbf{e}_3$,

$$\mathbf{e}_i \mathbf{e}_j = \delta_{ij} \tag{9.8}$$

Permutation symbol

The permutation symbol is defined as

$$\varepsilon_{ijk} = \begin{cases} 1 & \text{if } ijk \text{ is even permutation of 1, 2, 3} \\ -1 & \text{if } ijk \text{ is odd permutation of 1, 2, 3} \\ 0 & \text{otherwise (i.e. if two of the indices are equal)} \end{cases} \tag{9.9}$$

where the even and odd permutations are indicated as

Even permutation   Odd permutation

The following relations are also used in this chapter:

(1)   For the two vectors $\mathbf{a} = a_i\mathbf{e}_i$, and $\mathbf{b} = b_j\mathbf{e}_j$, then

$$\mathbf{a} \times \mathbf{b} = \varepsilon_{ijk}a_ib_j\mathbf{e}_k \tag{9.10}$$

where the unit vector $\mathbf{e}_k$ is normal to the plane containing both $\mathbf{e}_i$ and $\mathbf{e}_j$.

(2)   For the matrix $\mathbf{m} = [m_{ij}]$, its determinant is

$$\det(\mathbf{m}) = \begin{vmatrix} m_{11} & m_{12} & m_{13} \\ m_{21} & m_{22} & m_{23} \\ m_{31} & m_{32} & m_{33} \end{vmatrix}$$
$$= \tfrac{1}{6}\varepsilon_{ijk}\varepsilon_{pqr}m_{ip}m_{jq}m_{kr} \tag{9.11}$$

9.2.2   *Lagrangian and Eulerian descriptions*

Both the Lagrangian and Eulerian descriptions have been used in finite elasticity. The description of the relation between the undeformed and deformed configurations of a continuum can be considered as a 'mapping' between domains D_o and D (Fig. 9.3). To find the transformation relation, let X and x be two fixed rectangular Cartesian coordinates associated with the original and deformed configurations, respectively. The position of a generic particle P inside the domain D_o is defined by the position vector $\mathbf{X}$ and coordinates X_j ($j = 1,2,3$). After deformation, this particle assumes the location P' with the new position vector $\mathbf{x}$ and coordinates x_i ($i = 1,2,3$). Then, the 'mapping' or the deformation of the con-

Fig. 9.3. The undeformed configuration D_o and the deformed configuration D of an elastic body. (After Luo and Chou 1990b.)

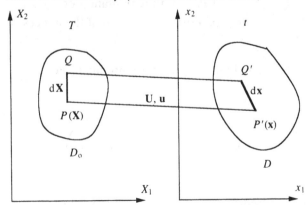

figuration can be described mathematically by the coordinate transformation between X_j and x_i.

The coordinate system X_1–X_2–X_3 is chosen as the reference system. The description of deformation, of which the independent variable is the particle position vector **X** in the original state, is known as the Lagrangian description. The reference system X is known as a Lagrangian coordinate. The transformation equation in terms of X_j is

$$x_i = x_i(X_1, X_2, X_3) \tag{9.12}$$

where the specified function x_i (X_1, X_2, X_3) is assumed to be continuous and differentiable. It follows, then:

$$dx_i = \frac{\partial x_i}{\partial X_j} dX_j \equiv x_{i,j} dX_j \tag{9.13}$$

where the dummy, or repeating, index, j, denotes a summation over its range. The matrix form of Eq. (9.13) can be written as $[dx] = [g][dX]$, where the deformation gradient matrix $[g]$ is

$$g_{ij} = \frac{\partial x_i}{\partial X_j} \tag{9.14}$$

The strain tensor associated with the Lagrangian system is called the *Lagrangian strain* (E_{ij}), also known as the *Green's* or *St. Venant's strain*, and it is defined in matrix form as

$$[E] = \tfrac{1}{2}([g]^\mathrm{T} - [g] - [\delta]) \tag{9.15}$$

where $[g]^\mathrm{T}$ is the transpose of the deformation gradient matrix $[g]$, and $[\delta]$ is the Kronecker delta. The explicit form of Eq. (9.15) is given by

$$E_{11} = \frac{1}{2}\left(\frac{\partial x_1}{\partial X_1}\frac{\partial x_1}{\partial X_1} + \frac{\partial x_2}{\partial X_1}\frac{\partial x_2}{\partial X_1} - 1 \right)$$

$$E_{22} = \frac{1}{2}\left(\frac{\partial x_1}{\partial X_2}\frac{\partial x_1}{\partial X_2} + \frac{\partial x_2}{\partial X_2}\frac{\partial x_2}{\partial X_2} - 1 \right) \tag{9.16}$$

$$E_{12} = E_{21} = \frac{1}{2}\left(\frac{\partial x_1}{\partial X_1}\frac{\partial x_1}{\partial X_2} + \frac{\partial x_2}{\partial X_1}\frac{\partial x_2}{\partial X_2} \right)$$

On the other hand, the coordinate system x_1–x_2–x_3 can be chosen as the reference system. Then the description, in which the independent variable is the particle position vector **x** in the deformed state, is known as the *Eulerian description*, and the reference system x is known as an *Eulerian coordinate*. The transformation equation in terms of x_i is $X_i = X_i(x_1, x_2, x_3)$. Then,

$$[dX] = [g]^{-1}[dx] \qquad (9.17)$$

where $[g]^{-1}$ is the inverse matrix of $[g]$ with the components

$$g_{ji}^{-1} = \frac{\partial X_j}{\partial x_i} = \frac{[co(g_{ij})]^T}{\det \mathbf{g}} \qquad (9.18)$$

where $[co(g_{ij})]^T$ is the transpose of the cofactor matrix of g_{ij}, and $\det \mathbf{g}$ is the determinant of $[g]$.

The strain tensor associated with the Eulerian system (in terms of the deformed configuration) is termed the *Eulerian strain* (e_{ij}), which is also known as *Almansi's strain* for large deformation and *Cauchy's strain* for infinitesimal deformation (Fung 1977). It is defined as

$$[e] = \tfrac{1}{2}\{[\delta] - ([g]^{-1})^T \cdot ([g]^{-1})\} \qquad (9.19)$$

or

$$2e_{11} = 1 - \left(\frac{\partial X_1}{\partial x_1}\frac{\partial X_1}{\partial x_1} + \frac{\partial X_2}{\partial x_1}\frac{\partial X_2}{\partial x_1} \right)$$

$$2e_{22} = 1 - \left(\frac{\partial X_1}{\partial x_2}\frac{\partial X_1}{\partial x_2} + \frac{\partial X_2}{\partial x_2}\frac{\partial X_2}{\partial x_2} \right) \qquad (9.20)$$

$$2e_{12} = 2e_{21} = -\left(\frac{\partial X_1}{\partial x_1}\frac{\partial X_1}{\partial x_2} + \frac{\partial X_2}{\partial x_1}\frac{\partial X_2}{\partial x_2} \right)$$

Equations (9.16) and (9.20) can be rewritten in terms of the displacement vectors $\mathbf{U}$ and $\mathbf{u}$, which are associated with the coordinate systems X and x (Fig. 9.3), respectively, and can be expressed as

$$\mathbf{U} = \mathbf{u} = \mathbf{x} - \mathbf{X} \qquad (9.21)$$

Then, the alternate formulas for Lagrangian and Eulerian strain tensors are

$$E_{ij} = \frac{1}{2}\left(\frac{\partial U_i}{\partial X_j} + \frac{\partial U_j}{\partial X_i} + \frac{\partial U_k}{\partial X_i}\frac{\partial U_k}{\partial X_j} \right) \qquad (9.22)$$

and

$$e_{ij} = \frac{1}{2}\left(\frac{\partial u_i}{\partial x_j} + \frac{\partial u_j}{\partial x_i} - \frac{\partial u_k}{\partial x_i}\frac{\partial u_k}{\partial x_j} \right) \qquad (9.23)$$

If the displacement gradients are sufficiently small, the quadratic terms in Eqs. (9.22) and (9.23) can be neglected in comparison with the linear terms. Then, the Lagrangian and Eulerian strain tensors are reduced to the linear forms, and both are equal to the strain (ε_{ij}) for infinitesimal deformation

$$\varepsilon_{ij} = \frac{1}{2}\left(\frac{\partial u_i}{\partial x_j} + \frac{\partial u_j}{\partial x_i}\right) \tag{9.24}$$

The force acting per unit area is known as *stress*. In the case of finite deformation, the area and normal direction of a surface of an undeformed element may be quite different from those of the same surface in the deformed state. Thus, the stress can be defined by the force per either undeformed or deformed area; the former is known as the *Piola–Kirchhoff stress* or *Lagrangian stress* (Π_{ij}), and the latter is known as the *Cauchy's stress* or *Eulerian stress* (σ_{ij}).

For the purpose of illustration, consider a rectangular element *ABCD* of unit thickness, which is deformed into *A'B'C'D'* under a uniaxial load *P* (Fig. 9.4), and neglect the dimensional change in the thickness direction. The Eulerian stress is defined as the force per unit deformed area,

$$\sigma_{xx} = \frac{P}{C'D'} \tag{9.25}$$

The Lagrangian stress is defined as the force, which is acting on the deformed surface, divided by the original surface area (correspond-

Fig. 9.4. A force *P* acting on a deformable body.

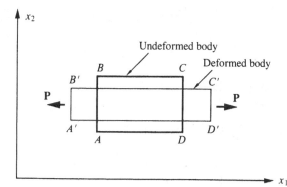

ing to the deformed area),

$$\Pi_{xx} = \frac{P}{CD} \tag{9.26}$$

From Eqs. (9.25) and (9.26),

$$\sigma_{xx} C'D' = \Pi_{xx} CD = P \tag{9.27}$$

The general relations between the two stress descriptions can be found by analyzing the deformation of a generic two-dimensional element as shown in Fig. 9.5. Let QR, with area δA and normal $\mathbf{N}$, be the edge surface of the element OQR in the undeformed state. The corresponding edge surface in the deformed state is $Q'R'$ with area δa and normal $\mathbf{n}$. The coordinates x_1–x_2 are fixed on the deformed element. Also let the surface force vector per unit area of the deformed surface $(Q'R')$ be $\mathbf{f}$ (traction), then the total surface force acting on $Q'R'$ is $\mathbf{f}\delta a$. The nominal traction, $\mathbf{F}$, acting on the undeformed edge surface (QR) is so defined that

$$\mathbf{F}\delta A = \mathbf{f}\delta a = \mathbf{P} \tag{9.28}$$

Fig. 9.5. The correspondence between Lagrangian stress and Eulerian stress.

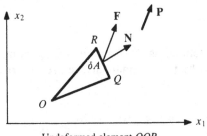

Undeformed element OQR

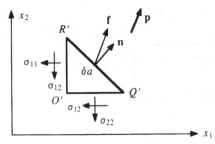

Deformed element $O'Q'R'$

where **P** is the actual force acting on the surface element $Q'R'$ in the deformed state. By neglecting the body force and assuming that the displacement rate with respect to time is very small, the following relation from force equilibrium can be obtained:

$$f_i = \sigma_{ji} n_j \tag{9.29}$$

where f_i and n_j denote the components of **f** and **n**, respectively. Similarly,

$$F_i = \Pi_{ji} N_j \tag{9.30}$$

where F_i and N_j denote the components of **F** and **N**, respectively.

Substituting Eqs. (9.29) and (9.30) into Eq. (9.28) and applying the relation

$$N_i = (\det \mathbf{g})^{-1} n_j \frac{\delta a}{\delta A} g_{ji} \tag{9.31}$$

the result is

$$\sigma = (\det \mathbf{g})^{-1} \mathbf{g} \Pi \tag{9.32}$$

or

$$\sigma_{ji} = (\det \mathbf{g})^{-1} g_{jk} \Pi_{ki} \tag{9.33}$$

where $\det \mathbf{g}$ is the determinant of **g**. The inversion of Eq. (9.32) gives

$$\Pi = (\det \mathbf{g}) \mathbf{g}^{-1} \sigma \tag{9.34}$$

or

$$\Pi_{ji} = (\det \mathbf{g}) g_{jk}^{-1} \sigma_{ki} \tag{9.35}$$

Note that the Eulerian stress tensor is symmetric (i.e. $\sigma_{ij} = \sigma_{ji}$) whereas the Lagrangian stress tensor (Eq. (9.34)) is not symmetric. However, the following quantity

$$P_{AB} = \frac{\partial X_A}{\partial x_i} \Pi_{Bi} = (\det \mathbf{g}) g_{Ai}^{-1} g_{Bj}^{-1} \sigma_{ji} \tag{9.36}$$

is symmetric, and P_{AB} is known as the *second Piola–Kirchhoff stress tensor,* or simply the *Kirchhoff stress*.

9.3 Constitutive relations based on the Lagrangian description

9.3.1 *Finite deformation of a composite lamina*

A basic element in a flexible composite is assumed to be a thin lamina consisting of straight, parallel continuous elastic fibers embedded in an elastic matrix which can sustain large deformation. It is also assumed that the lamina is homogeneous on a scale much larger than that of the inter-fiber spacing. Then, the flexible composite lamina can be treated as a homogeneous two-dimensional orthotropic elastic continuum. In this section, the constitutive equations for such an element under finite deformation are derived based on the Lagrangian description.

Figure 9.2(a) illustrates a unidirectional flexible composite lamina under a finite deformation, where the initial fiber orientation is at an angle θ_o with respect to the X_1 axis. The rectangular Cartesian coordinates l–t are along the initial fiber and transverse directions, respectively. Under loading, the rectangular element $ABCD$ in the undeformed lamina is deformed into a quadrilateral element $A'B'C'D'$ in the deformed lamina. There is an angle $\Delta\theta$ between AD and $A'D'$. Corresponding to this change, the current fiber orientation l' is at an angle θ with respect to the X_1 axis, and

$$\theta = \theta_o + \Delta\theta \tag{9.37}$$

Because of the low shear modulus of the matrix and the highly anisotropic nature ($E_{11} \gg E_{22}$) of flexible composites, $\Delta\theta$ may be quite large and the effective elastic properties of the composite become very sensitive to the fiber orientation. The geometric nonlinearity of a flexible composite is mainly caused by the reorientation of fibers. The material nonlinearity is also pronounced in elastomeric composites under large deformation.

The deformation of the basic element $ABCD$ (Fig. 9.2a) is further examined in Fig. 9.6. Let the rectangular Cartesian coordin-

Fig. 9.6. Deformation of a rectangular element of a composite lamina, referring to the principal material coordinate system. (After Luo and Chou 1990b.)

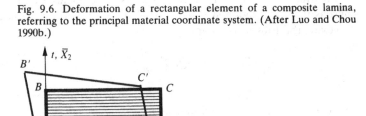

ate system $\bar{X}$ coincide with the initial principal material coordinates $l-t$, where the axis $\bar{X}_1$ is parallel to l. Here, a quantity with an over bar refers to the initial principal material coordinates. Then, the Lagrangian strain matrix in the system $\bar{X}$ is written as

$$[\bar{E}] = \tfrac{1}{2}([\bar{g}]^T[\bar{g}] - [\delta]) \tag{9.38}$$

The deformation of the element shown in Fig. 9.6 can be expressed in terms of these Lagrangian strain components. Let the line elements $AD = dl_o$ and $AB = dt_o$ in the undeformed lamina; also $A'D' = dl$ and $A'B' = dt$ in the deformed lamina. Then, the following relations can be found:

$$2\bar{E}_{11} = [(dl)^2 - (dl_o)^2]/(dl_o)^2$$
$$2\bar{E}_{22} = [(dt)^2 - (dt_o)^2]/(dt_o)^2 \tag{9.39}$$
$$2\bar{E}_{12} = -\sin(\Delta\phi)\sqrt{(1 + 2\bar{E}_{11})}\sqrt{(1 + 2\bar{E}_{22})}$$

where

$$\Delta\phi = \angle B'A'D' - \angle BAD = \phi - \pi/2 \tag{9.40}$$

9.3.2 *Constitutive equations for a composite lamina*

9.3.2.1 *Strain-energy function*
Rivlin (1959) made the following remarks concerning the strain-energy function: 'It was realized that the physical properties of an elastic material can be characterized by a strain-energy function and that this cannot depend on the nine displacement gradients in a completely arbitrary fashion. And also if the material has symmetry, the dependence of the strain-energy on these strain components cannot be arbitrary either.' Following R. S. Rivlin (private communications, 1986–7), the strain-energy density treated here is assumed to be a function of the Lagrangian strain components referring to the principal material coordinates. In the two-dimensional case, referring to Eq. (9.39), the strain-energy per unit volume is written as

$$W = W(\bar{E}_{11}, \bar{E}_{22}, \bar{E}_{12}) \tag{9.41}$$

Since W is unchanged by the following permutation, namely,

$$\bar{X}_1 \rightarrow -\bar{X}_1 \quad \text{and} \quad \bar{x}_1 \rightarrow -\bar{x}_1$$

or

$$\bar{X}_2 \rightarrow -\bar{X}_2 \quad \text{and} \quad \bar{x}_2 \rightarrow -\bar{x}_2 \tag{9.42}$$

the strain-energy function must be an even function of $\bar{E}_{12}$. Then, Eq. (9.41) is rewritten as

$$W = W(\bar{E}_{11}, \bar{E}_{22}, \bar{E}_{12}^2) \qquad (9.43)$$

Finally, the strain-energy per unit volume of the undeformed lamina is assumed in the following fourth-order polynomial form:

$$W = \tfrac{1}{2}C_{11}\bar{E}_1^2 + \tfrac{1}{3}C_{111}\bar{E}_1^3 + \tfrac{1}{4}C_{1111}\bar{E}_1^4 + C_{12}\bar{E}_1\bar{E}_2$$
$$+ \tfrac{1}{2}C_{22}\bar{E}_2^2 + \tfrac{1}{3}C_{222}\bar{E}_2^3 + \tfrac{1}{4}C_{2222}\bar{E}_2^4 + \tfrac{1}{2}C_{66}\bar{E}_6^2 + \tfrac{1}{4}C_{6666}\bar{E}_6^4$$

$$(9.44)$$

where C_{ij}, C_{ijk}, and C_{ijkl} are elastic constants. The short-hand notations are used, namely, $\bar{E}_1 = \bar{E}_{11}$, $\bar{E}_2 = \bar{E}_{22}$, and $\bar{E}_6 = 2\bar{E}_{12}$.

9.3.2.2 *General constitutive equations for a unidirectional lamina*

The stress matrix referring to the material principal coordinate system $\bar{X}_1$–$\bar{X}_2$, is given in terms of W (Rivlin 1970),

$$\bar{\Pi}_{ij} = \frac{\partial W}{\partial \bar{g}_{ji}}$$

$$(9.45)$$

$$\bar{\sigma}_{ij} = \frac{1}{\det \bar{\mathbf{g}}}\, \bar{g}_{ip}\, \frac{\partial W}{\partial \bar{g}_{jp}}$$

Using Eqs. (9.38) and (9.43), it follows that

$$\bar{\Pi}_{ji} = \frac{1}{2}\, \bar{g}_{ip} \left(\frac{\partial W}{\partial \bar{E}_{jp}} + \frac{\partial W}{\partial \bar{E}_{pj}} \right)$$

$$\bar{\sigma}_{ij} = \frac{1}{\det \bar{\mathbf{g}}} \{ W_{11}\bar{g}_{i1}\bar{g}_{j1} + W_{22}\bar{g}_{i2}\bar{g}_{j2} \qquad (9.46)$$
$$+ \tfrac{1}{2}W_{12}(\bar{g}_{i1}\bar{g}_{j2} + \bar{g}_{i2}\bar{g}_{j1}) \}$$

where

$$W_{11} = \frac{\partial W}{\partial \bar{E}_{11}} = C_{11}\bar{E}_{11} + C_{111}\bar{E}_{11}^2 + C_{1111}\bar{E}_{11}^3 + C_{12}\bar{E}_{22}$$

$$W_{22} = \frac{\partial W}{\partial \bar{E}_{22}} = C_{22}\bar{E}_{22} + C_{222}\bar{E}_{22}^2 + C_{2222}\bar{E}_{22}^3 + C_{12}\bar{E}_{11} \qquad (9.47)$$

$$W_{12} = \frac{\partial W}{\partial \bar{E}_{12}} = 4(C_{66}\bar{E}_{12} + 4C_{6666}\bar{E}_{12}^3)$$

To derive the general constitutive equations with reference axes other than the principal material directions, a two-dimensional rectangular Cartesian coordinate system X_1–X_2 is chosen in the plane of the lamina. The angle between X_i and $\bar{X}_i$ is θ_o (Fig. 9.7). Let $[a]$ be an orthogonal transformation matrix,

$$[a] = \begin{bmatrix} \cos\theta_o & -\sin\theta_o \\ \sin\theta_o & \cos\theta_o \end{bmatrix} \tag{9.48}$$

and $[X] = [a] \cdot [\bar{X}]$. Then, the transformation relations for the deformation gradient and Lagrangian strain between coordinate systems X and $\bar{X}$ are:

$$[\bar{g}] = [a]^{\mathrm{T}}[g][a] \tag{9.49}$$

and

$$[\bar{E}] = [a]^{\mathrm{T}}[E][a] \tag{9.50}$$

With Eq. (9.48), Eq. (9.50) yields

$$\bar{E}_{11} = \tfrac{1}{2}E_{11}(1 + \cos 2\theta_o) + E_{12}\sin 2\theta_o + \tfrac{1}{2}E_{22}(1 - \cos 2\theta_o)$$

$$\bar{E}_{22} = \tfrac{1}{2}E_{11}(1 - \cos 2\theta_o) - E_{12}\sin 2\theta_o + \tfrac{1}{2}E_{22}(1 + \cos 2\theta_o)$$

$$\bar{E}_{12} = \bar{E}_{21} = \tfrac{1}{2}(E_{22} - E_{11})\sin 2\theta_o + E_{12}\cos 2\theta_o$$

$$\tag{9.51}$$

The stress matrix referring to the coordinate system X is

$$[\sigma] = [a][\bar{\sigma}][a]^{\mathrm{T}}$$

$$[\Pi] = [a][\bar{\Pi}][a]^{\mathrm{T}} \tag{9.52}$$

With Eqs. (9.46) and (9.49), Eqs. (9.52) are expressed as

$$\sigma_{ij} = \frac{1}{\det \mathbf{g}} g_{ip}g_{jq}\{a_{p1}a_{q1}W_{11} + a_{p2}a_{q2}W_{22} + \tfrac{1}{2}(a_{p1}a_{q2} + a_{p2}a_{q1})W_{12}\} \tag{9.53}$$

and

$$\Pi_{ji} = g_{ip}\{a_{p1}a_{j1}W_{11} + a_{p2}a_{j2}W_{22} + \tfrac{1}{2}(a_{p1}a_{j2} + a_{p2}a_{j1})W_{12}\} \tag{9.54}$$

Then, from Eq. (9.48), Eq. (9.54) is given in the following explicit form:

$$\Pi_{11} = [g_{11}c^2 + g_{12}cs]W_{11} + [g_{11}s^2 - g_{12}cs]W_{22}$$
$$+ [-g_{11}cs + \tfrac{1}{2}g_{12}(c^2 - s^2)]W_{12}$$

$$\Pi_{22} = [g_{22}s^2 + g_{21}cs]W_{11} + [g_{22}c^2 - g_{21}cs]W_{22}$$
$$+ [g_{22}cs + \tfrac{1}{2}g_{21}(c^2 - s^2)]W_{12}$$

$$(9.55)$$

$$\Pi_{12} = [g_{22}cs + g_{21}c^2]W_{11} + [g_{21}s^2 - g_{22}cs]W_{22}$$
$$+ [-g_{21}cs + \tfrac{1}{2}g_{22}(c^2 - s^2)]W_{12}$$

$$\Pi_{21} = [g_{11}cs + g_{12}s^2]W_{11} + [g_{12}c^2 - g_{11}cs]W_{22}$$
$$+ [g_{12}cs + \tfrac{1}{2}g_{11}(c^2 - s^2)]W_{12}$$

where $c = \cos\theta_0$ and $s = \sin\theta_0$.

Equations (9.55) are the general constitutive equations for a composite lamina under finite deformation, where the deformation gradients, g_{ij}, represent the geometric nonlinearity influenced by the configuration changes of the lamina. The nonlinear expressions of W_{ij} (Eqs. (9.47)) represent the material nonlinearity of the composites. If the deformation of the composite lamina is infinitesimal (i.e. $g_{ij} = \delta_{ij}$) and only the linear terms (i.e. C_{ij}) remain in the expression of W_{ij}, Eqs. (9.55) can be easily reduced to the familiar linear stress–strain equation used for rigid composites.

For a specific deformation, the deformation gradient matrix, $[g]$, is calculated from Eq. (9.14); the Lagrangian strain referring to the principal material coordinates, $[\bar{E}]$, is obtained from Eqs. (9.15) and (9.51); and W_{ij} are obtained by introducing $[\bar{E}]$ into Eqs. (9.47). Then, the corresponding Lagrangian stresses, $[\Pi]$, can be determined from Eqs. (9.55). In the following Sections 9.3.2.3–9.3.2.5, this procedure is illustrated by some specific examples.

9.3.2.3 *Pure homogeneous deformation*

Consider the rectangular lamina of Fig. 9.7; its edges are parallel to the axes of the coordinate system X. The lamina is subjected to a pure homogeneous deformation with principal extension ratios λ_1 and λ_2 defined along the axes of the coordinate system X. The deformation is described by

$$x_1 = \lambda_1 X_1$$
$$x_2 = \lambda_2 X_2$$

$$(9.56)$$

Consequently, referring to Eqs. (9.14) and (9.15),

$$[g] = \begin{bmatrix} \lambda_1 & 0 \\ 0 & \lambda_2 \end{bmatrix} \tag{9.57}$$

and

$$2[E] = \begin{bmatrix} \lambda_1^2 - 1 & 0 \\ 0 & \lambda_2^2 - 1 \end{bmatrix} \tag{9.58}$$

From Eqs. (9.51) and (9.58), the following can be obtained:

$$\bar{E}_{11} = \tfrac{1}{4}[(\lambda_1^2 + \lambda_2^2 - 2) + (\lambda_1^2 - \lambda_2^2)\cos 2\theta_0]$$
$$\bar{E}_{22} = \tfrac{1}{4}[(\lambda_1^2 + \lambda_2^2 - 2) - (\lambda_1^2 - \lambda_2^2)\cos 2\theta_0] \tag{9.59}$$
$$\bar{E}_{12} = \tfrac{1}{4}(\lambda_2^2 - \lambda_1^2)\sin 2\theta_0$$

Then, the components of the Lagrangian stress are obtained from Eqs. (9.55) and (9.57) as

$$\Pi_{11} = \lambda_1(c^2 W_{11} + s^2 W_{22} - cs W_{12})$$
$$\Pi_{22} = \lambda_2(s^2 W_{11} + c^2 W_{22} + cs W_{12})$$
$$\Pi_{12} = \lambda_2[cs(W_{11} - W_{22}) + \tfrac{1}{2}(c^2 - s^2)W_{12}] \tag{9.60}$$
$$\Pi_{21} = \lambda_1[cs(W_{11} - W_{22}) + \tfrac{1}{2}(c^2 - s^2)W_{12}]$$

where W_{11}, W_{22} and W_{12} are given by Eqs. (9.47) and (9.59).

Fig. 9.7. Pure homogeneous deformation. (After Luo and Chou 1990b.)

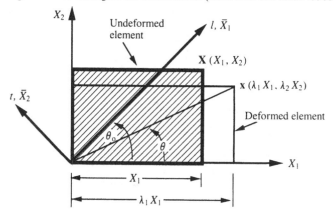

9.3.2.4 *Simple shear*

Suppose that the rectangular lamina, with its edges parallel to the axes of the coordinate system X, is subjected to a simple shear of amount K in the direction of the X_1 axis (Fig. 9.8). Then, the deformation is described by

$$x_1 = X_1 + KX_2$$
$$x_2 = X_2 \tag{9.61}$$

For this deformation, referring to Eqs. (9.14), (9.15) and (9.51),

$$[g] = \begin{bmatrix} 1 & K \\ 0 & 1 \end{bmatrix}, \qquad 2[E] = \begin{bmatrix} 0 & K \\ K & K^2 \end{bmatrix} \tag{9.62}$$

and

$$\bar{E}_{12} = \tfrac{1}{4}[K^2 + (2K \sin 2\theta_o - K^2 \cos 2\theta_o)]$$
$$\bar{E}_{22} = \tfrac{1}{4}[K^2 - (2K \sin 2\theta_o - K^2 \cos 2\theta_o)] \tag{9.63}$$
$$\bar{E}_{12} = \tfrac{1}{4}[K^2 \sin 2\theta_o + 2K \cos 2\theta_o]$$

Then, the components of the Lagrangian stress are obtained from Eqs. (9.55) and (9.62):

$$\Pi_{11} = [c^2 + Kcs]W_{11} + [s^2 - Kcs]W_{22}$$
$$\qquad + [-cs + \tfrac{1}{2}K(c^2 - s^2)]W_{12}$$
$$\Pi_{22} = s^2 W_{11} + c^2 W_{22} + cs W_{12} \tag{9.64}$$
$$\Pi_{12} = cs W_{11} - cs W_{22} + \tfrac{1}{2}(c^2 - s^2)W_{12}$$
$$\Pi_{21} = [cs + Ks^2]W_{11} + [Kc^2 - cs]W_{22} + [Kcs + \tfrac{1}{2}(c^2 - s^2)]W_{12}$$

Fig. 9.8. Simple shear deformation. (After Luo and Chou 1990b.)

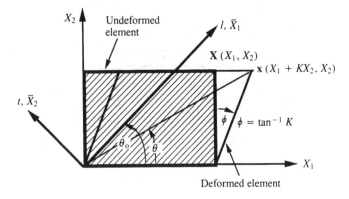

Considering the example of $\theta_o = 45°$, Eqs. (9.63) yield

$$\bar{E}_{11} = \tfrac{1}{2}K + \tfrac{1}{4}K^2$$
$$\bar{E}_{22} = -\tfrac{1}{2}K + \tfrac{1}{4}K^2 \qquad (9.65)$$
$$\bar{E}_{12} = \tfrac{1}{4}K^2$$

Then Eqs. (9.64) give

$$\Pi_{11} = \tfrac{1}{2}\{(1 + K)W_{11} + (1 - K)W_{22} - W_{12}\}$$
$$\Pi_{22} = \tfrac{1}{2}\{W_{11} + W_{22} + W_{12}\}$$
$$\Pi_{12} = \tfrac{1}{2}\{W_{11} - W_{22}\} \qquad (9.66)$$
$$\Pi_{21} = \tfrac{1}{2}\{(1 + K)W_{11} + (K - 1)W_{22} + KW_{12}\}$$

Figure 9.9 shows the theoretical prediction of the stress–strain relation (Eqs. (9.64)) for Kevlar/silicone elastomer laminae with various initial fiber orientations under simple shear deformation. The elastic constants used in the analysis are shown in Table 9.1 (Luo 1988). The result shows that the simple shear properties of a composite lamina under finite deformation are significantly influenced by the fiber orientation. Figure 9.10 gives the comparison between analytical predictions and experimental results of a 0° specimen under simple shear. Figure 9.11 shows the same comparison on a

Fig. 9.9. Theoretical predictions of simple shear deformation of Kevlar/silicone elastomer composite laminae for various initial fiber orientations. (After Luo and Chou 1990b.)

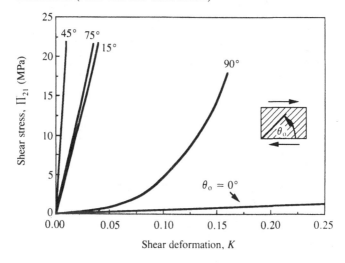

Table 9.1. *Elastic constants of Kevlar-49/silicone flexible composites (After Luo, 1988).*

S_{11}	$(MPa)^{-1}$	0.114×10^{-3}
S_{1111}	$(MPa)^{-3}$	0
S_{12}	$(MPa)^{-1}$	-69.9×10^{-6}
S_{22}	$(MPa)^{-1}$	0.306
S_{2222}	$(MPa)^{-3}$	0.563
S_{66}	$(MPa)^{-1}$	0.387
S_{6666}	$(MPa)^{-3}$	77.5×10^{-3}
S_{166}	$(MPa)^{-2}$	3.43×10^{-6}
S_{2266}	$(MPa)^{-3}$	56.3×10^{-3}
C_{11}	(MPa)	8.6×10^{3}
C_{1111}	(MPa)	0
C_{12}	(MPa)	-1.3
C_{22}	(MPa)	2.77
C_{2222}	(MPa)	-12.5
C_{66}	(MPa)	2.55
C_{6666}	(MPa)	-2.45

Fig. 9.10. Comparisons between theoretical predictions and experimental data of simple shear response of 0° Kevlar/silicone elastomer composite laminae. (After Luo and Chou 1990b.)

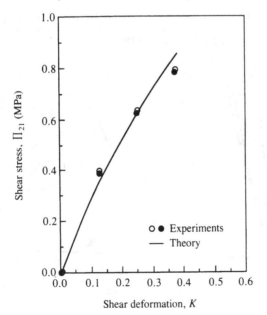

90° specimen; the experimental data, which are obtained from three-rail shear tests (Whitney, Daniel and Pipes 1982), are lower than the predicted values at large shear deformation. This is caused by the edge effects, and fiber pull-out from clamped edges for 90° specimens at large deformation.

9.3.2.5 *Simple shear superposed on simple extension*

A rectangular lamina, with the edges parallel to the axes of the rectangular Cartesian coordinate system X, is first subjected to the pure homogeneous deformation described by Eqs. (9.56), followed by a simple shear of magnitude K. There are two cases for the direction of the shear deformation: (1) parallel to the X_1 axis, and (2) parallel to the X_2 axis; both are discussed in the following:

Case 1

Figure 9.12(a) illustrates this deformation, which can be specified as

$$x_1 = \lambda_1 X_1 + K\lambda_2 X_2$$
$$x_2 = \lambda_2 X_2$$

(9.67)

Fig. 9.11. Comparisons between theoretical predictions and experimental data of simple shear response of 90° Kevlar/silicone elastomer composite lamina. (After Luo and Chou 1990b.)

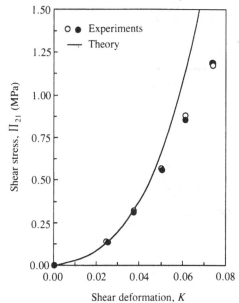

From Eqs. (9.14), (9.15) and (9.51),

$$[g] = \begin{bmatrix} \lambda_1 & K\lambda_2 \\ 0 & \lambda_2 \end{bmatrix}, \qquad 2[E] = \begin{bmatrix} \lambda_1^2 - 1 & K\lambda_1\lambda_2 \\ K\lambda_1\lambda_2 & \lambda_2^2(K^2+1) - 1 \end{bmatrix}$$

(9.68)

and

$$\bar{E}_{11} = \tfrac{1}{4}[(\lambda_1^2 + \lambda_2^2 - 2 + \lambda_2^2 K^2) + (\lambda_1^2 - \lambda_2^2 - \lambda_2^2 K^2)\cos 2\theta_o$$
$$+ 2K\lambda_1\lambda_2 \sin 2\theta_o]$$

$$\bar{E}_{22} = \tfrac{1}{4}[(\lambda_1^2 + \lambda_2^2 - 2 + \lambda_2^2 K^2) - (\lambda_1^2 - \lambda_2^2 - \lambda_2^2 K^2)\cos 2\theta_o$$
$$- 2K\lambda_1\lambda_2 \sin 2\theta_o]$$

(9.69)

$$\bar{E}_{12} = \tfrac{1}{4}[(\lambda_2^2 - \lambda_1^2 + \lambda_2^2 K^2)\sin 2\theta_o + 2K\lambda_1\lambda_2 \cos 2\theta_o]$$

The components of the Lagrangian stress are obtained from Eqs.

Fig. 9.12. Simple shear superposed on simple extension. (After Luo and Chou 1990b.)

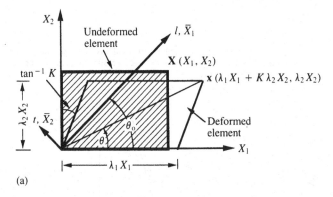

(a)

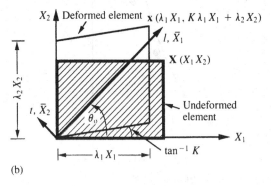

(b)

(9.55) and (9.68):

$$\Pi_{11} = [\lambda_1 c^2 + K\lambda_2 cs]W_{11} + [\lambda_1 s^2 - K\lambda_2 cs]W_{22}$$
$$+ [-\lambda_1 cs + \tfrac{1}{2}K\lambda_2(c^2 - s^2)]W_{12}$$

$$\Pi_{22} = \lambda_2[s^2 W_{11} + c^2 W_{22} + cs W_{12}]$$

$$\Pi_{12} = \lambda_2[cs W_{11} - cs W_{22} + \tfrac{1}{2}(c^2 - s^2)W_{12}] \tag{9.70}$$

$$\Pi_{21} = [\lambda_1 cs + K\lambda_2 s^2]W_{11} + [K\lambda_2 c^2 - \lambda_1 cs]W_{22}$$
$$+ [K\lambda_2 cs + \tfrac{1}{2}\lambda_1(c^2 - s^2)]W_{12}$$

Case 2

Figure 9.12(b) shows the deformation defined by

$$x_1 = \lambda_1 X_1$$
$$x_2 = K\lambda_1 X_1 + \lambda_2 X_2 \tag{9.71}$$

For this deformation, referring to Eqs. (9.14), (9.15) and (9.51),

$$[g] = \begin{bmatrix} \lambda_1 & 0 \\ K\lambda_1 & \lambda_2 \end{bmatrix}, \qquad 2[E] = \begin{bmatrix} \lambda_1^2(K^2 + 1) - 1 & K\lambda_1\lambda_2 \\ K\lambda_1\lambda_2 & \lambda_2^2 - 1 \end{bmatrix} \tag{9.72}$$

and

$$\bar{E}_{11} = \tfrac{1}{4}[(\lambda_1^2 + \lambda_2^2 - 2 + \lambda_1^2 K^2) + (\lambda_1^2 - \lambda_2^2 + \lambda_1^2 K^2)\cos 2\theta_{\mathrm{o}}$$
$$+ 2K\lambda_1\lambda_2 \sin 2\theta_{\mathrm{o}}]$$

$$\bar{E}_{22} = \tfrac{1}{4}[(\lambda_1^2 + \lambda_2^2 - 2 + \lambda_1^2 K^2) - (\lambda_1^2 - \lambda_2^2 + \lambda_1^2 K^2)\cos 2\theta_{\mathrm{o}}$$
$$- 2K\lambda_1\lambda_2 \sin 2\theta_{\mathrm{o}}] \tag{9.73}$$

$$\bar{E}_{12} = \tfrac{1}{4}[(\lambda_2^2 - \lambda_1^2 + \lambda_1^2 K^2)\sin 2\theta_{\mathrm{o}} + 2K\lambda_1\lambda_2 \cos 2\theta_{\mathrm{o}}]$$

The components of the Lagrangian stress obtained from Eqs. (9.55) and (9.72) are:

$$\Pi_{11} = \lambda_1[c^2 W_{11} + s^2 W_{22} - cs W_{12}]$$

$$\Pi_{22} = [\lambda_2 s^2 + K\lambda_1 cs]W_{11} + [\lambda_2 c^2 - K\lambda_1 cs]W_{22}$$
$$+ [\lambda_2 cs + \tfrac{1}{2}K\lambda_1(c^2 - s^2)]W_{12}$$

$$\Pi_{12} = [\lambda_2 cs + K\lambda_1 c^2]W_{11} + [K\lambda_1 s^2 - \lambda_2 cs]W_{22} \tag{9.74}$$
$$+ [-K\lambda_1 cs + \tfrac{1}{2}\lambda_2(c^2 - s^2)]W_{12}$$

$$\Pi_{21} = \lambda_1[cs W_{11} - cs W_{22} + \tfrac{1}{2}(c^2 - s^2)W_{12}]$$

Figure 9.13 illustrates an off-axis specimen under uniaxial tension. It is understood that the clamping of the specimen at the ends induces a local non-uniform strain field. However, if the length-to-width ratio of the specimen is sufficiently large, a uniform state of stress and strain prevails at the center of the specimen (Pagano and Halpin 1968), and the central lines of the specimen remain straight in the X_1 direction. Then, this deformation corresponds to Case 1, namely, 'simple shear superposed on simple extension'. The uni-axial loading condition can be described as $\Pi_{11} \neq 0$, $\Pi_{22} = 0$ and $\Pi_{21} = 0$. Then, from Eqs. (9.74),

$$\Pi_{11} = \lambda_1 [c^2 W_{11} + s^2 W_{22} - cs W_{12}]$$

$$0 = s^2 W_{11} + c^2 W_{22} + cs W_{12} \tag{9.75}$$

$$0 = cs W_{11} - cs W_{22} + \tfrac{1}{2}(c^2 - s^2) W_{12}$$

where W_{ij} are obtained by from Eqs. (9.68) and (9.47). The three unknowns λ_1, λ_2 and K in Eqs. (9.75) can be solved from these equations. Figure 9.14 shows the comparison between analytical predictions and experimental results for the off-axis response of Kevlar/silicone elastomer laminae under simple tension. The fiber initial orientations are 10°, 30° and 60°. The elastic constants used in the calculation are shown in Table 9.1.

Fig. 9.13. Off-axis specimens of flexible composite laminae (a) without loading, (b) with loading. The 15° one is tirecord/rubber, and the 10° and 30° ones are Kevlar/silicone elastomer. (After Luo and Chou 1988a.)

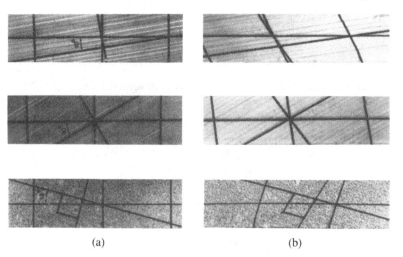

(a)   (b)

9.3.3 *Constitutive equations of flexible composite laminates*

9.3.3.1 *Constitutive equations*

The analytical methodology developed in Section 9.3.2 for composite laminae is applied to study the constitutive relations of laminated flexible composites (Fig. 9.15) under finite plane deformation (Luo and Chou 1989). The stress resultant in Lagrangian description (N_{ij}) is defined as

$$N_{ij} = \int_{-h/2}^{h/2} \Pi_{ij} \, dz \tag{9.76}$$

where h is the initial laminate thickness. N_{ij} so defined gives the total force in the i direction per unit length of the undeformed laminate.

Assume that the laminate is composed of n layers of unidirectional laminae. By neglecting the interlaminar shear deformation, the deformation gradient, g_{ij} (Eq. (9.14)), has the same value for all the layers; this is also true for E_{ij} (Eq. (9.15)). For an arbitrary kth lamina within the laminate, let $\theta_{o}^{(k)}$ be the fiber orientation angle

Fig. 9.14. Comparisons between theoretical predictions and experimental results on 10°, 30° and 60° off-axis stress–strain response of Kevlar/silicone elastomer composite laminae.

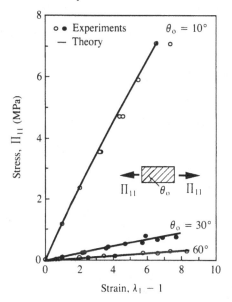

with respect to the coordinate X_1, and $a_{ij}^{(k)}$ the values given by Eq. (9.48) for $\theta_o = \theta_o^{(k)}$. Also, for the kth lamina, let $\bar{E}_{ij}^{(k)}$ be the values of $\bar{E}_{ij}$ given by Eqs. (9.51), and $W_{ij}^{(k)}$ the values of W_{ij} given by Eqs. (9.47). Then, from Eqs. (9.54) and (9.76) the following can be derived:

$$N_{ji} = g_{ip} \sum_{k=1}^{n} (h_k - h_{k-1})\{a_{p1}^{(k)}a_{j1}^{(k)}W_{11}^{(k)}$$
$$+ a_{p2}^{(k)}a_{j2}^{(k)}W_{22}^{(k)} + \tfrac{1}{2}(a_{p1}^{(k)}a_{j2}^{(k)} + a_{p2}^{(k)}a_{j1}^{(k)})W_{12}^{(k)}\} \qquad (9.77a)$$

If the laminae are identical in thickness, t, then

$$N_{ji} = tg_{ip} \sum_{k=1}^{n} \{a_{p1}^{(k)}a_{j1}^{(k)}W_{11}^{(k)}$$
$$+ a_{p2}^{(k)}a_{j2}^{(k)}W_{22}^{(k)} + \tfrac{1}{2}(a_{p1}^{(k)}a_{j2}^{(k)} + a_{p2}^{(k)}a_{j1}^{(k)})W_{12}^{(k)}\} \qquad (9.77b)$$

Equations (9.77) are the general constitutive equations for flexible composite laminates under finite deformations. The applications of the constitutive relations are exemplified in the following.

9.3.3.2 *Homogeneous deformation*

The homogeneous deformation of a composite laminate (Fig. 9.16) is defined by

$$x_1 = \lambda_1 X_1$$
$$x_2 = \lambda_2 X_2 \qquad (9.78)$$

Fig. 9.15. A composite laminate. (After Luo and Chou 1989.)

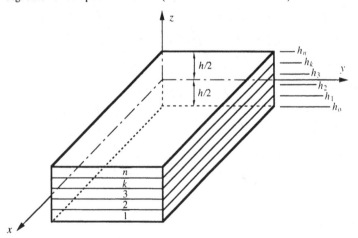

where λ_1 and λ_2 are the extension ratios in the X_1 and X_2 directions, respectively. Thus, referring to Eqs. (9.14) and (9.15), for all laminae:

$$[g] = \begin{bmatrix} \lambda_1 & 0 \\ 0 & \lambda_2 \end{bmatrix}, \qquad 2[E] = \begin{bmatrix} \lambda_1^2 - 1 & 0 \\ 0 & \lambda_2^2 - 1 \end{bmatrix} \tag{9.79}$$

The Lagrangian strains referring to the principal material coordinate for the kth lamina are obtained from Eqs. (9.51)

$$\bar{E}_{11}^{(k)} = \tfrac{1}{2}(\lambda_1^2 \cos^2 \theta_o^{(k)} + \lambda_2^2 \sin^2 \theta_o^{(k)} - 1)$$

$$\bar{E}_{22}^{(k)} = \tfrac{1}{2}(\lambda_1^2 \sin^2 \theta_o^{(k)} + \lambda_2^2 \cos^2 \theta_o^{(k)} - 1) \tag{9.80}$$

$$\bar{E}_{12}^{(k)} = \tfrac{1}{2}(\lambda_2^2 - \lambda_1^2) \sin \theta_o^{(k)} \cos \theta_o^{(k)}$$

Then, the components of the Lagrangian stress resultant are obtained from Eq. (9.77b)

$$N_{11} = \frac{t}{2} \lambda_1 \sum_{k=1}^{n} \{ W_{11}^{(k)}(1 + \cos 2\theta_o^{(k)})$$
$$+ W_{22}^{(k)}(1 - \cos 2\theta_o^{(k)}) - W_{12}^{(k)} \sin 2\theta_o^{(k)} \}$$

$$N_{22} = \frac{t}{2} \lambda_2 \sum_{k=1}^{n} \{ W_{11}^{(k)}(1 - \cos 2\theta_o^{(k)})$$
$$+ W_{22}^{(k)}(1 + \cos 2\theta_o^{(k)}) + W_{12}^{(k)} \sin 2\theta_o^{(k)} \} \tag{9.81}$$

$$N_{12} = \frac{t}{2} \lambda_2 \sum_{k=1}^{n} \{ (W_{11}^{(k)} - W_{22}^{(k)}) \sin 2\theta_o^{(k)} + W_{12}^{(k)} \cos 2\theta_o^{(k)} \}$$

$$N_{21} = \frac{t}{2} \lambda_1 \sum_{k=1}^{n} \{ (W_{11}^{(k)} - W_{22}^{(k)}) \sin 2\theta_o^{(k)} + W_{12}^{(k)} \cos 2\theta_o^{(k)} \}$$

Fig. 9.16. A symmetric flexible composite laminate under a uniaxial load. (After Luo and Chou 1989.)

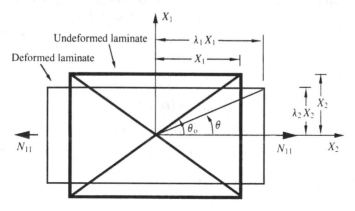

where $W_{11}^{(k)}$, $W_{22}^{(k)}$ and $W_{12}^{(k)}$ are obtained from Eqs. (9.47) and (9.80) with two variables λ_1 and λ_2.

9.3.3.3 *Simple extension of a symmetric composite laminate*

(A)   Tensile stress–strain relation

The state of homogeneous deformation is assumed for a symmetric composite laminate with fiber orientation sequences of $+\theta_o/-\theta_o/-\theta_o/+\theta_o$ under unidirectional tension. Because $\bar{E}_{11}$ and $\bar{E}_{22}$ are even functions of θ_o, and $\bar{E}_{12}$ is an odd function of θ_o (Eqs. (9.80)), Eqs. (9.47) become

$$W_{11}^{(\theta)} = W_{11}^{(-\theta)}$$
$$W_{22}^{(\theta)} = W_{22}^{(-\theta)} \qquad (9.82)$$
$$W_{12}^{(\theta)} = -W_{12}^{(-\theta)}$$

Then, Eqs. (9.81) can be reduced to

$$N_{11} = \frac{h}{2}\lambda_1\{W_{11}^{(\theta)}(1 + \cos 2\theta_o)$$
$$+ W_{22}^{(\theta)}(1 - \cos 2\theta_o) - W_{12}^{(\theta)}\sin 2\theta_o\}$$
$$N_{22} = \frac{h}{2}\lambda_2\{W_{11}^{(\theta)}(1 - \cos 2\theta_o) \qquad (9.83)$$
$$+ W_{22}^{(\theta)}(1 + \cos 2\theta_o) + W_{12}^{(\theta)}\sin 2\theta_o\}$$
$$N_{12} = N_{21} = 0$$

where h is the thickness of the laminate; $W_{ij}^{(\theta)}$, obtained from Eqs. (9.47) and (9.80), is a function of λ_1 and λ_2. With the uniaxial loading condition and the values of elastic constants, the two unknowns, λ_1 and λ_2, in Eqs. (9.83) can be solved.

For example, let $\theta_o = 45°$, from Eqs. (9.47), (9.80) and (9.83),

$$N_{11}/h = \lambda_1\{C_{66}(\lambda_1^2 - \lambda_2^2) + \tfrac{1}{4}C_{6666}(\lambda_1^2 - \lambda_2^2)^3\}$$
$$N_{22}/h = 0$$
$$= (D - 4C_{66})\lambda_1^2 - (D + 4C_{66})\lambda_2^2 \qquad (9.84)$$
$$+ \tfrac{1}{4}(C_{111} + C_{222})(\lambda_1^2 + \lambda_2^2 - 2)^2$$
$$+ \tfrac{1}{16}(C_{1111} + C_{2222})(\lambda_1^2 + \lambda_2^2 - 2)^3 + C_{6666}(\lambda_2^2 - \lambda_1^2)^3$$

where $D = C_{11} + 2C_{12} + C_{22}$.

Figure 9.17 shows the comparison between the theoretical predictions and experimental results of the stress–strain relation of $[\pm 45°]_s$ Kevlar/silicone elastomer composite laminates under uniaxial load. Reasonable agreement has been found.

(B)   Effective Poisson's ratio

The Poisson's ratio is defined as the negative ratio of the strain in the X_j direction to the strain in the X_i direction due to an applied stress in the X_i direction. The Poisson's ratio of a symmetric composite laminate was derived by Posfalvi (1977) based upon a finite deformation consideration. Although experimental results of large deformation were presented, the comparison of theory with experiments was still limited to the small deformation range.

From the above analysis the effective Poisson's ratio in the finite deformation range can be readily predicted. For example, for a $[+\theta_o / -\theta_o]_s$ laminate under unidirectional load, the effective Poisson's ratio at a given strain level can be determined from Eqs. (9.79) as

$$\frac{E_{22}}{E_{11}} = \frac{\lambda_2^2 - 1}{\lambda_1^2 - 1} \tag{9.85}$$

Fig. 9.17. Comparisons between theoretical predictions and experimental data of stress–strain response of a $[\pm 45°]_s$ Kevlar/silicone elastomer composite laminate under uniaxial load. (After Luo and Chou 1989.)

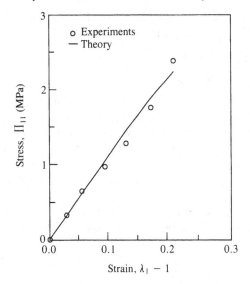

where the relation between λ_1 and λ_2 can be obtained from Eqs. (9.83) with $N_{22} = 0$.

The approximate order of the ratio E_{22}/E_{11} can be obtained by neglecting the non-linear terms (i.e. $C_{111}, \ldots, C_{6666}$, etc.) in the expressions of Eqs. (9.47) for W_{ij}. Then

$$-\frac{\lambda_2^2 - 1}{\lambda_1^2 - 1} = \frac{A}{B} \tag{9.86}$$

where

$$A = C_{11} \cos^2 \theta_o \sin^2 \theta_o + C_{12}(\sin^4 \theta_o + \cos^4 \theta_o)$$
$$+ C_{22} \cos^2 \theta_o \sin^2 \theta_o - 4C_{66} \cos^2 \theta_o \sin^2 \theta_o$$

$$B = C_{11} \sin^4 \theta_o + 2C_{12} \cos^2 \theta_o \sin^2 \theta_o + C_{22} \cos^4 \theta_o$$
$$+ 4C_{66} \cos^2 \theta_o \sin^2 \theta_o$$

For example, for $\theta_o = 45°$, Eq. (9.86) yields

$$\frac{A}{B} = \frac{(C_{11} + 2C_{12} + C_{22}) - 4C_{66}}{(C_{11} + 2C_{12} + C_{22}) + 4C_{66}} \tag{9.87}$$

Since the shear modulus C_{66} for flexible composites is relatively small, it can be assumed $A/B \approx 1$. Then, Eq. (9.85) becomes

$$\frac{E_{22}}{E_{11}} = -1 \tag{9.88}$$

Furthermore, if the flexible composite is very stiff in the fiber direction (i.e. $C_{11} \gg C_{ij}$, $ij \neq 11$) and $\theta_o \neq 0$, Eq. (9.86) becomes

$$\frac{E_{22}}{E_{11}} = \frac{\lambda_2^2 - 1}{\lambda_1^2 - 1} = -\frac{\cos^2 \theta_o}{\sin^2 \theta_o} \tag{9.89}$$

The results of Eq. (9.89) can also be derived by using the 'ideal fiber reinforced material theory' (Adkins and Rivlin 1955).

Figure 9.18 gives the comparison between theoretical predictions and experimental results of the ratio λ_2/λ_1 for $[\pm\theta_o]_s$ Kevlar/silicone elastomer composite laminates under uniaxial load. The initial fiber orientations are 15°, 30° and 45°. Very good agreement has been found.

Also, using the definition of Posfalvi (1977), the current Poisson's ratio at a given strain level can be derived from Eq. (9.86) as

$$\frac{d\lambda_2}{d\lambda_1} = -\frac{\lambda_1}{\lambda_2} \frac{A}{B} \tag{9.90}$$

Referring to Fig. 9.16, the current fiber orientation, $\theta^{(k)}$, of the kth lamina, can be expressed in terms of λ_1, λ_2 and the initial fiber orientation $\theta_o^{(k)}$ as

$$\tan \theta^{(k)} = \frac{\lambda_2 \sin \theta_o^{(k)}}{\lambda_1 \cos \theta_o^{(k)}} = \frac{\lambda_2}{\lambda_1} \tan \theta_o^{(k)} \tag{9.91}$$

where λ_1 and λ_2 are obtained by solving Eqs. (9.83).

9.3.4 *Determination of elastic constants*

In Section 9.3.2.1, the strain-energy per unit volume of an undeformed lamina is assumed in a polynomial form (Eq. (9.44)). The elastic constants in the strain-energy expression need to be determined experimentally. Some experimental methods for characterizing these constants are summarized below (Luo 1988).

9.3.4.1 *Tensile properties*

The constants C_{11}, C_{111}, C_{1111}, C_{22}, C_{222}, C_{2222}, and C_{12} are associated with the tensile behavior of flexible composites and are determined by unidirectional tensile tests. Consider a composite lamina under a unidirectional load (i.e. $\Pi_{11} \neq 0$, $\Pi_{22} = 0$ and

Fig. 9.18. Comparisons between theoretical predictions and experimental results of the ratio λ_2/λ_1 of $[\pm\theta_o]_s$ Kevlar/silicone elastomer composite laminates under uniaxial load. (After Luo and Chou 1989.)

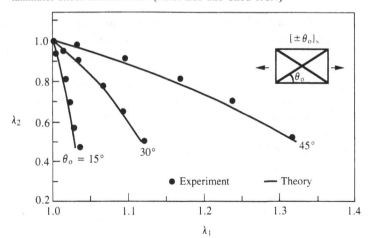

$\Pi_{12} = 0$). For $\theta_o = 0°$, Eqs. (9.47) and (9.60) yield

$$\Pi_{11}/\lambda_1 = C_{11}\left(\frac{\lambda_1^2 - 1}{2}\right) + C_{111}\left(\frac{\lambda_1^2 - 1}{2}\right)^2$$

$$+ C_{1111}\left(\frac{\lambda_1^2 - 1}{2}\right)^3 + C_{12}\left(\frac{\lambda_2^2 - 1}{2}\right)$$

$$\Pi_{22} = 0 = C_{22}\left(\frac{\lambda_2^2 - 1}{2}\right) + C_{222}\left(\frac{\lambda_2^2 - 1}{2}\right)^2 \qquad (9.92)$$

$$+ C_{2222}\left(\frac{\lambda_2^2 - 1}{2}\right)^3 + C_{12}\left(\frac{\lambda_1^2 - 1}{2}\right)$$

$$\Pi_{12} = \Pi_{21} = 0 = W_{12}$$

For $\theta_o = 90°$, Eqs. (9.47) and (9.60) yield

$$\Pi_{11}/\lambda_1 = C_{22}\left(\frac{\lambda_1^2 - 1}{2}\right) + C_{222}\left(\frac{\lambda_1^2 - 1}{2}\right)^2$$

$$+ C_{2222}\left(\frac{\lambda_1^2 - 1}{2}\right)^3 + C_{12}\left(\frac{\lambda_2^2 - 1}{2}\right)$$

$$\Pi_{22} = 0 = C_{11}\left(\frac{\lambda_2^2 - 1}{2}\right) + C_{111}\left(\frac{\lambda_2^2 - 1}{2}\right)^2 \qquad (9.93)$$

$$+ C_{1111}\left(\frac{\lambda_2^2 - 1}{2}\right)^3 + C_{12}\left(\frac{\lambda_1^2 - 1}{2}\right)$$

$$\Pi_{12} = \Pi_{21} = 0 = W_{12}$$

Π_{11}, λ_1 and λ_2 are measured experimentally from both $\theta_o = 0°$ and $90°$ unidirectional tensile tests. The constants C_{11}, C_{12}, and C_{22} in Eqs. (9.92) and (9.93) are related to the initial slope of these experimental curves of Π_{11}/λ_1 vs. $(\lambda_1^2 - 1)/2$. The constants C_{111} and C_{222} are the nonlinear terms associated with the bi-modulus properties of the composite; and the constants C_{1111} and C_{2222} are the fourth-order nonlinear terms in Eq. (9.44). C_{111}, C_{222}, C_{1111}, and C_{22222} are determined by fitting the theoretical curves of Eqs. (9.92) and (9.93) to the longitudinal and transverse experimental curves of Π_{11}/λ_1 vs. $(\lambda_1^2 - 1)/2$, respectively.

9.3.4.2 *Shear properties*

C_{66} and C_{6666} are the elastic constants associated with the shear properties. Two test methods have been used for characterizing the shear behavior: (1) three-rail 0° simple shear, and (2) simple tension of $[\pm 45°]_s$ specimen.

First, consider the simple shear test in which the applied shear force is parallel to the fiber direction. From Eqs. (9.64), for $\theta = 0$,

$$\Pi_{21} = KW_{22} + \tfrac{1}{2}W_{12}$$
$$= K(C_{22}\bar{E}_2 + C_{222}\bar{E}_2^2 + C_{2222}\bar{E}_2^3) + C_{66}\bar{E}_6 + C_{6666}\bar{E}_6^3$$
$$= K^3(C_{22} + \tfrac{1}{2}C_{222}K^2 + \tfrac{1}{8}C_{2222}K^4) + C_{66}K + C_{6666}K^3$$

(9.94)

Since the values of C_{22}, C_{222} and C_{2222} are already known from tensile property measurements, C_{66} and C_{6666} can be determined by fitting the experimental data of Π_{21} vs. K.

Next, consider the tensile test using $[\pm 45°]_s$ specimens. For a $[\pm 45°]_s$ laminate specimen under a tensile load, Eqs. (9.84) can be rewritten as

$$N_{11}/h\lambda_1 = C_{66}(\lambda_1^2 - \lambda_2^2) + \tfrac{1}{4}C_{6666}(\lambda_1^2 - \lambda_2^2)^3$$

(9.95)

By measuring h, N_{11}, λ_1 and λ_2, the curve of $N_{11}/h\lambda_1$ vs. $(\lambda_1^2 - \lambda_2^2)$ can be determined experimentally. Then a curve fitting method can be used to identify the constants C_{66} and C_{6666}. Experiments using both simple shear and tensile tests on Kevlar-49/silicone elastomer composites have yielded comparable results of the elastic constants as shown in Table 9.1.

The tensile experiment on $[\pm 45]_s$ specimens has been quite often used to determine the shear modulus of conventional rigid polymer composites (see ASTM Standard D 3518-76). The basic equation for this experiment is

$$\sigma_{xx}/2 = G_{12}(\varepsilon_{xx} - \varepsilon_{yy})$$

(9.96)

where the engineering stress (σ_{xx}) and strains (ε_{xx} and ε_{yy}) are measured experimentally.

In order to compare Eq. (9.95) with Eq. (9.96) the following relations are introduced:

$$\lambda_1 = 1 + \varepsilon_{xx}$$
$$\lambda_2 = 1 + \varepsilon_{yy}$$

(9.97)

$$\frac{N_{11}}{2h\lambda_1} = \frac{\Pi_{xx}}{2}\frac{1}{\lambda_1} = \frac{\sigma_{xx}}{2}\frac{\lambda_2}{\lambda_1}$$

Substitution of Eqs. (9.97) into Eq. (9.95) gives

$$\frac{\sigma_{xx}}{2} = \frac{1 + \varepsilon_{xx}}{1 + \varepsilon_{yy}} [C_{66}[(\varepsilon_{xx} - \varepsilon_{yy}) + (\varepsilon_{xx}^2 - \varepsilon_{yy}^2)]$$

$$+ C_{6666}[(\varepsilon_{xx} - \varepsilon_{yy}) + (\varepsilon_{xx}^2 - \varepsilon_{yy}^2)]^3 \qquad (9.98a)$$

For linear elastic materials (i.e. $C_{6666} = 0$) under small deformation, and by neglecting the higher order terms of strain, Eq. (9.98a) can be rewritten as

$$\frac{\sigma_{xx}}{2} = \frac{1 + \varepsilon_{xx}}{1 + \varepsilon_{yy}} [C_{66}(\varepsilon_{xx} - \varepsilon_{yy})] \qquad (9.98b)$$

Since the initial shear modulus $G_{12} = C_{66}$, the difference between Eqs. (9.96) and (9.98b) is the geometric factor $(1 + \varepsilon_{xx})/(1 + \varepsilon_{yy})$. Obviously, for infinitesimal deformation $(1 + \varepsilon_{xx})/(1 + \varepsilon_{yy}) = 1$ and Eqs. (9.96) and (9.98b) are identical. However, if the deformation is not infinitesimal, G_{12} determined from Eq. (9.96) may not be accurate because of the omission of the geometric factor. For instance, let the strain $\varepsilon_{xx} = 0.02$ and use the relation of Eq. (9.98a); the error is $(1 + \varepsilon_{xx})/(1 + \varepsilon_{yy}) = 4.1\%$.

The elastic constants of Kevlar-49/silicone elastomer obtained from the above methods are listed in Table 9.1. The higher order elastic constants (i.e. C_{iii} and C_{iiii}) are determined by a regression curve fitting to the experimental data. Thus, they are valid only within the strain level at which they are obtained experimentally.

9.4   Constitutive relations based on the Eulerian description

In the above, the Lagrangian system has been used to derive the closed form constitutive equations for flexible composites, based upon a strain-energy function, for both lamina and laminate. These equations can be used to predict the nonlinear elastic behavior of flexible composites under different cases of finite deformation. It should be mentioned that the Lagrangian stress, defined as force per undeformed area, is a nominal stress and the real force equilibrium is established in the deformed or contemporary configuration. Furthermore, the anisotropic elastic properties of the composite always refer to the deformed configuration. For example, the current Young's modulus describes the stiffness in the current fiber direction which rotates during deformation. Therefore, in some cases, it is convenient to use the deformed body as the reference to describe the constitutive relation.

In this section, a nonlinear constitutive relation has also been developed based upon the Eulerian description where the deformed configuration of the composite is used as the reference state (Luo and Chou 1988a). A stress-energy function, referring to the current principal material coordinate χ (Fig. 9.2b), provides the basis for deriving the constitutive relations.

9.4.1 *Stress-energy function*

In finite elasticity, the energy densities in terms of either the Eulerian or Lagrangian stresses are not unique referring to a fixed coordinate; this can be demonstrated through the consideration of a 'rigid-body rotation' (Fung 1969). As an example, consider a bar which is subjected to a simple tension and rotating about the z axis. At one instant, the bar is parallel to the x axis so that $\sigma_{xx} \neq 0$ and $\sigma_{yy} = 0$. At another instant, when the bar becomes parallel to the y axis, the stress state is given by $\sigma_{xx} = 0$ and $\sigma_{yy} \neq 0$. Thus a rigid-body rotation changes the stress tensor, even though the state of stress in the bar remains unchanged. A complementary energy function referring to a fixed coordinate may be defined based upon the second Piola–Kirchhoff stress tensor P_{AB}. However, as indicated in Eq. (9.36), the second Piola–Kirchhoff stress still involves the displacement gradient. Thus, the use of complementary energy in terms of P_{ij} does not really make the constitutive relation simpler.

In order to establish the stress-energy function, a moving Eulerian coordinate system is introduced in this section. Figure 9.2(b) illustrates a unidirectional flexible composite lamina under a finite deformation in the Eulerian system. Unlike Fig. 9.2(a), here the deformed configuration has been chosen as the reference state, and the rectangular element $A'E'F'D'$ in the deformed body is considered. The sides $A'D'$ and $A'E'$ coincide with the current principal material coordinate system $l'-t'$ or $\chi_1-\chi_2$, with l' referring to the current fiber direction. The underline of a quantity refers to the current principal material coordinates $\chi_1-\chi_2$. Thus, the element $AEFD$ corresponds to the element $A'E'F'D'$ in the undeformed state. One may assume that the rectangle $A'E'F'D'$ undergoes two stages of deformation in restoring to its initial shape $AEFD$. These stages are illustrated in Fig. 9.19. First, $A'E'F'D'$ becomes a smaller rectangle $A''E''F''D''$ by removing the normal stresses; then it reverses to $AEFD$ by removing the shear stress.

The deformations depicted in Fig. 9.19 can be related to the Eulerian strain components. Let the line elements $AD = \mathrm{d}l_o$ and $AE = \mathrm{d}t_o$ in the undeformed lamina (Fig. 9.19c); also define

$A'D' = dl$ and $A'E' = dt$ in the deformed lamina (Fig. 9.19a). Then, the physical significance of the Eulerian strains can be explained as

$$2\underset{\sim}{e}_{11} = [(dl)^2 - (dl_o)^2]/(dl)^2$$

$$2\underset{\sim}{e}_{22} = [(dt)^2 - (dt_o)^2]/(dt)^2 \tag{9.99}$$

$$2\underset{\sim}{e}_{12} = \sin \gamma_{12}(\sqrt{(1 - 2\underset{\sim}{e}_{11})}\sqrt{(1 - 2\underset{\sim}{e}_{22})})$$

where $\underset{\sim}{e}_{ij}$ are the Eulerian strains referring to the current principal material coordinates $\underset{\sim}{x}_1$–$\underset{\sim}{x}_2$, and γ_{12} is the angular deviation from a right-angle as shown in Fig. 9.19.

The stress-energy per unit area of the deformed lamina $(A'E'F'D')$ is assumed to be a function of the Eulerian stress components referring to the current principal material coordinate

Fig. 9.19. Illustration of the deformation of a rectangular element in the Eulerian system. (After Luo and Chou 1988a.)

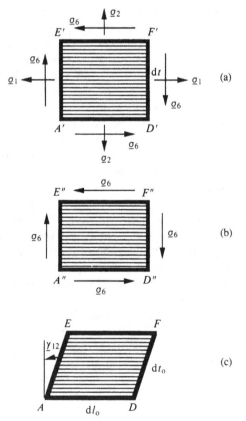

x_1-x_2, namely $W^* = W^*(\sigma_{11}, \sigma_{22}, \sigma_{12}^2)$. The following expression is adopted:

$$W^* = \tfrac{1}{2}S_{11}\sigma_1^2 + \tfrac{1}{3}S_{111}\sigma_1^3 + \tfrac{1}{4}S_{1111}\sigma_1^4$$
$$+ S_{12}\sigma_1\sigma_2 + \tfrac{1}{2}S_{22}\sigma_2^2 + \tfrac{1}{3}S_{222}\sigma_2^3$$
$$+ \tfrac{1}{4}S_{2222}\sigma_2^4 + \tfrac{1}{2}S_{66}\sigma_6^2 + \tfrac{1}{4}S_{6666}\sigma_6^4$$
$$+ S_{166}\sigma_1\sigma_6^2 + S_{2266}\sigma_2^2\sigma_6^2 \tag{9.100}$$

where σ_i are the Eulerian stresses referring to the current principal material coordinates x_1-x_2. Also, the short-handed notations are used, i.e. $\sigma_1 = \sigma_{11}$, $\sigma_2 = \sigma_{22}$ and $\sigma_6 = \sigma_{12}$. S_{ij}, S_{ijk} and S_{ijkl} are the compliance constants. Equation (9.100) is similar to the expressions of Hahn and Tsai (1973) in their mathematical forms. However, due to the finite deformation, it should be mentioned that: (1) The Eulerian coordinate x used here is a moving coordinate, which is chosen to coincide with the current fiber longitudinal and transverse directions, l' and t'. Therefore, the energy function satisfies the test of rigid-body rotation. (2) The Eulerian stresses, σ_{ij}, used in the energy function are the current stress state of the deformed lamina.

9.4.2 General constitutive equations

The complementary energy per unit volume of a deformed a lamina is $W^* = \sigma_{ij}\varrho_{ij} - W(\varrho_{ij})$. Here, $W(\varrho_{ij}) = \sigma_{ij}\delta\varrho_{ij}$ is the strain-energy density. Then,

$$\delta W^* = \sigma_{ij}\delta\varrho_{ij} + \varrho_{ij}\delta\sigma_{ij} - \frac{\partial W}{\partial\varrho_{ij}}\delta\varrho_{ij} \tag{9.101}$$

Since

$$\sigma_{ij} = \frac{\partial W}{\partial\varrho_{ij}}, \tag{9.102}$$

the following can be obtained from Eq. (9.101):

$$\delta W^* = \varrho_{ij}\delta\sigma_{ij} \tag{9.103}$$

or

$$\varrho_{ij} = \frac{\partial W^*}{\partial\sigma_{ij}} \tag{9.104}$$

Substituting Eq. (9.100) into Eq. (9.104), the Eulerian strain components referring to the coordinates x_1–x_2 are obtained as

$$\varrho_1 = S_{11}\sigma_1 + S_{111}\sigma_1^2 + S_{1111}\sigma_1^3 + S_{12}\sigma_2 + S_{166}\sigma_6^2$$

$$\varrho_2 = S_{22}\sigma_2 + S_{222}\sigma_2^2 + S_{2222}\sigma_2^3 + S_{12}\sigma_1 + 2S_{2266}\sigma_2\sigma_6^2 \qquad (9.105)$$

$$\varrho_6 = S_{66}\sigma_6 + S_{6666}\sigma_6^3 + 2S_{166}\sigma_1\sigma_6 + 2S_{2266}\sigma_2^2\sigma_6$$

Here $\varrho_1 = \varrho_{11}$, $\varrho_2 = \varrho_{22}$, $\varrho_6 = 2\varrho_{12}$. The choice of compliance constants in Eq. (9.100) is made on the following basis. First, S_{11}, S_{22}, S_{12} and S_{66} are associated with the linear deformation. Second, the terms S_{11} and S_{222} are adopted for representing the bi-modulus behavior in the axial and transverse directions, respectively. Third, the nonlinear deformations are represented by S_{1111}, S_{2222} and S_{6666}. Lastly, the greatest uncertainty involves the coupling terms between the normal and shear deformations. Unlike in rigid composites, the coupling effects may not be negligible in flexible composites. Two terms, S_{166} and S_{2266}, are retained to represent the interactions between axial and shear deformations in Eqs. (9.105).

Having established the constitutive relations with respect to the principal material coordinates x_1–x_2, the general constitutive relations of a composite lamina referring to the fixed coordinates x_1–x_2 (Fig. 9.2b) can be derived from Eq. (9.105) and the tensor transformation relation,

$$[e] = [T]^{\mathrm{T}}[\mathbf{S}][T][\sigma] = [\mathbf{S}^*][\sigma] \qquad (9.106)$$

where

$$[\mathbf{S}] = \begin{bmatrix} S_{11} + S_{111}\sigma_1 + S_{1111}\sigma_1^2 & S_{12} & S_{166}\sigma_6 \\ S_{12} & S_{22} + S_{222}\sigma_2 + S_{2222}\sigma_2^2 & 2S_{2266}\sigma_2\sigma_6 \\ 2S_{166}\sigma_6 & 2S_{2266}\sigma_2\sigma_6 & S_{66} + S_{6666}\sigma_6^2 \end{bmatrix}$$

$$\{e\} = \begin{Bmatrix} e_1 \\ e_2 \\ e_6 \end{Bmatrix}, \quad \{\sigma\} = \begin{Bmatrix} \sigma_1 \\ \sigma_2 \\ \sigma_6 \end{Bmatrix}, \quad \text{and} \quad [T] = \begin{bmatrix} c^2 & s^2 & 2cs \\ s^2 & c^2 & -2cs \\ -cs & cs & c^2 - s^2 \end{bmatrix}$$

Here, $c = \cos\theta$ and $s = \sin\theta$, where θ denotes the current fiber orientation angle. Also, e_i and σ_i are, respectively, the Eulerian stress and strain referring to the coordinates x_1–x_2. The full expression of $[\mathbf{S}^*]$ in Eq. (9.106) is

$$[\mathbf{S}^*] = \begin{bmatrix} \mathbf{S}_{11}^* & \mathbf{S}_{12}^* & \mathbf{S}_{16}^* \\ \mathbf{S}_{21}^* & \mathbf{S}_{22}^* & \mathbf{S}_{26}^* \\ \mathbf{S}_{61}^* & \mathbf{S}_{62}^* & \mathbf{S}_{66}^* \end{bmatrix}$$

$$= \begin{bmatrix} \begin{matrix} c^4 S_{11} + 2c^2 s^2 S_{12} \\ + s^4 S_{22} + c^2 s^2 S_{66} \end{matrix} & \begin{matrix} c^2 s^2 S_{11} + (c^4 + s^4) S_{12} \\ + c^2 s^2 S_{22} - c^2 s^2 S_{66} \end{matrix} \\ \begin{matrix} c^2 s^2 S_{11} + (c^4 + s^4) S_{12} \\ + c^2 s^2 S_{22} - c^2 s^2 S_{66} \end{matrix} & \begin{matrix} s^4 S_{11} + 2c^2 s^2 S_{12} \\ + c^4 S_{22} + c^2 s^2 S_{66} \end{matrix} \\ \begin{matrix} c^3 s S_{11} - cs(c^2 - s^2) S_{12} \\ - cs^3 S_{22} - \tfrac{1}{2} cs(c^2 - s^2) S_{66} \end{matrix} & \begin{matrix} cs^3 S_{11} + cs(c^2 - s^2) S_{12} \\ - c^3 s S_{22} + \tfrac{1}{2} cs(c^2 - s^2) S_{66} \end{matrix} \end{bmatrix}$$

$$\begin{matrix} 2c^3 s S_{11} - 2cs(c^2 - s^2) S_{12} \\ - 2cs^3 S_{22} - cs(c^2 - s^2) S_{66} \\ 2cs^3 S_{11} + 2cs(c^2 - s^2) S_{12} \\ - 2c^3 s S_{22} + cs(c^2 - s^2) S_{66} \\ 2c^2 s^2 S_{11} - 4c^2 s^2 S_{12} \\ + 2c^2 s^2 S_{22} + \tfrac{1}{2}(c^2 - s^2) S_{66} \end{matrix}$$

$$+ (S_{111}\sigma_1 + S_{111}\sigma_1^2) \begin{bmatrix} c^4 & c^2 s^2 & 2c^3 s \\ c^2 s^2 & c^2 s^2 & 2cs^3 \\ c^3 s & cs^3 & 2c^2 s^2 \end{bmatrix}$$

$$+ (S_{222}\sigma_2 + S_{222}\sigma_2^2) \begin{bmatrix} s^4 & c^2 s^2 & -2cs^3 \\ c^2 s^2 & c^2 s^2 & -2c^3 s \\ -cs^3 & -c^3 s & 2c^2 s^2 \end{bmatrix}$$

$$+ S_{6666}\sigma_6^2 \begin{bmatrix} c^2 s^2 & -c^2 s^2 & -cs(c^2 - s^2) \\ -c^2 s^2 & -c^2 s^2 & cs(c^2 - s^2) \\ -\tfrac{1}{2} cs(c^2 - s^2) & \tfrac{1}{2} cs(c^2 - s^2) & \tfrac{1}{2}(c^2 - s^2) \end{bmatrix}$$

$$+ S_{166}\sigma_6 \begin{bmatrix} -3c^3 s & c^3 s - 2cs^3 & c^4 - 5c^2 s^2 \\ 2c^3 s - 2cs^3 & 3cs^3 & -s^4 + 5c^2 s^2 \\ c^4 - 2c^2 s^2 & -s^4 + 2c^2 s^2 & 3c^3 s - 3cs^3 \end{bmatrix}$$

$$+ S_{2266}\sigma_2\sigma_6 \begin{bmatrix} -4cs^3 & 2cs^3 - 2c^3 s & -2s^4 + 6c^2 s^2 \\ 2cs^3 - 2c^3 s & 4c^3 s & 2c^4 - 6c^2 s^2 \\ -s^4 + 3c^2 s^2 & c^4 - 3c^2 s^2 & 4cs^3 - 4c^3 s \end{bmatrix}$$

$$\text{(9.107)}$$

The stresses in the current principal material directions, σ_i, are also obtained as

$$[\sigma] = [T][\sigma] \tag{9.108}$$

Referring to Fig. 9.2(b) the current fiber orientation angle is

$$\theta = \theta_o + \Delta\theta \tag{9.109}$$

The fiber reorientation angle $(\Delta\theta)$ due to finite deformation can be determined as follows. First, the angles DAD' and EAE' are defined as α and β, respectively. Then,

$$\Delta\theta = (\alpha + \beta)/2 + (\alpha - \beta)/2 \tag{9.110}$$

Here, the symmetric part of $\Delta\theta$, $(\alpha + \beta)/2$, equals $\gamma_{12}/2$; the antisymmetric part of $\Delta\theta$, $(\alpha - \beta)/2$, is defined as ω. It is understood that ω is the rigid-body rotation, which is independent of the coordinate system but dependent on the boundary conditions. If ω can be expressed in terms of the strain tensors, from Eqs. (9.99) and (9.110)

$$\Delta\theta = \frac{1}{2}\sin^{-1}\left(\frac{2\underline{e}_{12}}{\sqrt{(1 - 2\underline{e}_{11})}\sqrt{(1 - 2\underline{e}_{22})}}\right) + \omega(\underline{e}_{ij}) \tag{9.111}$$

Introducing Eq. (9.111) into Eq. (9.109), the current fiber orientation angle θ is expressed as a function in terms of the strain tensor. Then the general constitutive relations can be completely determined from Eqs. (9.106) and (9.111). The following are two illustrative examples.

9.4.3   *Pure homogeneous deformation*

The pure homogeneous deformation, with principal extension ratios λ_1 and λ_2 defined along the axes of the fixed coordinate system X, is shown in Fig. 9.7 and described by Eqs. (9.56). Referring to Eqs. (9.14) and (9.20),

$$[g] = \begin{bmatrix} \lambda_1 & 0 \\ 0 & \lambda_2 \end{bmatrix}, \qquad [g]^{-1} = \begin{bmatrix} \dfrac{1}{\lambda_1} & 0 \\ 0 & \dfrac{1}{\lambda_2} \end{bmatrix} \tag{9.112}$$

and

$$2[e] = \begin{bmatrix} 1 - \left(\dfrac{1}{\lambda_1}\right)^2 & 0 \\ 0 & 1 - \left(\dfrac{1}{\lambda_2}\right)^2 \end{bmatrix} \tag{9.113}$$

Eq. (9.113) gives

$$\lambda_1 = \frac{1}{\sqrt{(1 - 2e_{11})}}$$

$$\lambda_2 = \frac{1}{\sqrt{(1 - 2e_{22})}}$$

(9.114)

Referring to Fig. 9.7 the current fiber orientation can also be written as

$$\theta = \tan^{-1}\left\{\frac{\lambda_2 \sin \theta_o}{\lambda_1 \cos \theta_o}\right\} = \tan^{-1}\left\{\frac{\lambda_2}{\lambda_1} \tan \theta_o\right\}$$

(9.115)

Substituting Eqs. (9.114) into Eq. (9.115),

$$\theta = \tan^{-1}\left\{\frac{\sqrt{(1 - 2e_{11})}}{\sqrt{(1 - 2e_{22})}} \tan \theta_o\right\}$$

(9.116)

The substitution of Eq. (9.116) into Eq. (9.106) results in three independent equations. It is known from Eq. (9.113) that $e_{12} = 0$. Thus, by giving any two values of the following five variables in Eq. (9.106): stresses σ_1, σ_2, σ_6, and strains e_1, e_2 (or λ_1 and λ_2), the remaining three can be solved.

Finally, it is worth noting that from Eqs. (9.35) and (9.112), the Lagrangian stresses can be written in terms of the Eulerian stresses as

$$\Pi_{11} = \lambda_2 \sigma_{11}$$
$$\Pi_{22} = \lambda_1 \sigma_{22}$$
$$\Pi_{12} = \lambda_2 \sigma_{12}$$
$$\Pi_{21} = \lambda_1 \sigma_{12}$$

(9.117)

9.4.4 *Simple shear superposed on simple extension*

The deformation of 'simple shear superposed on simple extension' (Case 1) is shown in Fig. 9.12(a) and expressed by Eqs. (9.67). Using Eqs. (9.14) and (9.20), it can be found

$$[\mathbf{g}] = \begin{bmatrix} \lambda_1 & K\lambda_2 \\ 0 & \lambda_2 \end{bmatrix}, \qquad [\mathbf{g}]^{-1} = \begin{bmatrix} \dfrac{1}{\lambda_1} & -\dfrac{K}{\lambda_1} \\ 0 & \dfrac{1}{\lambda_2} \end{bmatrix}$$

(9.118)

and

$$2[e] = \begin{bmatrix} 1 - \left(\dfrac{1}{\lambda_1}\right)^2 & \dfrac{K}{\lambda_1^2} \\ \dfrac{K}{\lambda_1^2} & 1 - \left[\left(\dfrac{K}{\lambda_1}\right)^2 + \left(\dfrac{1}{\lambda_2}\right)^2\right] \end{bmatrix} \tag{9.119}$$

Invert the above equation to obtain

$$\lambda_1 = \frac{1}{\sqrt{(1 - 2e_{11})}}$$

$$\lambda_2 = \frac{\sqrt{(1 - 2e_{11})}}{\sqrt{[(1 - 2e_{11})(1 - 2e_{22}) - 4e_{12}^2]}} \tag{9.120}$$

$$K = \frac{2e_{12}}{1 - 2e_{11}}$$

Also, referring to Fig. 9.12(a), the current fiber orientation can be expressed as

$$\theta = \tan^{-1}\left(\frac{x_2}{x_1}\right) = \tan^{-1}\left(\frac{\lambda_2 \tan \theta_o}{\lambda_1 + \lambda_2 K \tan \theta_o}\right) \tag{9.121}$$

The substitution of Eq. (9.121) into Eq. (9.106), results in three independent equations. If the values are known for any three of the following six variables: stresses σ_1, σ_2, σ_6, and strains e_1, e_2, e_6 (or λ_1, λ_2 and K), the remaining three can be solved.

As mentioned in Section 9.3.2.5, for an off-axis specimen under uniaxial loading, with a length/width ratio $\gg 1$, the central lines of the middle section of the specimen remain straight in the loading direction. Then, this deformation can be referred to as 'simple shear superposed on simple extension (Case 1)'. Using the uniaxial loading conditions, $\sigma_{11} \neq 0$, $\sigma_{22} = \sigma_{21} = 0$, and Eqs. (9.106), (9.120) and (9.121), the deformation parameters λ_1, λ_2, and K (or e_{ij}) can be solved.

Figure 9.20 shows the comparison between analytical predictions and experimental results for the off-axis responses of Kevlar/elastomer composites under simple tension based upon the Eulerian approach. The fiber initial orientations are 10°, 30° and 60°. The same comparisons for tirecord/rubber composite specimens are shown in Fig. 9.21. The fiber initial orientations are 15°, 30° and 60° in this case. The predicted results are based upon Eqs. (9.106), (9.120) and (9.121) and an iterative calculation method. The elastic constants are shown in Table 9.1.

9.4.5 *Determination of elastic compliance constants*

The compliance constants in Eq. (9.100) can be determined experimentally (Luo and Chou 1988a). The second-order constants (S_{11}, S_{22}, S_{12} and S_{66}) are based on the linear behavior. The other constants are obtained by fitting the theoretical curves to experimental data. For example, for unidirectional tensile test in the x_1 direction (i.e. $\sigma_1 \neq 0$ and $\sigma_2 = \sigma_6 = 0$), Eq. (9.105) becomes

$$\underline{e}_1 = S_{11}\sigma_1 + S_{111}\sigma_1^2 + S_{1111}\sigma_1^3 \tag{9.122}$$

where the underline denotes the current principal material coordinate. Then, S_{11} is obtained from the initial slope of the experimental σ_1–$\underline{e}_1$ curve (i.e. $S_{11} = 1/$Young's modulus). S_{111} (which reflects bi-modulus behavior) and S_{1111} are determined by fitting the theoretical curves to experimental data in both tension and compression. With the unidirectional load applied in the x_2 direction, S_{22}, S_{222} and S_{2222}, can be determined by the same procedures as in the x_1 direction.

Fig. 9.20. Comparisons between theoretical predictions and experimental results of 10°, 30° and 60° off-axis stress–strain response of Kevlar/silicone elastomer composite laminae (Eulerian description). (After Luo and Chou 1988a.)

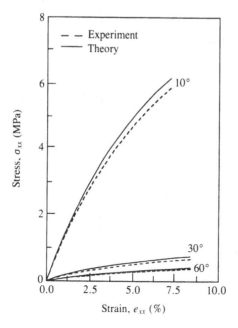

The remaining constants are related to shear (S_{66} and S_{6666}) and stretching–shear coupling (S_{166} and S_{2266}). If S_{166} and S_{2266} are negligibly small, the shear constants can be determined experimentally as described in Section 9.3.4.2, with proper stress and strain transformations from the Lagrangian system into the Eulerian system as described in Section 9.2.2.

The shear constants including the stretching–shear coupling listed in Table 9.1 are obtained by off-axis tensile tests at various fiber off-axis angles. For the unidirectional tensile condition ($\sigma_{11} \neq 0$, $\sigma_{22} = \sigma_{12} = 0$), Eq. (9.106) can be rewritten as

$$[c^4 S_{11} + 2c^2 s^2 S_{12} + s^4 S_{22}]\sigma_{11}/(cs) - e_{11}/(cs)$$
$$= S_{66}\sigma_6 + S_{6666}\sigma_6^3 - (3c/s)S_{166}\sigma_6^2 + (4s/c)S_{2266}\sigma_6^3 \quad (9.123)$$

where $c = \cos\theta$ and $s = \sin\theta$, $\sigma_6 = cs\sigma_{11}$. The fiber orientation angle, θ, and the stress–strain relations are measured experimentally. In Eq. (9.123), there are four unknown constants, S_{66}, S_{6666}, S_{166} and S_{2266}. The relations between σ_6 and the values of Eq. (9.123) which are determined by experimental measurements of σ_{11},

Fig. 9.21. Comparisons between theoretical predictions and experimental results of 15°, 30° and 60° off-axis stress–strain response of tirecord/rubber composite laminae (Eulerian description). (After Luo and Chou 1988a.)

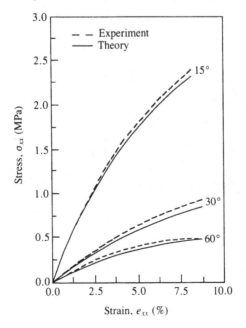

e_{11}, θ and the elastic constants related to the tensile properties (S_{11}, S_{12} and S_{22}). S_{66} is the initial slope of the stress–strain curve. Given sufficient experimental data (the number of specimens with different initial fiber orientations should be larger than the number of unknown constants), the remaining compliance constants S_{6666}, S_{166} and S_{2266} can be determined by a regression technique to fit the theoretical curve of Eq. (9.123) to the experimental curves.

9.5 Elastic behavior of flexible composites reinforced with wavy fibers

9.5.1 *Introduction*

In Chapter 8, an iso-phase model for flexible composites containing sinusoidally shaped fibers (Fig. 8.13) is presented and the analysis of the elastic behavior of such composites is based upon a step-wise incremental technique and the classical lamination theory. Being a well established analytical technique, the lamination theory does provide a convenient tool for describing the basic characteristics of flexible composites. However, because of the use of super-position techniques for nonlinear finite deformation problems, the limitation of incremental analysis is obvious. In an effort to provide a rigorous treatment, Luo and Chou (1988a&b, 1990b) applied the constitutive models based upon the Lagrangian (Section 9.3) and Eulerian descriptions (Section 9.4) to study the nonlinear elastic behavior of flexible composites with wavy fibers under finite deformation.

The deformation of the iso-phase flexible composite (see Fig. 8.13) is best understood by examining a representative element which contains a full wavelength of the sinusoidal curve (Fig. 9.22). This element is further divided into sub-elements along the x_1 axis. Each sub-element of the composite between x_1 and $x_1 + \Delta x_1$ is approximated by an off-axis unidirectional fiber composite, in which fibers are inclined at an angle $\theta_0^{(n)}$ to the x_1 axis. Referring to Eq. (8.14), the initial fiber orientation of sub-element (n), for example, is given as

$$\theta_0^{(n)} = \frac{1}{2}\left[\tan^{-1}\left(\frac{2\pi a}{\lambda}\cos\frac{2\pi x_1}{\lambda}\right)\right.$$

$$\left. + \tan^{-1}\left(\frac{2\pi a}{\lambda}\cos\frac{2\pi(x_1 + \Delta x_1)}{\lambda}\right)\right] \qquad (9.124)$$

It is also assumed that the stress and strain of a sub-element are homogeneous under axial loading. This assumption is supported by the photoelastic analysis (Luo and Chou 1988a). Figure 9.23 is a photoelastic view of a flexible composite sample under longitudinal loading; it shows that relatively uniform strain is maintained in distinct regions along the longitudinal direction. It should be noted that although all the experimental data collected are based upon a Kevlar-49/silicone elastomer system, the photograph shown in Fig. 9.23 is based upon graphite fiber as a reinforcement materials, so better contrast between the fiber and matrix in the photograph can be achieved.

Based upon the above assumptions, the analysis for the iso-phase model consists of two steps: (1) The constitutive relation of an off-axis sub-element under finite deformation is examined based upon the analysis developed in Sections 9.3 and 9.4. (2) The total deformation of the composite is the summation of the deformations of all these sub-elements.

9.5.2   *Longitudinal elastic behavior based on the Lagrangian approach*

Under the uniaxial tensile force F_1 in the x_1 direction, the following plane stress condition of the flexible composite is

Fig. 9.22. Deformation of a sub-element of a flexible composite containing sinusoidally shaped fibers under longitudinal tension. (After Luo and Chou 1990b.)

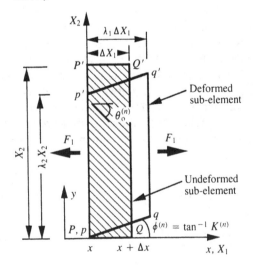

assumed:

$$\Pi_{11} = F_1/A_o$$

$$\Pi_{22} = 0 \tag{9.125}$$

$$\Pi_{21} = 0$$

where A_o is the initial cross-sectional area perpendicular to the x_1 axis.

Figure 9.22 shows the deformation of a typical sub-element $PQQ'P'$. $pqq'p'$ represents the configuration in the deformed state. Due to the iso-phase fiber arrangement, the edge qq' remains perpendicular to the X_1 axis. Let $x_i^{(n)}$ be the coordinates of an arbitrary particle in $PQQ'P'$, and $x_i^{(n)}$ be the corresponding local coordinates of this particle in $pqq'p'$. Here, the superscript refers to the sub-element (n). This deformation is specified as

$$x_1^{(n)} = \lambda_1^{(n)} X_1^{(n)}$$

$$x_2^{(n)} = k^{(n)} \lambda_1^{(n)} X_1^{(n)} + \lambda_2^{(n)} X_2^{(n)} \tag{9.126}$$

Equations (9.126) specify a deformation equivalent to Case 2 of Section 9.3.2.5, namely 'simple shear superposed on simple extension'. Using Eqs. (9.74) and the stress boundary condition of Eqs.

Fig. 9.23. A photoelastic view of a flexible composite lamina under longitudinal tension. (After Luo and Chou 1988a.)

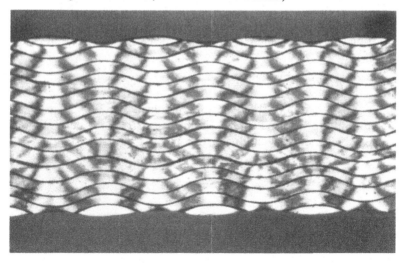

(9.125), the following can be obtained:

$$\Pi_{11} = \lambda_1^{(n)}[c^2 W_{11} + s^2 W_{22} - cs W_{12}]$$

$$0 = \Pi_{22} = s^2 W_{11} + c^2 W_{22} + cs W_{12} \tag{9.127}$$

$$0 = \Pi_{21} = cs W_{11} - cs W_{22} + \tfrac{1}{2}(c^2 - s^2) W_{12}$$

Here, W_{ij} is a function of $K^{(n)}$, $\lambda_1^{(n)}$ and $\lambda_2^{(n)}$, and it is given by Eqs. (9.47) and (9.73). The initial fiber orientation $\theta_0^{(n)}$ is given by Eq. (9.124). Then, the three unknowns, $K^{(n)}$, $\lambda_1^{(n)}$ and $\lambda_2^{(n)}$ can be solved from Eqs. (9.127). It is interesting to note that Π_{12} does not vanish, and it can be found from Eqs. (9.74)

$$\Pi_{12} = K^{(n)} \Pi_{11} \tag{9.128}$$

The current fiber orientation, $\theta^{(n)}$, of the sub-element (n) (Fig. 9.22) is

$$\theta^{(n)} = \tan^{-1}\left[K^{(n)} + \frac{\lambda_2^{(n)}}{\lambda_1^{(n)}} \tan \theta_0^{(n)} \right] \tag{9.129}$$

The average extension ratio of the wavelength in the X_1 direction can be derived as

$$\lambda_1 = \frac{\Delta x}{\lambda} \sum_{n=1}^{m} \lambda_1^{(n)} \tag{9.130}$$

9.5.3   Longitudinal elastic behavior based on the Eulerian approach

Under the uniaxial tension force F_1 in the longitudinal direction, the following stress states in the Eulerian system are assumed:

$$\sigma_{11} = F_1/A, \qquad \sigma_{22} = \sigma_{12} = 0 \tag{9.131}$$

where A is the area of the section perpendicular to the longitudinal direction in the deformed state. From Eqs. (9.126) the deformation of the sub-element (n) can be written in the Eulerian system as

$$X_1 = \frac{1}{\lambda_1^{(n)}} x_1$$

$$X_2 = -\frac{K^{(n)}}{\lambda_2^{(n)}} x_1 + \frac{1}{\lambda_2^{(n)}} x_2 \tag{9.132}$$

Using Eqs. (9.18) and (9.20), it can be found

$$[\mathbf{g}]^{-1} = \begin{bmatrix} \dfrac{1}{\lambda_1^{(n)}} & 0 \\ -\dfrac{K^{(n)}}{\lambda_2^{(n)}} & \dfrac{1}{\lambda_2^{(n)}} \end{bmatrix} \tag{9.133}$$

and

$$2[\mathbf{e}] = \begin{bmatrix} 1 - \left[\left(\dfrac{1}{\lambda_1^{(n)}}\right)^2 - \left(\dfrac{K^{(n)}}{\lambda_2^{(n)}}\right)^2\right] & \dfrac{K^{(n)}}{(\lambda_2^{(n)})^2} \\ \dfrac{K^{(n)}}{(\lambda_2^{(n)})^2} & 1 - \left(\dfrac{1}{\lambda_2^{(n)}}\right)^2 \end{bmatrix} \tag{9.134}$$

Invert the above equation to obtain

$$\lambda_1^{(n)} = \frac{\sqrt{(1 - 2e_{22}^{(n)})}}{\sqrt{[(1 - 2e_{11}^{(n)})(1 - 2e_{22}^{(n)}) - 4(e_{12}^{(n)})^2]}}$$

$$\lambda_2^{(n)} = \frac{1}{\sqrt{(1 - 2e_{22}^{(n)})}} \tag{9.135}$$

$$K^{(n)} = \frac{2e_{12}^{(n)}}{1 - 2e_{22}^{(n)}}$$

where $\lambda_1^{(n)}$ and $\lambda_2^{(n)}$ are the extension ratios of the sub-element (n) in the longitudinal and transverse directions, respectively; and $K^{(n)} = \tan \Phi^{(n)}$ as shown in Fig. 9.22. The current fiber orientation is obtained from Eq. (9.129). Then, $\lambda_1^{(n)}$, $\lambda_2^{(n)}$, and $K^{(n)}$ can be determined by an iterative calculation from Eqs. (9.106), (9.131) and (9.134). Also, the average extension of a wavelength in the longitudinal direction, λ_1, can be determined by Eq. (9.130).

The predictions of the longitudinal constitutive relations based upon the Lagrangian and Eulerian approaches are compared to experimental results and an incremental analysis in the finite deformation range (Luo, Kuo and Chou 1988). The model composite system consists of silicone elastomer reinforced with sinusoidally shaped Kevlar fibers ($a/\lambda = 0.09$). Due to fiber waviness, the volume fraction may vary among the sub-elements. An average fiber volume fraction $V_f = 9\%$ is used in the calculation. Also, the elastic constants are given in Table 9.1. Figure 9.24 compares the analytical predictions with experimental results. The heavy solid line indicates theoretical predictions of the Lagrangian approach (Luo and Chou 1988b, 1990b); the thin solid line indicates theoretical predictions based upon the Eulerian approach (Luo and Chou 1988a, 1990a); and the dotted line is from the incremental analysis (Kuo, Takahashi and Chou 1988). Experimental results are also presented.

Furthermore, the local strains in the sub-element can be predicted directly from Eqs. (9.127). The current fiber orientation angle of the sub-element is given by Eq. (9.129). These results show that the maximum local tensile strain of the fiber occurs at the

Fig. 9.24. Stress–strain relations of Kevlar/silicone elastomer composite laminae containing sinusoidally shaped fibers for $a/\lambda = 0.09$. (After Luo and Chou 1990b.)

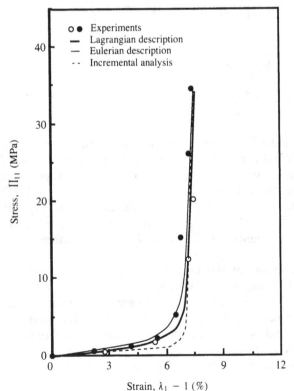

region where the initial fiber orientation angle equals zero (i.e. $X_1 = \pm\lambda/4$). The maximum local shear strain of the composites occurs in the region where the initial fiber orientation is a maximum (i.e. $X_1 = 0$, $\lambda/2$). Hence, the strength of the flexible composites may be determined by the maximum tensile strain at $X_1 = \lambda/4$ and the maximum shear strain at $X_1 = 0$.

References

Chapter 1

Ashby, M. F. (1987) 'Technology on the 1990s: advanced material and predictive design', *Phil. Trans. R. Soc. London*, **A322**, 393–407.

Bunsell, A. R. and Harris, B. (1974) 'Hybrid carbon and glass fiber composites', *Composites*, **5**, 157–64.

Chou, T. W. (1989) 'Flexible composites', *J. Mat. Sci.*, **24**, 761–83.

Chou, T. W. and Kelly, A. (1976) 'What we do not know about fiber composites', *Mat. Sci. Engr.*, **25**, 35.

Chou, T. W., Kelly, A. and Okura, A. (1985) 'Fiber reinforced metal matrix composites', *Composites*, **16**, 177.

Chou, T. W., McCullough, R. L. and Pipes, R. B. (1986) 'Composites', *Sci. Am.*, **255**, 192–203.

Clark, J. P. and Flemings, M. C. (1986) 'Advanced materials and the economy', *Sci. Am.*, **255**, 50–7.

Compton, W. D. and Gjostein, N. A. (1986) 'Materials for ground transportation', *Sci. Am.*, **255**, 92–100.

Congress of the United States, Office of Technology Assessment (1988) *New Structural Materials Technologies*, OTA-E-352, Washington, D.C.

Gordon, J. E. (1988) *The Science of Structures and Materials*, Scientific American Library, New York.

Humphrey, J. D. and Yin, F. C. P. (1987) 'A new constitutive formulation for characterizing the mechanical behavior of soft tissues', *Biophys. J.*, **52**, 563–70.

Kelly, A. (1985) 'Composites in context', *Comp. Sci. Tech.*, **23**, 171–200.

Kelly, A. (1987a) 'An outline of trends in materials science and processing', *Mat. Sci. Engr.*, **85**, 1–13.

Kelly, A. (1987b) 'Composites for the 1990's', *Phil. Trans. R. Soc. London*, **A322**, 409–423.

McCullough, R. L. (1985) 'Generalized combining rules for predicting transport properties of composite materials', *Comp. Sci. Tech.*, **22**, 3–21.

Mignery, L. A., Tan, T. M. and Sun, C. T. (1985) *The use of stitching to suppress delamination in laminated composites*, ASTM STP876, American Society for Testing and Materials, Philadelphia, PA, pp. 371–85.

Mody, P. B. and Majidi, A. P. (1987) 'Metal and ceramic matrix composites: the heat is on', *Composites in Manufacturing*, **3**, 1–5.

Nardone, V. C. and Prewo, K. M. (1988) 'Tensile performance of carbon-reinforced glass', *J. Mat. Sci.*, **23**, 168–80.

Port, O., King, R. W. and Hawkins, C. (1988) 'Materials that think for themselves', *Business Week*, December 5, pp. 166–7.

Rogers, C. A., ed. (1988) *Smart Materials, Structures, and Mathematical Issues*, Technomic Pub. Co., Lancaster.

Sousa, L. J. (1988) *Problems and Opportunities in Metals and Materials: An Integrated Perspective*, US Department of the Interior, Washington, D.C.

Steinberg, M. A. (1986) 'Materials for aerospace', *Sci. Am.*, **255**, 66–72.

Sun, C. T. (1989) 'Intelligent tailoring of composite laminates', *Carbon*, **27**, 679–87.

Takagi, T. (1989) 'A concept of intelligent materials in Japan', in *Proceedings of International Workshop on Intelligent Materials*, The Society of Non-Traditional Technology, Tokyo, Japan, pp. 1–10.

Vinson, J. R. and Chou, T. W. (1975) *Composite Materials and Their Use in Structures*, Elsevier-Applied Science, London.

Chapter 2

Aköz, A. Y. and Tauchert, T. R. (1972) 'Thermal stresses in an orthotropic elastic semispace', *J. Appl. Mech.*, **39**, 87–90.

Aköz, A. Y. and Tauchert, T. R. (1978) 'Thermoelastic analysis of a finite orthotropic slab'. *J. Mech. Eng. Sci.*, **20**, 65–71.

Ashton, J. E., Halpin, J. C. and Petit, P. H. (1969) *Primer on Composite Materials: Analysis*, Technomic, Westport, Connecticut.

Carlsson, L. A. and Pipes, R. B. (1987) *Experimental Characterization of Advanced Composite Materials*, Prentice-Hall, Englewood Cliffs, New Jersey.

Carslaw, H. S. and Jaeger, J. C. (1959) *Conduction of Heat in Solids*, Clarendon Press, Oxford.

Chamis, C. C. (1983) *NASA Tech. Memo 83320* (presented at the 38th Annual Conference of the Society of Plastics Industry (SPI), Houston, TX, Feb. 1983).

Chang, Y. P. (1977) 'Analytical solution for heat conduction of anisotropic media in infinite, semi-infinite and two-plane bounded regions', *Int. J. Heat and Mass Transfer*, **20**, 1019.

Chawla, K. K. (1987) *Composite Materials Science and Engineering*, Springer-Verlag, New York.

Cheng, C. M. (1951) 'Resistance to thermal shock', *J. Am. Rocket Soc.*, **21**, 147–53.

Chou, T. W. (1989a) 'Flexible composites', *J. Mat. Sci.*, **24**, 761–83.

Chou, T. W. (1989b) 'Elastic properties of laminates', *Concise Encyclopedia of Composite Materials*, Pergamon Press, Oxford, p. 159.

Christensen, R. M. (1979) *Mechanics of Composite Materials*, Wiley-Interscience, New York.

Chu, H. S., Weng, C. I. and Chen, C. K. (1983) 'Transient response of a composite straight fin', *ASME J. Heat Transfer*, **105**, 307–11.

Fukunaga, H. and Chou, T. W. (1988a) 'On laminate configurations for simultaneous failure', *J. Comp. Mat.*, **22**, 271.

Fukunaga, H. and Chou, T. W. (1988b) 'Simplified design techniques for laminated cylindrical pressure vessels under stiffness and strength constraints', *J. Comp. Mat.*, **22**, 1156–69.

Halpin, J. C. (1984) *Primer on Composite Materials Analysis*, Technomic Pub. Co., Lancaster, Pennsylvania.

Halpin, J. C. and Tsai, S. W. (1967) *Environmental Factors in Composite Materials Design*, Air Force Materials Laboratory Technical Report 67-423.

Hsu, P. W. and Herakovich, C. T. (1977) 'A perturbation solution for interlaminar stresses in bidirectional laminates', *Composite Materials Testing and Design (4th Conference)*, ASTM STP 617, American Society for Testing and Materials, Philadelphia, pp. 296–316.

Hsu, P. W. and Herakovich, C. T. (1977) 'Edge effects in angle-ply composite laminates', *J. Comp. Mat.*, **11**, 422–8.

Huang, S. C. and Chang, Y. P. (1980) 'Heat conduction in unsteady, periodic, and steady states in laminated composites', *ASME J. Heat Transfer*, **102**, 742–8.

528   References

Jones, R. M. (1975) *Mechanics of Composite Materials*, McGraw-Hill, New York.

Jost, W. (1960) *Diffusion*, Academic Press, New York.

Katayama, K., Saito, A. and Kobayashi, N. (1974) 'Transient heat conduction in anisotropic solids', *Proceedings of the International Conference on Heat and Mass Transfer*, Tokyo, p. 137.

Kingery, W. D. (1955) 'Factors affecting thermal stress resistance of ceramic materials', *J. Am. Ceramic. Soc.*, **38**, 3–5.

Lo, K. H., Christensen, R. M. and Wu, E. M. (1977a) 'A higher-order theory of plate deformation. Part 1: Homogeneous plates', *J. Appl. Mech.*, **44**, 663–8.

Lo, K. H., Christensen, R. M. and Wu, E. M. (1977b) 'A higher-order theory of plate deformation. Part 2: Laminated plates', *J. Appl. Mech.*, **44**, 669–76.

Luo, J. and Sun, C. T. (1989) 'Global–local methods for thermoelastic stress analysis of thick fiber-wound cylinders', *Proceedings of the Fourth Technical Conference on Composite Materials, American Society for Composites*, Technomic Pub. Co., Lancaster, Pennsylvania, pp. 535–44.

Mossakowska, Z. and Nowacki, W. (1958) 'Thermal stresses in transversely isotropic bodies', *Archiv. Mech. Stosow*, **10**, (4), 569–603.

Noda, N. (1983) 'Transient thermal stress problem in a transversely isotropic finite circular cylinder under three-dimensional temperature field', *J. Thermal Stresses*, **6**, 57–71.

Nomura, S. and Chou, T. W. (1986) 'Heat conduction in composite materials due to oscillating temperature field', *Int. J. Engr. Sci.*, **24**, 643.

Ozisik, M. N. (1980) *Heat Conduction*, John Wiley and Sons, Inc., New York.

Pipes, R. B., Vinson, J. R. and Chou, T. W. (1976) On the hygrothermal response of laminated composite systems. *J. Comp. Mat.*, **10**, 129–48.

Poon, K. C. and Chang, Y. P. (1978) 'Transformation of heat conduction problems from anisotropic to isotropic', *Heat and Mass Transfer*, **5**, 215.

Reddy, J. N. (1984) 'A simple higher-order theory for laminated composite plates', *J. Appl. Mech.*, **51**, 745–52.

Reissner, E. (1945) 'Transverse shear deformation on the bending of elastic plates', *J. Appl. Mech.*, **2**, (2), 69–77.

Rosen, B. W. (1973). 'Stiffness of fibre composite materials', *Composites*, **4**, 16–25.

Sharma, B. (1958). 'Thermal stresses in transversely isotropic semi-infinite elastic solid', *J. Appl. Mech.*, **25**, 86–8.

Singh, A. (1960) 'Axisymmetric thermal stresses in transversely isotropic bodies', *Archiv. Mech. Stosow.*, **39**, (3), 287–304.

Stein, M. and Jegley, D. C. (1987) 'Effects of transverse shearing on the cylindrical bending, vibration, and buckling of laminated plates', *AIAA J.*, **25**, (1), 123–9.

Sugano, Y. (1979) 'Transient thermal stress in a transversely isotropic finite circular cylinder due to an arbitrary internal heat-generation', *Int. J. Engr. Sci.*, **17**, 729–39.

Sun, C. T. and Li, S. (1988) 'Three-dimensional effective elastic constants for thick laminates', *J. Comp. Mat.*, **22**, 629–39.

Takahashi, K. and Chou, T. W. (1988) 'Transverse elastic moduli of unidirectional fiber composites with interfacial debonding', *Met. Trans. AIME*, **19A**, 129.

Takeuti, Y. and Noda, N. (1978). 'A general treatise on the three-dimensional thermoelasticity of curvilinear aeolotropic solids', *J. Thermal Stresses*, **1**, 25–39.

Tauchert, T. R. and Aköz, A. Y. (1974) 'Thermal stresses in an orthotropic elastic slab due to prescribed surface temperatures', *J. Appl. Mech.*, **41**, 222–8.

Tsai, S.-W. and Hahn, H. T. (1980) *Introduction to Composite Materials*, Technomic, Westport, Connecticut.

Van Dyke, M. (1975) *Perturbation Methods in Fluid Mechanics*, The Parabolic Press, Stanford, California.

Vinson, J. R. and Chou, T. W. (1975) *Composite Materials and Their Use in Structures*, Elsevier-Applied Science, London.

Wang, H. S. and Chou, T. W. (1985) 'Transient thermal stress analysis of a rectangular orthotropic slab', *J. Comp. Mat.*, **19**, 424–42.

Wang, H. S. and Chou, T. W. (1986) 'Transient thermal behavior of a thermally and elastically orthotropic medium', *AIAA J.*, **24**, 664–72.

Wang, H. S., Pipes, R. B. and Chou, T. W. (1986) 'Thermal transient stresses due to rapid cooling in thermally and elastically orthotropic medium', *Met. Trans. A*, **17A**, 1051–5.

Wang, Y. R. and Chou, T. W. (1988) 'Three-dimensional analysis of transient interlaminar thermal stress of laminated composites', *Symposium on Mechanics of Composite Materials ASME AMD*, **92**, 185–92.

Wang, Y. R. and Chou, T. W. (1989) 'Three-dimensional analysis of transient interlaminar thermal stress of laminated composites', *J. Appl. Mech.*, **56**, 601.

Wang, Y. R. and Chou, T. W. (1991). 'Thermal shock resistance of laminated ceramic matrix composites', *J. Mat. Sci.* in press.

Whitney, J. M. (1972) 'Stress analysis of thick laminated composite and sandwich plates', *J. Comp. Mat.*, **6**, 426–40.

Whitney, J. M. and Pagano, N. J. (1970) 'Shear deformation in heterogeneous anisotropic plates', *J. Appl. Mech.*, **37**, (4), 1031–6.

Whitney, J. M. and Sun, C. T. (1973) 'A higher order theory for extensional motion of laminated composites', *J. Sound and Vibration*, **30**, 85–97.

Chapter 3

Aveston, J. and Kelly, A. (1973) 'Theory of multiple fracture of fibrous composites', *J. Mat. Sci.*, **8**, 352–62.

Aveston, J. and Kelly, A. (1980) 'Tensile first cracking strain and strength of hybrid composites and laminates', *Phil. Trans. Royal Soc. London Series A*, **294**, 519–34.

Aveston, J., Cooper, G. and Kelly, A. (1971) 'Single and multiple fracture', in *The Properties of Fibre Composites*, Conference Proceedings, National Physical Laboratory, IPC Science and Technology Press Ltd., pp. 15–26.

Bader, M. G., Bailey, J. E., Curtis, P. T. and Parvizi, A. (1979) 'The mechanism of initiation and development of damage in multi-axial fibre-reinforced plastics on laminates', in *Mechanical Behavior of Materials*, K. J. Miller and R. F. Smith, eds., Pergamon Press, Oxford, pp. 227–39.

Bailey, J. E., Curtis, P. T. and Parvizi, A. (1979) 'On the transverse cracking and longitudinal splitting behavior of glass and carbon fibre reinforced epoxy cross-ply laminates and the effect of Poisson and thermally generated strain', *Proc. Royal Soc. London, Series A*, **366**, 599–623.

Bailey, J. E. and Parvizi, A. J. (1981) *J. Mat. Sci.*, **16**, 649.

Bjeletich, J. G., Crossman, F. W. and Warren, W. J. (1979) 'The influence of stacking sequence on failure modes in quasi-isotropic graphite-epoxy laminates', in *Failure Modes in Composites – IV*, J. R. Cornie and F. W. Crossman, eds., American Institute of Mining, Metallurgical and Petroleum Engineers, New York, p. 118.

Bradley, W. L. and Cohen, R. N. (1985) 'Matrix deformation and fracture in graphite reinforced epoxies', in *Delamination and Debonding of Materials*, W. S. Johnson, ed., ASTM STP 876, pp. 389–410.

Budiansky, B., Hutchinson, J. W. and Evans, A. G. (1986) 'Matrix fracture in fiber-reinforced ceramics', *J. Mech. Phys. Sol.*, **34**, 167–89.

Burgel, B., Perry, A. J. and Scheider, W. R. (1970) 'On the theory of fiber strengthening', *J. Mech. Phys. Sol.*, **18**, 101–14.

Carrara, A. S. and McGarry, F. J. (1968) 'Matrix and interface stresses in a discontinuous fiber composite model', *J. Comp. Mat.*, **2**, 222–43.

Chen, C. H. (1973) 'Tension of a composite bar with fibre discontinuities and soft inter-fibre material', *Fibre Sci. Tech.*, **6**, 1.

Chen, P. E. (1971) 'Strength properties of discontinuous fiber composites', *Polymer Eng. Sci.*, **11**, 51–6.

Chi, Z. F. and Chou, T. W. (1983) 'An experimental study of the effect of prestressed loose carbon strands on composite strength', *J. Comp. Mat.*, **17**, 196–209.

Chi, Z. F., Chou, T. W. and Shen, G. (1984) 'Determination of single fibre strength distribution from fibre bundle testings', *J. Mat. Sci.*, **19**, 3319–24.

Coleman, B. D. (1958) 'On the strength of classical fibres and fibre bundles', *J. Mech. Phys. Sol.*, **7**, 60.

Crossman, F. W. and Wang, A. S. D. (1982) *ASTM Symposium on Damages in Composite Materials: Basic Mechanisms, Accumulation, Tolerance, and Characterization*, K. L Reifsnider, ed., ASTM STP 775, ASTM, Philadelphia.

Crossman, F. W., Warren, W. J., Wang, A. S. D. and Law, G. E. (1980) 'Initiation and growth of transverse cracks and edge delamination in composite laminates, II. Experimental correlation', *J. Comp. Mat.*, **14**, 88–108.

Daniels, H. E. (1945) 'The statistical theory of the strength of bundles of threads I', *Proc. Royal Soc. London Series, A*, **183**, 405.

Dhingra, A. K. (1980) 'Alumina fibre FP', *Phil. Trans. R. Soc. London, (A)*, **294**, 411–17.

Epstein, B. (1948) 'Statistical aspects of fracture problems', *J. Appl. Phys.*, **19**, 140.

Fichter, B. W. (1969) 'Stress concentration around broken filaments in a filament-stiffened sheet', NASA TN D-5453.

Fichter, B. W. (1970) 'Stress concentrations in filament-stiffened sheets of finite length', NASA TN D-5947.

Fukuda, H. and Kawata, K. (1976a) 'On the stress concentration factor in fibrous composites', *Fibre Sci. Tech.*, **9**, 189.

Fukuda, H. and Kawata, K. (1976b) 'Strength estimation of unidirectional composites', *Trans. Japan Soc. Comp. Mat.*, **2**, 59.

Fukuda, H. and Kawata, K. (1977) 'On the strength distribution of unidirectional fibre composites', *Fibre Sci. Tech.*, **10**, 53.

Fukuda, H. and Kawata, K. (1980) 'Stress distribution of laminates including discontinuous layers', *Fibre Sci. Tech.*, **13**, 255–67.

Fukunaga, H., Peters, P. W. M., Schulte, K. and Chou, T. W. (1984) 'Probabilistic failure strength analyses of graphite/epoxy cross-ply laminates', *J. Comp. Mat.*, **18**, 339.

Garrett, K. W. and Bailey, J. E. (1977a) *J. Mat. Sci.*, **12**, 157.

Garrett, K. W. and Bailey, J. E. (1977b) *J. Mat. Sci.*, **12**, 2189.

Goree, J. G. and Gross, R. S. (1979) 'Analysis of a unidirectional composite containing broken fibers and matrix damage', *Eng. Fracture Mech.*, **13**, 563–78.

Goree, J. G. and Gross, R. S. (1980) 'Stresses in a three-dimensional unidirectional composite containing broken fibers', *Eng. Fracture Mech.*, **13**, 395–405.

Gucer, D. E. and Gurland, J. (1962) 'Comparison of the statistics of two fracture modes', *J. Mech. Phys. Sol.*, **10**, 365.

Harlow, D. G. (1979) 'Properties of the strength distribution for composite materials', *Composite Materials: Testing and Design (Fifth conference)*, ASTM STP 674, S. W. Tsai, ed., American Society for Testing and Materials, pp. 484–501.

Harlow, D. G. and Phoenix, S. L. (1978a) 'The chain-of-bundles probability model for the strength of fibrous materials I: analysis and conjectures', *J. Comp. Mat.*, 12, 195.

Harlow, D. G. and Phoenix, S. L. (1978b) 'The chain-of-bundles probability model for the strength of fibrous materials II: a numerical study of convergence', *J. Comp. Mat.*, 12, 314.

Harlow, D. G. and Phoenix, S. L. (1979) 'Bounds on the probability of failure of composite materials', *Int. J. Fracture*, 15, 321–36.

Harlow, D. G. and Phoenix, S. L. (1981a) 'Probability distributions for the strength of composite materials I: two level bounds', *Int. J. Fracture*, 17, 347–72.

Harlow, D. G. and Phoenix, S. L. (1981b) 'Probability distributions for the strength of composite materials II: a convergent sequence of tight bounds', *Int. J. Fracture*, 17, 601–30.

Hedgepeth, J. M. (1961) 'Stress concentrations in filamentary structures', NASA TN D-882.

Hedgepeth, J. M. and Van Dyke, P. (1967) 'Local stress concentrations in imperfect filamentary composite materials', *J. Comp. Mat.*, 1, 294.

Henstenburg, R. B. and Phoenix, S. L. (1989) 'Interfacial shear strength studies using the single-filament-composite Test II: a probability model and Monte-Carlo simulation', *Polymer Composites*, 10, 389–408.

Hikami, F. and Chou, T. W. (1984a) 'A probabilistic theory for the strength of discontinuous fiber composites', *J. Mat. Sci.*, 19, 1805.

Hikami, F. and Chou, T. W. (1984b) 'Statistical treatment of transverse crack propagation in aligned composites', *AIAA J.*, 22, 1485.

Hikami, F. and Chou, T. W. (1990) 'Explicit crack problem solutions of unidirectional composites: elastic stress concentrations', *AIAA J.*, 28, 499–505.

Ji, X. (1982) 'On the hybrid effect and fracture mode of interlaminated hybrid composites', *Proceedings of the Fourth International Conference on Composite Materials*, Tokyo, p. 1137.

Ji, X., Liu, X. R. and Chou, T. W. (1985) 'Dynamic stress concentration factors in unidirectional composites', *J. Comp. Mat.*, 19, 269–75.

Kelly, A. (1973) *Strong Solids*, Clarendon Press, Oxford.

Kelly, A. (1976) 'Composites with brittle matrices', in *Frontiers in Materials Science*, L. E. Murr and C. Stein, eds., Marcel Dekker Inc., New York, pp. 335–64.

Kelly, A. and Nicholson, R. B. (eds.) (1971) *Strengthening Methods in Crystals*, Elsevier, London.

Kies, J. A. (1962) US Naval Research Laboratory, Report No. 5752.

Kirkpatrick, E. G. (1974) *Introductory Statistics and Probability for Engineering, Science and Technology*, Prentice-Hall, Englewood Cliffs, New Jersey.

Kulkarni, S. V., Rosen, B. W. and Zweben, C. (1973) 'Load concentration factors for circular holes in composite laminates', *J. Comp. Mat.*, 7, 387.

Lei, S. C. (1986) 'A stochastic model for the damage growth during the transverse cracking process in composite laminates', Ph.D. Thesis, Drexel University.

Lipson, S. G. and Lipson, H. (1981) *Optical Physics*, 2nd edn., Cambridge University Press.

McCartney, L. N. (1987) 'Mechanics of matrix cracking in brittle-matrix fibre-reinforced composites', *Proc. Royal Soc. London, Series A*, 409, 329–50.

Manders, P., Bader, M. and Chou, T. W. (1982) 'Monte Carlo simulation of the strength of composite fiber bundles', *Fiber Sci. Tech.*, **17**, 183.

Manders, P. W. and Chou, T. W. (1983a) 'Variability of carbon and glass fibers, and the strength of aligned composites', *J. Reinforced Plastics & Composites*, **2**, 43.

Manders, P. W. and Chou, T. W. (1983b) 'Enhancement of strength in composites reinforced with previously-stressed fibers', *J. Comp. Mat.*, **17**, 26.

Manders, P. W., Chou, T. W., Jones, F. R. and Rock, J. W. (1983) 'Statistical analysis of multiple fracture in $0°/90°/0°$ glass fibre/epoxy resin laminates', *J. Mat. Sci.*, **18**, 2876–89.

Metcalfe, A. G. and Schmitz, G. K. (1964) 'Effect of length on the strength of glass fibers', *Proc. ASTM*, **64**, 1075.

Mileiko, S. T. (1969) 'The tensile strength and ductility of continuous fibre composites', *J. Mat. Sci.*, **4**, 974.

Mills, G. J. and Dauksys, R. J. (1973) 'Effect of prestressing boron/epoxy prepreg on composite strength properties', *AIAA J.*, **11**, 1459.

Netravali, A. N., Henstenburg, R. B., Phoenix, S. L. and Schwartz, P. (1989) 'Interfacial shear strength studies using the single-filament-composite Test I: experiments on graphite fibers in epoxy', *Polymer Composites*, **10**, 226–41.

Oh, K. P. (1979) 'A Monte Carlo study of the strength of unidirectional fiber-reinforced composites', *J. Comp. Mat.*, **13**, 311.

Pagano, N. J. and Pipes, R. B. (1971) 'The influence of stacking sequence on laminate strength', *J. Comp. Mat.*, **5**, 50–7.

Parratt, N. J. (1960) 'Defects in glass fibers and their effects on the strength of plastic mouldings', *Rubber and Plastics Age*, March 1960.

Parvizi, A. (1979). 'Transverse cracking in glass fibre reinforced plastic composites', Ph.D. Thesis, University of Surrey.

Parvizi, A. and Bailey, J. E. (1978) 'On multiple transverse cracking in glass fiber epoxy cross-ply laminates', *J. Mat. Sci.*, **13**, 2131.

Parvizi, A., Garrett, K. W. and Bailey, J. E. (1978) 'Constrained cracking in glass fiber-reinforced epoxy cross-ply laminates', *J. Mat. Sci.*, **13**, 195.

Peters, P. W. M. and Chou, T. W. (1987) 'On cross-ply cracking in glass- and glass-epoxy laminates', *Composites*, **18**, 40.

Phoenix, S. L. (1974) 'Probabilistic strength analysis of fiber bundle structures', *Fibre Sci. Tech.*, **7**, 15.

Phoenix, S. L. (1979) 'Statistical aspects of failure of fibrous composites', *Composite Materials: Testing and Design (Fifth Conference)*, ASTM STP 674, S. W. Tsai, ed., American Society for Testing and Materials, pp. 455–83.

Phoenix, S. L., Schwartz, P. and Robinson IV, H. H. (1988) 'Statistics for the strength and lifetime in creep-rupture of model carbon/epoxy composites', *Comp. Sci. Tech.*, **32**, 81–120.

Phoenix, S. L. and Smith, R. L. (1983) 'A comparison of probabilistic techniques for the strength of fibrous materials under local load-sharing among fibers', *Int. J. Sol. Structures*, **19**, 479–96.

Phoenix, S. L. and Taylor, H. M. (1973) 'The asymptotic strength distribution of a general fiber bundle', *Adv. Appl. Prob.*, **5**, 200.

Pipes, R. B. and Pagano, N. J. (1970) 'Interlaminar stresses in composite laminates under uniaxial extension', *J. Comp. Mat.*, **4**, 538–48.

Reifsnider, K. L., Henneke, E. G., Stinchcomb, W. W. and Duke, J. C. (1983) 'Damage mechanics and NDE of composite laminates', in *Mechanics of Composite Materials – Recent Advances* (Z. Hashin and C. T. Herakovich, eds.), Pergamon, New York, pp. 399–420.

Rosen, B. W. (1964) 'Tensile failure of fibrous composites', *AIAA J.*, **2**, 1985.

Rosen, B. W. (1970) 'Thermomechanical properties of fibrous composites', *Proc. Royal Soc. London, Series A*, **319**, 79–94.

Russell, A. J. and Street, K. N. (1985) 'Moisture and temperature effects on the mixed-mode delamination fracture of unidirectional graphite-epoxy', in *Delamination and Debonding of Materials*, W. S. Johnson, ed., ASTM STP 876, pp. 349–70.

Scop, P. M. and Argon, A. S. (1967) 'Statistical theory of strength of laminated composites', *J. Comp. Mat.*, **1**, 92.

Scop, P. M. and Argon, A. S. (1969) 'Statistical theory of strength of laminated composites II', *J. Comp. Mat.*, **3**, 30.

Smith, R. L. (1980) 'A probability model for fibrous composites with local load-sharing', *Proc. Royal Soc. London, Series A*, **372**, 539–53.

Smith, R. L. (1982) 'A note on a probability model for fibrous composites', *Proc. Royal Soc. London, Series A*, **382**, 179–82.

Smith, R. L. and Phoenix, S. L. (1981) 'Asymptotic distributions for the failure of fibrous materials under series–parallel structure and equal load sharing', *J. Appl. Mech.*, **48**, 75–82.

Smith, R. L., Phoenix, S. L., Greenfield, M. R., Henstenburg, R. B. and Pitt, R. E. (1983) 'Lower-tail approximations for the probability of failure of three-dimensional fibrous composites with hexagonal geometry', *Proc. Royal Soc. London, Series A*, **388**, 353–91.

Spiegel, M. R. (1961) *Statistics*, Schaum's Outline Series, McGraw-Hill, New York.

Takao, Y., Taya, M. and Chou, T. W. (1981) 'Stress field due to a cylindrical inclusion with constant axial eigenstrain in an infinite elastic body', *ASME J. Appl. Mech.*, **48**, 853–8.

Talreja, R. (1985) 'A continuum mechanics characterization of damage in composite materials', *Proc. Royal Soc., London, Series A*, **399**, 195–216.

Talreja, R. (1986) 'Stiffness properties of composite laminates with matrix cracking and interior delamination', *Eng. Fract. Mech.*, **25**, 751–62.

Talreja, R. (1987) *Fatigue of Composite Materials*, Technomic Pub. Co., Lancaster, Pennsylvania.

Talreja, R. (1989) 'Fatigue of composites', in *Concise Encyclopedia of Composite Materials*, A. Kelly, ed., Pergamon Press, Oxford, pp. 77–81.

Van Dyke, P. and Hedgepeth, J. M. (1969) 'Stress concentrations from single-filament failures in composite materials', *Textile Res. J.*, July, p. 618.

Vinson, J. R. and Chou, T. W. (1975) *Composite Materials and Their Use in Structures*, Applied Science Publishers, London.

Wang, A. S. D. (1984) 'Fracture mechanics of sublaminate cracks in composite materials', *Comp. Techl. Rev.*, **6**, 45–62.

Wang, A. S. D. (1987) 'Strength, failure, and fatigue analysis of laminates', *Engineering Materials Handbook*, **1**, 236–51, ASM International, Metals Park, Ohio.

Wang, A. S. D., Chou, P. C. and Lei, S. C. (1984) 'A stochastic model for the growth of matrix cracks in composite laminates', *J. Comp. Mat.*, **18**, 239–54.

Wang, A. S. D. and Crossman, F. W. (1977) 'Some new results on edge effects in symmetric composite laminates', *J. Comp. Mat.*, **11**, 92–102.

Wang, A. S. D. and Crossman, F. W. (1980) 'Initiation and growth of transverse cracks and edge delamination, I. An energy method', *J. Comp. Mat.*, **14**, 71-87.

Wang, A. S. D., Kishore, N. N. and Li, C. A. (1985) 'On crack development in graphite-epoxy $[0_2/90_n]_s$ laminates under uniaxial tension', *Comp. Sci. Tech.*, **23**, 1–31.

Wang, A. S. D., Slomiana, M. and Bucinell, R. B. (1985) 'Delamination crack growth in composite laminates', in *Delamination and Debonding of Materials*, W. S. Johnson, ed., ASTM STP 876, pp. 135–67.

Weibull, W. (1939a) 'A statistical theory of the strength of materials', *Ing. Vetenskaps Akad. Handl.*, no. 151.

Weibull, W. (1939b) 'The phenomenon of rupture in solids', *Ing. Vetenskaps Akad. Handl.*, no. 153.

Weibull, W. (1951) 'A statistical distribution function of wide applicability', *J. Appl. Mech.*, **18**, 293.

Zender, G. W. and Deaton, J. W. (1963) 'Strength of filamentary sheets with one or more fibers broken', NASA TN D-1609.

Zweben, C. (1968) 'Tensile failure of fiber composites', *AIAA J.*, **6**, 2325.

Zweben, C. (1974) 'An approximate method of analysis for notched unidirectional composites', *Eng. Frac. Mech.*, **6**, 1.

Zweben, C. and Rosen, B. W. (1970) 'A statistical theory of material strength with application to composite materials', *J. Mech. Phys. Sol.*, **18**, 189.

Chapter 4

Akasaka, T. (1974) 'A practical method of evaluating the isotropic elastic constants of glass mat reinforced plastics', *Comp. Mat. Struct. (Japan)*, **3**, 21–2.

Anderson, R. M. and Lavengood, R. E. (1968) 'Variables affecting strength and modulus of short fiber composites', *Soc. Plastic. Engrs. J.*, **24**, 20.

Arridge, R. G. C. (1963) 'Orientation effects in fibre reinforced composites where the modulus of the fibres is no more than an order of magnitude greater than that of the matrix', *Proc. 18th Ann. Tech. Conf. Reinf. Plastics Div., Soc. Plastics Industry*, Sec. 4-A, February 1963.

Bader, M. G., Chou, T. W. and Quigley, J. (1979) 'On the strength of discontinuous fiber composites with polymeric matrices', in *New Developments and Applications in Composites*, D. Wilsdorf, ed., TMS-AIME, New York.

Baker, R. M. and MacLaughlin, T. F. (1971) 'Stress concentrations near a discontinuity in fibrous composites', *J. Comp. Mat.* **5**, 492.

Bert, C. W. (1979) 'Composite-material mechanics: prediction of properties of planar-random fiber composites'. Presented at the 34th Annual Conference of the Reinforced Plastics/Composites Institute, New Orleans, Louisiana, January 29–February 2.

Beran, M. (1965) *Nuovo Cimento Ser. X*, **35**, 771.

Beran, M. and Molyneax, J. (1966) *Quart. Appl. Math.*, **24**, 107.

Blumentritt, B. F., Vu, B. T. and Cooper, S. L. (1974) 'The mechanical properties of oriented discontinuous fibre-reinforced thermoplastics I. Unidirectional fiber orientation', *Polymer Eng. Sci.*, **14**, 633–45.

Blumentritt, B. G., Vu, B. T. and Cooper, S. L. (1975) 'Mechanical properties of discontinuous fiber reinforced thermoplastics, II. Random-in-plane fiber orientation', *Polymer Eng. Sci.*, **15**, 428–36.

Bowyer, W. H. and Bader, M. G. (1972) 'On the reinforcement of thermoplastics by imperfectly aligned discontinuous fibers', *J. Mat. Sci.*, **7**, 1315.

Budiansky, B. (1965) 'On the elastic moduli of some heterogeneous materials', *J. Mech. Phys. Sol.*, **13**, 223–7.

Budiansky, B. (1970) *J. Comp. Mat.*, **4**, 286.

Burgel, B., Perry, A. J. and Schneider, W. R. (1970) 'On the theory of fibre strengthening', *J. Mech. Phys. Sol.*, **18**, 101–14.

Carrara, A. S. and McGarry, F. J. (1968) 'Matrix and interface stresses in a discontinuous fiber composite model', *J. Comp. Mat.*, **2**, 222–43.

Chamis, C. C. and Sendeckyj, G. P. (1968) 'Critique on theories predicting thermoelastic properties of fibrous composites', *J. Comp. Mat.*, **2**, 232.

Chang, C. I., Conway, H. D. and Weaver, T. C. (1972) 'The elastic constants and bond stresses for a three-dimensional composite reinforced by discontinuous fibers', *Fibre Sci. Tech.*, **5**, 143–62.

Chen, P. E. (1971) 'Strength properties of discontinuous fiber composites', *Polymer Eng. Sci.*, **11**, 51.

Chen, P. E. and Lavengood, R. E. (1969) 'Stress fields around multiple inclusions', Monsanto/Washington University, ONR/ARPA Association, HPC 68-60, AD 846907, January, 1969.

Chen, P. E. and Lewis, T. B. (1970) 'Stress analysis of ribbon reinforced composites', *Polymer Eng. Sci.*, **10**, 43.

Chou, T. W. and Kelly, A. (1976) 'Fiber composites', in 'Challenges and opportunities in materials science and engineering', *Mat. Sci. Eng.*, **25**, 35.

Chou, T. W. and Kelly, A. (1980) 'Mechanical properties of composites', *Ann. Rev. Mat. Sci.*, **10**, 229.

Chou, T. W. and Nomura, S. (1980) 'On the thermoelastic behavior of short fiber and hybrid composites', *Proceedings of the Third International Conference on Composite Materials*, Pergamon Press, New York, pp. 69–80.

Chou, T. W. and Nomura, S. (1981) 'Fiber orientation effects on the thermoelastic properties of short-fiber composites', *Fibre Sci. Tech.*, **14**, 279.

Chou, T. W., Nomura, S. and Taya, M. (1980) 'A self-consistent approach to the elastic stiffness of short-fiber composites', *J. Comp. Mat.*, **14**, 178.

Christensen, R. M. (1971) *Theory of Viscoelasticity*, Academic Press, New York.

Christensen, R. M. and Waals, F. M. (1972) 'Effective stiffness of randomly oriented fibre composites', *J. Comp. Mat.*, **6**, 518–32.

Christensen, R. M. (1979) *Mechanics of Composite Materials*, Wiley-Interscience, New York.

Conway, H. D. and Chang, C. I. (1971) 'The effective elastic constants and bond stresses for a fiber reinforced elastic sheet', *Fibre Sci. Tech.*, **5**, 249–60.

Cook, J. (1968) 'The elastic constants of an isotropic matrix reinforced with imperfectly oriented fibres', *Brit. J. Appl. Phys. (J. Phys. D)*, Series 2, **1**, 799–812.

Cox, H. L. (1952). 'The elasticity and strength of paper and other fibrous materials', *Brit. J. Appl. Phys.*, **3**, 72–9.

Curtis, P. T., Bader, M. G. and Bailey, J. E. (1978) 'The stiffness and strength of a polyamide thermoplastic reinforced with glass and carbon fibres', *J. Mat. Sci.*, **13**, 377.

Dederichs, P. H. and Zeller, R. (1973). 'Variational treatment of the elastic constants of disordered materials', *Z. Physik*, **259**, 103.

Edwards, H. and Evans, N. P. (1980) 'A method for the production of high quality aligned short fibre mats and their composites', *Proceedings of the Third International Conference on Composite Materials*, Pergamon Press, New York, pp. 1620–35.

Eimer, C. Z. (1971) 'The viscoelasticity of multi-phase media'. *Arch. Mech.*, **23**, 3–15.

Eshelby, J. D. (1957) 'The determination of the elastic field of an ellipsoidal inclusion and related problems', *Proc. Royal Soc.*, **A241**, 376–96.

Eshelby, J. D. (1961) 'Elastic inclusions and inhomogeneity', in *Progress in Solid Mechanics*, I. N. Sneddon and R. Hill, eds., vol. 2, North-Holland, Amsterdam, p. 89.

Favre, J. P. and Perrin, J. (1972) 'Carbon fibre adhesion to organic matrices', *J. Mat. Sci.*, **7**, 1113.

Friedrich, K. (1985) 'Microstructural efficiency and fracture toughness of short fiber/thermoplastic matrix composites', *Comp. Sci. Tech.*, **22**, 43–74.

Friedrich, K. (1989) 'Fractographic analysis of polymer composites', in *Application of Fracture Mechanics to Composite Materials*, Composite Material Series, vol. 6, Klaus Friedrich, ed., Elsevier, Amsterdam, p. 425.

Friedrich, K. and Karger-Kocsis, J. (1989) 'Fractography and failure mechanisms of unfilled and short fiber reinforced semi-crystalline thermoplastics', in *Fractography and Failure Mechanisms of Polymers and Composites*, A. C. Roulin-Moloney, ed., Elsevier-Applied Science, London, pp. 437–94.

Friedrich, K., Schulte, K., Horstenkamp, G. and Chou, T. W. (1985) 'Fatigue behavior of aligned short carbon fiber reinforced polyimide and polyethersulfone composites', *J. Mat. Sci.*, **20**, 3353.

Fukuda, H. and Chou, T. W. (1981a) 'An advanced shear-lag model applicable to discontinuous fiber composites', *J. Comp. Mat.*, **15**, 79.

Fukuda, H. and Chou, T. W. (1981b) 'A probabilistic theory for the strength of short-fiber composites', *J. Mat. Sci.*, **16**, 1088.

Fukuda, H. and Chou, T. W. (1982) 'A probabilistic theory of the strength of short-fiber composites and variable fiber length and orientation', *J. Mat. Sci.*, **17**, 1003.

Fukuda, H. and Kawata, K. (1974) 'On Young's modulus of short fibre composites', *Fibre Sci. Tech.*, **7**, 207–22.

Fukuda, H. and Kawata, K. (1977) 'On the strength distribution of unidirectional fibre composites', *Fibre Sci. Tech.*, **10**, 53.

Haener, J. and Ashbaugh, N. (1967) 'Three-dimensional stress distribution in a unidirectional composite', *J. Comp. Mat.*, **1**, 54–63.

Hahn, H. T. (1978) 'Stiffness and strength of discontinuous fiber composites', in *Composite Materials in the Automotive Industry*, S. V. Kulkarni, C. H. Zweben and R. B. Pipes, eds., ASME, New York, pp. 85–109.

Hale, D. K. and Kelly, A. (1972) 'Strength of fibrous composite materials', in *Annual Review of Materials Science*, vol. 2, R. A. Huggins, ed., Annual Review, Inc., Palo Alto, California, p. 405.

Halpin, J. C. (1969) 'Stiffness and expansion estimates for oriented short fiber composites', *J. Comp. Mat.*, **3**, 732–4.

Halpin, J. C. (1984) *Primer on Composite Materials: Analysis*, Technomic Publishing Co., Inc., Lancaster, Pennsylvania.

Halpin, J. C., Jerine, K. and Whitney, J. M. (1971) 'The laminate analogy for 2 and 3 dimensional composite materials', *J. Comp. Mat.*, **5**, 36–49.

Halpin, J. C. and Pagano, N. J. (1969) 'The laminate approximations for randomly oriented fibrous composites', *J. Comp. Mat.*, **3**, 720–4.

Hancock, P. and Cuthbertson, J. (1970) 'The effect of fibre length and interfacial bond in glass fibre-epoxy resin composites, *J. Mat. Sci.*, **15**, 762–8.

Hashin, Z. (1965a) 'On elastic behavior of fibre reinforced materials of arbitrary transverse phase geometry', *J. Mech. Phys. Sol*, **13**, 179.

Hashin, Z. (1965b) 'Viscoelastic behavior of heterogeneous media', *J. Appl. Mech.*, **32**, 630–6.

Hashin, Z. (1968) 'Assessment of the self-consistent scheme approximation: conductivity of particulate composites', *J. Comp. Mat.*, **2**, 284–300.

Hashin, Z. (1969) 'The inelastic inclusion problem', *Int. J. Eng. Sci.*, **7**, 11–36.

Hashin, Z. (1972) *Theory of Fiber Reinforced Materials*, NASA CR-1974.

Hashin, Z. and Rosen, B. W. (1964) 'The elastic moduli of fiber-reinforced materials', *J. Appl. Mech.*, **31**, 223–32.

Hashin, Z. and Shtrikman, S. (1962) *J. Appl. Phys.*, **30**, 3125.

Hashin, Z. and Shtrikman, S. (1963) 'A variational approach to the theory of the elastic behavior of multiphase materials', *J. Mech. Phys. Sol*, **11**, 127–40.

Hermans, J. J. (1967) 'The elastic properties of fibre reinforced materials when the fibers are aligned', *Proc. Konigl. Nederl, Akad, van Weteschappen Amsterdam, Series B*, **70**, 1–9.

Hikami, F. and Chou, T. W. (1984a) 'A probabilistic theory for the strength of discontinuous fiber composites', *J. Mat. Sci.*, **19**, 1805.

Hikami, F. and Chou, T. W. (1984b) 'Statistical treatment of transverse crack propagation in aligned composites', *AIAA J.*, **22**, 1485–90.

Hikami, F. and Chou, T. W. (1990) 'Explicit crack problem solutions of unidirectional composites: elastic stress concentrations', *AIAA J.*, **28**, 499–505.

Hill, R. (1952) 'The elastic behavior of a crystalline aggregate', *Proc. Phys. Soc.*, **A65**, 349.

Hill, R. (1965a) 'Theory of mechanical properties of fiber-strengthened materials – III. Self-consistent model', *J. Mech. Phys. Sol*, **13**, 189–98.

Hill, R. (1965b) 'A self-consistent mechanics of composite materials', *J. Mech. Phys. Sol*, **13**, 213–25.

Hori, M. and Yonezawa, F. (1975) 'Statistical theory of effective electrical, thermal and magnetic properties of random heterogeneous materials IV', *J. Math. Phys.*, **16**. 352.

Hsu, P. L., Yau, S. S. and Chou, T. W. (1986) 'Stress-corrosion cracking and its propagation in aligned short-fiber composites', *J. Mat. Sci.*, **21**, 3703.

Ishikawa, H., Chou, T. W. and Taya, M. (1982) 'Prediction of failure modes in unidirectional short fiber composites', *J. Mat. Sci.*, **17**, 832.

Jackson, P. W. and Cratchley, D. (1966) *J. Mech. Phys. Sol*, **14**, 49.

Kacir, L. and Narkis, M. (1975) *Polymer Eng. Sci.*, **15**, 525.

Kardos, J. (1973) 'Structure property relations in short-fiber reinforced plastics', *CRC Crit. Rev. Solid State Sci.*, August, pp. 419–50.

Kelly, A. (1971) 'Reinforcement of structural materials by long strong fibres', *Met. Trans.*, **3**, 2313.

Kelly, A. (1973) *Strong Solids*, 2nd edn., Clarendon Press, Oxford.

Kelly, A. and Davies, G. J. (1965) 'The principles of the fibre reinforcement of metals', *Met. Rev.*, **10**, 1.

Kelly, A. and Tyson, W. R. (1965a) 'Fibre-strengthened materials', in *High Strength Materials, V*, F. Zackay, ed., J. Wiley and Sons, Inc., New York, p. 578.

Kelly, A. and Tyson, W. R. (1965b) 'Tensile properties of fiber-reinforced metals: copper/tungsten and copper/molybdenum', *J. Mech. Phys. Sol*, **13**, 329.

Kerner, E. H. (1956) 'The elastic and thermoelastic properties of composite media', *Proc. Phys. Soc.*, **B69**, 808–13.

Kilchinskii, A. A. (1965) 'On the model for determining thermoelastic characteristics of fiber reinforced materials', *Prikladnaia Mekhanika*, **1**, 1.

Kilchinskii, A. A. (1966) 'Approximate method for determining the relation between the stresses and strains for reinforced materials of the fiber glass type',

Thermal Stresses in Elements of Construction, Naukova Kumka, Kiev,
6, 123.

Knibbs, R. H. and Morris, J. B. (1974) 'The effects of fibre orientation on the physical properties of composites', *Composites*, **5**, 209–18.

Kröner, E. (1958) 'Berechnung der Elastichen Konstanten des Vielkristalls aus den Konstanten des Eikristalls', *Z. Physik*, **151**, 504.

Kröner, E. (1967) 'Elastic moduli of perfectly disordered composite materials', *J. Mech. Phys. Sol.*, **15**, 319.

Kröner, E. (1972) *Statistical Continuum Mechanics*, CISM Courses and Lectures no. 92, Udine, Springer-Verlag, Wien.

Kröner, E. (1977) 'Bounds for effective elastic moduli of disordered materials', *J. Mech. Phys. Sol*, **25**, 137.

Lavengood, R. E. (1972) 'Strength of short-fiber reinforced composites', *Polymer Eng. Sci.*, **12**, 48.

Laws, N. (1973) 'On the thermostatics of composite materials', *J. Mech. Phys. Sol*, **21**, 9.

Laws, N. and McLaughlin, R. (1978) 'Self-consistent estimates for the viscoelastic creep compliances of composite materials', *Proc. Royal Soc.*, **A359**, 251–73.

Lee, L. H. (1969) 'Strength-composition relationships of random short glass fiber–thermoplastics composites', *Polymer Eng. Sci.*, **9**, 213–24.

Lees, J. K. (1968) 'A study of the tensile modulus of short fiber reinforced plastics', *Polymer Eng. Sci.*, **8**, 186–94

Levin, V. M. (1967) *Inzh. Zh. Mekk. Tverd. Tela*, No. 1, p. 88.

MacLaughlin, T. F. (1966) 'Effect of fiber geometry on stress in fiber reinforced composite materials', *Exp. Mech.*, **6**, 481–92.

MacLaughlin, T. F. (1968) 'A photoelastic analysis of fiber discontinuities in composite materials', *J. Comp. Mat.*, **2**, (1), 44.

McNally, D. (1977) 'Short fiber orientation and its effect on the properties of thermoplastic composite materials', *Polymer Plast. Tech. Eng.*, **8**, 101–54.

Mandell, J. F., Grande, D. H., Tsiang, T.-H. and McGarry, F. J. (1986) 'Modified microdebonding test for direct in situ fiber/matrix bond strength determination in fiber composites', in *Composite Materials: Testing and Design (Seventh Conference)*, ASTM STP 893, J. M. Whitney, ed., American Society for Testing and Materials, Philadelphia, pp. 87–108.

Manders, P. W. and Chou, T. W. (1982) 'The strength of aligned short-fiber carbon, glass, and hybrid carbon/glass composites', *Proceedings of the Fourth International Conference on Composite Materials*, The Japan Society for Composite Materials, Tokyo, pp. 1075–82.

Manera, M. (1971) 'Elastic properties of randomly oriented short fiber-glass composites', *J. Comp. Mat.*, **11**, 235–47.

Miller, B., Muri, P. and Rebenfeld, L. (1987) *Comp. Sci. Tech.*, **28**, 17.

Muki, R. and Sternberg, E. (1969) 'On the diffusion of an axial load from an infinite cylindrical bar embedded in an elastic medium', *Int. J. Sol. Struct.*, **5**, 587.

Muki, R. and Sternberg, E. (1970) 'Elastostatic load-transfer to a half-space from a partially embedded axially loaded rod', *Int. J. Sol. Struct.*, **6**, 69.

Muki, R. and Sternberg, E. (1971) 'Load-absorption by a discontinuous filament in a fiber-reinforced composite', *Z. Angew. Math. Phys.*, **22**, 809.

Mura, T. (1982) *Micromechanics of Defects in Solids*, Martinus Nijhoff Publishers, The Hague.

Nicolais L. (1975) 'Mechanics of composites', *Polymer Eng. Sci.*, **15**, 137–49.

Nielsen, L. E. and Chen, P. E. (1968) 'Young's modulus of composites filled with randomly oriented fibers', *J. Mat.*, **3**, 352–8.

Nomura, S. and Chou, T. W. (1980) 'Bounds of effective thermal conductivity of short-fiber composites', *J. Comp. Mat.*, **14**, 120.

Nomura, S. and Chou, T. W. (1981) 'Effective thermoelastic constants of short-fiber composites', *Int. J. Eng. Sci.*, **19**, 1.

Nomura, S. and Chou, T. W. (1984) 'Bounds of elastic moduli of multiphase short-fiber composites', *J. Appl. Mech.*, **51**, 540.

Nomura, S. and Chou, T. W. (1985) 'The viscoelastic behavior of short-fiber composite materials', *Int. J. Eng. Sci.*, **23**, 193.

Outwater, J. O., Jr. (1956) 'The mechanics of plastics reinforcement in tension', *Mod. Plast.*, **33**, 156–62.

Pakdemirli, E. and Williams, J. G. (1969) 'Metal fibre reinforced thermoplastics and the role of adhesion efficiency', *J. Mech. Eng. Sci.*, **11**, 68–75.

Piggott, M. R. (1987) 'Debonding and friction at fibre-polymer interfaces I: Criteria for failure and sliding', *Comp. Sci. Tech.*, **30**, 295–306.

Piggott, M. R. and Dai, S. R. (1988) 'Debonding and friction at fibre-polymer interfaces II: microscopic model experiments, *Comp. Sci. Tech.*, **31**, 15–24.

Reuss, A. (1929) *Zeit Angew Math. U. Mech.*, **9**, 49.

Richter, H. (1980) 'Single fibre and hybrid composites with aligned discontinuous fibres in polymer matrix', *Proceedings of the Third International Conference on Composite Materials,* Pergamon Press, New York, pp. 387–98.

Riley, V. R. and Reddaway, J. L. (1968) *J. Mat. Sci.*, **3**, 41.

Rosen, B. W. (1964) 'Tensile failure of fibrous composites', *AIAA J.*, **2**, 1985.

Rosen, B. W. and Shu, L. S. (1971) 'On some symmetry conditions for three dimensional fibrous composites', *J. Comp. Mat.*, **5**, 279.

Rosen, W. and Hashin, Z. (1970) 'Effective thermal expansion coefficients and specific heats of composite materials', *Int. J. Eng. Sci.*, **8**, 157–73.

Schapery, R. A. (1967) 'Stress analysis of viscoelastic composite materials', *J. Comp. Mat.*, **1**, 228–67.

Schapery, R. A. (1968) 'Thermal expansion coefficients of composite materials based on energy principles', *J. Comp. Mat.*, **2**, 380–404.

Schapery, R. A. (1974) *Composite Materials*, G. P. Sendeckyj, ed., Vol. 2, Academic Press, New York.

Schierding, R. G. and Deex, O. D. (1969) 'Factors influencing the properties of whisker–metal composites', *J. Comp. Mat.*, **3**, 618–29.

Stowell, E. Z. and Liu, T. S. (1961) *J. Mech. Phys. Sol*, **9**, 242.

Takao, T., Chou, T. W. and Taya, M. (1982) 'Effective longitudinal Young's modulus of misoriented short fiber composites', *J. Appl. Mech.*, **49**, 536.

Takao, Y., Taya, M. and Chou, T. W. (1981) 'Stress field due to cylindrical inclusion with constant axial eigenstrain in an infinite body', *J. Appl. Mech.*, **48**, 853.

Takao, Y., Taya, M. and Chou, T. W. (1982) 'Effects of fiber-end cracks on the stiffness of aligned short-fiber composites', *Int. J. Sol. Struct.*, **8**, 723.

Taya, M. and Chou, T. W. (1981) 'On two kinds of ellipsoidal inhomogeneities in infinite elastic body: an application to a hybrid composite', *Int. J. Sol. Struct.*, **17**, 553.

Taya, M. and Chou, T. W. (1982) 'Prediction of the stress–strain curve of a short fiber reinforced thermoplastics', *J. Mat. Sci.*, **17**, 2801.

Tsai, S. W. and Pagano, N. J. (1968) 'Invariant properties of composite materials', in *Composite Materials Workshop*, S. W. Tsai, J. C. Halpin and N. J. Pagano, eds., Technomic Pub. Co., Stamford, CT, pp. 233–53.

van de Poel, C. (1958) 'On the rheology of concentrated dispersions', *Rheol. Acta*, **1**, 198–205.

Vinson, J. R. and Chou, T. W. (1975) *Composite Materials and Their Use in Structures*, Elsevier-Applied Science, London.

Voigt, W. (1889) *Ann. Phys.*, **33**, 573.

Wadsworth, N. J. and Spilling, I. (1968) *Brit. J. Appl. Phys.*, **1**, 1049.

Warren, F. and Norris, C. B. (1953) 'Mechanical properties of laminate design to be isotropic', Forest Products Laboratory, Madison, WI, Report No. 1841, May 1953.

Wilczynki, A. P. (1978) 'Random directional reinforcement theory', *Fibre Sci. Tech.*, **11**, 19–22.

Willis, J. R. (1977) 'Bounds and self-consistent estimates for the overall properties of anisotropic composites', *J. Mech. Phys. Sol.*, **25**, 185–202.

Wu, C.-T. D. and McCullough, R. L. (1977) 'Constitutive relationships for heterogeneous materials', in *Developments in Composite Materials—1*, G. S. Holister, ed., Applied Science Publishers, London, p. 119.

Yates, B., Overy, M. J., Sargent, J. P., McCalla, B. A., Kingston-Lee, D. M., Phillips, L. N. and Rogers, K. F. (1978) *J. Mat. Sci.*, **13**, 433.

Yau, S. S. and Chou, T. W. (1989) 'Low temperature performance of short fiber reinforced thermoplastics', in *Test Methods and Design Allowables for Fiber Composites*, vol. 2, C. C. Chamis, ed., ASTM STP 1003, p. 45.

Zeller, R. and Dederichs, P. H. (1973) 'Elastic constants of polycrystals', *Phys. Stat. Sol.*, **b55**, 831.

Chapter 5

Adam, T., Fernando, G., Dickson, R. F., Reiter, H. and Harris, B. (1989) 'Fatigue life prediction for hybrid composites', *Int. J. Fatigue*, **11**, 233–7.

Adams, D. F. (1975) 'A scanning electron microscopic study of hybrid composite impact response', *J. Mat. Sci.*, **10**, 1591–602.

Adams, D. F. and Miller, A. K. (1975) 'An analysis of the impact behavior of hybrid composite materials', *Mat. Sci. Eng.*, **19**, 245–60.

Adams, D. F. and Miller, A. K. (1976) 'The influence of transverse shear on the static flexure and Charpy impact response of hybrid composite materials', *J. Mat. Sci.*, **11**, 1697–710.

Adams, D. F. and Zimmerman, R. S. (1986) 'Static and impact performance of PE fiber/graphite fiber hybrid composites', *SAMPE J.*, Nov./Dec., pp. 10–16.

Arrington, M. and Harris, B. (1978) 'Some properties of mixed fibre CFRP', *Composites*, **9**, 149–52.

Aveston, J. and Kelly, A. (1980) 'Tensile first cracking strain and strength of hybrid composites and laminates', *Phil. Trans. Royal Soc. London*, **A294**, 519–34.

Aveston, J. and Sillwood, J. M. (1976) 'Synergistic fibre strengthening in hybrid composites', *J. Mat. Sci.*, **11**, 1877.

Bader, M. G. and Manders, P. W. (1978) 'Failure strain enhancement in carbon/glass fiber hybrid composites', *Proceedings of the Third International Conference on Composite Materials*, vol. 3, Toronto.

Bader, M. G. and Manders, P. W. (1981a) 'The strength of hybrid glass/carbon fiber composites, part I, failure strain enhancement and failure mode', *J. Mat. Sci.*, **16**, 2233–45.

Bader, M. G. and Manders, P. W. (1981b) 'The strength of hybrid glass/carbon fiber composites, part II, a statistical model', *J. Mat. Sci.*, **16**, 2246–56.

Beaumont, P. W. R., Riewald, P. G. and Zweben, C. (1974) 'Methods for improving the impact resistance of composite materials', in *Foreign Object Impact Damage to Composites*, ASTM STP 568, American Society of Testing and Materials, Philadelphia, pp. 134–58.

Bucci, R. J., Mueller, L. N., Schultz, R. W. and Prohaska, J. L. (1987) 'ARALL laminates – results from a cooperative test program', in *Advanced Materials Technology 87, Proceedings 32nd International SAMPE Symposium*, vol. 32, Society for the Advancement of Material and Process Engineering, Corina, CA, pp. 902–16.

Bunsell, A. R. (1976) Letter to the Editor, *Composites*, **7, 158**.

Bunsell, A. R. and Harris, B. (1974) 'Hybrid carbon and glass fibre composites', *Composites*, **5**, 157–64.

Chamis, C. C., Hanson, M. P. and Serafini, T. T. (1972) 'Impact resistance of unidirectional fiber composites', *Composite Materials: Testing and Design (Second Conference)*, ASTM STP 497, American Society for Testing and Materials, Philadelphia, pp. 324–49.

Chamis, C. C. and Lark, R. F. (1978) 'Non-metallic hybrid composites: analysis, design, application and fabrication', in *Hybrids and Selected Metal-Matrix Composites: A State-of-the-Art Review*, W. J. Renton, ed., AIAA, New York, pp. 13-51.

Chamis, C. C. and Sinclair, J. H. (1979) 'Micromechanics of intraply hybrid composites: elastic and thermal properties', in *Modern Developments in Composite Materials and Structures*, J. R. Vinson, ed., The American Society of Mechanical Engineers, New York, pp. 253–67.

Chan, W. S., Rogers, C. and Aker, S. (1976) 'Improvement of edge delamination strength using adhesive layers', in *Composite Materials Testing and Design*, 7th Conference, J. M. Whitney, ed., ASTM STP893, American Society for Testing and Materials, Philadelphia, pp. 266–85.

Chen, J. L. and Sun, C. T. (1989) 'Modeling of orthotropic elastic–plastic properties of ARALL laminates', *Comp. Sci. Tech.*, **36**, 321–38.

Chou, T. W. and Kelly, A. (1980a) 'Mechanical properties of fiber composite materials', in *Annual Review of Materials Science*, vol. 10, Annual Review, Inc., Palo Alto, pp. 229–59.

Chou, T. W. and Kelly, A. (1980b) 'The effect of transverse shear on the compressive strength of fiber composites', *J. Mat. Sci.*, **15**, 327.

Chou, T. W., Nomura, S. and Taya, M. (1980) 'A self-consistent approach to the elastic stiffness of short-fiber composites', *J. Comp. Mat.*, **14**, 178.

Chou, T. W., Steward, B. and Bader, M. G. (1979) 'On the compression strength of glass–epoxy composites', in *New Developments and Applications in Composites*, D. Wilsdorf, ed., TMS-AIME, New York.

Dorey, G., Sidey, G. R. and Hutchings, J. (1978) 'Impact properties of carbon fibre/Kevlar 49 fibre hybrid composites', *Composites*, **9**, 25–32.

Eshelby, J. D. (1957) 'The determination of the elastic field of an ellipsoidal inclusion, and related problem', *Proc. Royal Soc. London*, **241**, 376–96.

Fernando, G., Dickson, R. F., Adam, T., Reiter, H. and Harris, B. (1988) 'Fatigue behaviour of hybrid composites: I carbon/Kevlar hybrids', *J. Mat. Sci.*, **23**, 3732–43.

Fischer, S. and Marom, G. (1987) 'The flexural behavior of aramid fiber hybrid composite materials', *Comp. Sci. Tech.*, **28**, 1–24.

Fukuda, H. (1983a) 'Mechanics of hybrid composites, part I', *Trans. Japan Soc. Composite Mat.*, **9**, 76–80.

Fukuda, H. (1982b) 'Mechanics of hybrid composites, part II', *Trans. Japan Soc. Composite Mat.*, **9**, 118–23.

Fukuda, H. (1983c) 'Mechanics of hybrid composites, part III'. *Trans. Japan Soc. Composite Mat.*, **9**, 153–9.

Fukuda, H. and Chou, T. W. (1981) 'Stress concentrations around a discontinuous fiber in a hybrid composite sheet', *Trans. Japan Soc. Composite Mat.*, **7**, 37–42.

Fukuda, H. and Chou, T. W. (1982a) 'Monte Carlo simulation of the strength of hybrid composites', *J. Comp. Mat.*, **16**, 357.

Fukuda, H. and Chou, T. W. (1982b) 'A statistical approach to the strength of hybrid composites', *Proceedings of the Fourth International Conference on Composite Materials*, Japan Society for Composite Materials, Tokyo, pp. 1145–52.

Fukuda, H. and Chou, T. W. (1983) 'Stress concentration in a hybrid composite sheet', *J. Appl. Mech.*, **50**, 845–8.

Fukunaga, H., Chou, T. W. and Fukuda, H. (1984) 'Strength of intermingled hybrid composites', *J. Reinforced Plastics Composites*, **3**, 145–60.

Fukunaga, H., Chou, T. W. and Fukuda, H. (1989) 'Probabilistic strength analyses of interlaminated hybrid composites', *Comp. Sci. Tech.*, **35**, 331.

Fukunaga, H., Chou, T. W., Peters, P. M. W. and Schulte, K. (1984a) 'Probabilistic failure strength analyses of graphite/epoxy cross-ply laminates', *J. Comp. Mat.*, **18**, 339–56.

Fukunaga, H., Chou, T. W., Schulte, K. and Peters, P. M. W. (1984b) 'Probabilistic initial failure strength of hybrid and non-hybrid laminates', *J. Mat. Sci.*, **19**, 3546.

Gruber, M. B., Overbeeke, J. L. and Chou, T. W. (1982) 'A reusable sandwich beam concept for composite compression test', *J. Comp. Mat.*, **16**, 162–71.

Gruber, M. B. and Chou, T. W. (1983) 'Elastic properties of intermingled hybrid composites', *Polymer Composites*, **4**, 265–9.

Hancox, N. L. and Wells, H. (1973) 'Izod impact properties of carbon-fibre/glass-fibre sandwich structures', *Composites*, **4**, 26–30.

Harlow, D. G. (1983) 'Statistical properties of hybrid composites', *Proc. Royal Soc.*, **A389**, 67–100.

Harlow, D. G. and Phoenix, S. L. (1978a) 'The chain-of-bundles probability model for the strength of fibrous materials I: analysis and conjectures', *J. Comp. Mat.*, **12**, 195.

Harlow, D. G. and Phoenix, S. L. (1978b) 'The chain-of-bundles probability model for the strength of fibrous materials II: a numerical study of convergence', *J. Comp. Mat.*, **12**, 314.

Harris, S. J. and Bradley, P. D. (1976) *Proceedings of the First International Conference on Composite Materials*, pp. 327–35.

Harris, B. and Bunsell, A. R. (1975) 'Impact properties of glass-fibre/carbon fibre hybrid composites', *Composites*, **6**, 197–201.

Hayashi, T. (1972) 'On the improvement of mechanical properties of composites by hybrid composition', *Proceedings of the Eighth International Reinforced Plastics Conference*, paper 22.

Hedgepeth, J. M. (1961) 'Stress concentration in filamentary structures', NASA TN D-882.

Jang, B. Z., Chen, L. C., Wang, C. Z., Lin, H. T. and Zee, R. H. (1989) 'Impact resistance and energy absorption mechanisms in hybrid composites', *Comp. Sci. Tech.*, **34**, 305–35.

Ji, X. (1982) 'On the hybrid effect and fracture mode of interlaminated hybrid composites', in *Proceedings of the Fourth International Conference on Composite Materials*, T. Hayashi, K. Kawata and S. Umekawa, eds., Japan Society for Composite Materials, Tokyo, pp. 1137–44.

Ji, X., Hsiao, G. C. and Chou, T. W. (1981) 'A dynamic explanation of the hybrid effect', *J. Comp. Mat.*, **15**, 443–61.

Kalnin, I. L. (1972) 'Evaluation of unidirectional glass–graphite fiber/epoxy resin composites', *Composite Materials Testing and Design (Second Conference)*, ASTM STP 497, American Society for Testing and Materials, Philadelphia, pp. 551–63.

Kenaga, D., Doyle, J. F. and Sun, C. T. (1987) 'The characterization of boron/aluminum composite in nonlinear range as an orthotropic elastic–plastic material', *J. Comp. Mat.*, **21**, 516–31.

Kirk, J. N., Munro, M. and Beaumont, P. W. R. (1978) 'The fracture energy of hybrid carbon and glass composites', *J. Mat. Sci.*, **13**, 2197–204.

Kretsis, G. (1987) 'A review of the tensile, compressive, flexural and shear properties of hybrid fiber-reinforced plastics', *Composites*, **18**, 13–23.

McColl, I. R. and Morley, J. G. (1977) 'Crack growth in hybrid fibrous composites', *J. Mat. Sci.*, **12**, 1165–75.

McCullough, R. L. and Peterson, J. M. (1977) 'Property optimization analysis for multicomponent (hybrid) composites', in *Developments in Composite Materials – 1*, G. S. Holister, ed., Applied Science Publisher, London.

Marissen, R. (1984) 'Flight simulation behavior of aramid reinforced aluminum laminates (ARALL)', *Eng. Fracture Mech.*, **19**, 261–77.

Marissen, R., Trautmann, K. H., Foth, J. and Nowack, H. (1984) 'Microcrack growth in aramid reinforced aluminum laminates (ARALL)', in *Fatigue 84, Proceedings 2nd International Conference on Fatigue and Fatigue Thresholds*, C. J. Beevers, ed. vol. II. EMAS Ltd, Warley, UK, pp. 1081–91.

Marom, G. and Chen, E. J. H. (1987) 'Asymmetric hybrid composite: a design concept to improve flexural properties of Kevlar aramid composites', *J. Comp. Sci. Tech.*, **29**, 161–8.

Marom, G., Fischer, S., Tuler, F. R. and Wagner, H. (1978) 'Hybrid effects in composites', *J. Mat. Sci.*, **13**, 1419–26.

Mori, T. and Tanaka, K. (1973) 'An average stress in matrix and average elastic energy of materials with misfitting inclusions', *Acta Met.*, **21**, 571–4.

Mueller, L. N., Prohaska, J. L. and Davis, J. W. (1985) 'ARALL (aramid aluminum laminates): introduction of a new composite material', *Proceedings AIAA Aerospace Engineering Conference*, AIAA, New York, AIAA paper no. 85-0846.

Nomura, S. and Chou, T. W. (1984) 'Bounds of elastic moduli of multiphase short-fiber composites', *J. Appl. Mech.*, **51**, 540.

Phillips, L. N. (1976) 'The hybrid effect–does it exist?' *Composites*, **7**, 7–8.

Pitkethly, M. J. and Bader, M. G. (1987) 'Failure modes of hybrid composites consisting of carbon fiber bundles dispersed in a glass fiber epoxy resin matrix', *J. Phys. D: Appl. Phys.*, **20**, 315–22.

Renton, W. J. (ed.) (1978) *Hybrids and Selected Metal–Matrix Composites: A State-of-the-Art Review*, AIAA, New York.

Rosen, B. W. (1964) 'Tensile failure of fibrous composites', *AIAA J.*, **2**, 1985.

Rybicki, E. and Kanninen, M. (1978) 'Fracture mechanics of non-metallic hybrid composites', in *Hybrids and Selected Metal–Matrix Composites: A State-of-the-Art Review*, W. J. Renton, ed., AIAA, New York, pp. 53–65.

Summerscales, J. and Short, D. (1978) 'Carbon fibre and glass fibre hybrid reinforced plastics', *Composites*, **9**, 157–66.

Sun, C. T. and Luo, J. (1985) 'Failure loads for notched graphite/epoxy laminates with a softening strip', *Comp. Sci. Tech.*, **22**, 121–33.

Sun, C. T. and Norman, T. L. (1988) 'Design of laminated composite with controlled-damage concept', *Proceedings of American Society for Composites, 3rd Technical Conference*, Technomic Pub. Co., Lancaster, PA, pp. 485–9.

Sun, C. T. and Rechak, S. (1988) 'Effect of adhesive layers on impact damage in composite laminates', in *Composite Materials Testing and Design, 8th Conference*, ASTM STP 972, J. D. Whitcomb, ed., American Society for Testing and Materials, Philadelphia.

Takahashi, K. and Chou, T. W. (1987) 'Non-linear deformation and failure behavior of carbon/glass hybrid laminate', *J. Comp. Mat.*, **21**, 396–420.

Taya, M. and Chou, T. W. 'On two kinds of ellipsoidal inhomogeneities in an infinite elastic body: an application to a hybrid composite', *Int. J. Sol. Struct.*, **17**, 553–63.

Vogelesang, L. B. and Gunnink, J. W. (1986) 'ARALL, a materials challenge for the next generation of aircraft', *Mat. & Design*, **7**, 287–300.

Wagner, H. D. and Marom, G. (1982) 'On composition parameters for hybrid composite materials', *Composites*, **13**, 18.

Walton, P. L. and Majumdar, A. J. (1975) *Composites*, **6**, 209–16.

Wells, H. and Hancox, N. L. (1971) 'Stiffening and strengthening GRP beams with CFRP', *Composites*, **2**, 147–51.

Yau, L. N. and Chou, T. W. (1989) 'Analysis of hybrid effect in unidirectional composites under longitudinal compressions', *Composite Structures*, **12**, 27–37.

Zweben, C. (1977) 'Tensile strength of hybrid composites', *J. Mat. Sci.*, **12**, 1325–37.

Chapter 6

Bishop, S. M. (1989) 'Strength and failure of woven carbon-fibre reinforced plastics for high performance applications', in *Textile Structural Composites*, T. W. Chou and F. K. Ko, eds., Elsevier Science Publishers B.V., Amsterdam, pp. 173–207.

Byun, J. H. and Chou, T. W. (1989) 'Modeling and characterization of textile structural composites: a review', *J. Strain Analysis*, **24**, 253–62.

Chang, L. W., Yau, S. S. and Chou, T. W. (1987) 'Notched strength of woven fabric composites with moulded-in holes', *Composites*, **18**, 233–41.

Chou, T. W. (1985) 'Characterization and modeling of textile structural composites: an overview', *Proceedings of the European Conference on Composite Materials*, AEMC, Bordeaux, pp. 133–7.

Chou, T. W. (1986) 'Strength and failure behavior of textile structural composites', *Proceedings of the American Society for Composites First Technical Conference*, Technomic Pub. Co., Lancaster, PA, p. 104.

Chou, T. W. (1989a) 'Properties of woven fabric composites', in *Encyclopedia of Materials Science and Engineering* and *Concise Subject Encyclopedias*, Pergamon Press, Oxford, p. 292.

Chou, T. W. (1989b) 'Mechanics of two-dimensional woven fabric composites', in *Mechanical Behavior and Properties of Composite Materials*, Technomic Pub. Co., Lancaster, PA, pp. 131–50.

Chou, T. W. and Ishikawa, T. (1989) 'Analysis and modeling of two-dimensional fabric composites', in *Textile Structural Composites*, T. W. Chou and F. K. Ko, eds., Elsevier Science Publishers B.V., Amsterdam, pp. 209–64.

Chou, T. W. and Ko, F. K. (1989) *Textile Structural Composites*, Elsevier Science Publishers B.V., Amsterdam.

Curtis, P. T. and Bishop, S. M. (1984) 'An assessment of the potential of woven carbon fibre-reinforced plastics for high performance applications', *Composites*, **15**, 259–65.

Douglass, W. A. (1964) *Braiding and Braiding Machinery*, Centrex Pub. Co., Eindhoven.

Dow, N. F. (1969) Triaxial Fabric, U.S. Patent 3446251, May 1969.

Dow, N. F. (1982) 'Studies of woven fabric reinforced composites for automotive applications', *Tech. Final Rep., MSC TFR 1301/8101*, Materials Science Corp., Springhouse, Pennsylvania.

Dow, N. F. and Tranfield, G. (1970) 'Preliminary investigations of feasibility of weaving triaxial fabrics (Doweave)', *Text. Res. J.*, **40**, 986–98.

Ghasemi Nejhad, M. N. and Chou, T. W. (1990a) 'Compression behavior of woven carbon fibre-reinforced epoxy composites with moulded-in and drilled holes', *Composites*, **21**, 33–40.

Ghasemi Nejhad, M. N. and Chou, T. W. (1990b) 'A model for the prediction of compressive strength reduction of composite laminates with molded-in holes', *J. Comp. Mat.* **24**, 236–55.

Hahn, H. T. and Tsai, S. W. (1973) 'Nonlinear elastic behavior of unidirectional composite laminate', *J. Comp. Mat.*, **7**, 102–18.

Hearle, J. W. S. (1989) 'Mechanics of yarns and nonwoven fabrics', in *Textile Structural Composites*, T. W. Chou and F. K. Ko, eds., Elsevier Science Publishers B.V., Amsterdam, pp. 27–65.

Hearle, J. W. S., Grosberg, P. and Backer, S. (1969) *Structural Mechanics of Fibers, Yarns, and Fabrics*, vol. 1, Wiley-Interscience, New York.

Ishikawa, T. (1981) 'Anti-symmetric elastic properties of composite plates of satin weave cloth', *Fiber Sci. Tech.*, **15**, 127–45.

Ishikawa, T. and Chou, T. W. (1982a) 'Elastic behavior of woven hybrid composites', *J. Comp. Mat.*, **16**, 2–19.

Ishikawa, T. and Chou, T. W. (1982b) 'Stiffness and strength behavior of woven fabric composites', *J. Mat. Sci.*, **17**, 3211–20.

Ishikawa, T. and Chou, T. W. (1982c) 'Stiffness and strength properties of woven fabric composites', in *Proceedings of the Fourth International Conference on Composite Materials*, Japan Society for Composite Materials, Tokyo, pp. 489–96.

Ishikawa, T. and Chou, T. W. (1983a) 'In-plane thermal expansion and thermal bending coefficients of fabric composites', *J. Comp. Mat.*, **17**, 92–104.

Ishikawa, T. and Chou, T. W. (1983b) 'One-dimensional analysis of woven fabric composites', *AIAA J.*, **21**, 1714.

Ishikawa, T. and Chou, T. W. (1983c) 'Nonlinear behavior of woven fabric composites', *J. Comp. Mat.*, **17**, 399–413.

Ishikawa, T. and Chou, T. W. (1983d) 'Thermoelastic analysis of hybrid fabric composite', *J. Mat. Sci.*, **18**, 2260–8.

Ishikawa, T., Koyama, K. and Kobayashi, S. (1977) 'Elastic moduli of carbon-epoxy composites and carbon fibers', *J. Comp. Mat.*, **11**, 332–44.

Ishikawa, T., Matsushima, M., Hayashi, Y. and Chou, T. W. (1985) 'Experimental confirmation of the theory of elastic moduli of fabric composites', *J. Comp. Mat.*, **19**, 443–58.

Jones, R. M. (1975) *Mechanics of Composite Materials*, McGraw-Hill, New York.

Kimpara, I., Hamamoto, A. and Takehana, M. (1977) *Trans. Japan Soc. Comp. Mat.*, **3**, 21.

Lekhnitskii, S. G. (1963) *Theory of Elasticity of an Anisotropic Elastic Body*, Holden-Day, San Francisco.

Lord, P. R. and Mohamed, M. H. (1982) *Weaving: Conversion of Yarn to Fabric*, 2nd edn, Merrow Publishing Company, Durham, UK.

Mody, P. B., Chou, T. W. and Friedrich, K. (1988) 'Effect of testing conditions and microstructure on the sliding wear of graphite fiber/PEEK matrix composites', *J. Mat. Sci.*, **23**, 4319–30.

Mody, P. B., Chou, T. W. and Friedrich, K. (1989) 'Abrasive wear behavior of unidirectional and woven graphite fiber/PEEK composites', in *Test Methods and Design Allowables for Fiber Composites*, ASTM STP 1003, C. C. Chamis, ed., American Society for Testing and Materials, Philadelphia, Pennsylvania, p. 75.

Rogers, K. F., Kingston-Lee, D. M., Phillips, L. N., Yates, B., Chandra, M. and Parker, S. F. H. (1981) 'The thermal expansion of carbon fibre-reinforced plastics, part 6. The influence of fibre weave in fabric reinforcement', *J. Mat. Sci.*, **16**, 2803–18.

Rogers, K. F., Phillips, L. N., Kingston-Lee, D. M., Yates, B., Overy, M. J., Sargent, J. P. and McCalla, B. A. (1977) 'The thermal expansion of carbon fibre-reinforced plastics, part 1. The influence of fibre type and orientation', *J. Mat. Sci.*, **12**, 718–34.

Rosen, B. W., Chatterjee, S. N. and Kibler, J. J. (1977) 'An analysis model for spatially oriented fiber composites', ASTM STP 617, *Composite Materials: Testing and Design (Fourth Conference)*, American Society for Testing and Materials, Philadelphia, pp. 243–54.

Scardino, F. L. (1989) 'An introduction to textile structures and their behavior', in *Textile Structural Composites*, T. W. Chou and F. K. Ko, eds., Elsevier Science Publishers B.V., Amsterdam, pp. 1–26.

Scardino, F. L. and Ko, F. (1981) 'Triaxial woven fabrics Part I: behavior under tensile, shear, and burst deformation', *Text. Res. J.*, **51**, 80–9.

Schulte, K., Reese, E. and Chou, T. W. (1987) 'Fatigue behavior and damage development in woven fabric and hybrid fabric composites', *Proceedings of the Sixth International Conference and Second European Conference on Composite Materials*, vol. 4, Elsevier Applied Science, London, pp. 89–99.

Schwartz, P. (1984) A mathematical analysis of a fabric having non-orthogonal interlacings using strain energy methods', *Fiber Sci. Tech.*, **20**, 273–82.

Schwartz, P., Fornes, R. E. and Mohamed, M. H. (1982) 'An analysis of the mechanical behavior of triaxial fabrics and the equivalency of conventional fabrics', *Text Res. J.*, **52**, 388–94.

Schwartz, P., Rhodes, T. and Mohamed, M. H. (1982) *Fabric Forming Systems*, Noyes Publications, Park Ridge, New Jersey.

Skelton, J. (1971) Triaxially woven fabrics – their structure and properties, *Text. Res. J.*, **41**, 637–47.

Thomas, D. G. B. (1971) *An Introduction to Warp Knitting*, Merrow, Watford, UK.

Wray, G. R. and Vitols, R. (1982) 'Advances in stitch-bonding, warp- and weft-knitting systems, and automated knitwear manufacture', in *Contemporary Textile Engineering*, F. Happey, ed., Academic Press, London, pp. 375–409.

Yang, J. M. and Chou, T. W. (1986) 'Performance optimization of woven fabric composites for printed circuit boards', in *Electronic Packaging Materials Science, II*, Symposia Proceedings vol. 72, Materials Research Society, Pittsburgh, pp. 163–73.

Yang, J. M. and Chou, T. W. (1987) 'Performance maps of textile structural composites', *Proceedings of the Sixth International Conference on Composite Materials*, vol. 5, Elsevier Applied Science, London, p. 579.

Yang, J. M. and Chou, T. W. (1989) 'Thermo-elastic analysis of triaxial woven fabric composites', in *Textile Structural Composites*, T. W. Chou and F. K. Ko, eds., Elsevier Science Publishers B.V., Amsterdam, pp. 265–77.

Yates, B., Overy, M. J., Sargent, J. P., McCalla, B. A., Kingston-Lee, D. M., Phillips, L. N. and Rogers, K. F. (1978) 'The thermal expansion of carbon fibre-reinforced plastics, part 2. The influence of fibre volume fraction', *J. Mat. Sci.*, **13**, 433–40.

Yau, S. S. and Chou, T. W. (1988) 'Strength of woven fabric composites with drilled and molded holes', in *Composite Materials Testing and Design (Eighth Conference)*, ASTM STP 972, J. D. Whitcomb, ed., ASTM, Philadelphia, pp. 423–37.

Yokoyama, A., Fujita, A., Kobayashi, H., Hamada, H. and Maekawa, Z. (1989) 'A new braiding process – robotised braiding mechanism', in *Materials and Processing – Move into the 90's*, S. Benson, T. Cook, E. Trewin and R. M. Turner, eds., Elsevier, Amsterdam, pp. 87–99.

Zweben, C. and Norman, J. C. (1976) 'Kevlar 49/Thornel 300 hybrid fabric composites for aerospace application', *SAMPE Quarterly*, **1**, 1–10.

Chapter 7

Annual Book of ASTM Standards, Part 10, Practice E399, 'Plane-strain fracture toughness of metallic materials', American Society for Testing and Materials, Philadelphia, PA.

Ashby, M. F., Gandi, C. and Taplin, D. M. R. (1979) 'Fracture-mechanism maps and their construction for F.C.C. metals and alloys', *Acta Metall.*, **27**, 699–729.

ASTM Standards and Literature References for Composite Materials, D 4255-83, 'In-plane shear properties of composite laminates', American Society for Testing and Materials, Philadelphia, PA.

ASTM Standards and Literature References for Composite Materials, D 2344-84, 'Apparent interlaminar shear strength of parallel fiber composites by short-beam method', American Society for Testing and Materials, Philadelphia, PA.

Brunnschweiler, D. (1954) 'The structure and tensile properties of braids', *J. Text. Instrum.*, **45**, T55–T77.

Byun, J. H., Leach, B. S., Stroud, S. S. and Chou, T. W. (1990a) 'Structural characteristics of three-dimensional angle-interlock woven fabric preforms', in *Processing of Polymers and Polymeric Composites*, ASME, MD-vol. 19 American Society for Mechanical Engineers, New York, pp. 177.

Byun, J. H., Du, G. W. and Chou, T. W. (1991) 'Analysis and modeling of 3-D textile structural composites', ACS Symposium Series **457**, American Chemical Society, Washington D.C., pp. 22–33.

Byun, J. H., Gillespie, J. W. and Chou, T. W. (1989) 'Mode II delamination of three-dimensional textile structural composites', *Proceedings of the American Society for Composites, 4th Technical Conference*, Technomic Pub. Co., Lancaster, PA, pp. 287–96.

Byun, J. H., Gillespie, J. W. and Chou, T. W. (1990b) 'Mode I delamination of a three-dimensional fabric composite', *J. Comp. Mat.*, **24**, 497.

Byun, J. H., Whitney, T. J., Du, G. W. and Chou, T. W. (1991) 'Analytical characterization of two-step braided composites' *J. Comp. Mat.* in press.

Chou, T. W. (1989) 'Structure–performance maps', in *Encyclopedia of Materials Science and Engineering and Concise Subject Encyclopedias*, Pergamon Press, Oxford, p. 261.

Chou, T. W., McCullough, R. L. and Pipes, R. B. (1986) 'Composites', *Sci. Am.*, **254**, 193–203.

Chou, T. W. and Yang, J. M. (1986) 'Structure–performance maps of polymeric, metal and ceramic matrix composites', *Metall. Trans. A*, **17A**, 1547–9.

Cole, P. M. (1988) 'Three-dimensional structures of interlocked strands', U.S. Patent 4,737,399.

Cooper, G. A. and Kelly, A. J. (1967) 'Tensile properties of fiber-reinforced metals: fracture mechanics', *Mech. Phys. Solids*, **15**, 279–97.

Crane, R. M. and Camponeschi, E. T. (1986) 'Experimental and analytical characterization of multidimensionally braided graphite/epoxy composites', *Exp. Mech.*, **26**, 259.

Dexter, H. B. and Funk, J. G. (1986) 'Impact resistance and interlaminar fracture toughness of through-the-thickness reinforced graphite epoxy', AIAA Paper 86-1020-CP, pp. 700–9.

Dhingra, A. K., Champion, A. R. and Krueger, W. H. (1975) 'Fiber FP reinforced aluminum and magnesium composites', in *Proceedings of the First Metal Matrix Composite Workshop*, Paper Number C-501, Institute for Defense Analyses, Washington, D.C., September.

Dow, N. F. (1984) *Proceedings of the Fiber Society/SAMPE Conference on High Performance Textile Structures*, Philadelphia College of Textile and Science, Philadelphia, PA.

Du, G. W., Popper, P. and Chou, T. W. (1989) 'Analysis and automation of two-step braiding', *FIBER-TEX 88*, NASA Conference Publication no. 3038, pp. 217–33.

Du, G. W., Popper, P. and Chou, T. W. (1991) 'Analysis of 3D textile preform for multidirectional reinforcement of composites', *J. Mat. Sci.* in press.

Florentine, R. (1982) 'Apparatus for weaving a three-dimensional article', U.S. Patent 4,312,261.

Fowser, S. W. (1986) 'The behavior of orthogonal fabric composites', M.S. thesis, University of Delaware.

Fowser, S. W. and Chou, T. W. (1989) 'Simplified Green's functions for mode I and II cracks', *Int. J. Fracture*, **39**, 301–21.

Fowser, S. W. and Chou, T. W. (1990a) 'Integral equations solution for reinforced mode I cracks opened by internal pressure', *J. Appl. Mech.*, in press.

Fowser, S. W. and Chou, T. W. (1990b) 'Numerical integration of Green's functions for an edge-loaded infinite strip', *Computers and Structures*, in press.

Frost, H. J. and Ashby, M. F. (1982) *Deformation-Mechanism Maps*, Pergamon Press, Oxford.

Gandi, C. and Ashby, M. F. (1979) 'Fracture mechanism maps for materials which cleave: F.C.C. and H.C.P. metals and ceramics', *Acta Metall.*, **27**, 1565–1602.

Guénon, V. A., Chou, T. W. and Gillespie, J. W. (1987) 'Interlaminar fracture toughness of a three-dimensional fabric composite', *Proceedings of the Society of*

Manufacturing Engineers, EM87-551, 1–17, Society of Manufacturing Engineers, Dearborn, Michigan.

Guénon, V. A., Chou, T. W. and Gillespie, J. W. (1989) 'Toughness properties of a three-dimensional carbon–epoxy composite', *J. Mat. Sci.,* **24,** 4168–75.

Guess, T. R. and Reedy, Jr., E. D. (1985) 'Comparison of interlocked fabric and laminated fabric Kevlar 49/epoxy composites', *J. Comp. Tech. Res.,* **7,** 136–42.

Hearle, J. W. S., Grosberg, P. and Backer, S. (1969) *Structural Mechanics of Fibers, Yarns, and Fabrics,* vol. 1, Wiley-Interscience, New York, p. 80.

Hunston, D. H. (1984) *Comp. Tech. Rev.,* **6,** 176.

Iosipescu, N. (1967) 'New accurate procedures for single shear testing of metals', *J. Mat.,* **2,** 537–66.

Ishikawa, T. and Chou, T. W. (1982) 'Stiffness and strength behavior of woven fabric composites', *J. Mat. Sci.,* **17,** 3211–20.

Kelly, A. and Macmillan, N. H. (1986) *Strong Solids,* 3rd edn, Clarendon Press, Oxford.

Kies, J. A. (1962) 'Maximum strains in the resin of fiberglass composites', U.S. Naval Research Laboratory Report NRL 5752.

Ko, F. K. (1989a) Three-dimensional fabrics for composites', in *Textile Structural Composites,* T. W. Chou and F. K. Ko, eds., Elsevier Science Publishers B.V., Amsterdam, pp. 129–71.

Ko, F. K. (1989b) 'Preform fiber architecture for ceramic–matrix composites', *Ceramic Bull.,* **68,** 401–14.

Ko, F. K. (1986) 'Tensile strength and modulus of a 3-D braided composite' in *Composite Materials: Testing and Design,* ASTM STP 893, American Society for Testing and Materials, Philadelphia, PA, p. 392.

Ko, F. K. and Pastore, C. M. (1985) 'Structure and properties of an integrated 3-D fabric for structural composites', ASTM STP 864, American Society for Testing and Materials, Philadelphia, PA, p. 428.

Ko, F. K., Pastore, C. M., Yang, J. M. and Chou, T. W. (1986) 'Structure and properties of multilayer, multidirectional warp knit fabric reinforced composites', in *Composites '86: Recent Advances in Japan and the United States,* Japan Society for Composite Materials, Tokyo.

Ko, F. K., Soebroto, H. B. and Lei, C. (1988) '3-D net shaped composites by the 2-step braiding process', *Proceedings of 33rd International SAMPE Symposium,* vol. 33, SAMPE International Business Office, Covina, California, pp. 912–21.

Kregers, A. F. and Teters, G. A. (1982) 'Structural model of deformation of anisotropic three-dimensionally reinforced composites', *Mech. Comp. Mat.,* **1,** 14.

Lekhnitskii, S. G. (1963) *Theory of Elasticity of an Anisotropic Elastic Body,* Holden-Day, San Francisco.

Li, W. and El Shiekh, A. (1988) 'The effect of processes and processing parameters on 3-D braided preforms for composites', *SAMPE Quarterly,* **19,** 22–8.

Li, W., Kang, T. J., and El Shiekh, A. (1988) 'Structural mechanics of 3-D braided preforms for composites, part I: Geometry of fabric produced by 4-step process', in *Proceedings of Fiber-Tex'87 Conference,* NASA Conference Publication.

Liu, C. H. and Chou, T. W. (1989) 'Mode II interlaminar fracture toughness of three-dimensional textile structural composites', *Proceedings of the 4th Japan–U.S. Conference on Composite Materials,* Technomic Pub. Co., 981.

Ma, C. L., Yang, J. M. and Chou, T. W. (1986) 'Elastic stiffness of three-dimensional braided textile structural composites', in *Composite Materials, Testing* and Design (Seventh Conference), ASTM STP 893, American Society for Testing and Materials, Philadelphia, PA, pp. 404–21.

Majidi, A. P. and Chou, T. W. (1986) 'Impact tolerance of braided alumina fiber reinforced aluminum composites', *Proceedings of the 31st International SAMPE Symposium*, SAMPE International Business Office, Covina, California.

Majidi, A. P. and Chou, T. W. (1987) 'Structure-reliability studies of three-dimensionally braided metal matrix composites', in *Proceedings of the Sixth International and Second European Conference on Composite Materials*, vol. 2, Elsevier Applied Science, London.

Majidi, A. P., Rémond, O. G. and Chou, T. W. (1987) 'The effect of fiber architecture on the mechanical performance of metal matrix composites', *Proceedings of the 2nd Annual Conference of the American Society for Composites*, Technomic Pub. Co., Lancaster, Pennsylvania, p. 371.

Majidi, A. P., Yang, J. M. and Chou, T. W. (1986) 'Toughness characteristics of three-dimensionally braided Al2O3/Al–Li composites', in *Interfaces in Metal–Matrix Composites*, A. K. Dhingra and S. G. Fishman, eds., The Metallurgical Society, Warrendale, PA, pp. 27–44.

Majidi, A. P., Yang, J. M. and Chou, T. W. (1988) 'Mechanical behavior of three-dimensional braided metal–matrix composites', *Testing Technology of Metal Matrix Composites*, ASTM STP 964, American Society for Testing and Materials, Philadelphia, PA, p. 31.

Majidi, A. P., Yang, J. M., Pipes, R. B. and Chou, T. W. (1985) 'Mechanical behavior of three-dimensional woven fiber composites', in *Proceedings of the Fifth International Conference on Composite Materials*, The Metallurgical Society of AIME, Warrendale, PA, pp. 1247–65.

Masters, J. E. (1987) 'Characterization of impact development in graphite epoxy laminates', ASTM, STP 948, American Society for Testing and Materials, Philadelphia, PA, pp. 238–58.

Mignery, L. A., Tan, T. M. and Sun, C. T. (1985) 'The use of stitching to suppress delamination in laminated composites', ASTM STP 876, American Society for Testing and Materials, Philadelphia, PA, pp. 371–85.

Ogo, Y. (1987) 'The effect of stitching on in-plane and interlaminar properties of carbon–epoxy fabric laminates', M.S. Thesis, University of Delaware,

Peirce, F. T. (1937) 'The geometry of cloth structure', *J. Text. Instrum.*, **28**, T45–T96.

Popper, P. and McConnell, R. F. (1988) 'Complex shaped braided structures', U.S. Patent 4,719,837.

Rémond, G. O. (1987) 'Characterization and modeling of 3-D braided metal matrix composites', M.S. Thesis, University of Delaware.

Rybicki, E. F. and Kanninen, M. F. (1977) 'A finite element calculation of stress intensity factors by a modified crack closure integral', *Eng. Fracture Mech.*, **9**, 931–8.

Simonds, R. A., Stinchcomb, W. and Jones, R. M. (1988) 'Mechanical behavior of braided composite materials', ASTM STP 972, American Society for Testing and Materials, Philadelphia, PA, p. 438.

Steeger, USA, Inc. (1989) Production Program, Spartanburg, South Carolina.

Takahashi, K. and Chou, T. W. (1986) 'Modeling of the interfacial behavior of flexible composites', in *Interfaces in Metal–Matrix Composites*, A. K. Dhingra and S. G. Fishman, eds., The Metallurgical Society, Warrendale, PA, pp. 45–59.

Tattersall, H. G. and Tappin, G. J. (1966) *J. Mat. Sci.* **1**, 296.

Walrath, D. E. and Adams, D. F. (1983a) 'The Iosipescu shear test as applied to composite materials', *Exp. Mech.*, **23**, 105–10.

Walrath, D. E. and Adams, D. F. (1983b) 'Analysis of the stress state in an Iosipescu shear test specimen', Technical Report UWME-DR-301-102-1, University of Wyoming, Laramie, Wyoming.

Weller, R. D. (1985) 'AYPEX: a new method of composite reinforcement braiding, *3-D Composite Materials*, NASA Conference Publication 2420.

Whitcomb, J. D. (1989) 'Three dimensional stress analysis of plain weave composites', NASA Technical Memorandum 101672.

Whitney, J. M., Browning, C. E. and Hoogsteden, W. (1982) *J. Reinf. Plastics Comp.*, **1**, 297.

Whitney, T. J. (1988) 'Analytical characterization of 3-D textile structural composites using plane stress and 3-D lamination analogies', M.S. Thesis, University of Delaware.

Whitney, T. J. and Chou, T. W. (1988) 'Modeling of elastic properties of 3-D textile structural composites', *Proceedings of the American Society for Composites, Third Technical Conference*, Technomic Pub. Co., Lancaster, Pennsylvania, p. 427.

Whitney, T. J. and Chou, T. W. (1989) 'Modeling of 3-D angle-interlock textile structural composites', *J. Comp. Mat.*, **23**, 890–911.

Yang, J. M. and Chou, T. W. (1989) 'Thermo-elastic analysis of triaxial woven fabric composites', in *Textile Structural Composites*, T. W. Chou and F. K. Ko, eds., Elsevier Science Publishers B.V., Amsterdam, pp. 265–77.

Yang, J. M., Ma, C. L. and Chou, T. W. (1986) 'Fiber inclination model of three-dimensional textile structural composites', *J. Comp. Mat.*, **20**, 472–84.

Yau, S. S., Ko, F. and Chou, T. W. (1986) 'Flexural and axial compressive failures of three-dimensionally braided composite I-beams', *Composites*, **17**, 227.

Yoshida, H., Ogasa, T. and Hayashi, R. (1986) 'Statistical approach to the relationship between ILSS and void content of CFRP', *Comp. Sci. Tech.*, **25**, 3–18.

Chapter 8

Akasaka, T. (1959–64) Various Reports/Bulletins, Faculty of Science and Engineering, Chuo University, Tokyo.

Akasaka, T. (1989) 'Flexible composites', in *Textile Structural Composites* T. W. Chou and F. Ko, eds., Elsevier Science Publishers, Amsterdam, pp. 279–330.

Akasaka, T. and Hirano, M. (1972) *Comp. Mat. Struc.*, **1**, 70.

Akasaka, T. and Yoshida, N. (1972) *Proc. Intl. Conf. Mech. Behavior of Mater.*, Kyoto, Japan, vol. 5, pp. 187–97.

Alley, V. A. and Fairslon, R. W. (1972) 'Experiment investigation of strains in a fabric under biaxial and shear forces', *J. Aircraft*, **9**, 55.

Bert, C. W. and Kumar, M. (1981) 'Experiments on highly nonlinear elastic composites', *Proc. NCKU/AAS Int-Sym. in Eng. Sci. and Mech.*, National Chen Kung Univ., Taiwan, vol. 2, pp. 1269–83.

Bert, C. W. and Reddy, J. N. (1982) 'Mechanics of bimodular composite structures', in *Mechanics of Composite Materials: Recent Advances*, Proceedings of the IUTAM Symposium, Virginia Polytechnic Institute, pp. 323–37.

Biderman, V. I., Gusliter, R. L., Sakharov, S. P., Nenakhov, B. V., Seleznev, I. I. and Tsukerberg, S. M. (1963) 'Automobile tires, construction, design, testing and usage', NASA, TT F-12, 382, 1969. (Original publication in Russian, State Scientific and Technical Press for Chemical Literature, Moscow).

Bohm, F. (1966) 'Mechanik des Gürtelreifens', *Ing-Arch.*, **35**, 82–101.

Chamis, C. C. (1984) 'Simplified composite micromechanics equations for hygral, thermal and mechanical properties', *SAMPE Quarterly*, **15**, 14–23.

Chou, T. W. (1985) 'Characterization and modeling of textile structural composites: an overview', *Proceedings of First European Conference on Composite Materials*, A.E.M.C., Bordeaux, France, pp. 133–8.

Chou, T. W. (1989) 'Review: flexible composites', *J. Mat. Sci.*, **24**, 761–83.

Chou, T. W. (1990) 'Flexible composites', in *International Encyclopedia of Composites*, VCH Publishers, New York.

Chou, T. W. and Takahashi, K. (1987) 'Nonlinear elastic behavior of flexible fiber composites', *Composites*, **18**, 25.

Chou, T. W. and Yang, J. M. (1986) 'Structure-performance maps of textile structural composites in polymeric, metal and ceramic matrices', *Met. Trans. A*, **17A**, 1547.

Clark, S. K. (1963a) 'The plane elastic characteristics of cord–rubber laminates', *Textile Res. J.*, **33**, 295–313.

Clark, S. K. (1963b) 'Internal characteristics of orthotropic laminates', *Textile Res. J.*, **33**, 935–53.

Clark, S. K. (1964) 'A review of cord–rubber elastic characteristics', *Rubber Chem. Tech.*, **37**, 1365–90.

Clark, S. K. (1980) 'The role of textiles in pneumatic tires', in *Mechanics of Flexible Fibre Assembles*, J. W. S. Hearle, J. J. Thwaites and J. Amirbayat, eds., Sijthoff and Noordhoff, The Netherlands.

Clark, S. K. and Dodge, R. N. (1969) 'A load transducer for tire cord', SAE Paper 690521, Society of Automotive Engineers, Warrendale, PA.

Gough, V. E. (1968) 'Stiffness of cord and rubber constructions', *Rubber Chem. Tech.*, **41**, 988–1021.

Hearle, J. W. S., Grosberg, P. and Backer, S. (1969) *Structural Mechanics of Fibers, Yarns and Fabrics*, vol. 1, Wiley-Interscience, New York.

Ishikawa, T. and Chou, T. W. (1983) 'Nonlinear behavior of woven fabric composites', *J. Comp. Mat.*, **17**, 399.

James, H. M. and Guth, E. (1943) 'Theory of the elastic properties of rubber', *J. Chem. Phys.*, **11**, 455–81.

Jones, R. E. (1975) *Mechanics of Composite Materials*, McGraw-Hill, New York.

Kuo, C. M., Takahashi, K. and Chou, T. W. (1988) 'Effects of fiber waviness on the nonlinear elastic behavior of flexible composites', *J. Comp. Mat.*, **12**, 1004.

Lou, A. Y. C. and Walter, J. D. (1978) 'Interlaminar shear strain measurements in cord–rubber composites', paper presented at SESA meeting, Wichita, Kansas, May.

Luo, S. Y. and Chou, T. W. (1988) 'Finite deformation and nonlinear elastic behavior of flexible composites', *J. Appl. Mech.*, **55**, 149–55.

Modern Plastics Encyclopedia, Engineering Data Bank, McGraw-Hill, Inc., New York.

Patterson, R. G. (1969) 'The measurement of cord tensions in tires', *Rubber Chem. Echnology*, **42**, 812.

Petit, P. H. and Waddoups, M. E. (1969) 'A method of predicting the nonlinear behavior of laminated composites', *J. Comp. Mat.*, **3**, 2–19.

Reinhardt, H. W. (1976) 'On the biaxial testing and strength of coated fabrics', *Exp. Mech.*, **11**, 71.

Skelton, J. (1971) 'The biaxial stress–strain behavior of fabrics for air-supported tents', *J. Mat., J.M.L.S.A.*, **6**, 656.

Stubbs, N. (1988) 'Elastic and inelastic response of coated fabrics to arbitrary loading paths', in *Textile Structural Composites* T. W. Chou and F. K. Ko, eds., Elsevier Science Publishers, Amsterdam, pp. 331–54.

Stubbs, N. and Thomas, S. (1984) 'A nonlinear elastic constitutive model for coated fabrics', in *Mechanics of Material*, S. Nernat-Nasser, eds., vol. 3, Elsevier Science Publishers, BV, Amsterdam, pp. 157–68.

Takahashi, K. and Chou, T. W. (1986) 'Modeling of the interfacial behavior of flexible composites', in *Interfaces in Metal–Matrix Composites*, A. K. Dhingra and S. G. Fishman, eds., The Metallurgical Society, Warrendale, PA, pp. 45–59.

Takahashi, K., Kuo, C. M. and Chou, T. W. (1986) 'Nonlinear elastic constitutive equations of flexible fiber composites', in *Composites '86: Recent Advances in Japan and the United States*, Japan Society for Composite Materials, Tokyo, p. 389.

Takahashi, K., Yano, T., Kuo, C. M. and Chou, T. W. (1987) 'Effect of fiber waviness on elastic moduli of fiber composites', *Trans. Japan Fiber Soc.*, **43**, 376.

Treloar, L. R. G. (1973) 'The elasticity and related properties of rubbers', *Rep. Prog. Phys.*, **36**, 755–826.

Walter, J. D. (1978) 'Cord–rubber tire composites: theory and application', *Rubber chem. Tech.*, **51**, 524.

Walter, J. D. and Hall, G. L. (1969) 'Cord load characteristics in bias and belted-bias tires', SAE Paper 690522, Society of Automotive Engineers, Warrendale, PA.

Walter, J. D. and Patel, H. P. (1979) ''Approximate expressions for the elastic constants of cord–rubber laminates', *Rubber Chem. Tech.*, **52**, 710–24.

Chapter 9

Adkins, J. E. and Rivlin, R. S. (1955) 'Large elastic deformation of isotropic materials X. Reinforcements by inextensible cords', *Phil. Trans. Royal Soc. London*, (*A*), **248**, 201–23.

Aspden, R. M. (1986) 'Relation between structure and mechanical behavior of fibre-reinforced composite materials at large strains', *Proc. Royal Soc. London*, (*A*), **406**, 287–98.

ASTM Standard D3518-76 (1982) 'Practice for in-plane shear stress–strain response of unidirectional reinforced plastics' American Society for Testing and Materials, Philadelphia.

Chou, T. W. (1989) 'Flexible composites', *J. Mat. Sci.*, **24**, 761–83.

Ericksen, J. L. and Rivlin, R. S. (1954) 'Large elastic deformations of homogeneous anisotropic materials', *J. Rational Mech. Analysis*, 3 (3) 281–301.

Fung, Y. C. (1965) *Foundations of Solid Mechanics*, Prentice-Hall Inc., Englewood Cliffs, NJ.

Fung, Y. C. (1977) *A First Course in Continuum Mechanics*, Prentice-Hall, Inc., Englewood Cliffs, N.J.

Fung, Y. C. (1981) *Biomechanics: Mechanical Properties of Living Tissues*, Springer Verlag, New York.

Hahn, H. T. (1973) 'Nonlinear behavior of laminated composites', *J. Comp. Mat.*, **7**, 257–71.

Hahn, H. T. and Tsai, S. W. (1973) 'Nonlinear elastic behavior of unidirectional composite laminae', *J. Comp. Mat.*, **7**, 102–18.

Humphrey, J. D. and Yin, F. C. P. (1987) 'A new constitutive formulation for characterizing the mechanical behavior of soft tissues', *Biophys. J.*, **52**, 563–70.

Ishikawa, T. and Chou, T. W. (1983) 'Nonlinear behavior of woven fabric composites', *J. Comp. Mat.*, **17**, 399–413.

Jones, R. S. and Morgan, H. S. (1977) 'Analysis of nonlinear stress–strain behavior of fiber-reinforced composite materials', *AIAA J.*, **15**, 1669–76.

Kuo, C. M., Takahashi, K. and Chou, T. W. (1988) 'Effect of fiber waviness on the nonlinear elastic behavior of flexible composites', *J. Comp. Mat.*, **12**, 1004.

Lai, W. M., Rubin, D. and Krempl, E. (1978) *Introduction to Continuum Mechanics*, Pergamon Press, Oxford.

Luo, S. Y. (1988) 'Theoretical modeling and experimental characterization of flexible composites', Ph.D. Dissertation, University of Delaware, Newark, Delaware.

Luo, S. Y. and Chou, T. W. (1988a) 'Finite deformation and nonlinear elastic behavior of flexible composites', *J. Appl. Mech.*, **55**, 149–55.

Luo, S. Y. and Chou, T. W. (1988b) 'Constitutive relations of flexible composites under finite elastic deformation', in *Mechanics of Composite Materials – 1988*, G. J. Dvorak and N. Laws, eds., ASME, AMD, New York, vol. **92**, pp. 209–16.

Luo, S. Y. and Chou, T. W. (1989) 'Elastic behavior of laminated flexible composites under finite deformation', in *Micromechanics and Inhomogeneity – The Toshio Mura Anniversary Volume*, G. J. Weng, M. Taya and H. Abe, eds., Springer-Verlag, New York, pp. 243–56.

Luo, S. Y. and Chou, T. W. (1990a) 'Modeling of the nonlinear elastic behavior of elastomeric flexible composites', in *Composites: Chemical and Physicochemical Aspects*, T. L. Vigo and B. J. Kinzig eds., VCH Publishers, New York, in press.

Luo, S. Y. and Chou, T. W. (1990b) 'Finite deformation of flexible composites', *Proc. Royal Soc. London*, (*A*), **429**, 569–86.

Luo, S. Y., Kuo, C. M. and Chou, T. W. (1988) 'Theoretical modeling and experimental characterization of flexible composites', *Proceedings of the Fourth Japan–United States Conference on Composite Materials*, Technomic Pub. Co., Lancaster, Pennsylvania, pp. 885–74.

Malvern, L. E. (1969) *Introduction to the Mechanics of a Continuous Medium*, Prentice-Hall, Inc., Englewood Cliffs.

Pagano, N. J. and Halpin, J. C. (1968) 'Influence of end constraint in the testing of anisotropic bodies', *J. Comp. Mat.*, **2**, 18–31.

Petit, P. H. and Waddoups, M. E. (1969) 'A method of predicting the nonlinear behavior of laminated composites', *J. Comp. Mat.*, **3**, 2–19.

Pipkin, A. C. and Rogers, T. G. (1971) 'Plane deformation of incompressible fiber-reinforced materials', *J. Appl. Mech.*, **38**, 634–40.

Posfalvi, O. (1977) 'The Poisson ratio for rubber–cord composites', *Rubber Chem. Tech.*, **50**, 224–32.

Rivlin, R. S. (1948a) 'Large elastic deformation of isotropic materials I. Fundamental Concepts', *Phil. Trans. Royal Soc. London*, (*A*), **240**, 459–90.

Rivlin, R. S. (1948b) 'Large elastic deformation of isotropic materials IV. Further developments of the general theory', *Phil. Trans. Royal Soc. London*, (*A*), **241**, 379–97.

Rivlin, R. S. (1959) 'Mathematics and rheology, the 1958 Bingham Medal Address', *Phys. Today*, **12**, (5), 32–6.

Rivlin, R. S. (1964) 'Networks of inextensible cords', in *Nonlinear Problems of Engineering*, W. F. Ames, ed., Academic Press, New York.

Rivlin, R. S. (1970) 'Nonlinear continuum theories in mechanics and physics and their applications', in *Centro Internazionale Matematico Estivo*, Ciclo, II and R. S. Rivlin, eds., Edizioni Cremonese, Roma.

Rivlin, R. S. and Saunders, D. W. (1951) 'Large elastic deformation of isotropic materials XII. Experiments on the deformation of rubber', *Phil. Trans. Royal Soc. London, (A)*, **243**, 251–98.

Spencer, A. J. M. (1972) *Deformation of Fibre Reinforced Materials*, Clarendon Press, Oxford.

Truesdell, C. (1966) *Elements of Continuum Mechanics*, Springer-Verlag, New York.

Whitney, J. M., Daniel, I. M. and Pipes, R. B. (1982) *Experimental Mechanics of Fiber Reinforced Composite Materials*, The Society for Experimental Stress Analysis, Brookfield Center, Connecticut.

Author index

Adam, T. 277, 278, 540, 541
Adams, D. F. 277, 431, 540, 551
Adkins, J. E. 475, 504, 553
Akasaka, T. 191, 444, 448, 449, 452, 456, 534, 551
Aker, S. 275, 541
Aköz, A. Y. 54, 55, 78, 527, 528
Alley, V. A. 444, 551
Anderson, R. M. 183, 534
Argon, A. S. 115, 118, 533
Arridge, R. G. C. 187, 534
Arrington, M. 231, 540
Ashbaugh, N. 170, 536
Ashby, M. F. 1, 2, 419, 526, 547, 548
Ashton, J. E. 29, 527
Aspden, R. M. 476, 553
Aveston, J. 82, 84, 134, 135, 136, 231, 248, 251, 254, 256, 257, 258, 276, 529, 540, 541

Backer, S. 286, 390, 392, 447, 545, 549, 552
Bader, M. G. 80, 98, 133, 136, 144, 204, 205, 207, 208, 216, 222, 226, 230, 231, 261, 275, 278, 529, 534, 535, 540, 541, 543
Bailey, J. E. 136, 137, 138, 139, 140, 144, 149, 207, 216, 230, 529, 530, 532, 535
Baker, R. M. 229, 534
Beaumont, P. W. R. 276, 277, 541, 543
Beran, M. 195, 534
Bert, C. W. 191, 456, 534, 551
Biderman, V. I. 444, 552
Bishop, S. M. 371, 372, 544, 545
Bjeletich, J. G. 160, 529
Blumentritt, B. F. 182, 225, 534
Bohm, F. 453, 552
Bowyer, W. H. 204, 205, 216, 226, 534
Bradley, P. D. 231, 542
Bradley, W. L. 165, 529
Browning, C. E. 435, 551
Brunnschweiler, D. 392, 547
Bucci, R. J. 275, 541
Bucinell, R. B. 158, 160, 165, 534
Budianski, B. 136, 178, 198, 530, 534, 535
Bunsell, A. R. 21, 231, 254, 256, 268, 526, 541, 542
Burgel, B. 85, 170, 530, 535
Byun, J. H. 302, 384, 386, 414, 415, 417, 418, 422, 427, 437, 438, 439, 440, 544, 547, 548

Camponeschi, E. T. 442, 548
Carlsson, L. A. 29, 527
Carrara, A. S. 85, 170, 530, 535
Carslaw, H. S. 75, 527
Chamis, C. C. 33, 177, 231, 249, 250, 251, 277, 469, 527, 535, 541, 552
Champion, A. R. 428, 548
Chan, W. S. 275, 541
Chandra, M. 333, 546
Chang, C. I. 182, 535
Chang, L. W. 372, 544
Chang, Y. P. 54, 527, 528
Chatterjee, S. N. 362, 546
Chawla, K. K. 29, 527
Chen, C. H. 94, 530
Chen, C. K. 54, 527
Chen, E. J. H. 278, 543
Chen, J. L. 275, 541
Chen, L. C. 276, 277, 543
Chen, P. E. 85, 94, 176, 187, 224, 225, 229, 530, 535, 539
Cheng, C. M. 73, 527
Chi, Z. F. 106, 107, 108, 109, 110, 111, 112, 114, 166, 167, 168, 530
Chou, P. C. 150, 533
Chou, T. W. 1, 6, 8, 9, 10, 19, 21, 29, 33, 40, 46, 48, 54, 55, 58, 62, 65, 66, 67, 68, 69, 72, 73, 74, 76, 77, 78, 80, 81, 85, 88, 90, 91, 92, 93, 94, 96, 97, 103, 106, 107, 108, 109, 110, 111, 112, 113, 114, 133, 136, 145, 146, 147, 148, 149, 150, 166, 167, 169, 171, 172, 174, 175, 176, 178, 179, 180, 181, 184, 191, 193, 195, 197, 198, 199, 200, 201, 202, 203, 206, 207, 208, 209, 210, 211, 212, 213, 214, 215, 216, 217, 222, 223, 225, 226, 229, 230, 231, 232, 233, 234, 238, 239, 241, 242, 243, 244, 246, 247, 248, 249, 251, 258, 259, 261, 262, 263, 265, 266, 267, 268, 269, 271, 272, 278, 285, 288, 300, 302, 303, 307, 308, 313, 314, 316, 317, 318, 319, 320, 321, 322, 324, 325, 326, 330, 332, 334, 335, 336, 337, 338, 341, 342, 344, 346, 347, 348, 349, 350, 353, 354, 355, 357, 358, 359, 360, 364, 365, 366, 367, 368,

556

Subject index